Probabilistic Safety Assessment
in the Chemical and Nuclear Industries

Probabilistic Safety Assessment in the Chemical and Nuclear Industries

Ralph R. Fullwood

Boston Oxford Auckland Johannesburg Melbourne New Delhi

Copyright © 2000 by Butterworth–Heinemann

 A member of the Reed Elsevier group

All rights reserved.

No part of this publication may be reproduced, stored in a retrieval system, or transmitted in any form or by any means, electronic, mechanical, photocopying, recording, or otherwise, without the prior written permission of the publisher.

 Recognizing the importance of preserving what has been written, Butterworth–Heinemann prints its books on acid-free paper whenever possible.

 Butterworth–Heinemann supports the efforts of American Forests and the Global ReLeaf program in its campaign for the betterment of trees, forests, and our environment.

Library of Congress Cataloging-in-Publication Data

Fullwood, R. R.
    Probabilistic safety assessment in the chemical and nuclear industries / Ralph R. Fullwood.
        p. cm.
    Rev. edition of: Probabilistic risk assessment in the nuclear power industry, 1988.
        ISBN 0-7506-7208-0 (alk. paper)
        1. Nuclear power plants—Risk assessment. 2. Chemical plants—Risk assessment. I. Title.
TK9153.F853 1999   99-37453
363.17'992—dc21   CIP

British Library Cataloguing-in-Publication Data
A catalogue record for this book is available from the British Library.

The publisher offers special discounts on bulk orders of this book.
For information, please contact:
Manager of Special Sales
Butterworth-Heinemann
225 Wildwood Avenue
Woburn, MA 01801-2041
Tel: 781-904-2500
Fax: 781-904-2620

For information on all Butterworth–Heinemann publications available, contact our World Wide Web home page at: http://www.bh.com

10 9 8 7 6 5 4 3 2 1

Printed in the United States of America

Disclaimer

The information in this book is collected from published and unpublished literature. Responsibility for the accuracy of this material is disclaimed, however, responsibility is accepted for the selection, organization, and presentation. The vastness of the information necessitates selectivity in the attempt to make a comprehensive and cohesive presentation. The material is selected to illustrate a procedure or principle not advocacy. Every effort toward objectivity was made to balance human health and safety, environment, economic welfare, and civilization.

Neither I, Brookhaven National Laboratory from which I am retired, nor the publisher are responsible for the materials presented here.

RALPH R. FULLWOOD

Upton, NY
October 1998

Contents

Chapter 1 Protecting the Public Health and Safety 1
   1.1 Historical Review 1
      1.1.1 Beginnings 1
      1.1.2 Industrial Revolution 2
      1.1.3 This Century 3
   1.2 Risk Assessment Objectives 5
   1.3 Risk, Hazard, and other Terms 6
   1.4 Quantitative Aspects of Risk 6
      1.4.1 Actuarial or Linear Risk 6
      1.4.2 Shortcomings of Linear Risk 7
      1.4.3 Presentation of Risk 8
      1.4.4 Public Perception of Risk 12
   1.5 Safety Goals 13
   1.6 Emergency Planning Zones 15
   1.7 Use of PSA by Government and Industry 17
   1.8 Regulation of Nuclear Power 18
      1.8.1 Regulations 18
      1.8.2 Regulatory Structure 18
      1.8.3 Licensing Process 19
      1.8.4 Public Participation 20
      1.8.5 Advisory Committee on Reactor Safety (ACRS) 20
      1.8.6 Inspection 21
      1.8.7 Decommissioning 21
      1.8.8 Accident Severity Criteria 21
      1.8.9 PSA Requirements 22
   1.9 Regulation of Chemical Processing and Wastes 22
      1.9.1 Environmental Law 22
      1.9.2 Occupational Risk Protection: the PSM Rule 27
   1.10 Summary 33
   1.11 Problems 34

Chapter 2. Mathematics for Probabilistic Safety 35
   2.1 Boolean Algebra 35
   2.2 Venn Diagram and Mincuts 37
   2.3 Probability and Frequency 39
   2.4 Combining Probabilities 41
      2.4.1 Intersection or Multiplication 41
      2.4.2 Union or Addition 41
      2.4.3 M-Out-of-N-Combinations 42
   2.5 Distributions 42
      2.5.1 Discrete Distributions 43

      2.5.2 Continuous Distributions 44
      2.5.3 Confidence Limits 47
      2.5.4 Markov Modeling 48
      2.5.5 Summary of Functions and their Generating Functions 49
2.6 Bayesian Methods 50
      2.6.1 Bayes' Equation 50
      2.6.2 Bayes Conjugates for Including New Information 51
      2.6.3 Constant Failure Rate Model 52
      2.6.4 Failure on Demand Model 54
      2.6.5 Interpretations of Bayes Statistics 55
2.7 Uncertainty Analysis 56
      2.7.1 Convolution 56
      2.7.2 Moments Method 57
      2.7.3 Taylor's Series 57
      2.7.4 Monte Carlo 59
      2.7.5 Discrete Probability Distribution (DPD) 60
2.8 Sensitivity Analysis and Importance Measures 61
      2.8.1 Sensitivity Analysis 61
      2.8.2 Importance Measures 62
      2.8.3 Relationships between the Importance Measures 63
      2.8.4 Interpretation and Usage 64
2.9 Summary 65
2.10 Problems 66

Chapter 3 Chemical and Nuclear Accident Analysis Methods 67
3.1 Guidance from the PSM Rule 67
      3.1.1 Rule Objectives 67
      3.1.2. Employee Involvement in Process Safety 67
      3.1.3. Process Safety Information 68
3.2 Process Hazard Analysis 70
      3.2.1 Overview 70
      3.2.2 Operating Procedures and Practices 71
      3.2.3. Employee Training 71
      3.2.4 Contractors 71
      3.2.5 Pre-Startup Safety 72
      3.2.6 Mechanical Integrity 72
      3.2.7 Nonroutine Work Authorization 73
      3.2.8 Managing Change 73
      3.2.9 Investigation of Incidents 74
      3.2.10 Emergency Preparedness 74
      3.2.11 Compliance Audits 75
3.3 Qualitative Methods of Accident Analysis 76
      3.3.1 Checklist 77

      3.3.2 What-If Analysis 81
      3.3.3 What-If/Checklist Analysis 84
      3.3.4 Hazard and Operability (HAZOP) 86
      3.3.5 Failure Mode and Effects Analysis 94
3.4 Quantitative Methods of Accident Analysis 97
      3.4.1 Parts Count 98
      3.4.2 FMEA/FMECA 99
      3.4.3 Reliability Block Diagram (RBD) 100
      3.4.4 Fault Tree Analysis 101
      3.4.5 Event Trees 111
      3.4.6 Alternatives to Fault Tree Analysis 119
3.5 Common Cause of Failure 123
      3.5.1 Known Deterministic Coupling 124
      3.5.2 Known Stochastic Coupling 125
      3.5.3 Unknown Stochastic Coupling 125
      3.5.4 Modeling Known Dependencies 125
      3.5.5 Geometric Mean 126
      3.5.6 Beta Factor 126
      3.5.7 Example of the Beta-Factor Method: Emergency Electric Power 127
      3.5.8 Common Cause Multiparameter Models 127
3.6 Computer Codes for System Reliability Analysis 128
      3.6.1 Codes for Finding Minimal Cutsets and Tree Quantification 130
      3.6.2 Truncation of a Fault Tree 133
      3.6.3 Time Dependence 134
      3.6.4 Uncertainty Analysis 134
      3.6.5 Importance Calculations 134
      3.6.6 Processing Cutset Information 135
      3.6.7 System Analysis Code Usage in Past PSAs 136
      3.6.8 Logistics of Acquiring Codes 136
3.7 Code Suites 136
      3.7.1 SAPHIRE 136
      3.7.2 PSAPACK 141
      3.7.3 RISKMAN 143
      3.7.4 R&R Workstation 144
      3.7.5 WinNUPRA, NUCAP+, and SAFETY MONITOR 145
3.8 Summary 147
3.9 Problems 148

Chapter 4 Failure Rates, Incidents and Human Factors Data 151
4.1 Databases 151
      4.1.1 Background 151
      4.1.2 Some Reliability Data Compilations 151
4.2 Incident Reports 158

- 4.3 Database Preparation 160
  - 4.3.1 How to Estimate Failure Rates 160
  - 4.3.2 Making a Database 160
  - 4.3.3 Test Reports and Procedures 161
  - 4.3.4 Maintenance Reports 162
  - 4.3.5 Operating Procedures 162
  - 4.3.6 Control Room Log 162
  - 4.3.7 Information Flow in Plant Data Collecting 163
  - 4.3.8 Data Sources Used in Past PSAs 163
- 4.4 Human Reliability Analysis 163
  - 4.4.1 Human Error in System Accidents 163
  - 4.4.2 Lack of Human Error Considerations 166
  - 4.4.3 Cases Involving Human Error 168
- 4.5 Incorporating Human Reliability into a PSA 173
  - 4.5.1 Human Reliability Analysis Models 173
  - 4.5.2 Quantifying Human Error Probabilities (HEPs) 175
  - 4.5.3 Human Factors Data 179
  - 4.5.4 Example of Human Error Analysis 180
  - 4.5.5 HRA Event Tree (NUREG/CR-1278) 181
  - 4.5.6 Comparison of Human Factors in PSAs 183
- 4.7 Summary 184
- 4.8 Problems 184

Chapter 5 External Events 185
- 5.1 Seismic Events 185
  - 5.1.1 Overview 185
  - 5.1.2 Richter Magnitude - Frequency of Occurrence Distribution 188
  - 5.1.3 Ground Coupling with Attenuation 190
  - 5.1.4 Shaking Model of the Plant 190
  - 5.1.5 Fragility Curves 192
  - 5.1.6 System Analysis 194
- 5.2 Fires 195
  - 5.2.1 Introduction 195
  - 5.2.2 Procedures for Fire Analysis 196
  - 5.2.3 Screening Analysis 197
  - 5.2.4 Fire Frequencies 197
  - 5.2.5 Fire Growth Modeling 198
  - 5.2.6 Fragility Curves 199
  - 5.2.7 Systems Analysis 199
- 5.3 Flood 200
  - 5.3.1 Overview of Internal Flooding 200
  - 5.3.2 Internal Flooding Incidents 200
  - 5.3.3 Internal Flood Modeling 202

        5.3.4 External Flooding 203
    5.4 Summary 204

Chapter 6 Analyzing Nuclear Reactor Safety Systems 205
    6.1 Nuclear Power Reactors 205
        6.1.1 U.S. Light Water Reactors 206
        6.1.2 Pressurized Water Reactors 208
        6.1.3 Boiling Water Reactors 211
        6.1.4 Advanced Light Water Reactors 213
    6.2 TMI-2 and Chernobyl Accidents 221
        6.2.1 The TMI-2 Accident 221
        6.2.2 The Chernobyl Accident 223
    6.3 Preparing a Nuclear Power Plant PSA 227
        6.3.1 Overview of the Probabilistic Safety Process Using Event Trees 228
        6.3.2 Preparing for the PSA 228
        6.3.3 PSA Construction 236
    6.4 Example: Analyzing an Emergency Electric Power System 238
        6.4.1 The Problem: Emergency Electric Power 239
        6.4.2 Overview of FTAPSUIT 239
    6.5 Summary 243
    6.6 Problems 243

Chapter 7 Analyzing Chemical Process Safety Systems 245
    7.1 Chemical Process Accidents 245
        7.1.1 The Nature of Process Accidents 245
        7.1.2 Some Deadly and Severe Chemical Accidents 247
    7.2 Chemical Processes 261
        7.2.1 The Topography of Chemical Processes 261
        7.2.2 Inorganic Chemicals 262
        7.2.3 Fertilizer Production 264
        7.2.4 Halogens and Their Compounds 266
        7.2.5 Organic Chemicals 269
        7.2.6 Explosives 272
        7.2.7 Plastics and Resins 276
        7.2.8 Paints and Varnishes 283
        7.2.9 Petrochemical Processing 286
    7.3 Chemical Process Accident Analysis 293
        7.3.1 Scoping Analysis 295
        7.3.2 Performing a Detailed Probabilistic Safety Analysis 299
    7.4 Example: Analyzing a Chemical Tank Rupture 304
        7.4.1 Defining the Problem with Fault Tree Analysis 304
        7.4.2 Applying FTAPSUIT to Chemical Process Tank Rupture 305
    7.5 Summary 307

7.6 Problems 308
Chapter 8 Nuclear Accident Consequence Analysis 309
   8.1 Meltdown Process 309
      8.1.1 Defense in Depth Barriers Providing Public Protection 309
      8.1.2 Qualitative Description of Core Melt 310
   8.2 Source Terms for In-Plant Radionuclide Transport 314
      8.2.1 Background on the Source Term 314
      8.2.2 Computer Codes for Fission Product Release and In-Plant Transport .. 316
      8.2.3 Comparison with WASH-1400 320
   8.3 Ex-Plant Transport of Radionuclides 321
      8.3.1 Atmospheric Transport Models 321
      8.3.2 Health Effects 323
      8.3.3 Radiation Shielding and Dose 325
      8.3.4 Computer Codes for Consequence Calculation 329
      8.3.5 Aquatic Transport 331
   8.4 Summary 331
   8.5 Problems 331

Chapter 9 Chemical Process Accident Consequence Analysis 333
   9.1 Hazardous Release 333
      9.1.1 Pipe and Vessel Rupture 333
      9.1.2 Discharge through a Pipe 335
      9.1.3 Gas Discharge from a Hole in a Tank 337
      9.1.4 Liquid Discharge from a Hole in a Tank 338
      9.1.5 Unconfined Vapor Cloud Explosions (UVCE) 338
      9.1.6 Vessel Rupture (Physical Explosion) 342
      9.1.7 BLEVE, Fireball and Explosion 343
      9.1.8 Missiles 345
   9.2 Chemical Accident Consequence Codes 346
      9.2.1 Source Term and Dispersion Codes 347
      9.2.2 Explosions and Energetic Events 362
      9.2.3 Fire Codes 365
   9.3 EPA's Exposure Model Library and Integrated Model Evaluation System 368
      9.3.1 IMES 368
      9.3.2 PIRANHA 371
      9.3.3 REACHSCN 371
      9.3.4 SIDS 371
      9.3.5 THERDCD 372
   9.4 Summary 373

Chapter 10 Assembling and Interpreting the PSA 375
   10.1 Putting It Together 375
      10.1.1 Integrated and Special PSAs 375

      10.1.2 Assembling the PSA 375
  10.2 Insights and Criticisms 377
      10.2.1 Insights from Past PSAs 378
      10.2.2 Criticism of Past PSAs 378

Chapter 11 Applications of PSA 383
  11.1 U.S. Commercial Nuclear PSAs 383
      11.1.1 Commercial Nuclear PSAs before IPE 383
      11.1.2 ATWS 384
      11.1.3 Risk-Based Categorization of NRC Technical and Generic Issues ... 385
      11.1.4 Use of PSA in Decision Making 385
      11.1.5 Value-Impact Analysis (VIA) 386
      11.1.6 Response to the TMI-2 Accident 386
      11.1.7 Special Studies 386
      11.1.8 Individual Plant Evaluation PSAs 392
      11.1.9 Risk-Based Regulation 400
      11.1.10 Practical Application of PSA: Utility Experience and
             NRC Perspective 402
  11.2 PSA of the CANDU (Heavy Water Power Reactor) 404
      11.2.1 Reactor Description 404
      11.2.2 CANDU-2 PSA 405
      11.2.3 CANDU-6 PSA 406
      11.2.4 CANDU-9 PSA 407
  11.3 Research and Production Reactor PSAs 408
      11.3.1 Advanced Test Reactor PSA 408
      11.3.2 High Flux Beam Reactor PSA 411
      11.3.3 HFIR PSA 414
      11.3.4 K-Reactor PSA 416
      11.3.5 N-Reactor PSA 422
      11.3.6 Omega West Reactor PSA 427
  11.4 Chemical Process PSAs 428
      11.4.1 Canvey Island 428
      11.4.2 PSA of a Butane Storage Facility (Oliveira, 1994) 438
      11.4.3 Comparative Applications of HAZOP, Facility Risk Review, and
             Fault Trees 440
      11.4.4 Probabilistic Safety Analysis of an Ammonia Storage Plant 445
  11.5 Problems 449

Chapter 12 Appendix: Software on the Distribution Disk 451
  12.1 ANSPIPE 451
  12.2 BETA 452
  12.3 BNLDATA 453
  12.4 FTAPSUIT 453

  12.5 Lambda 457
  12.6 UNITSCNV 458

Chapter 13 Glossary of Acronyms and Unusual Terms 459

Chapter 14 References 467
  14.1 Nuclear Regulatory Commission Reports Identified by NUREG Numbers 467
  14.2 Nuclear Regulatory Commission Contractor Reports Identified by
     NUREG/CR Numbers 468
  14.3 Electric Power Research Institute Reports Identified by NP Number 471
  14.4 References by Author and Date 472

Chapter 15 Answers to Problems 493
  15.1 Chapter 1 493
  15.2 Chapter 2 495
  15.3 Chapter 3 496
  15.4 Chapter 4 500
  15.5 Chapter 6 501
  15.6 Chapter 7 504
  15.7 Chapter 8 504
  15.8 Chapter 11 506

Index 509

List of Figures[a]

1.4.2-1 Curve Enveloping Accident Data and a Linear Envelope for a Looser Straight-Line Bound 7
1.4.2-2 Exceedance Probability (accidents 1-6) and composite envelope 8
1.4.3-1 Fatality Frequency for Man-Caused Events (from WASH-1400) 9
1.4.3-2 Confidence Intervals: CCDF for Latent Cancer Fatalities Internal and External Seismicity 10
1.4.3-3 Latent Cancer Risk at Indian Point site if the Reactor were one of the five types shown 10
1.4.3-4 Early fatality risk if the Indian Point Plant had been at one of these sites 10
1.4.3-5 Risk compared as volumes or masses 11
1.4.3-6 Relative contributions to the Zion and ANO-1 core melt frequencies 12
1.4.4-1 Revealed and risk-benefit relationships 13
2.2-1 Venn diagram of two sets 37
2.2-2 Mincut Explanation by Venn Diagrams 39
2.3-1 Probability of Scores in Craps 40
2.4-1 Venn Diagram of Sets A and B 41
2.5-1 The discrete distribution of probability of crap scores has impressed on it a continuous distribution with quite a different meaning 43
2.5-2 Conceptual Life-Cycle of a Component 46
2.5-3 Transitions between States 48
2.7-1 ERMA simulation of a low pressure coolant injection system 59
2.7-2 Histogram of probability distribution generated by Monte Carlo simulation 60
2.8-1 Risk worth ratios with respect to core melt for Sequoyah safety systems 65
2.8-2 Risk worth ratios with respect to core melt for Oconee safety systems. 65
2.10-1 Control Rod Pattern 66
3.1.3-1 Example of a Block Flow Diagram 68
3.1.3-2 Process Flow Diagram 69
3.3.1-1 Hydrogen Fluoride Supply System 76
3.3.1-2 Cooling Tower Water 77
3.3.4-1 The HAZOP Process 88
3.4.1-1 Illustrative Cooling Systems 98
3.4.3-1 Reliability Block Diagram of Figure 3.4.1-1 100

---

[a] The first number is the chapter number, the second number is the section number and the third number (if present) is the subsection number. The number after the dash is the figure number.

3.4.4-1 Symbols of Fault Tree Analysis 101
3.4.4-2 Sketch of an Internal Combustion Engine 102
3.4.4-3 Fault Tree of Engine 103
3.4.4-4 Example Fault Tree 104
3.4.4-5 Mincut Fault Tree of Figure 3.4.4-4 105
3.4.4-6 Simple injection system 109
3.4.4-7 Failure to Inject Fault Tree 109
3.4.4-8 Boolean Reduction of "Failure to Inject Fault Tree" 109
3.4.4-9 Fault Tree from the Mincut Representation of Figure 3.4.4-8 Fault Tree 110
3.4.4-10 Success Tree Transformation of Figure 3.4.4-9 Fault Tree 110
3.4.5-1 Event Tree Scenario Depiction 111
3.4.5-2 Large LOCA as a Function Event Tree 113
3.4.5-3 Fault Trees at the Nodes of an Event Trees to Determine the Probability 114
3.4.5-4 Large LOCA as a System Event Tree 115
3.4.5-5 Large LOCA Event Tree in the LESF style 116
3.4.5-6 Containment Event Tree 119
3.5.7-1 Fault Tree diagram of the emergency electric power system 127
3.6-1 The Saphire Main Menu and the Fault Tree Editor Menu 139
3.7-1 Fault Tree of a doubly redundant ECCS 148
3.7-2 Time-sequential testing 149
 3.7-3 Staggered testing 149
4.3-1 Plant Data Processing for PSA 163
4.4-1 Releases based on plant status 165
4.4-2 Releases based on cause code 165
4.4-1 Venn Diagram of Error Cause 167
4.5-1 Phases of Human Reliability Analysis 172
4.5-2 General incorporation of HRA into the PSA process 174
4.5-3 Failure vs Problem-Solving Time 180
4.5-4 HRA Event Tree 180
4.5-5 HRA Event Tree for Actions Outside the Control Room 181
4.5-6 Example of Spread in HEPs from IPE for BWRs 183
5.1-1 Procedures for Analyzing External Events 187
5.1-2 Comparison of the world frequency-magnitude shape with data from the CMMG 189
5.1-3 Seismic hazard curves for a hypothetical site 191
5.1-4 Plan view of nuclear steam supply system 191
5.1-5 Nodal diagram of a nuclear steam supply system 191
5.1-6 Large LOCA Seismic Event Tree for a PWR 195
5.3.1 Fault Tree Identification of Flood-Critical Equipment 202
5.3-2 Flood data plotted lognormally 204
6.1-1 Sketch of a Pressurized Water Reactor (PWR) 207
6.1-2 Sketch of a Boiling Water Reactor (BWR) 212
6.1-3 The In-Containment Passive Safety Injection System 215

6.1-4 AP500 Passively Cooled containment 216
6.1-5 Sketch of the PIUS Principle 218
6.1-6 SWBR LOCA Response. 219
6.2-1 The Damaged TMI-2 Core 221
6.2-2 Schematic of the RBMK-1000 224
6.3-1 Overview of the PSA process 227
6.3-2 Possible PSA Organization Chart 229
6.3-3 Core Damage Master Logic Diagrams for Indian Point 2 and 3 Nuclear Power
       Stations 233
6.3-4 Organizing the Analysis via Event Tree 236
6.3-5 Relative core melt contributions 238
6.4-1 Emergency Electric Power 238
6.4-2 Procedure for Using FTAPSUIT 240
6.4.-3 Menu for FTAPSUIT 241
6.4-4 FTAPlus Menu 241
6.4-5 Post Processor Menu 241
7.1-1 Increasing Cost of Process Accidents per 4 Year Period 246
7.1-2 Average Loss by Type of Plant 246
7.1-3 Average Loss by Loss Mechanism 246
7.1-4 Average Loss by Loss Cause 246
7.1-5 Average Loss by Type of Equipment 246
7.1-6 Temporary Repair at Flixborough 250
7.1-7 Common Headers to the MIC Storage Tanks at Bhopal 252
7.1-8 MIC Storage Tank 253
7.2-1 Fertilizers and Uses 264
7.2-2 Uses for Benzene 269
7.2-3 Unsaturated Hydrocarbons from Petroleum 270
7.2-4 Diagram of a Fractional Distillation Column 287
7.2-5 Schematic of a Catalytic Cracking Unit 289
7.3-1 Process Scoping Steps 296
7.3-2 A Detailed PSA 299
7.4-1 Fault Tree for Tank Rupture 304
7.4-2 FTAP Input file 305
7.4-3 FTAP Punch output 305
7.4-4 Appearance of File 305
7.4-5 Edited Output of IMPORTANCE 306
7.4-6 Monte Carlo Calculation of the Chemical Process Tank Rupture 307
8.1-1 Decay Heat from a 3000 MWe Core 310
8.1-2 Total heat evolved from fuel and absorbed by concrete 310
8.1-3a Molten droplets and rivulets begin to flow downward 311
8.1-3b Formation of a local blockage in colder rod regions 311
8.1-3c Formation of the molten pool 312
8.1-3d Radial and axial growth of themolten pool 312

8.1-3e Melt migrates to the side and fails the core barrel 312
8.1-3f Melt migrates downward into remaining water 312
8.1-3g Downward progress of coherent molten mass as the below core structure weakens ... 312
8.1-3h Failure of support 312
8.1-3i Two possible modes of accident progression 313
8.1-3j Three possible failure modes 313
8.1-3k Concept of solidified core in concrete after meltdown 313
8.2-1 BMI-2104 suite of codes used in the source term assessment 317
8.2-2 Meltdown model in the MARCH code 318
8.2-3 Complementary cumulative distribution for latent cancer fatalities 320
8.3-1 Lateral diffusion parameter vs distance from the source 322
8.3-2 Vertical diffusion parameter vs distance from the source 322
8.3-3 Correction factors for σ_y by atmospheric stability class 322
8.3-4 Isotropic Radiation 325
8.3-5 Common Doses from Various Sources 329
9.1-1 Diagram for Hoop Stress 334
9.1-2 Viscous Sliding 335
9.1-3 Diagram for Viscous Flow in a Pipe 335
9.1-4 Shock Wave Parameters for a Hemispherical TNT Surface Explosion at Sea level. 341
9.1-5 Shock Wave Parameters for a Spherical TNT Surface Explosion at Sea 341
9.1-6 Range of LPG Missiles 345
9.1-7 Scaled Fragment Range vs Scaled Force 346
9.2-1 Case Study Refinery and Town 358
9.2-2 Input to Phast 358
9.2-3 Calculated Toxic Plumes on Map 360
10.1-1 PSA Assembly Process 376
11.1-1 Summary of BWR and PWR CDFs from IPEs 395
11.1-2 Conditional Containment Failure Probabilities from the IPEs 395
11.2-1 Sketch of the CANDU 404
11.3-1 Omega West Results 428
11.4-1 Map of the Canvey Island/Thurrock Area 430
11.4-2 Alkylation Process 441
11.4-3 Proposed New Chemical Reactor 444
11.4-4 Case Studies of QRA Results 444
12.1-1 Main Screen for ANSPIPE 451
12.1-2 Main Beta Screen 452
12.4-1 Procedure for Using FTAPSUIT 453
12.4-1 Menu for FTAPSUIT 455
12.4-2 FTAPlus Menu 455
12.4-3 PostProcessor Menu 455
12.4-4 Menu for Lambda 457
12.4-5 Menu for the Units Conversion Program 457

15.1.2-1 Redraw of Figure 1.4.3-1 493
15.1.3-1 Comparing by Area 493
15.2.1-1 Control Rod Pattern 495
15.3.1-1 Slow Melt Fault Tree 497
15.3.1-2 Melt Fault Tree 497
15.3.2-1 RBD of a Simple Injection System 497
15.3.5-1 Fault Tree of Emergency Electric Power 498
15.4.3-1 HRA Tree for Driving to Work 501
15.5.2-1 The Percolator Scheme 503
15.5.2-2 Using Residual Heat 503
15.6.1-1 Bhopal Accident Event Tree 504
15.6.2-1 Failure to Inject Fault Tree Input 504
15.7.2-1 A Radiation Fallout Field 505

List of Tables[b]

1.4.3-1 Illustrative Cutset Listing 9
1.4.3-2 Expected Annual Consequences (risk) from Five LWR Designs if Sited at
 Indian Point 11
1.4.3-3 Estimated Frequencies of Severe Core Damage 11
1.6-1 Chronology of Federal Emergency Regulations 15
1.6-2 Emergency Action Levels and Responses 17
1.8-1 Some Federal Regulations Governing the Environment 18
1.9-1 List of Highly-Hazardous Chemicals, Toxics, and Reactives (Mandatory) 28
2.1-1 Comparison of Ordinary and Boolean Algebra 36
2.5-1 Values of the Inverse Cumulative Chi-Squared Distribution in terms of percentage
 confidence with M failures 47
2.5-2 Mean Variance and Moment-Generating Functions for Several Distributions 49
2.6-1 Bayes Conjugate Self-Replication 52
2.6-2 Effect of Bayes Conjugation on a Gamma Prior 52
2.8-1 Relationships between Importance Measures 64
3.3.1-1 Possible Checklist Main Headings 77
3.3.1-2 Possible Headings and Subheadings 78
3.3.1-3 Simplified Process Hazards Analysis Checklist 79
3.3.1-4 Checklist Analysis of Dock 8 HF Supply 80
3.3.1-5 Checklist of Cooling Tower Chlorination 80
3.3.1-6 Estimated Time for a Checklist 81
3.3.2-1 What-If Analysis of Dock 8 HF Supply System 82
3.3.2-2 What-If Analysis of the Cooling Tower Chlorination System 83
3.3.2-3 Estimated Time for a What-If 84
3.3.3-1 Estimated Time for a What-If/Checklist 85
3.3.3-2 What-If/Checklist Analysis of Dock 8 HF Supply System 85
3.3.3-3 What-If/Checklist Analysis of Cooling Tower Chlorination System 85
3.3.4-1 Guide Words for HAZOP 90
3.3.4-2 HAZOP Process Parameters and Deviations 91
3.3.4-3 HAZOP of the Cooling Tower Chlorination System 92

---

[b] The first number is the chapter number, the second number is the section number and the third number (if present) is the subsection number. The number after the dash is the figure number.

3.3.4-4 HAZOP of the Dock 8 HF Supply System 93
3.3.4-5 Estimated Time for a HAZOP 93
3.3.5-1 FMEA Consequence Severity Categories 94
3.3.5-2 Partial FMEA for the Cooling Water Chlorination System 96
3.3.5-3 Partial FMEA for the Dock 8 HF Supply System 97
3.3.5-4 Estimated Time for an FMEA 97
3.4.2-1 Failure Modes Effects and Criticality Analysis Applied to System B of
      Figure 3.4.1-1 99
3.4.4-1 Hypothetical Engine Operating Experience 103
3.4.4-2 Example Format for an Auxiliary Feedwater System Interaction FMEA 107
3.4.5-1 Sequence of Events after LOCA 114
3.4.6-1. Summary of Other Methods 120
3.4.6-2 Attributes of GO 121
3.5.6-1 Generic Beta Factors for Reactor Components 126
3.6-1 Computer Codes for Qualitative Analysis 129
3.6-2 Computer Codes for Quantitative Analysis 131
3.6-3 Computer Codes for Uncertainty Analysis 132
3.6-4 Computer Codes for Dependent-Failure Analysis 133
3.6-5 Code Usage in Utility PSAs 135
3.6-6 Code Usage in PSA Studies 136
3.6-7 The Saphire Suite 137
3.6-8 Applications of the R&R Workstation 145
4.1-1 Partial History of Reliability Data Collection 152
4.1-2 Data Collections 154
4.1-3 NPDRS Contents 154
4.1-4 Sample of NUCLARR Comonent Data 155
4.1-5 NUREG 1150 Nuclear Plant Reliability Data 156
4.2-1 Internet Web Sites Relevant to Chemical Safety 158
4.2-3 Some Environmental Internet Addresses 159
4.3-1 Sources of Data 161
4.3-2 Contents of Maintenance Reports 162
4.3-3 Reliability Data Sources for Big Rock Point and Zion PSAs 164
4.4-1 Studies of Human Error in the CPI Magnitude of the Human Error Problem 165
4.5-1 HRA Analysis Methods 173
4.5-2 Requirements for Confusion Matrix 175
4.5-3 Requirements for Direct Numerical Estimation 176
4.5-4 Requirements for LER-HEP 176
4.5-5 Requirements for MAPPS 176
4.5-6 Requirements for MSF 177
4.5-7 Requirements for OAT 177
4.5-8 Requirements for Paired Comparisons 177
4.5-9 Requirements for SLIM-MAUD 178

4.5-10 Requirements for THERP	178
4.5-11 Sample of NUCLARR Human Error Probability Data	179
4.5-12 Estimated Human Error	179
4.5-13 Comparison of Human Interaction Evaluations in Utility and NRC PSA Studies	182
5-1 Natural and Man-caused Risks	186
5.1-1 Modified Mercali Scale (Wood-Neumann Scale) Perceived Intensity	187
5.1-2 Richter Magnitudes, Energy, Effects and Frequencies	188
5.1-3 Motor-Operated Valve Failure Modes	192
5.1-4 Differences between Seismic and Internal Events Analysis	194
5.1-5 Initiating Events Used in the SSMRP	195
5.2-1 Statistical Evidence of fires in Light Water Nuclear Power Plants[*] (<May 1978)	197
5.2-2 Distribution of the Frequency of Fires[a]	198
5.3-1 Turbine Building Flooding in U.S. Nuclear Power Plants	201
5.3-2 Flooding Frequencies for Turbine and Auxiliary Buildings	201
6.1-1 List of PWR Transient Initiating Events	209
6.1-2 Typical Frontline Systems for PWRs	209
6.1-3 ANO-1 Frontline vs Support and Support vs Support Dependencies	210
6.1-4 List of BWR Transient Initiating Events	213
6.1-5 Typical BWR Frontline Systems	214
6.1-6 Attributes for Second-Generation Reactors	214
6.1-7 Comparison Showing Reductions in AP-600 Compared with a Comparable First-Generation Plant	215
6.2-1 Precursors to the TMI-2 Accident	222
6.2-2 Summary of the Chernobyl Accident	226
6.3-1 Comparison of Estimated and Actual PSA Resources (person-months)	229
6.3.2 Zion Principal Information Sources	231
6.3.3 Qualitative Categorizations from MIL-STD-882A	232
6.3-4 Sample Preliminary Hazards Analysis	232
6.3-5 Initiating Event Summaries	234
6.3-6 Inspection Importance Ranking with Health Effects for Indian Point 3 Accident Initiators	235
7.1-1 Safety Enhancement	245
7.1-2 Ammonium Perchlorate Use in Rockets (million lb/year)	258
7.2-1 Chemical Process Steps	262
7.3-1 Hazard Criteria	293
7.3-2 Objectives for Chemical PSA	294
7.3.1-1 A Spread Sheet for Scoping Analysis Data	296
7.3.1-2 Some Toxicology Sources	297
7.3.2-1 Investigations in a Detailed PSA (from CCPS, 1989)	301
7.4-1 Parameters for Figure 7.4-1	304
8.1-1 Melting and Boiling Points of Material	311
8.2-1 Reactor Safety Study Accident Release Categories	315
8.3-1 Units of Radioactivity	327

8.3-2 Units of Energy Deposition 327
8.3-3 Units of Health Effects 327
8.3-4 Q Factors for Various Radiation 327
8.3-5 Some Maximum Doses from NCRP-91 329
8.3-6 Codes for Nuclear Accident Consequences 331
9.1-1 Yield Strength of Some Metals 333
9.1-2 Temperature Dependent Viscosity of Some Gases and Liquids ... 336
9.1-3 Gamma for Several Gases 337
9.1-4 Some Heats of Combustion 339
9.1-5 Damage Produced by Blast Overpressure 341
9.3-1 Computer Models on the EPA Disk 371
11.1-1 PSA Studies of U.S. Plants Completed before IPE 384
11.1-2 Shutdown CDF (Mid-Loop Operation) and Full Power CDF 390
11.1-3 IPE Submittals 393
11.1-4 IPE Submittals' Containment Types 396
11.1-5 Findings from the IPEs 397
11.1-6 Plants Improvements from IPE from Drouin, 1996 399
11.2-1 CANDU-2 PSA Results 405
11.2-2 CANDU-6 PSA Results 406
11.3-1 ATR Initiating Events 409
11.3-2 Quantification of HFBR PSA 412
11.3-3 HFIR Risk 416
11.3-4 SRS Methodology 417
11.3-5 SRS Accident Classes 418
11.3-6 Point Estimate CDFs for the K-Reactor 421
11.4-1 Frequency /1000 years of Exceeding Maximum Casualties for Existing Facilities 437
11.4-2 Risk Reductions and Costs 439
11.4-3 HF Alkylation HAZOP Results 440
11.4-4 Recommendations from FRR of Mining Operation 442
11.4-5 Frequency (1/y) vs Assumptions 445
11.4-6 Comparison of Some PSA Methods 446
15.3.4-1 Failure Modes Effects Analysis for Valve A in Figure 3.4.4-6 ... 498
15.5.1-1 What-If Analysis of Critical Assembly 501
15.6.1-2 Critical Assembly Accident Initiators 502

Foreword

The last three decades have seen the development of a new science to help us better understand the risk of events about which there is often very little information. The reason there is interest in such a science is that there are a great many societal benefits from activities that involve risk; risk that if properly managed through better understanding can greatly benefit the quality of all life on the planet earth, both plant and animal. That science is quantitative risk assessment, also known by such names as probabilistic risk assessment and probabilistic safety assessment, the latter being the preferred name for this text. Probabilistic safety assessment divides the risk question into three questions "What can go wrong?" "How likely is it?" and "What are the consequences?"

Probabilistic safety assessment has had its greatest push in relation to the assessment of risk associated with nuclear power plant operation as documented in the author's previous book. This new book, besides updating and reorganizing the nuclear portions of the previous text, ventures into the safety assessment of chemical facilities, another important industry driver of probabilistic safety assessment methods and applications.

As expected, the nuclear sections of this book have much greater scope (breadth and depth) than the chemical sections. This is a direct result of the difference in the experience of the two industries in the development and application of PSA methods. The great appeal of extending the text to the chemical field is that it indicates not only the growth of the discipline, but the contribution to PSA that can come from different industries applying such methods. For example, the chemical field is extremely cost and competitively driven and has been very innovative in coupling safety assessment with plant performance considerations, a factor not nearly as visible in nuclear safety analysis. Other advantages of adding the chemical industry safety assessment practices to this text is the extensive operating experience base involved and the efforts of the chemical safety experts to employ more abbreviated probabilistic safety assessment methods.

This book, for the most part, is a stand-alone text. It addresses not only the fundamentals of PSA as a science, but insights on the regulatory framework affecting its development and application. In particular, it provides the basic methods of analysis that can be employed, available databases, an excellent set of examples, software resources, chapter summaries that facilitate comprehension, and problem sets that are very well connected to the theory. While much has been written about probabilistic safety assessment over the last three decades, this is the most comprehensive attempt so far to provide a much needed college level textbook for the education of risk and safety professionals. It also provides a valuable reference for any individual curious enough about the risk and safety sciences to want to become much more informed.

B. John Garrick, Ph.D, P.E.

Preface

This book aims at a unified presentation of probabilistic safety assessment as it is applied to the chemical processing and nuclear electrical generation industries. As John Garrick points out in the foreword, probabilistic safety assessment has developed over the latter part of this century to assess and thereby enhance safety in industries that have a remote possibility of affecting many people. PSA's genesis was in the space industry to achieve better equipment reliability. Its application to safety came with the U.S. governments "Plowshare" program for peaceful use of nuclear energy for electric power generation. The electric utility industry was reluctant to adopt this energy source because of the unknown liabilities. The Price-Anderson Act provided the insurance to protect the industry from an unknown risk. A worst case analysis (WASH-740) indicated a large potential hazard with unknown probability of occurrence.

There had been small-scale probabilistic risk studies, but the first in-depth study was initiated by the U.S. Atomic Energy Commission in September 1972 and completed by the Nuclear Regulatory Commission (NRC). This was known as the "Reactor Safety Study," (WASH-1400, October 1975) that set the pattern for subsequent PSAs not only nuclear, but chemical and transportation. PSA had it beginnings in nuclear power because of the unknown risk and the large amounts of funds for the investigation.

There is a close kinship between the chemical process industry and the nuclear electric power industry. In fact once the physics of nuclear reaction was established the rest is chemistry and heat transfer. The word "reactor" is from chemistry for the location the reaction takes place. A nuclear reactor consists of a vessel in which a nuclear reaction heats water to make steam to drive a turbine to generate electricity. Thus the primary components are pipes, valves, pumps heat exchangers, and water purifiers similar to the components found in a chemical plant. Following the success of WASH-1400, PSA was used to analyze the chemical processing of nuclear fuel and waste preparation for disposal.

A leader in applying PSA to other parts of the chemical process industry has been the AIChE's Center for Chemical Process Safety. A major difference between PSA for nuclear power and PSA for chemical processing has been the lack of government regulations that require risk analysis for chemical processes. A primary impetuous has been the Occupational Safety and Health Administration's (OSHA) PSM rule that defines the application of PSA to the chemical industry for the protection of the public and workers. In addition, the Environmental Protection Agency (EPA) regulates waste disposal.

This book describes the evolution of PSA.

WASH-1400 provided the first comprehensive estimate of the risk of two nuclear power plants which were assumed to be generic, hence, taken to represent the industry in the U.S. The NRC, the agency responsible for the regulation of nuclear power in the U.S. used and extended WASH-1400 for regulating the plants and making decision on retrofitting, inspection, maintenance and other purposes. Thus, the legal requirements for a nuclear power plant license drives the use of PSA for this purpose. While not an NRC licensee, the Department of Energy applied PSA to its nuclear power plants and process facilities and adopted many of NRC's safety criteria.

Considerations for preparing this book were:

- Show the long historical usage of PSA by government and industry for protecting health safety, environment and the infrastructure of civilization. PSA has only been known recently by this term, but its processes have long been practiced.
- Present risk from its basis in insurance. This is the natural basis that says that risk is the product of probability and consequences.
- Show the complex iterations between government laws and regulations and the PSA response to not only comply but to protect the process industry. The real impact of the accident at the Three-Mile Island nuclear plant was not radiation, which was within regulations but financial losses to the utility and the acceptance of nuclear electrical power in the United States. The effects of the Bhopal accident were in human life but it also had a profound effect on the chemical industry: financially, and its acceptability and growth.
- Present the mathematics used in PSA in one chapter to be skipped, studied, or referred to according to the readers needs.
- Provide the reader with a computer disk containing programs that I have found useful and to provide FTAP and its associated programs. Computer codes that calculate fault trees are either proprietary or expensive or both. FTAP and associated codes are the only code suite that is in the public domain.[c] The disk proves them grouped as FTAPSUIT as a "bat" file in association with the FTAPLUS program to aid in formatting. FTAPSUIT is a sophisticated set of programs that finds the cutsets, calculates their probabilities, handles dependencies, and calculates importances and uncertainties. It was written long ago before user-friendly interfaces but this fact may assist understanding and learning.
- As far as I know, this is the only book that merges nuclear and chemical PSA. Although both processes are similar, nuclear PSA places emphasis on reliability and the probabilistic calculation other than the consequence calculation. Chemical PSA places more emphasis on the consequences than on the probability. Chapter 3 presents methods used by both to determine the reliability of systems and the probability of failure. It begins with OSHA's PSM rule that outlines process for use in analyzing PSA. Qualitative methods are presented before quantitative. The chapter ends by describing computer codes to aid in the calculations and results presentation. Chapter 4 presents data and human factors analysis that are used by both industries. Chapter 5 outlines methods for analyzing the effects of "externalities" on plants. These include earthquake, wind, fire and flood. Chapter 6 describes nuclear reactors, and how they are analyzed to determine the probability of an accident. The only major accidents: Three-Mile Island and Chernobyl are described as to cause and response. Chapter 8 describes how the consequences of such an accident are calculated. Chapter 7 describes the scope of the chemical process industry, major recent accidents that have occurred, including Bhopal. It goes on to describe hazardous chemical processes and how to analyze the probability of an accident. Nuclear reactors and how they are analyzed to determine the probability of an accident; Chapter 9 describes how the consequences of such an accident are calculated by hand and computer codes that are available for more detailed calculations. Chapter 10 describes how the accident probabilities and consequences are

---

[c] These codes are provided by Dr. Howard Lambert of LLNL and FTA Associates, 3728 Brunell Dr., Oakland CA 94602.

assembled for a risk assessment. Chapter 11 describes how PSA is applied in both industries. More examples of nuclear than chemical applications are given because many of the chemical applications are proprietary. Chapter 12 describes codes on the distribution disk. Chapter 13 is an extensive glossary to define acronyms for the reader that is unfamiliar with them and allows me to use them as shorthand for those that are familiar to them. Chapter 14 provides references from which some information was extracted and some references are provided for further study by the reader. Chapter 15 provides answers to the problems that are at the end of some of the chapters. These problems are not so much exercises as providing an extension of the text. My answers are not simply stated but are worked out to show the reasoning.

I am a physicist who switched to nuclear engineering for my Ph D. My introduction to PSA was as an original participant in the Reactor Safety Study in 1972. Material for this book was first gathered in 1974 for a workshop on what to expect in WASH-1400 (the results of the Reactor Safety Study). Materials were gathered over the years for EPRI, Savannah River Laboratory, and other workshops. A culmination was in 1988 with "Probabilistic Risk Assessment in the Nuclear Power Industry" with Robert Hall as coauthor. This book updates these materials and adds material on PSA in the chemical process industry. I prepared the material for printing using a word processor.

Ralph Fullwood
Brookhaven National Laboratory, retired, 1998.

Acknowledgments

I am greatly indebted to many people over the years who have taught and advised me. To Professor Rasmussen who introduced me to the subject and colleagues on the Reactor Safety Study: Dr. William Vesely, John Kelly, Fred Leverenz, and Abel Garcia. To Dr. Robert Erdman for whom I worked at Science Applications International Corporation and other friends in that organization including Kartik Majumdar, Eugene Hughes Dr. Edward Burns, and Dr. Bryce Johnson. To Professor N. J. McCormick at the University of Washington whose book was an inspiration for this one.

At Brookhaven National Laboratory, I wish to acknowledge the support, assistance, discussions and suggestions with Robert Hall (coauthor on the first book), and Drs. Robert Bari, Ali Azarm, Robert Bari, John Boccio, Lewis Chu, John Lehner, Semyon Shteyngart, and Trevor Pratt. Helen Todosow was very helpful for library assistance.

I am particularly indebted to Dr. Howard Lambert of Lawrence Livermore Laboratory and FTA Associates for providing the PC computer codes FTAP (fault tree code), POSTPR (post processor), IMPORT (importance calculation), MONTE (Monte Carlo error determination), and supporting documentation and instructions.

Dr. Richard Walentowicz provided the EPA CD-ROM disk entitled "Exposure Models Library and Integrated Model Evaluation System" with other reference material. Lester Wittenberg of the Center for Chemical Process Safety, AIChE was particularly helpful in providing a chemical industry perspective and reference material as was Dr. Steven Arendt of JBF Associates, Inc. Drs. David Hesse of Battelle Columbus Laboratories and Vinod Mubayi of Brookhaven National Laboratory were very helpful in providing material on the chemical consequence codes.

For the foreword, I would like to thank Dr. John Garrick, who began PSA work in the 1960s and innovated many of the procedures.

Lastly, I wish to thank my publisher Butterworth-Heinemann and staff for their patient, accurate and prompt work in preparing the manuscript.

Ralph R. Fullwood

Chapter 1

Protecting the Public Health and Safety

Probabilistic Safety Assessment (PSA) is an analytic method for protecting the public health and safety. The institutional methods are traced from early times to the present to show a progressive understanding of risk, related terms and of methods for anticipating safety concerns before they are manifest as death and injury. Besides presenting common definitions of terms related to safety, mathematical representations of risk are presented for calculating risk. Short-comings of the mathematical forms are discussed as well as differences between calculated risk and public perception of risk. Safety goals for nuclear power are discussed as well as the regulatory bases for nuclear power and chemical processing.

1.1 Historical Review

1.1.1 Beginnings

State intervention in man's activities to protect the health of the inhabitants goes back to prehistory. The motivation may not have been altogether altruistic; the king acted to protect his subjects because he regarded them as his property. Public health protection began for disease control. With industrialization, came the need for control of even more hazardous forces and substances. This extended protection became technological in accident analysis and response. Present efforts in controlling risk, such as from nuclear power, are a continuation of this development.

Safety, as it relates to public protection from disease, has a history extending to early history. Ruins in the Indus Valley reveal that as early as 400 B.C., building codes and sanitary engineering was in effect. The Egyptians from the middle kingdom (approximately 2000 B.C.) had bathroom and sewage facilities, as did the Incas. The Greeks formulated principles of hygiene and attempted to show a causal relationship between environmental factors and disease. Indeed, the basic text on epidemiology for 2,000 years was "Air, Waters, and Places" from the Hippocratic collection. The Romans perceived a relationship between swamps and malaria and drained many swamps. They also devised dust respirators for workers, built sewage systems, public baths and great aqueducts. Officials were empowered to destroy impure foodstuffs and regulate public baths, brothels and burial grounds. Justinian I of Byzantine, to combat one of the worst plagues in history (532 A.D.), set up quarantine posts and required certificates of health for admission to Constantinople.

Health protection is a very old concept that has also been incorporated into state religion. The Biblical books: Leviticus and Numbers established dietary laws, rules of hygiene, precautions against contagious diseases and prohibitions against consanguineous marriages. These rules are similar to those imposed by modern states concerning food preservation, epidemic control, and hereditary diseases.

While health care declined after the fall of the Roman Empire, England used the common-law concept of public nuisance to protect the public from flagrant cases of polluting the waters. In France, Germany, and Italy, tanners were prohibited from washing skins in the water supply. London, from 1309, had ordinances regulating cesspools and sewers. The Florentines forbade the sale of meat on Monday that had been slaughtered on Friday.

National health legislation came into being in the 19th century primarily in the form of laws that governed the conditions of child labor and eventually prohibited it. In Germany, medical police were organized to make and enforce health and safety regulations. Both France and Germany became committed to the proposition that government had a positive duty to provide for the health, safety and welfare of workers and citizens.

The coming of industrialization intensified existing problems and created new ones. With the Clearances in England, came migration of farm labor to the cities as well as improvements in agricultural productivity to support the increasing urban population and consequent increase in communicable diseases. Smallpox was the most widespread disease in the 18th century. Peak years in London occurred between 1723 and 1796, with a periodicity of about five years. Each outbreak took over three thousand lives. In the 1740s, 75 % of London's infants died before the age of five. The diseases of typhus and scarlet fever were also major contributors.

Victorian England led the world to better health by actions improving nutrition and working conditions. The Public Health Act of 1848, established Local Boards of Health specifying educational levels of the district health officers and empowering them to enforce sanitation requirements.

1.1.2 Industrial Revolution

The other effects on safety brought by industrialization resulted from new and more powerful energy sources. Although water and wind power was used in the Middle Ages, these forces were "natural" and believed to be understood. However, the steam engine was something new. The original condensation engine was sub-atmospheric, but with Watt's invention and Carnot's theory, the quest for higher steam pressure and temperature began.

The original steam generators were simple pressure vessels that were prone to catastrophic failures and loss of life. Due to better boiler design, tube-fired boilers, and boiler inspections, the incidence of catastrophic failure is now to a rare event (about once every 100,000 vessel-years). In Great Britain in 1866, there were 74 steam boiler explosions causing 77 deaths. This was reduced to 17 explosions and 8 deaths in 1900 as a result of inspections performed by the Manchester Steam User Association. In the United States, the American Society of Mechanical Engineers established the ASME Pressure Vessel Codes with comparable reductions.

The development of steam and later the internal combustion engine made possible transportation by rail, road and air at speeds never before experienced. In all cases, the regulations, inspections, and design standards were imposed after the hazards had been exhibited by many deaths and injuries. Nuclear power has attempted, rather successfully, to anticipate the risks before they occur and avoid them through design, control and regulation. PSA is an essential analytical tool for accomplishing this result.

1.1.3 This Century

The discovery of nuclear fission made possible a far more concentrated energy source than ever before. Its hazards were recognized from the beginning, and for the first time, a commitment by government, to safely bring a technology on line without the deadly learning experiences that occurred to safely use earlier technologies. During World War II, experience was acquired in the operation of plutonium production and experimental reactors. Shortly after passage of the Atomic Energy Act of 1948, the Reactor Safeguards Committee was formed (1947) which was to merge with the Industrial Committee on Reactor Location Problems (1951) to become the ACRS (Advisory Committee on Reactor Safeguards). The Atomic Energy Act of 1954 made industrial nuclear power possible, and the first plant began operation at Shippingport, PA, in 1957. The risk posed by a nuclear power plant at this time was unknown, hence the Price-Anderson Act was passed to limit the financial risk.

The first report on nuclear power plant accidents, WASH-740, was issued by Brookhaven National Laboratory (1957). The consequences predicted were unacceptable, but it was believed that the probability of such an accident was very small. This report and the technically untenable Maximum Credible Accident method in licensing gave rise, during the 1960s, to probabilistic approaches to siting (Farmer, 1967; Otway and Erdmann, 1970) and to accident analysis (Garrick et al., 1967; Salvatori, 1970; Brunot, 1970; Otway et al., 1970; Crosetti, 1971; and Vesely, 1971). The most ambitious of the pre-Reactor Safety Studies was Mulvihill, 1966, which consisted of a fault tree probability analysis followed by consequence analysis of the postulated accidents at a nuclear power plant.

Reactor Safety Study

The Reactor Safety Study (RSS) directed by Professor Norman Rasmussen of MIT may have had its beginnings in a letter from Senator Pastore to James Schlesinger, AEC Chairman, requesting risk information for the Price-Anderson renewal. The RSS study began in September 1972 with Saul Levine, full-time staff director assisted by John Bewick and Thomas Murley (all AEC).

A significant development of the study was the use of event trees to link the system fault trees to the accident initiators and the core damage states as described in Chapter 3. This was a response to the difficulties encountered in performing the in-plant analysis by fault trees alone. Nathan Villalva and Winston Little proposed the application of decision trees, which was recognized by Saul Levine as providing the structure needed to link accident sequences to equipment failure.

The Reactor Safety Study was the most important development in PSA because it:

- Established a pattern for performing a PSA of a nuclear plant;
- Provided a basis for comparison;
- Identified transients and small LOCAs as the major risk contributors, rather than the previous emphasis on a large LOCAs;
- Showed that the radiological risk of a nuclear power plant is small compared with other societal risks;
- Originated the event tree for linking initiators, systems, and consequences, and introduced the fault tree to a large audience;
- Compiled a database;
- Showed that human error is a major contributor;
- Showed the impact of test and maintenance; and
- Showed the importance of common mode interactions.

The work was published as draft WASH-1400 in August 1974 and extensively reviewed. The revised report was published as WASH-1400 (FINAL) in October 1975.

Following the release of WASH-1400, the techniques were disseminated by the authors and interpreters through publications, lectures, and workshops. Many organizations set up in-house PSA groups, and the nucleus of the organization that had produced the Reactor Safety Study continued at the NRC.

Critique of the Reactor Safety Study (RSS)

WASH-1400 (FINAL), Appendix XI presents comments and responses on the draft report. Some of these resulted in changes that were incorporated in the final report. Only a few critiques of the final report have been published. Two of these are NUREG/CR-0400 (Lewis Report) and Kendall, 1977. Of these, the Lewis Report's comments are the most objective (Leverenz and Erdmann, 1979, provide a review of the Lewis review). Some comments are summarized:

1. Despite its shortcomings, WASH-1400 provides at this time (1978) the most complete single picture of accident probabilities associated with nuclear reactors. The fault tree/event tree approach coupled with an adequate database is the best method available to quantify these probabilities.

2. The Committee is unable to determine whether the absolute probabilities of accident sequences in WASH-1400 are high or low, but it is believed that the error bounds on those estimates are, in general, greatly understated. This is due in part to an inability to quantify common cause failures, and in part to some questionable methodological and statistical procedures.

3. It should be noted that the dispersion model for radioactive material developed in WASH-1400 for reactor sites as a class cannot be applied to individual sites without significant refinement and sensitivity tests.

4. The biological effects models should be updated and improved in the light of new information.

5. After having studied the peer comments about some important classes of initiating events, we remain unconvinced of the WASH-1400 conclusion that they contribute negligibly to the overall risk. Examples include fires, earthquakes, and human accident initiation.

6.   It is conceptually impossible to be complete in a mathematical sense in the construction of event trees and fault trees. What matters is the approach to completeness and the ability to demonstrate with reasonable assurance that only small contributions are omitted. This inherent limitation means that any calculation using this methodology is always subject to revision and to doubt as to its completeness.

7.   The statistical analysis in WASH-1400 suffers from a spectrum of problems, ranging from lack of data on which to base input distributions to the invention and the use of wrong statistical methods. Even when correct, the analysis is often presented in so murky a way as to be very hard to decipher.

8.   For a report of this magnitude, confidence in the correctness of the result can only come from a systematic and deep peer review process. The peer review process of WASH-1400 was defective in many ways and the review was inadequate.

9.   Lack of scrutability is a major failing of the report, impairing both its usefulness and the quality of possible peer review.

These criticisms only are partially addressed in subsequent work.

1.2 Risk Assessment Objectives

The assessment of risk with respect to nuclear power plants is intended to achieve the following general objectives:

- Identify initiating events and event sequences that might contribute significantly to risk,
- Provide realistic quantitative measures of the likelihood of the risk contributors,
- Provide a realistic evaluation of the potential consequences associated with hypothetical accident sequences, and
- Provide a reasonable risk-based framework for making decisions regarding nuclear plant design, operation, and siting.

One of the products of a nuclear power plant PSA is a list of plant responses to initiating events (accident starters) and the sequences of events that could follow. By evaluating the significance of the identified risk contributors, it is possible to identify the high-risk accident sequences and take actions to mitigate them.

Although the consequences of the high-risk accident sequences may vary from one PSA to another, all PSAs attempt to evaluate realistically, the consequences of hypothetical accident sequences. Depending on the scope of the PSA, these evaluations may include an estimation of the number of latent cancers, the number of immediate fatalities, the probability of core damage, or a number of other consequence measures.

When used to identify and evaluate significant risk contributors, as well as to assess the consequences of accident sequences, the PSA provides a comprehensive framework for making many types of decisions regarding reactor design, operation, and siting. These and other applications can be facilitated by the rational evaluation of the risks associated with a particular installation.

Protecting the Public Health and Safety

1.3 Risk, Hazard, and other Terms

This text is concerned with quantified risk. To treat any subject mathematically, precise definitions are necessary for a common understanding. Risk is related to safety, danger, hazard, loss, injury, death, toxicity, and peril but it has two meanings that may cause confusion. The first definition concerns "hazard, peril, and exposure to injury or loss," which suggests an unrealized potential for harm. If the danger becomes real, then it is no longer risk but becomes "injury, loss, or death." The second definition is more explicit: "Risk is the <u>chance</u> of loss, injury, or death." Chance, likelihood, and probability are all related words for a random process.

1.4 Quantitative Aspects of Risk

Risk is a nebulous concept, but when low risk equipment leads to major consequences, the public feels that something is wrong - especially after the media perform their work. Putting risk on a mathematical foundation is a first step in setting a number to risk.

1.4.1 Actuarial or Linear Risk

To convert the words of the second definition of risk into mathematics, let "chance" be probability, "loss" be consequences, and "of" be multiplication. Expressed in words, this is: Risk = Probability times Consequences which is expressed by equation 1.4-1, using R for risk, p for probability, and C for consequence.

$$R = p * C \quad (1.4\text{-}1)$$

A useful example of equation 1.4-1 is insurance, which was invented by the 15th century Genoese to protect against individual catastrophic shipping losses by sharing the risk. To derive the risk equation, suppose the insurer collects premium R for insuring N ships per year of which n are lost and an award C is paid for each lost ship.

$$N * R = n * C \quad (1.4\text{-}2)$$
$$R = (n/N) * C \quad (1.4\text{-}3)$$

The insurance company receives N*R in premiums; it pays out n*C and breaks even (neglecting insurance company expenses) when these are equal (equation 1.4-2). Solving equation 1.4-2 for risk gives equation 1.4-3. As N becomes very large, the ratio, n/N, approaches probability, p (Section 2.3). Thus, R = p * C, as stated in equation 1.4-1.

To illustrate, if you are in an age group with 1% probability of dying per year and the insurance pays $10,000, your annual premium must be at least $100 for the insurance company to break-even without considering the insurance company's expenses.

A risk equation for nuclear power may be derived by imagining a world with a very large nuclear power plant population. All plants are identical with the same demography and meteorology. The plants are separated such that one does not affect the other. Each year, n_i plants fail in the ith failure mode, causing a population dose d_i. If the effects are additive, the population dose (other risk measures could be used) is linearly proportional to the number failing (Equation 1.4-4), where c_i is

Quantitative Aspects of Risk

a proportionality constant that is interpreted as the average population dose per plant caused by the ith failure mode. The total population dose D is the sum over the doses from M failure modes (equation 1.4-5). The probability of occurrence of the ith failure mode in the limit of large N is equation 1.4-6. The population dose per year per plant, and hence the plant risk, is given by equation 1.4-7 which says that risk is the expected consequence in the same sense that an insurance premium is the expected consequence of the awards.

$$d_i = c_i * n_i \quad (1.4\text{-}4)$$

$$D = \sum_{i=1}^{M} d_i = \sum_{i=1}^{M} c_i * n_i \quad (1.4\text{-}5)$$

$$p_i = \lim_{N \to \infty} \frac{n_i}{N} \quad (1.4\text{-}6)$$

$$R = \lim_{N \to \infty} \frac{D}{N} = \sum_{i=1}^{M} c_i * p_i \quad (1.4\text{-}7)$$

In conclusion, *risk is probability times consequences* which is the expectation value of the consequences. In analogy to insurance, *risk is the premium paid by society for the use of a technology.*

$$R = \sum_{i=1}^{M} c_i^{\nu} * p_i \quad (1.4\text{-}8)$$

1.4.2 Shortcomings of Linear Risk

Equation 1.4-7 is unsatisfactory because the risk from a large number of small accidents is the same as from a small number of large accidents if the total number of effects, say fatalities, is the same for each case. It is hypothesized that the perceived risk of a large accident is greater than the equivalent risk from many small accidents because of human nature and the emphasis of the news services on the unusual (50,000 traffic deaths per year is not newsworthy, but a single accident killing 50,000 is very newsworthy).

To address this nonlinearity, it has been proposed that the risk equation be modified as shown in equation 1.4-8 where the consequences are raised to the ν-power to account for the effects of perception. Unfortunately, a physical basis for the value of ν has not been established, but a suggested value is 1.2 (NUREG-0739). If ν were set to 1, then risk would be linear and not allow for perception.

The perception problem may be avoided if risk is considered to be an ordered pair composed of probability and consequences without a relationship between the members of the pair. Figure 1.4.2-1 illustratively shows points representing the probability and consequence of accidents, associated with an activity plotted as a log-log graph. Encompassing the points are a smooth curve through the maximum points, and a tangent to this curve that defines at the maximum risk (p*C product). *Notice that neither curve bounds the total risk* they bound the p, C ordered pair combinations. That is, given p, the curves will indicate the maximum consequence with the curve giving a tighter (less conservative) bound than the straight line. The linear curve has the equation: p*c = k, where k is the constant risk value. (The slope of the constant risk curve will be 45° if the abscissa and ordinate have equal size decades.)

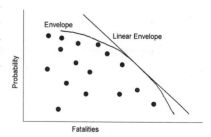

Fig. 1.4.2-1 Curve Enveloping Accident Data and a Linear Envelope for a Looser Straight-Line Bound

7

Although discussion up to this point has treated probability and consequences as precise quantities, there are uncertainties associated with both. It is common practice to present uncertainties as a "bell-shaped" curve called a probability distribution, probability density function (pdf), or just distribution (Figure 2.5-1). Such bell-shaped curves, however, are not probability but are the rate of change of probability (probability density). Probability is obtained by integrating over portions of the probability density. An exceedance probability (CCDF - complementary cumulative distribution function) is the integral from x to ∞ (equation 1.4-9). This is the way that WASH-1400 presents most of its results. The next section discusses several ways of presenting PRA results and this method in particular. Figure 1.4.2-2 illustrates this type of presentation. Instead of showing the accident (p, C) combinations as dots, the accidents are presented as exceedance plots, i.e., the probability that a consequence x will be exceeded.

$$CCD = \int_x^\infty pdf(x) * dx \qquad (1.4\text{-}9)$$

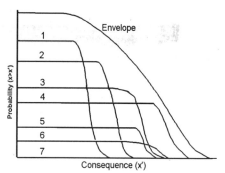

Fig. 1.4.2-2 Exceedance Probability (accidents 1-6) and composite envelope. Each curve represents the probability that the consequences of the accident will be greater than some value x_o

As a final comment on inadequacies of mathematical representations of risk, those who bear the risk are not necessarily those who receive the benefit. While unequal distribution of risk and benefit may not be fair, it is difficult to redress the inequity.

1.4.3 Presentation of Risk

Abstract calculations of risk have meaning only when their meaning is understood by people. Their significance may be communicated quantitatively and qualitatively.

1.4.3.1 Qualitative Results

System models, even unquantified, provide valuable insight into the robustness of the plant design. Because of the uncertainties in the data used to quantify the PSA, some people say that information in the cutsets is the most valuable result of a PSA. The cutsets are the groupings of things that must fail for an undesirable event to occur. Such an event could be some degree of core melt or it could be some system failing to perform its design function. Table 1.4.3-1 shows a typical cutset listing, grouped according to the order of combinations of failed components (single, double, triple, etc.) that must fail to cause system failure. For example, if failures A and D ("*" is the symbol for logical "AND"), occur concurrently, the system will fail. Furthermore, paired failures G*H, I*J, etc. will fail the system as will the triple combinations P*R*S, P*R*T, etc. Clearly, singlet components are more important than doublet which are more important than triplet

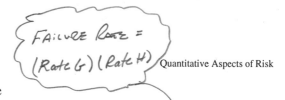

Quantitative Aspects of Risk

components. Prioritization of test, maintenance and inspection activities may be prioritized according to their order in a cutset listing. A cutset listing also indicates systems that may be taken out for repair when the system is operating. For example, G may be taken out as long as H is functioning, but if L is taken out, the failure of M, N or O will fail the system, thus giving L a higher importance than G.

This ordering by singles, doubles and triples takes added meaning when the failure rates of the active and passive components (Table 1.4.3-1) are included. A doublet has a failure frequency that is the product of the two failure rates; a triplet is the product of three failure rates.

Table 1.4.3-1 Illustrative Cutset Listing (letters represent the probability of system failing. The ordering has no significance)

Singlet	Doublet	Triplet
A	G*H	P*R*S
B	I*J	P*R*T
C	I*K	U*W*T
D	L*M	V*W*T
	L*N	X*Y*Z
	L*O	X*W*S

Another qualitative result, obtained from quantitative analysis, is the ordering of accident sequences, according to their fractional contribution to the risk. The order of the sequences is insensitive to data uncertainties (unless they are extreme).

The uncertainties in the data should be carried through the analysis and sensitivity studies performed. It is important to recognize that results can be useful despite large uncertainties because the order of importance may not be strongly affected by the uncertainties.

An important product of the analysis is the framework of engineering logic generated in constructing the models. The numerical estimates of frequencies need only be sufficiently accurate to distinguish risk-significant plant features.

The patterns, ranges, and relative behavior obtained can be used to develop insights into design and operation - insights that can be gained only from an integrated, consistent approach such as PSA. These insights are applicable to regulation and minimizing regulation impact.

Thus, PSA techniques serve as a valuable adjunct to the methods currently used in decision making by both industry and government. Although not yet developed to the point where they can be used without caution, they provide a framework of integrated engineering logic that can be used to identify and evaluate critical areas that influence economics and safety.

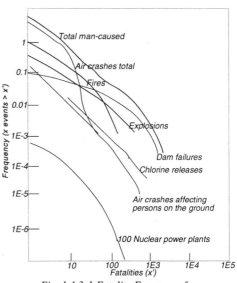

Fig. 1.4.3-1 Fatality Frequency for Man-Caused Events (from WASH-1400)

1.4.3.2 Quantitative Results

Figure 1.4.3-1 from WASH-1400 compares the risk of 100 nuclear plants with other man-caused risks. This is a CCDF that gives the frequency per year that accidents will exceed a value on the abscissa. For example, for 100 fatalities, the frequency that 100 nuclear power plants could do this is 1E-4, air crashes to persons on the ground: 1E-2, chlorine releases: 1.1E-2, dam failures: 7E-2, explosions: 8E-2, fires: 1.1E-1, air crashes (total): 5E-1, and total man-caused: 9E-1.

Some comments regarding Figure 1.4.3-1 are:

1. CCDF plots are difficult for many people to interpret.
2. Many activities are presented but the benefits of each are not the same. For example, there is no viable alternative to air travel, but there are alternatives to producing electricity with nuclear power plants. A better comparison would be between alternative methods for producing the same quantity. This was not done because the authors of WASH-1400 wanted to relate the risk of national nuclear power usage to risks with which the public is more familiar.
3. CCDFs may be compared if they have the same shape. If not, a line of constant risk may be drawn (Figure 1.4.2-1) and the comparison made by comparing the envelopes of constant risk.
4. Figure 1.4.3-1 does not reflect the uncertainties in the analysis; Figure 1.4.3-2 addresses this deficiency by presenting envelopes at the 5, 50, and 95% confidence levels. Of course, including confidence intervals on all curves, e.g., Figure 1.4.3-1 would be confusing.

Figure 1.4.3-3 is an example from the Indian Point hearings showing an effective use of CCDF. The purpose of this presentation was to assure that the reactor at the Indian Point site is about as safe as any that could be sited there; hence, comparisons are made of the risk

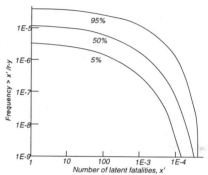

Fig. 1.4.3-2 Confidence Intervals: CCDF for Latent Cancer Fatalities Internal and External Seismicity

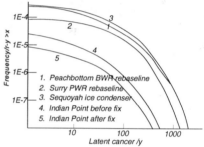

Fig. 1.4.3-3 Latent Cancer Risk at Indian Point site if the Reactor were one of the five types shown.

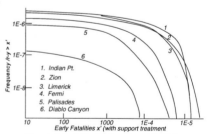

Fig. 1.4.3-4 Early fatality risk if the Indian Point Plant had been at one of these sites

Quantitative Aspects of Risk

Table 1.4.3-2 Expected Annual Consequences (Risk) from Five LWR Designs if Sited at Indian Point				
Situation	Early Deaths	Early Injuries	Latent Cancers	Damage k$
IP after fix	2.2E-4	2.7E-4	1.6E-5	199
IP before fix	6.3E-4	9.5E-4	4.4E-5	700
Surry rebaselined	6.1E-3	1.5E-2	5.4E-4	9,550
Sequoyah ice condenser	2.7E-3	2.2E-2	1.2E-3	14,800
Peach Bottom BWR rebaselined	1.7E-2	3.1E-3	1.1E-3	13,500

Table 1.4.3-3 Estimated Frequencies of Severe Core Damage		
Reactor	Type	Freq./r-yr
Surry	3-loop PWR	6E-5
Peach Bottom	BWR (Mark 1)	3E-5
Sequoyah	4-loop PWR (ice condenser)	4E-5
Oconee	2-loop PWR	2E-4
Calvert Cliffs	Calvert Cliffs	2E-4
Crystal River-3	2-loop PWR	3E-4
Biblis	4-loop PWR	4E-5
Indian Point	4-loop PWR	1E-5

posed by other reactors at the same site. Figure 1.4.3-4 is presented to show that the Indian Point site is reasonable by comparison with some other sites. This figure illustrates the point that CCDFs cannot be compared if the curves have different shapes.

Logarithmic scales are frequently used in presenting risk results but most of the public do not understand logarithmic scales. Fullwood and Erdman, 1983 circumvent this problem by comparing risk as cubes in which linear dimensions are the cube root of the volume/risk. Figure 1.4.3-5 compares the risks associated with nuclear fuel reprocessing, refabrication and waste disposal with non-nuclear risks.

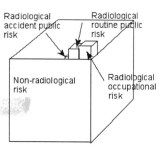

Fig. 1.4.3-5 Risk compared as volumes or masses

Presenting information as a cumulative integral (CCDF) may be confusing. Tables 1.4.3-2 and 1.4.3-3, corresponding to the preceding figures, present considerably more information in terms of expected effects (probability times consequences). Table 1.4.3-2, also from the Indian Point hearings, compares the PSA-assessed probabilities of severe core damage for various reactors.

Perhaps the most readily perceived type of presentation is a "pie" chart of the major contributors as determined in the Zion and Arkansas Nuclear One (ANO)-l PSAs and presented in Figure 1.4.3-6.

1.4.4 Public Perception of Risk

The preceding section shows the rather abstract, mathematical methods used to present the results of PSA. While the "public" is a very diverse group, the majority do not receive information in this fashion but receive it through the news media. It is not the nature of news to attempt a balanced presentation but to emphasize the unusual. To illustrate, the fact that 50,000 people die each year in traffic accidents, involving a few deaths at a time, is not newsworthy. Furthermore the latency of a hazard greatly affects risk perception. If smoking resulted in immediate death, however unlikely, the public attitude would be greatly changed because the cause-effect relationship would be apparent. There is similarity between the risk of smoking and that of a nuclear power accident in that both result in an increase in the probability of cancer developing with a latency period of about 20 years, yet the hazard of nuclear power is perceived quite differently from that of smoking.

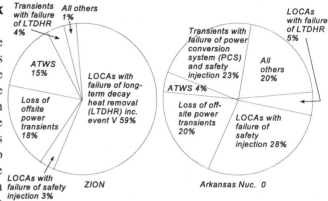

Fig. 1.4.3-6 Relative contributions to the Zion and ANO-1 core melt frequencies. Reprinted with permission of the Electric Power Research Institute USA

Equation 1.4-7 showed that as long as consequences are small enough, that effects are linear, i.e., not so catastrophic as to affect the perpetuation of civilization, the number of injuries or fatalities in either case are the same whether or not there are many small accidents or a few large accidents. But the public, either because of information sources or for more fundamental reasons does not see it this way. Since the public affects government, public perception is a concern.

The literature on this subject is so large that it cannot be encompassed in a brief review. NUREG/CR-1930 (1981) is a bibliographical survey of 123 references; Covello, 1981 lists 148 references. Since no risk should be tolerated if it has no benefits, most of the papers address the question "How safe is safe enough," by comparisons with acceptable risks. (In many cases these "acceptable" risks are really "tolerated" risks in that the cost of reduction does not seem to be warranted.)

Starr, 1969 approached this by investigating the "revealed preferences exhibited in society as the result of trial and error. (Similar to the "efficient market theory" in the stock market.) Starr conjectured that the risk of death from disease appears to determine a level of acceptable voluntary risk but that society requires a much lower level for involuntary risk. He noted that individuals seem to accept a much higher risk (by about 1000 times) if it is voluntary, e.g., sky-diving or mountain climbing, than if it is imposed, such as electric power or commercial air travel, by a correlating with the perceived benefit. From this study, a "law" of acceptable risk was found concluding that risk acceptability is proportional cube of the benefits. Figure 1.4.4-1 from Starr, 1972 shows these relationships. One aspect of revealed preferences is that these preferences do not necessarily remain constant (Starr et al., 1976). In Starr et al., 1976, it is shown that while nuclear power has the least risk of those activities compared, it also has the least perceived benefit. Clearly the public thinks that alternative ways of producing electricity have less risk despite risk comparisons of various ways of electric power production such as Inhaber, 1979, Okrent, 1980 or Bolten, 1983.

Another approach to public perception of risk is to simply ask the public. This was the approach of Fischoff et al., 1977 in which a survey of the League of Women Voters in Eugene, Oregon was taken. Later studies covered 40 college students at the University of Oregon, 25 Eugene businessmen and 15 national experts in risk analysis. Thomas, 1981 reports a survey of 224 selected Austrians.

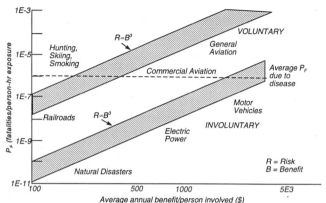

Fig. 1.4.4-1 Revealed and risk-benefit relationships (from Starr, 1971). Reprinted with permission from Perspectives on Benefit-Risk Decision Making, National Academy of Engineering, U.S.A.

Cross-comparing the risks of various activities is difficult because of the lack of a common basis of comparison, however Cohen and Lee, 1979 provide such a comparison on the basis of loss of life expectancy. Solomon and Abraham, 1979 used an index of harm in a study of 6 occupational harms - three radiological and three nonradiological to bracket high and low estimates of radiological effects. The index of harm consists of a weighting factor for parametric study: the lost time in an industry and the worker population at risk. The conclusions were that the data are too imprecise for firm conclusions but it is possible for a radiation worker under pessimistic health effects assumptions to have as high index of harm as the other industries compared.

To conclude, this sampling of the literature of risk perception, the comments of Covello, 1981 may be summarized. Surveys have been of small specialized groups - generally not representative of the population as a whole. There has been little attempt to analyze the effects of ethnicity, religion, sex, region age, occupation and other variables that may affect risk perception. People respond to surveys with the first thing that comes to mind and tend to stick to this answer. They provide an answer to any question asked even when they have no opinion, do not understand the question or have inconsistent beliefs. Surveys are influenced by the order of questions, speed of response, whether a verbal or numerical response is required and by how the answer is posed. Few studies have examined the relationships between perceptions of technological hazards and behavior which seems to be influenced by several factors such as positive identification with a leader, efficacy of social and action, physical proximity to arenas of social conflict.

1.5 Safety Goals

During the 1970s, a number of safety requirements were imposed that required backfitting existing nuclear power plants that increased the costs of new plants. It was believed that codifying a level of safety would stabilize the licensing and possibly aid in the public acceptance of nuclear power. The Three Mile Island-2 accident gave further impetus to establishing safety goals because the requirements of 10CFR100 (accidents must be less than 25 rem whole body or 300 rem thyroid dose to any individual) were not even closely approached. Nevertheless the public response essentially invalided the rule. The NRC, in responding to the President's Commission on the Accident at Three Mile Island, stated that it was "prepared to move forward with an explicit policy statement on safety philosophy and the role of safety-cost tradeoffs in the NRC safety decisions." The objective of the policy statement was to establish goals defining a acceptable level of radiological public accident risk imposed on the public by the operation of nuclear electric power plants. Hearings on both coasts were held and suggestions were solicited. Some of the papers that were published in response to this request were: Mattson et al., 1980; Starr, 1980; Salem et al., 1980; O'Donnell; 1982 and Cox and Baybutt, 1982. The NRC issued a "Discussion Paper on Safety Goals for Nuclear Power Plants" (NUREG-880, 1983) with qualitative goals and a suggested a quantitative goal of less than one core melt per 10,000 years.

Discussions continued and in August 1986 the formal statement was issued by the NRC (Federal Register 51.162 p30029 - 30033). The two qualitative safety goals are:

- "Individual members of the public should be provided a level of protection from the consequences of nuclear power plant operation such that individuals bear no significant additional risk to life and health.
- "Societal risks to life and health from nuclear power plant operation should be comparable to or less than the risks of generating electricity by viable competing technologies and should not be a significant addition to other societal risks."

The statement goes on to acknowledge the contribution of the Reactor Safety Study (WASH-1400) to risk quantification but points out that safety goals were not the study objectives and that the uncertainties make it unsuitable for such a purpose. After pointing out that the death of any individual is not "acceptable," it states two quantitative objectives:

- "The risk to an average individual in the vicinity of a nuclear power plant of prompt fatalities that might result from reactor accidents should not exceed 0.1% of the sum of prompt fatality risks from other accidents to which members of the U.S. population are generally exposed."
- "The risk to the population in the area near a nuclear power plant of cancer fatalities that might result from nuclear power plant operation should not exceed 0.1% of the sum of cancer fatality risk resulting from all other causes."

The fraction 0.1% is chosen to be so low that individuals living near a nuclear plant should have no special concern because of the closeness. Uncertainties in the analysis of risk are not caused by the "quantitative methodology" but are highlighted by it. Uncertainty reduction will be achieved by methodological improvements; mean values should be calculated. As a guideline for regulatory implementation, the following is recommended:

"...the overall mean frequency of a large release of radioactive materials to the environment should be less than 1 in 1,000,000 years of reactor operation."

These safety goals are not meant to supersede any of the present conservative design and operational practices such as, defense in depth, low population zone siting or emergency response capabilities. The goals may be considered as a factor in the licensing decision. Commissioner Asselstine of the USNRC provided an additional view by saying that the nuclear industry has been trying to distance itself from the Chernobyl accident on the basis of containment performance (the Russian plant had no containment structure) and recommended "a mean frequency of containment failure in the event of a severe core damage accident should be less than 1 in 100 severe core damage accidents." Commissioner Asselstine also recommends that the large release criterion of less than once per 1,000,000 reactor years be adopted with clarification as to the meaning of a "large release."

Commissioner Bernthal, felt that the statement does not address the question "How safe is safe enough?" or assure the public concerning the Chernobyl accident, while preserving the nuclear power option. A fool-proof containment, while protecting the public would not satisfy this last requirement. He furthermore decries the absence of population density consideration in the 0.1% goals which would allow a plant to be located in "New York's Central Park." Furthermore he points out that the 0.1% incremental societal health risk standard that is adopted is a purely subjective assessment of public acceptance.

Table 1.6-1 Chronology of Federal Emergency Regulations	
Date	Description
1970	10CFR50.34 required a discussion of plans for coping with emergencies.
1975	The format and content of onsite emergency plans was given in Reg. Guide 1.101. Offsite emergency planning was required for licensing purposes for a low-zone (EPZ) within a 3 mile radius about the plant.
1978	NUREG-0396 defined two EPZ at radii of 10 miles to provide protection from direct radiation by evacuating or sheltering the public and at 50 miles within which food and water interdiction would protect from this dose pathway.
1979	President Carter directs the Federal Emergency Management Administration (FEMA) responsibility of offsite emergency planning. The NRC retained jurisdiction over plant licensing and operation with respect to onsite emergency preparedness.
1980	NUREG-0654/FEMA-REP-1 and 10CFR50 Appendix E gave bases for both onsite and offsite emergency planning. It requires joint utility, state and local participation in an annual simulated accident exercise as a condition for an operating license.
1981	The NUREG-0654, etc., requirements were modified such that the annual exercise is needed only for operation above 5% power.
1983	FEMA relaxed the rule for offsite exercises to every two years. The onsite exercise frequency remains the same.

1.6 Emergency Planning Zones

The philosophy of public health protection used by the AEC and pursued ever since, is the use of multiple independent barriers, each a significant shield for the public. The last barrier involves the removal of people from the area over which the radioactive plume is expected to pass, interdiction of food supplies and the use of prophylaxis to reduce the iodine dose. Blood transfusions have also been used to save the lives of people lethally exposed. Emergency planning to protect the public in case of a nuclear power accident is in a much more advanced state than the emergency planning for more likely and potentially more dangerous accidents such as poisonous chemical releases or dam failure because it is required by law. It does present some vexing problems regarding legal responsibilities. The Atomic Energy Act of 1954 made the Federal Government responsible for nuclear power but emergency actions may require the participation of several states and many local jurisdictions. The lack of cooperation from any of these may be used to oppose Federal policy.

Table 1.6-1 from Buzelli, 1986 gives a chronology of the Federal regulations.

Evacuation in case of a nuclear accident had always been considered a possibility to be considered on an ad hoc basis much as it is for releases of hazardous material that could affect the public. The Reactor Safety Study (WASH-1400, 1975) showed the possibility of some immediate fatalities,[a] and set in motion the regulatory process shown in the table. The TMI-2 accident, during which Governor Thornburg, concerned about the possibility of a hydrogen-bubble explosion, recommended an evacuation, did much to increase the concern for emergency planning.

In the aftermath of TMI-2, NUREG-0654 and regulations were promulgated that strongly codified the emergency planning process. NUREG-0654 presents bases for emergency planning and its relationship to the EPZ. Sixteen standards are presented to be used in evaluating the plant's emergency plan. The criteria for emergency response are presented in Table 1.6-2.

The nuclear power industry has generally opposed these emergency planning requirements partially on the basis of cost but primarily because of the poor public image it gives these plants by involving the public in the drills not required for any other industry. Reducing the size of the EPZ is much to their liking. The technical basis for the size of the EPZ was WASH-1400 which was a basis for the criteria presented in NUREG-0396. Besides demonstrating an evacuation, TMI-2 also demonstrated that the quantities of radioactive releases used in WASH-1400 were grossly conservative for many isotopes. The industry response was the IDCOR program and the NRC response to a reexamination of the "source terms" is contained in NUREG-0956 (discussed in Section 7.2). Hazzan and Warman, 1986 using these and other sources recommend that the EPZ be reduced to 2 miles as does Kaiser, 1986. Solomon and Kastenburg, 1985 reviewed 4 PSA studies of the Shoreham plant - one of the most affected plants by emergency planning and concluded that plant features would allow a reduction to 10 miles.

In concluding this brief discussion on emergency planning, it may be pointed out that the planning criteria are not dependent on the size of the plant nor on special plant features. For example, a plant could have such a strong containment that nothing could escape, even in the most severe accident, and the EPZ would be the same as it is for any other plant. Another anomaly is that the 50 mile ingestion pathway from the Canadian nuclear power plants reach into the U.S. without the same requirements being applied.

---

[a] Immediate in emergency planning means a death within a year of exposure. Long-term deaths are a different problem because of the indistinguishability of a cancer death initiated by radiation and other causes, e.g.., smoking. The effect of a nuclear accident would be a very small contributor to a large background of other causes.

Table 1.6-2 Emergency Action Levels and Responses

Emergency class	Plant criteria	Release of activity	Response
Unusual event (UE)	Indication of potential plant degradation	None	Notification and information
Alert (A)	Actual or potential degradation of plant safety	Minor	Notification and information, partial activation of Emergency Operations Center (EOC)
Site area emergency (SAE)	Actual or likely failure of plant safety	Small - within Federal guidelines	Activation of EOC, deployment of emergency monitoring and communication teams
General emergency (GE)	Actual or imminent substantial core degradation loss of containment possible	Releases reasonably expected to exceed Federal guidelines	Recommendations to evacuate/shelter for a 2-mile radius, shelter downwind expected to sectors to five miles

1.7 Use of PSA by Government and Industry

Interpreting PSA as any risk assessment using accident probabilities and consequences, government and industry have prepared many examples. These are discussed as: Public Risk, Specialized Analyses, and Performance Improvement.

Initially PSAs were used, primarily, to evaluate risk imposed on the public or workers. This includes the use of PSA in licensing and/or review procedures to assure protection of public health and safety. Examples by user are:

- Department of Defense - requires that a PSA be performed according to MIL-STD-882A for any major activity or undertaking, e.g., analyses of the transportation of nuclear weapons and deactivation of chemical weapons.
- Environmental Protection Agency - has sponsored work on the risk of chemical manufacture and transportation, the risk of reprocessing nuclear fuel, and the risk of nuclear waste disposal.
- Department of Transportation - has sponsored work on air, ground, rail, and water transportation, using PSA methods.
- Department of Energy - has sponsored analyses of its reactors and process facilities, the risks of the breeder reactor, the risk of nuclear material transportation and disposal, and the risks of several fuel cycles.
- Chemical Industry - has risks comparable to or possibly greater then those of the nuclear power industry, but no risk studies of chemical plants in the U.S. have been published. Great Britain, on the other hand, has been active in this area, e.g., the Canvey Island Study (Section 11.4.1 and Green, 1982).

- Nuclear Regulatory Commission - has been a prime mover in the use of PSA and in requiring its use by applicants. Although the Reactor Safety Study (RSS) originated under the Atomic Energy Commission, it was completed by the NRC. Its initial purpose was to provide information for the, at that time, forthcoming Price-Anderson renewal hearings regarding the risk of commercial nuclear power plants.

1.8 Regulation of Nuclear Power

1.8.1 Regulations

The regulations concerning nuclear installations in the United States are governed by the Atomic Energy Act of 1954 as amended, and the Energy Reorganization Act of 1974, as amended (which created the NRC, and the DOE). The NRC administers statutes on the licensing of nuclear installations in the United States. In addition, other applicable statutes are given in Table 1.8-1

Nuclear installations in the United States must be licensed by the NRC, but some Federal government facilities are exempt from licensing such as the Department of Energy.

Table 1.8-1 Some Federal Regulations Governing the Environment

▸National Environmental Policy Act of 1969	▸Low Level Radioactive Waste Policy Amendments Act of 1985
▸Resource Conservation and Recovery Act (RCRA) of 1976	▸Administrative Procedure Act
▸Toxic Substances Control Act; Clean Air Act of 1977	▸Coastal Zone Management Act
	▸Endangered Species Act
▸Clean Water Act (CWA) of 1977	▸Federal Advisory Committee Act Federal Water Pollution Control Act
▸Uranium Mill Tailings Radiation Control Act of 1978	▸Freedom of Information Act
▸Comprehensive Environmental Response, Compensation, and Liability Act (CERCLA) of 1980	▸Government in the Sunshine Act
	▸National Historic Preservation Act
	▸Privacy Act
▸Federal Facilities Compliance Act of 1990	▸Wild and Scenic Rivers Act
▸West Valley Demonstration Project Act of 1980	
▸Nuclear Waste Policy Act of 1982	

1.8.2 Regulatory Structure

The NRC delegates the authority for reactor licensing to its Office of Nuclear Reactor Regulation (NRR). Fuel facility, nuclear waste storage and disposal are delegated to the Office of Nuclear Material Safety and Safeguards (NMSS). In cases where public hearings are required, the decision to license, and the condition of the license, rest with an Atomic Safety and Licensing Board (ASLB). For a construction permit, the ASLB decides all issues; in operating license proceedings, the ASLB decides controversial matters. These decisions are subject to review by an Appeal Board and the NRC Commissioners. In these cases, the NRC staff cannot communicate directly with the Board and the Commission on the merits of the proceedings until a final Commission decision is rendered. Applications for a license to construct and operate an enrichment facility are handled by NMSS; if public hearings are held on such applications, decisions are subject to review by the Commission. Thus, in the United States, it is necessary to distinguish between actions, positions and decisions of the NRC staff, the Boards, and those of the Commission itself. Additional support for the licensing activities of NRR and NMSS comes from the Office of Nuclear Regulatory Research and Regional Offices.

1.8.3 Licensing Process

The licensing process consists of two steps: construction and operating license that must be completed before fuel loading. Licensing covers radiological safety, environmental protection, and antitrust considerations. Activities not defined as production or utilization of special nuclear material (SNM), use simple one-step, Materials Licenses, for the possession of radioactive materials. Examples are: uranium mills, solution recovery plants, UO_2 fabrication plants, interim spent fuel storage, and isotopic separation plants.

An applicant for a construction permit files a Preliminary Safety Analysis Report (PSAR) presenting design criteria and preliminary design information, hypothetical accident analyses, safety features, and site data. An Environmental Report (ER) must be submitted to evaluate the environmental impact of the proposed facility, and information must be submitted to the Attorney General and the NRC staff for antitrust review.

The application is subjected to an acceptance review to determine whether it contains sufficient information for a detailed review, if insufficient, additional information may be requested. When minimum material is submitted, the application is docketed in the Public Document Room, near the project, and an announcement is published in the Federal Register with appropriate officials being informed. When the PSAR is submitted, a substantive review and inspection of the applicant's quality assurance program, covering design, construction and procurement is conducted. The NRC staff reviews the application for undue risk to the health and safety of the public. If such is found, plant modifications may be requested. Additional information for the review are: demography, site characteristics, engineered safety features, design, fabrication, testing, response to transients, accident consequences, PSA conduct of operations, and quality assurance. The NRC staff and its consultants conduct this review over about two to three years using previous evaluations of other licensed reactors.

When the review progresses to the point that the staff concludes that the documentation is acceptable, a Safety Evaluation Report is prepared that represents a summary of the review and evaluation of the application by the staff.

The National Environmental Policy Act (NEPA) review is performed concurrently by the staff. After completion, a Draft Environmental Statement (DES) is issued and circulated for review and comments by the appropriate Federal, State and local agencies, individuals and public. After receipt of comments and their resolution, the Final Environmental Statement (FES) is issued.

When the final design information and plans for operation are ready, a Final Safety Analysis Report (FSAR) and the operating Environmental Report are submitted. The final design of the facility, and the operational and emergency plans are given in the FSAR. Amendments to the application and reports may be submitted from time to time. Each license for operation of a nuclear reactor contains Technical Specifications (Tech. Specs.) and an Environmental Protection Plan assuring the protection of health and the environment.

For an early site permit, the application must describe the number, type, thermal power level, and contain a plan for site restoration if the site preparation activities are performed, but the permit expires without being used. The application must identify physical characteristics that could pose a significant impediment to the development of emergency plans. The issues presented in an early site permit proceeding are mostly environmental, but if they involve significant safety issues, they are reported to the ACRS on the permit application.

Certified Standard Designs allow preapproval of the NSSS to reduce the time required for power reactor licensing. Applications for certification of a design must contain a level of detail comparable to that required for a final design approval. Combined Construction Permits and Conditional Operating Licenses streamline the licensing process by referencing a standard design and early permitted site.

1.8.4 Public Participation

Public hearings are required at the construction permit stage; the ASLB may grant a public hearing at the operating stage if requested by: the NRC, applicant, or a member of the public. The public hearing is conducted by a three-member Atomic Safety and Licensing Board (Board) appointed from the NRC's Atomic Safety and Licensing Board Panel. The Board is composed of a lawyer (chairman), and two technically-qualified persons. The Safety Evaluation, its supplements and the Final Environmental Statement are offered as evidence by the staff at the public hearing. The hearing may be a combined safety and environmental hearing or separate hearings can take place. If the initial decision regarding NEPA and safety matters is favorable, a construction permit is issued to the applicant by the Director of NRR. The Board's initial decision is, in the case of a reactor license application, subject to review by an Atomic Safety and Licensing Appeal Board.

1.8.5 Advisory Committee on Reactor Safety (ACRS)

The ACRS is an independent statutory committee that advises the Commission on reactor safety, and reviews each application for a construction permit or an operating license. It is composed of a maximum of fifteen members appointed by the Commission for terms of four years each.

As soon as an application for a construction permit is docketed, copies of the PSAR are provided to the ACRS for assignment to a project subcommittee consisting of four to five members. During the Staff review, the ACRS is kept informed of requests for additional information, meetings and potential design changes. If the plant is a "standard design" and the site appears acceptable, the subcommittee review does not begin until the staff has nearly completed its detailed review. Otherwise, the ACRS subcommittee begins its formal review early in the process. The Staff's Safety

Evaluation Report and the ACRS subcommittee's report form the basis for ACRS's consideration of a project. The full ACRS meets, publicly, at least once with the staff and applicant and issues a letter report to the Commission and public.

1.8.6 Inspection

Construction inspections encompass all safety-related construction activities at the facility site. They consist of observation of work performance, quality assurance, testing, examination, inspection, records and maintenance. Inspection of pre-operational testing and operational readiness verifies the operability of systems, structures and components related to safety and whether the results of such tests demonstrate that the plant is ready for operation as specified in the SAR. It also verifies that the licensee has an operating organization and procedures consistent with the SAR; it verifies whether the tests conducted under both transient and operating conditions are consistent with the SAR. It also verifies that licensee management controls for the test program are consistent with NRC requirements and commitments.

Operations phase inspections verify, through direct observation, personnel interviews, and review of facility records and procedures that the licensee's management control system is effective and the facility is being operated safely and in conformance with the regulatory requirements. The NRC inspection program assigns resident inspectors at power reactors that are under construction or already operating. The licensee must give any resident inspector immediate, uninhibited access to the facility. In addition to regular inspections, staff members from several NRC offices investigate any significant incident and determine any hazards. Enforcement involves plant shutdowns, corporation and individual fines, and incarceration.

1.8.7 Decommissioning

The regulations require, before a license can be terminated, that the NRC must determine that the licensee's decommissioning activities have been carried out in accordance with the approved decommissioning plan, and the NRC order authorizing decommissioning with a final radiation survey demonstrating that the premises are suitable for release for unrestricted use. The regulations require licensee funding and completion of the decommissioning in a manner that protects public health, safety and environment.

1.8.8 Accident Severity Criteria

10 Code of Federal Regulations (10CFR) part 100 provides reactor siting criteria. It specifies that the fission product release calculated for major hypothetical accidents shall produce a whole

body dose less than 25 rem or thyroid dose less than 300 rem to the public (people outside of the site boundary fence). These criteria are not accepted limits but are intended as reference values.[b]

1.8.9 PSA Requirements

There are no PSA requirements for licensing the current generation of nuclear power plants.[c] Nevertheless PSA (PRA to the NRC) has been a tool for regulatory decisions by the NRC and has been used for compliance demonstration and modification justification by licensees.

10CFR50.54(f) states that the licensee must submit individual plant examinations (IPE) of significant safety issues to justify continuing operation of a reactor facility. The NRC issued Generic Letter stating that utilities with existing PSAs or similar analysis use these results, provided they are updated and certified that they reflect the actual design, operation, maintenance and emergency operations of the plant. The methodologies allow the user to determine dominant accident sequences and to assess the core damage preventive and mitigative capabilities of their plant. The IPEs are needed to confirm the absence of any plant unique vulnerabilities to severe accidents. This is to confirm vulnerabilities, identified by PRAs that were not previously identified by traditional methods.

The Generic Letter does not specify methodology or contents of an acceptable PSA. A satisfactory analysis is to be determined by the NRC staff, although NUREG-1560, volume 2 part 4 lists attributes of a quality PSA.

1.9 Regulation of Chemical Processing and Wastes

Chemical regulation is primarily concerned with the consequences to the public that may result from chemical releases to the environment. Note there is less concern with the probability than with the consequences of release. The major U. S. environmental laws are as follows.

1.9.1 Environmental Law

1.9.1.1 Clean Air Act (CAA) 42 U.S.C. s/s 7401 et seq. (1970)

The Clean Air Act is the comprehensive Federal law that regulates air emissions from area, stationary, and mobile sources. This law authorizes the U.S. Environmental Protection Agency (EPA) to establish National Ambient Air Quality Standards (NAAQS) to protect public health and the environment. The goal of the Act was to set and achieve NAAQS in every state by 1975. This setting of maximum pollutant standards was coupled with directing the states to develop state

---

[b] Author's note - The accident at TMI-1 released radioactivity far below 10CFR criteria but still the licensee was subject to severe regulatory action and suffered severe financial loss.

[c] 10CFR52 requires a PSA for the next generation of nuclear power plants.

implementation plans (SIPs) applicable to appropriate industrial sources in the state. The Act was amended in 1977 to set new goals for achieving NAAQS, since many areas of the country had failed to meet the deadlines. The 1990 amendments to the Clean Air Act in large part were intended to meet unaddressed or insufficiently addressed problems such as acid rain, ground level ozone, stratospheric ozone depletion, and air toxics.

1.9.1.2 Clean Water Act (CWA) 33 U.S.C. s/s 121 et seq. (1977)

The Clean Water Act is a 1977 amendment to the Federal Water Pollution Control Act of 1972, that set the basic structure for regulating discharges of pollutants to waters of the United States. This law gave the EPA the authority to set effluent standards on an industry-by-industry basis and continued the requirements to set water quality standards for all contaminants in surface waters. The CWA makes it unlawful for any person to discharge any pollutant from a point source into navigable waters unless a permit (NPDES) is obtained under the Act. The 1977 amendments focused on toxic pollutants. In 1987, the CWA was reauthorized and again focused on toxic substances, authorized citizen suit provisions, and funded sewage treatment plants (POTWs) under the Construction Grants Program. The CWA provides for the delegation by EPA of many permitting, administrative, and enforcement aspects, of the law to state governments. In states with the authority to implement CWA programs, EPA still retains oversight responsibilities.

1.9.1.3 Comprehensive Environmental Response, Compensation, and Liability Act (CERCLA or Superfund) 42 U.S.C. s/s 9601 et seq. (1980)

CERCLA (pronounced "serk-la") provides a Federal "Superfund" to clean up uncontrolled or abandoned hazardous waste sites as well as accidents, spills, and other emergency releases of pollutants and contaminants into the environment. Through the Act, EPA was given power to seek out those parties responsible for any release and assure their cooperation in the cleanup. EPA cleans up orphan sites when potentially responsible parties (PRPs) cannot be identified or located, or when they fail to act. Through various enforcement tools, EPA obtains private party cleanup through orders, consent decrees, and other small party settlements. EPA also recovers costs from financially viable individuals and companies once a response action has been completed. It is authorized to implement the Act in all 50 states and U.S. territories. Superfund site identification, monitoring, and response activities in states are coordinated through the state environmental protection or waste management agencies.

1.9.1.4 Emergency Planning & Community Right-to-Know Act (EPCRA) 42 U.S.C. 11011 et seq. (1986)

This is also known as Title III of SARA, EPCRA was enacted by Congress as the national legislation on community safety. This law was designed to help local communities protect public health, safety, and the environment from chemical hazards. To implement EPCRA, Congress required each state to appoint a State Emergency Response Commission (SERC). The SERCs were required to divide their states into Emergency Planning Districts and to name a Local Emergency Planning Committee (LEPC) for each district. Broad representation by firefighters, health officials, government and media representatives, community groups, industrial facilities, and emergency managers ensures that all necessary elements of the planning process are represented.

1.9.1.5 Endangered Species Act 7 U.S.C. 136; 16 U.S.C. 460 et seq. (1973)

The Endangered Species Act provides a program for the conservation of threatened and endangered plants and animals and the habitats in which they are found. The U.S. Fish and Wildlife Service (FWS) of the Department of Interior maintains the list of 632 endangered species (326 are plants) and 190 threatened species (78 are plants). Species include birds, insects, fish, reptiles, mammals, crustaceans, flowers, grasses, and trees. Anyone can petition FWS to include a species on this list or to prevent some activity, such as logging, mining, or dam building. The law prohibits any action, administrative or real, that results in a "taking" of a listed species, or adversely affects habitat. Likewise, import, export, interstate, and foreign commerce of listed species are all prohibited.

EPA's decision to register a pesticide is based in part on the risk of adverse effects on endangered species as well as the environmental fate (how a pesticide will affect the habitat). Under FIFRA, EPA can issue emergency suspensions of certain pesticides to cancel, or restrict their use if an endangered species will be adversely affected. Under a new program, EPA, FWS, and USDA are distributing hundreds of county bulletins which include habitat maps, pesticide use limitations, and other actions required to protect listed species.

In addition, EPA enforces regulations under various treaties, including the Convention on International Trade in Endangered Species of Wild Fauna and Flora (CITES). The U.S. and 70 other nations have established procedures to regulate the import and export of imperiled species and their habitat. The Fish and Wildlife Service works with U.S. Customs agents to stop the illegal trade of species, including the Black Rhino, African elephants, tropical birds and fish, orchids, and various corals.

1.9.1.6 Federal Insecticide, Fungicide and Rodenticide Act (FIFRA) 7 U.S.C. s/s 135 et seq. (1972)

The primary focus of FIFRA provides Federal control of pesticide distribution, sale, and use. EPA was given authority under FIFRA not only to study the consequences of pesticide usage but also to require users (farmers, utility companies, and others) to register when purchasing pesticides. Through later amendments to the law, users also must take examinations for certification as applicators of pesticides. All pesticides used in the U.S. must be registered (licensed) by EPA. Registration assures that pesticides will be properly labeled and that, if used in accordance with specifications, will not cause unreasonable harm to the environment.

1.9.1.7 Freedom of Information Act (FOIA) U.S.C. s/s 552 (1966)

The Freedom of Information Act provides specifically that "any person" can make requests for government information. Citizens who make requests are not required to identify themselves or explain why they want the information they have requested. The position of Congress in passing FOIA was that the workings of government are "for and by the people" and that the benefits of government information should be made available to everyone. All branches of the Federal government must adhere to the provisions of FOIA with certain restrictions for work in progress (early drafts), enforcement, confidential information, classified documents, and national security information.

1.9.1.8 National Environmental Policy Act (NEPA) 42 U.S.C. s/s 4321 et seq. (1969)

The National Environmental Policy Act was one of the first laws written to establish the broad national framework for protecting the environment. NEPA's basic policy is to assure that all branches of government give proper consideration to the environment prior to undertaking any major Federal action which significantly affects the environment. NEPA requirements are invoked when airports, buildings, military complexes, highways, parkland purchases, and other Federal activities are proposed. Environmental Assessments (EAs) and Environmental Impact Statements (EISs) that assess the likelihood of impacts from alternative courses of action, are required from all Federal agencies and are the most visible NEPA requirements.

1.9.1.9 Occupational Safety and Health Act 29 U.S.C. 61 et seq. (1970)

Congress passed the Occupational and Safety Health Act to ensure worker and workplace safety. Their goal was to make sure employers provide their workers a place of employment free from recognized hazards to safety and health, such as exposure to toxic chemicals, excessive noise levels, mechanical dangers, heat or cold stress, or unsanitary conditions. In order to establish standards for workplace health and safety, the Act also created the National Institute for Occupational Safety and Health (NIOSH) as the research institution for the Occupational Safety and Health Administration (OSHA). OSHA is a division of the U.S. Department of Labor which oversees the administration of the Act and enforces Federal standards in all 50 states.

1.9.1.10 Pollution Prevention Act 42 U.S.C. 13101 and 13102, s/s 6602 et. seq. (1990)

The Pollution Prevention Act focused industry, government, and public attention on reducing the amount of pollution produced through cost-effective changes in production, operation, and raw materials use. Opportunities for source reduction are often not realized because existing regulations, and the industrial resources required for compliance, focus on treatment and disposal. Source reduction is fundamentally different and more desirable than waste management or pollution control. Pollution prevention also includes other practices that increase efficiency in the use of energy, water, or other natural resources, and protect our resource base through conservation. Practices include recycling, source reduction, and sustainable agriculture.

1.9.1.11 Resource Conservation and Recovery Act (RCRA) 42 U.S.C. s/s 321 et seq. (1976)

RCRA (pronounced "reck-rah") gave EPA the authority to control hazardous waste from "cradle-to-grave" including generation, transportation, treatment, storage, and disposal of hazardous waste. RCRA also set forth a framework for the management of non-hazardous solid wastes. The 1986 amendments to RCRA enabled EPA to address environmental problems that could result from underground tanks storing petroleum and other hazardous substances. RCRA focuses only on active and future facilities and does not address abandoned or historical sites (see CERCLA).

The 1984 Federal Hazardous and Solid Waste Amendments (HSWA, pronounced "hiss-wa") to RCRA requires phasing-out land disposal of hazardous waste. Some of the other mandates of this law include increased enforcement authority for EPA, more stringent hazardous waste management standards, and a comprehensive underground storage tank program.

1.9.1.12 Safe Drinking Water Act (SDWA) 43 U.S.C. s/s 300f et seq. (1974)

The Safe Drinking Water Act protects the quality of drinking water in the U.S. This law focuses on all waters actually or potentially designated for drinking use, whether above or below ground. The Act authorized EPA to establish safe standards of purity and required all owners or operators of public water systems to comply with primary (health-related) standards. State governments, that assume this power from EPA, also encourage attainment of secondary standards (nuisance-related).

1.9.1.13 Superfund Amendments and Reauthorization Act (SARA) 42 U.S.C. 9601 et seq. (1986)

The Superfund Amendments and Reauthorization Act of 1986 reauthorized CERCLA to continue cleanup activities around the country. Several site-specific amendments, definitions, clarifications, and technical requirements were added to the legislation, including additional enforcement authorities. Title III of SARA also authorized the Emergency Planning and Community Right-to-Know Act (EPCRA).

1.9.1.14 Toxic Substances Control Act (TSCA) 15 U.S.C. s/s 2601 et seq. (1976)

The Toxic Substances Control Act of 1976 was enacted by Congress to test, regulate, and screen all chemicals produced or imported into the U.S. Many thousands of chemicals and their compounds are developed each year with unknown toxic or dangerous characteristics. To prevent tragic consequences, TSCA requires that any chemical that reaches the consumer market be tested for possible toxic effects prior to commercial manufacture. Any chemical that poses health and environmental hazards is tracked and reported under TSCA. Procedures are authorized for corrective action under TSCA in cases of cleanup of toxic materials contamination. TSCA supplements other Federal statutes, including the Clean Air Act and the Toxic Release Inventory under EPCRA.

1.9.2 Occupational Risk Protection: The PSM Rule

The chemical industry is not licensed by the Federal Government. Plants are built to the Uniform Building Code, Fire Protection Code and other applicable codes as needed for insurance purposes. They are subject to EPA regarding effluents, and OSHA for worker safety but there is no watchdog for public safety like the NRC. However OSHA in the PSM Rule (process safety management rule) takes a major step in requiring PSA analysis of operating facilities. The following is a condensation of the PSM Rule. Reference should be made to 29CFR1910.119 for the full text.

The PSM Rule provides requirements for preventing or minimizing the consequences of catastrophic releases of toxic, reactive, flammable, or explosive chemicals. These releases may result in toxic, fire or explosion hazards. The section applies to processes involving chemical, in excess of the threshold quantities listed in Table 1.9-1, flammable liquids or gases at one location in excess of 10,000 lbs, hydrocarbon fuels for workplace consumption, and flammable liquids stored in atmospheric tanks kept below their normal boiling points without refrigeration. It does not apply to: retail facilities, oil or gas well drilling or servicing operations, or normally unoccupied remote facilities.

1.9.2.1 Information

Safety Plan

Employers shall develop a written action plan that implements employee participation in developing and conducting process hazards analyses and other elements of process safety management.

The written process safety information shall be prepared by the employer before conducting process hazard analysis. It shall include information on hazards of the hazardous chemicals used or produced by the process, information pertaining to the technology of the process, and information pertaining to the equipment in the process.

Hazardous Chemical and Process Information

Materials information includes: toxicity, permissible exposure limits, physical properties, reactivity, corrosivity, thermal and chemical and hazardous effects of inadvertent mixing of different materials.[d] Process information consists of: 1) process flow diagrams, 2) process chemistry descriptions, 3) maximum amounts of chemicals, 4) safe ranges for temperatures, pressures, flows or 5) evaluation of the consequences of deviations.

---

[d] Note: Material Safety Data Sheets meeting the requirements of 29 CFR 1910.1200(g) may provide this information. They are available from, e.g., website http://hazard.com/.

Protecting the Public Health and Safety

Table 1.9-1 List of Highly-Hazardous Chemicals, Toxics, and Reactives (Mandatory)					
Chemical	CAS[a]	TQ[b]	Chemical	CAS[a]	TQ[b]
Acetaldehyde	75-07-0	2500	Chlorodiethylaluminum (also called Diethylaluminum Chloride)	96-10-6	5000
Acrolein (2-Propenal)	107-02-8	150	1-Chloro-2, 4-Dinitrobenzene	97-00-7	5000
Acrylyl Chloride	814-68-6	250	Chloromethyl Methyl Ether	107-30-2	500
Allyl Chloride	107-05-1	1000	Chloropicrin	76-06-2	500
Allylamine	107-11-9	1000	Chloropicrin and Methyl Bromide mixture	None	1500
Alkylaluminums	Varies	5000	Chloropicrin and MethylChloride mixture	None	1500
Ammonia, Anhydrous	7664-41-7	10000	Commune Hydroperoxide	80-15-9	5000
Ammonia solutions (greaterthan 44% ammonia by weight)	7664-41-7	15000	Cyanogen	460-19-5	2500
Ammonium Perchlorate	7790-98-9	7500	Cyanogen Chloride	506-77-4	500
Ammonium Permanganate	7787-36-2	7500	Cyanuric Fluoride	675-14-9	100
Arsine (also called Arsenic Hydride)	7784-42-1	100	Diacetyl Peroxide (concentration greater than 70%)	110-22-5	5000
Bis(Chloromethyl) Ether	542-88-1	100	Diazomethane	334-88-3	500
Boron Trichloride	10294-34-5	2500	Dibenzoyl Peroxide	94-36-0	7500
Boron Trifluoride	7637-07-2	250	Diborane	19287-45-7	100
Bromine	7726-95-6	1500	Dibutyl Peroxide (Tertiary)	110-05-4	5000
Bromine Chloride	13863-41-7	1500	Dichloro Acetylene	7572-29-4	250
Bromine Pentafluoride	7789-30-2	2500	Dichlorosilane	4109-96-0	2500
Bromine Trifluoride	7787-71-5	15000	Diethylzinc	557-20-0	10000
3-Bromopropyne (also called Propargyl Bromide)	106-96-7	100	Diisopropyl Peroxydicarbonate	105-64-6	7500
Butyl Hydroperoxide(Tertiary)	75-91-2	5000	Dilauroyl Peroxide	105-74-8	7500
Butyl Perbenzoate(Tertiary)	614-45-9	7500	Dimethyldichlorosilane	75-78-5	1000
Carbonyl Chloride(see Phosgene)	75-44-5	100	Dimethylhydrazine, 1,1-	57-14-7	1000
Carbonyl Fluoride	353-50-4	2500	Dimethylamine, Anhydrous	124-40-3	2500
Cellulose Nitrate (concentration greater than 12.6% nitrogen	9004-70-0	2500	2,4-Dinitroaniline	97-02-9	5000
Chlorine	7782-50-5	1500	Ethyl Nitrite	109-95-5	5000
Chlorine Dioxide	10049-04-4	1000	Ethyl Methyl Ketone Peroxide (also Methyl Ethyl KetonePeroxide; concentration greater than 60%)	1338-23-4	5000
Chlorine Pentrafluoride	13637-63-3	1000	Ethylamine	75-04-7	7500
Chlorine Trifluoride	7790-91-2	1000	Ethylene Fluorohydrin	371-62-0	100

Table 1.9-1 List of Highly-Hazardous Chemicals, Toxics, and Reactives (Mandatory continued)

Chemical	CAS[a]	TQ[b]	Chemical	CAS[a]	TQ[b]
Ethylene Oxide	75-21-8	5000	Methyl Ethyl Ketone Peroxide (concentration greater than 60%)	1338-23-4	5000
Ethyleneimine	151-56-4	1000	Methyl Fluoroacetate	453-18-9	100
Fluorine	7782-41-4	1000	Methyl Fluorosulfate	421-20-5	100
Formaldehyde (Formalin)	50-00-0	1000	Methyl Hydrazine	60-34-4	100
Furan	110-00-9	500	Methyl Iodide	74-88-4	7500
Hexafluoroacetone	684-16-2	5000	Methyl Isocyanate	624-83-9	250
Hydrochloric Acid, Anhydrous	7647-01-0	5000	Methyl Mercaptan	74-93-1	5000
Hydrofluoric Acid, Anhydrous	7664-39-3	1000	Methyl Vinyl Ketone	79-84-4	100
Hydrogen Bromide	10035-10-6	5000	Methyltrichlorosilane	75-79-6	500
Hydrogen Chloride	7647-01-0	5000	Nickel Carbonly (Nickel Tetracarbonyl)	13463-39-3	150
Hydrogen Cyanide, Anhydrous	74-90-8	1000	Nitric Acid (94.5% by weight or greater)	7697-37-2	500
Hydrogen Fluoride	7664-39-3	1000	Nitric Oxide	10102-43-9	250
Hydrogen Peroxide (52% by weight or greater)	7722-84-1	7500	Nitroaniline (para Nitroaniline)	100-01-6	5000
Hydrogen Selenide	7783-07-5	150	Nitrogen Dioxide	10102-44-0	250
Hydrogen Sulfide	7783-06-4	1500	Nitrogen Oxides (NO; NO_2; N_2O_4; N_2O_3)	10102-44-0	250
Hydroxylamine	7803-49-8	2500	Nitrogen Tetroxide (also called Nitrogen Peroxide)	10544-72-6	250
Iron, Pentacarbonyl	13463-40-6	250	Nitrogen Trifluoride	7783-54-2	5000
Isopropylamine	75-31-0	5000	Nitrogen Trioxide	10544-73-7	250
Ketene	463-51-4	100	Oleum (65% to 80% by weight; also called Fuming Sulfuric Acid)	8014-94-7	1000
Methacrylaldehyde	78-85-3	1000	Osmium Tetroxide	20816-12-0	100
Methacryloyl Chloride	920-46-7	150	Oxygen Difluoride (Fluorine Monoxide)	7783-41-7	100
Methacryloyloxyethyl Isocyanate	30674-80-7	100	Ozone	10028-15-6	100
Methyl Acrylonitrile	126-98-7	250	Pentaborane	19624-22-7	100
Methylamine, Anhydrous	74-89-5	1000	Peracetic Acid (concentration greater than 60% Acetic Acid; also called Peroxyacetic Acid)	79-21-0	1000
Methyl Bromide	74-83-9	2500	Nitromethane	75-52-5	2500
Methyl Chloride	74-87-3	15000	Perchloric Acid (concentration greater than 60% by weight)	7601-90-3	5000
Methyl Chloroformate	79-22-1	500	Perchloromethyl Mercaptan	594-42-3	150

Table 1.9-1 List of Highly-Hazardous Chemicals, Toxics, and Reactives (Mandatory continued)					
Chemical	CAS[a]	TQ[b]	Chemical	CAS[a]	TQ[b]
Perchloryl Fluoride	7616-94-6	5000	Sulfur Pentafluoride	5714-22-7	250
Peroxyacetic Acid (concentration greater than 60% Acetic Acid;also called PeraceticAcid)	79-21-0	1000	Sulfur Tetrafluoride	7783-60-0	250
Phosgene (also called Carbonyl Chloride)	75-44-5	100	Sulfur Trioxide (also called Sulfuric Anhydride)	7446-11-9	1000
Phosphine (HydrogenPhosphide)	7803-51-2	100	Sulfuric Anhydride (also called Sulfur Trioxide)	7446-11-9	1000
Phosphorus Oxychloride (also called Phosphoryl Chloride)	10025-87-3	1000	Tellurium Hexafluoride	7783-80-4	250
Phosphorus Trichloride	7719-12-2	1000	Tetrafluoroethylene	116-14-3	5000
Phosphoryl Chloride (also called Phosphorus Oxychloride)	10025-87-3	1000	Tetrafluorohydrazine	10036-47-2	5000
Propargyl Bromide	106-96-7	100	Tetramethyl Lead	75-74-1	1000
Propyl Nitrate	627-3-4	2500	Thionyl Chloride	7719-09-7	250
Sarin	107-44-8	100	Trichloro (chloromethyl) Silane	1558-25-4	100
Selenium Hexafluoride	7783-79-1	1000	Trichloro (dichlorophenyl) Silane	27137-85-5	2500
Stibine (Antimony Hydride)	7803-52-3	500	Trichlorosilane	10025-78-2	5000
Sulfur Dioxide (liquid)	7446-09-5	1000	Trifluorochloroethylene	79-38-9	10000

a CAS is- Chemical Abstract Service, Internet: http//\info.cas.org, part of the American Chemical Society, 1155 16th Street, NW, Washington DC, 20036, Internet: http:\\www.acs.org.
b TQ is Threshold Quantity in pounds

Trade secrets may be used in developing the information for the process hazard analysis emergency planning and responses, and compliance audits. Materials that are developed involving trade secrets may be treated as proprietary and may require signed statements for their protection.

Equipment information includes: 1) materials of construction, 2) piping and instrument diagrams (P&ID's), 3) electrical, 4) relief system design and design basis, 5) ventilation system design, 6) design codes and standards, 7) material and energy balances, and 8) safety systems (e.g., interlocks, detection or suppression systems).

1.9.2.2 Process Hazard Analysis

An initial process hazard analysis (PrHA) is performed on the processes, appropriate to the complexity to identify, evaluate and control the hazards. Employers determine the priority for conducting process hazard analyses based on a rationale which includes such considerations as extent of the process hazards, number of potentially affected employees, age of the process, and operating history of the process. The process hazard analysis is conducted as soon as possible to scope out the work.

PrHA Team

A process hazard analysis team includes one or more employees that are process experts and one or more that are process hazard experts in the methodology being used.

The employer establishes a system to promptly address the team's results, timely resolve recommendations, schedule completion, and communicate the activities to affected personnel. Every five years after the completion of the initial process hazard analysis, it is equivalently updated and revalidated. Employers retain the required process hazards analyses for the life of the process.

1.9.2.3 Safe Operation

Operating Procedures

The employer is responsible for developing and implementing written operating procedures that provide clear instructions for safely conducting activities involved in each covered process consistent with the process safety information by addressing: requirements, responsibilities, and procedures for:
- Initial startup
- Normal operations
- Temporary operations
- Emergency shutdown
- Emergency operations
- Normal shutdown
- Startup following a process modification or after an emergency shutdown

Operating Limits
- Consequences of deviations
- Avoidance of deviations

Safety and Health Considerations
- Process chemicals and their hazards
- Precautions: engineering controls, administrative controls, and protective equipment
- Exposure mitigation
- Inventory control
- Special hazards

Safety Systems and Functions
- Accessibility of operating procedures
- Currency of operating procedures - annual employer certification of accuracy
- Safe work practices for hazards control: lockout/tagout, confined space entry, procedures for opening process boundaries and entrance control for maintenance

Training

Initial training: All employees involved in operating a process are trained in a process overview and in the operating procedures. The training includes emphasis on the specific safety and health hazards under all conditions. In lieu of initial training for employees already involved in operating a process, an employer may certify, in writing, that the employee has the required knowledge, skills, and abilities to safely carry out the duties and responsibilities as specified in the operating procedures.

Refresher training at least every three years, to each employee involved in operating a process to assure understanding and adherence to current operating procedures.

Training for process maintenance employees are trained by the employer in maintaining the integrity of process equipment. This training includes an overview of the process, its hazards and the procedures for safe operation.

Documentation records that each employee involved in operating a process has received and understood the training required by this paragraph. The employer prepares a record which contains the identity of the employee, the date of training, and the means used to verify that the employee understood the training.

Inspection and Testing

Inspections and tests are performed on the process equipment following recognized and accepted engineering practices. The inspection and test frequency is consistent with manufacturers' recommendations, good engineering practices, and prior operating experience.

The inspections and tests are documented to include the date, the name of the responsible person, identification of the equipment, a description of the inspection or test and the results.

Equipment deficiencies that are outside acceptable limits (defined by the process safety information) are corrected before further use or in a safe and timely manner.

Quality Assurance

In the construction of new plants and equipment, the employer is responsible that the equipment is suitable for the process application. Checks and inspections are performed to assure that equipment is installed properly, and consistent with design specifications and the manufacturer's instructions. Maintenance materials, instructions, spare parts and equipment will be available.

Hot Work Permit

The employer issues a hot work permit for hot work operations conducted on or near a covered process. The permit documents that the fire prevention and protection requirements in 29 CFR 1910.252(a) have been implemented prior to beginning the hot work operations; it indicates the date(s) authorized for hot work; and identifies the object on which hot work is performed. The permit is kept on file until completion of the hot work operations.

Management of Change

The employer manages changes (except for "replacements in kind") with written procedures for changes in process chemicals, technology, equipment, procedures, and facilities. The procedures assure the following before any change:

- The technical basis for the proposed change,
- Impact of change on safety and health,
- Modifications to operating procedures,
- Time required,
- Authorization requirements for the proposed change,
- Notification of affected employees,
- Training needed prior to start-up of the process involving the change, and
- Updating process safety information and operating procedures.

Incident Investigation

The employer investigates incidents that result in, or could result in, a catastrophic release of highly hazardous chemicals. An incident investigation is initiated as soon as possible, but before 48 hours following the incident. An incident investigation team is established to consist of one or more experts in the process involved, and accident investigation. The report prepared at the conclusion of the investigation includes at a minimum:

- Date of incident,
- Date investigation began,
- Description of the incident,
- Contributing factors,
- Recommendations resulting from the investigation,
- Resolution the report findings and recommendations,
- Documentation of the resolutions and corrective actions,
- Report review by all affected personnel, and
- Incident report retention for five years.

Emergency Planning and Response

An emergency action plan is established and implemented for the entire plant in accordance with the provisions of 29 CFR 1910.38(a) and may be subject to the hazardous waste and emergency response provisions contained in 29 CFR 1910.120 (a), (p) and (q).

Compliance Audits

Employers certify compliance with 29CFR1910.119 every three years. The compliance audit is conducted by one or more persons knowledgeable in the process. A report of the audit findings is prepared. The employer documents the response to each of the findings and their correction. The two most recent compliance audits are retained.

1.10 Summary

This chapter provided a common basis for understanding the assigning of numerical values to "risk." in the context of probability as the behavior of an ensemble of plants. Predictions of short-term behavior are subject to statistical fluctuations and may be very misleading. Qualitative goals may be established; quantitative safety goals, while more precise, do not enhance the public's understanding of risk choices. Many examples were given of how PSAs results may be presented. Critiques of past PSA were presented with the view of addressing these criticisms and improving upon them in the future. An overview of U.S. regulations that govern radiological and chemical materials to protect the public was given. The chapter concluded with the PSM rule that provides guidelines for safe operation especially of chemical process facilities.

1.11 Problems

1. A certain insurance company requires a 30% overhead on the premiums. If the payment to your beneficiary is $100,000 and you pay $1,500/yr in premiums, what is your probability of dying in the year?

2. What is the mean frequency of deaths from nuclear power presented by the distribution from WASH-1400 presented in Figure 1.4.3-1?

3. Few people understand graphs plotted logarithmically. (a) Devise a method for compressing widely varying data into a convenient size - like a logarithm does, but more familiar. (b) Apply your method to the Zion data in Figure 1.4.3-6.

4. 10CFR50 Appendix I gives the value of a person-rem as $1,000. The report Biological Effects of Ionizing Radiation (BEIR, 1972) assesses about 10,000 person-rem per death (statistically not exposed to an individual). (a) Given this information, what is the cost of a death from radiation? (b) At Pickett's charge at Gettysburg, approximately 30,000 soldiers died in 30 minutes. What is the expenditure rate per minute if the same life equivalence were used?

5. It has been traditional to subject workers to higher risks than the public. (a) Write a short discussion of the equities of this practice with consideration for the fact the worker may leave employment that imposed a long term risk. (b) The Nevada Test Site has been selected as a possible location for a high level waste repository. For purposes of the risk assessment, the workers in the repository will be treated as workers with subsequent higher allowable risk. There are many other workers at NTS. Discuss whether these should be treated as workers or public. (c) If they are treated as workers in one calculation and public in the other where is the logic in this?

6. Discuss why the EPZ should be independent of plant size, safety features, demography and meteorology.

7. The death rate by automobile is about 1/10 million passenger miles. At TMI-2 Governor Thornburg recommended that everyone susceptible to radiation within a 5 mile radius to evacuate. Assume that 25,000 people evacuate and the distance each travels is 20 miles, compare the risk of evacuation with the risk of not evacuating, assuming their exposure is 0.5 mrem-person/hr and they evacuate for two days.

8. The government is charged by the Constitution with a responsibility for the public welfare. (a) Discuss the government's responsibility in controlling the risks in society. (b) Tobacco is not subject to the Delaney Amendment that bans material that can be demonstrated to cause cancer in animals regardless of the quantity of material used in the tests. Why aren't all materials subject to the same rules?

Chapter 2

Mathematics for Probabilistic Safety

Probabilistic safety is concerned with evaluating harmful effects of an artificial construct on man. This construct may be called an airplane, plant, facility, or system depending on the degree of its complexity.[a] Whatever, it is made of components that must work together, as designed, for the system to accomplish its intended function (to work). A logical system model consists of the important components of a system and the effects that component failure has on the system operability. This model treats each components as either working or not working,[b] hence, the state of the system may be represented by a logical equation composed of the states of the components. This, itself, is valuable information but it also may be used to calculate the probability of system failure by replacing each components Boolean state with the probability that that component will fail. A system's risk is the product of its failure probability and the consequences of failure. Hence, the subjects of this chapter are *logic modeling and probabilities*. Logic modeling uses the algebra of two-state variables called Boolean algebra. Other subjects are: meanings of probability, combining probabilities, calculating failure rates from inspection and incident data by classical and Bayes statistics, treating of uncertainties as distributed variables, calculating confidence intervals, and the importance of components to system operability.

2.1 Boolean Algebra

State Vector

It is convenient to consider the operability of a system as a function of its components $f=f(c_1,c_2,....c_n)$. If a components operability is identified as "1" for operating and "0" for failed, the status of the components at any time may be represented by a system state vector: $\psi = (1, 1, 1, 0, 0)$ meaning that components *1*, *2*, and *3* are operating and components *4* and *5* have failed. Requirements for system operability may be represented by a matrix $|O|$ that has 1s where components are required and 0s where they are non-essential; the result is: $\phi = |O|\psi$, where the rules

---

[a] The term "system" is used in this book to mean an aggregate of components; the term "component" refers to an entity that is considered as a unit and not decomposed for analysis.

[b] Degraded states (partially nonworking) can be in a model by defining a discrete degrading state and considering the component to be either in that state or not.

of matrix multiplication apply. The presence of a 1 in the ϕ vector indicates an essential condition; This matrix approach is valuable for the concept that the configuration of a plant at any time is represented by a vector whose elements represent the status of the components by a 1 for operable and a 0 for non-operable. A more refined and developed method is Boolean algebra.

Boolean Algebra

Boolean Algebra is the algebra of two-state variables invented by George Boole to provide a mathematical structure to logical reasoning. Consider an equation $C = A+B$ in which the variables are not continuous but may only have two values "0" or "1." Logically, something OR nothing is something; e.g., if $A = 1$, $B = 0$ then $C = 1$ ($1 = 1+0$). If you have

Table 2.1-1 Comparison of Ordinary and Boolean Algebra

Property	Ordinary Algebra	Boolean Algebra
Commutative	$A+B = B+A$ $A*B = B*A$	Same
Associative	$A+(B+C) = (A+B)+C$ $A*(B*C) = (A*B)*C$	Same
Distributive	$A*(B+C) = A*B+A*C$	Same
Idempotency	NA NA	$A*A = A$ $A+A = A$
Completeness	$A*\overline{A} = A-A^2$ $A+\overline{A} = 1$	$A+\overline{A} = 1$
Unity	$A+1 = A+1$ $A*1 = A$	$A+1 = 1$ $A*1 = 1$
Absorption	$A*(A+B) = A^2+A*B$ $A+(A*B) = A*(1+B)$	$A*(A+B) = A$ $A+(A*B) = A$
de Morgan's theorem	NA	$\overline{A*B} = \overline{A}+\overline{B}$ $\overline{A+B} = \overline{A}*\overline{B}$
Useful relationships	$A+(\overline{A}*B) = A+B-A*B$ $\overline{A}*(A+\overline{B}) = \overline{B}+A*(B-A)$	$A+(\overline{A}*B) = A+B$ $\overline{A}*(A+\overline{B}) = \overline{A}*\overline{B}$

both: $A = 1$ and $B = 1$, the requirement is over satisfied and $C = 1$ ($1 = 1+1$) as before. It may seem strange that $1 + 1 = 1$ but that is the meaning of logic and that is the way Boolean algebra works. The AND condition requires both, represented as $C = A*B$. If you have A but do not have B then the requirement is not satisfied so $C = 1*0 = 0$.

Table 2.1-1 compares the ordinary algebra of continuous variables with the Boolean algebra of 1s and 0s. This table uses the symbols "*" and "+" for the operations of "intersection" (AND) and "union" (OR) which mathematicians represent by "∩" and "∪" respectively. The symbols "*" and "+" which are the symbols of multiplication and addition, are used because of the similarity of their use to AND and OR in logic.

In the table, the rules of commutation, association and distribution are the same for both algebras. Idempotency, unique to Boolean algebra, relates to redundancy. Having "A" and "A" is the same as only having "A"; "A" or "A" is superfluous and equals "A." Complementation is introduced in the next rule. The universe is represented by "1". Completeness includes everything in this world and not in this world, hence $A+\overline{A}=1$ where not A is: $\overline{A} = 1-A$ which is the meaning of complementation. With this understanding, it is impossible to be both A and not A. Similarly A or not A is complete (the universe). Under unity, A is included in the universe (1) so $A+1 = 1$. For this reason $A*1 = 1$. Absorption follows from previous rules; de Morgan's theorem corresponds to the rules of double negatives in English. The last relationships may be derived from previous.

Systems may be modeled using Boolean algebra with two-state representation of component operability. Systems are not usually modeled as equations is not usually done because of preference by engineers for the more schematic methods that are presented in the next chapter.

2.2 Venn Diagram and Mincuts

Venn Diagram

A Venn Diagram is a graphical depiction of logic. It relates the universal set ("1") to a system's status represented by "sets."[c] Set theory is a branch of mathematics with its own symbols and language. For PSA, the graphical procedure known as a Venn diagram serves to illustrate the method.

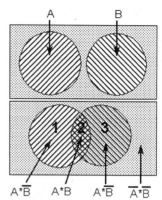

Fig. 2.2-1 Venn diagram of two sets. Upper figure: set A and set B separated but in a universe (background). Lower figure: sets A and B partially merged

Consider two sets A and B (upper Figure 2.2-1). The two sets are isolated and unrelated. They have no intersection (AND), but the union (OR) operation includes the two circles. In the lower operation, union is the two circles (A+B the fact that they overlap does not matter). The left partial moon (1) is $A*\overline{B}$; the right partial moon (3) is $\overline{A}*B$. The common lens (2) is $A*B$. These are depictions of equation 2.2-1, are equivalent to equation 2.2-2 by using complementation and multiplication.

$$A+B = (A*\overline{B})+(\overline{A}*B)+(A*B) \quad (2.2\text{-}1)$$
$$(A+B) = A+B-(A*B) \quad (2.2\text{-}2)$$

Shannon's Method for Expressing Boolean Functions[d]

Shannon's method,[e] expands a Boolean function of n variables in minterms consisting of all combinations of occurrences and non-occurrences of the events of interest. Consider a function of n Boolean variables: $f(X_1, X_2, ... X_n)$ which may be expanded about X_1 as shown in Equation 2.2-3 where $f(1, X_2, ..., X_n)$ where 1 replaces X_1. This says that a function of Boolean variables equals the function with a variable set to 1 plus the product of NOT the variable times the function with the variable set to 0. By extending Equation 2.2-3, a Boolean function may be expanded about all of its

---

[c] A set is a collection of related elements.

[d] This material was adapted from NUREG-0492.

[e] Claude Shannon won the Nobel Prize for relating the maximum information transfer to bandwidth using entropy.

arguments. For example, Equation 2.2-4 is an expansion about X_1 and X_2. It should be noted that the variables multiplying the function on the right are the state variables (minterms) for all possible combinations of the function that mirrors the state configuration. This process is continued for all variables. The complete expansion has 2^n minterms consisting of all possible combinations of occurrences and nonoccurrences of the X variables.

$$f(X_1,X_2,\ldots,X_n) = [X_1 * f(1,X_2,\ldots,X_n)] + [\overline{X_1} * f(0,X_2,\ldots,X_n)] \quad (2.2\text{-}3)$$

$$\begin{aligned}f(X_1,X_2,X_3,\ldots,X_n) &= [X_1 * X_2 * f(1,1,X_3,\cdots X_n)] \\ &+ X_1 * \overline{X_2} * f(1,0,X_3,\ldots,X_n)] \\ &+ \overline{X_1} * X_2 * f(0,1,X_3,\ldots,X_n)], \\ &+ \overline{X_1} * \overline{X_2} * f(0,0,X_3,\cdots,X_n)]\end{aligned} \quad (2.2\text{-}4)$$

$$\begin{aligned}(X*Y)+(\overline{X}*Z)+(Y*Z) &= [X*Y*Z*f(1,1,1)] \\ &+ [X*Y*\overline{Z}\cdot f(1,1,0)] + [X*\overline{Y}*\overline{Z}*f(1,0,0] \\ &+ [\overline{X}*Y*Z*f(0,1,1)] + [\overline{X}*Y*\overline{Z}*f(0,1,0)] \\ &+ [\overline{X}*\overline{Y}*Z*f(0,0,1)] + [\overline{X}*\overline{Y}*\overline{Z}*f(0,0,0)]\end{aligned} \quad (2.2\text{-}5)$$

$$f(1,1,1) = (1*1)+(0*1)+(1*0) = 1 \quad (2.2\text{-}6a)$$
$$f(1,1,0) = (1*1)+(0*0)+(1*1) = 1 \quad (2.2\text{-}6b)$$
$$f(1,0,1) = (1*0)+(0*1)+(0*1) = 0 \quad (2.2\text{-}6c)$$
$$f(1,0,0) = (1*0)+(0*0)+(0*0) = 0 \quad (2.2\text{-}6d)$$
$$f(0,1,1) = (0*1)+(1*1)+(1*1) = 1 \quad (2.2\text{-}6e)$$
$$f(0,1,0) = (0*1)+(1*0)+(1*0) = 0 \quad (2.2\text{-}6f)$$
$$f(0,0,1) = (0*0)+(1*1)+(0*1) = 1 \quad (2.2\text{-}6g)$$
$$f(0,0,0) = (0*0)+(1*0)+(0*0) = 0$$

Each minterm expansion is disjoint[f] from all the others. Thus, in Equation 2.2-4 the 4 terms: $X_1*X_2*f(1,1,X_3,\cdots X_n)$, $X_1*\overline{X_2}*f(1,0,X_3,\ldots,X_n)$, $\overline{X_1}*X_2*f(0,1,X_3,\ldots,X_n)$ and $\overline{X_1}*\overline{X_2}*f(0,0,X_3,\cdots,X_n)$, $X_1X_2f(l,1,X_3,\cdot,X_n)$ are disjoint.[g] Another reason for representing a Boolean function in minterms is their uniqueness. Such an expansion, then, provides a general technique for determining the equality of two Boolean expressions by identical minterms.

Shannon's expansion is demonstrated for two, 3-variable functions in Equation 2.2-5 which equates the first function to its expansion. Equations 2.2-6a-g evaluate the functions for all combinations of values for X, Y, and Z. Equation 2.2-7 equates the second function to its expansion, and equations 2.2-8a-g evaluate the functions for all combinations of values for X, Y, and Z.

Notice that equations 2.2-6a-g produce the same results as equations 2.2-8a-g. Furthermore, only those equations that result in "1" have relevance.

This means that Boolean equation 2.2-5 is the same as 2.2-7 and the third term in equation 2.2-5 is superfluous. Figure 2.2-2 interprets these equations as Venn diagrams.

Figures 2.2-2a-d represent Equations 2.2-6 a-d, or 2.2-8a-d. These are barriers bounded by lines but not intersected by any lines. Thus, they are elemental units called "minimal cutsets." The term $X*Y$ is represented by Figure 2.2-2e. Notice the area is cut by the line that is the boundary of X. This area is the

$$\begin{aligned}(X*Y)+(\overline{X}*Z) &= [X*Y*Z*f(1,1,1)] + \\ &[X*Y*\overline{Z}\cdot f(1,1,0)] + [X*\overline{Y}*\overline{Z}*f(1,0,0] \\ &+ [\overline{X}*Y*Z*f(0,1,1)] + [\overline{X}*Y*\overline{Z}*f(0,1,0)] \\ &+ [\overline{X}*\overline{Y}*Z*f(0,0,1)] + [\overline{X}*\overline{Y}*\overline{Z}*f(0,0,0)]\end{aligned} \quad (2.2\text{-}7)$$

$$f(1,1,1) = (1*1)+(0*1) = 1 \quad (2.2\text{-}8a)$$
$$f(1,1,0) = (1*1)+(0*0) = 1 \quad (2.2\text{-}8b)$$
$$f(1,0,1) = (1*0)+(0*1) = 0 \quad (2.2\text{-}8c)$$
$$f(1,0,0) = (1*0)+(0*1) = 0 \quad (2.2\text{-}8d)$$
$$f(0,1,1) = (0*1)+(1*1) = 1 \quad (2.2\text{-}8e)$$
$$f(0,1,0) = (0*1)+(1*0) = 0 \quad (2.2\text{-}8f)$$
$$f(0,0,1) = (0*0)+(1*1) = 1 \quad (2.2\text{-}8g)$$
$$f(0,0,0) = (0*0)+(1*0) = 0$$

---

[f] Events are disjoint if the have no intersection, i.e., they are unrelated.

[g] This is an important property when converting to probability to avoid double counting.

combination of Figures 2.2-2a-b. The term $\overline{X}*Z$ is represented by Figure 2.2-2f which is the combination of 2.2-2c-d. It is cut by the boundary line of y. The combination $X*Y+\overline{X}*Z$ (2.2-2e-f) is represented by Figure 2.2-2h. The Venn diagram shows that the term $Y*Z$ in equation 2.2-5, shown as Figure 2.2-2g, is superfluous because it is contained in parts of Figures 2.2-2e-f.

A reasonable question is "what is the significance of superfluence in the logical expression?" As logic, there is nothing wrong, the overstatement is extraneous. But when

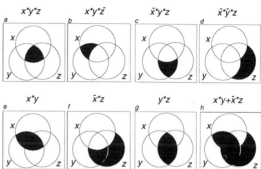

Fig. 2.2-2 Mincut Explanation by Venn Diagrams

system logic is converted to the probability of system failure by replacing component variables by the probability of failure, the superfluous elements result in double counting, hence, erroneous results. For example, the area $X*Y*Z$ is contained in both $X*Y$ and $Y*Z$.

Minimal Cut Sets and Minimal Path Sets

Boolean equations show how component failures can fail a system. *A minimal cut set is the smallest combination of component failures that can fail a system.* It is the set of non-superfluous components, such as in the previous example, with the superfluous combination $Y*Z$ $(X*Y)+(\overline{X}*Z)$ excluded. If they all occurred they would cause the top event to occur. One-component minimal cut sets, if there are any, are single failures that cause system failure. Two-component minimal cutsets are pairs of components, if they occur together cause system failure. Triple-components minimal cutsets are sets of three components that, if they fail together cause system failure, and so on to higher cutsets

A minimal pathset *is a smallest combination of component successes that can result in system success.* A minimal pathset is related to a minimal cutset via DeMorgan's theorem (Table 2.1-1).

2.3 Probability and Frequency

Frequency with the dimensions of per unit time, ranges from zero to infinity and means the number of occurrences per time interval. *Probability* is dimensionless, ranges from zero to one, and has several definitions. The confusion between frequency and probability arises from the need to determine the probability that a given system will fail in a year. Such a calculation of probability explicitly considers the time interval and, hence, is frequency. However, considerable care must be used to ensure that calculations are dimensionally correct as well as obeying the appropriate algebra. Three interpretations of the meaning of probability are:

Mathematics for Probabilistic Safety

Laplacian Probability

Equation 2.3-1 expresses the Laplacian meaning of probability. It is applicable when the number of results are countable and each outcome is equally likely (a true die is equally likely to land on any of its six faces). For example, with two dies you can throw *four* by throwing: 2, 2, or 1, 3, or 3, 1. Hence, there are *three* ways of throwing *four*. Since a die has *six*, presumably identical faces, the total number of possible outcomes with two dice is: 6*6 = 36 and the probability is 3/36 = 1/12 of making four (Figure 2.3-1). Notice that it is impossible for probability to exceed one.

$$Probability = \frac{Number\, of\, ways\, a\, result\, can\, occur}{The\, total\, number\, of\, ways\, all\, results\, can\, occur} \quad (2.3\text{-}1)$$

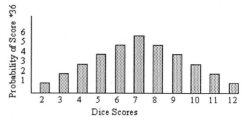

Fig. 2.3-1 Probability of Scores in Craps (note the multiplier)

The disadvantage of Laplacian probability is its use is limited to calculating the probability of processes for which all outcomes are known and equally probable. This eliminates the use of Laplacian probability for determining the probability of process system failure.

von Misesian Probability

This is experimental probability determined from operating experience by counting the number of results of a particular type divided by the number of trials. Laplacian probability is calculated from geometry. von Misesian probability is approximated as the ratio of the number of times, N, that a particular result occurs (say snake eyes) to a total number of throws N_o. As N_o becomes very large, the ratio approaches the true probability (equation 2.3-2). This can be demonstrated easily with a computer program that simulates dice throwing. In the limit of a very large number of throws, the von Misesian probability will approach the Laplacian result. The advantage of von Misesian probability is that it is correct answer even if the dice are loaded. It is used to determine the probability of component failure using operating experience, but it may be a poor approximation if the operating experience is limited. In practice, PSA data are limited and the probability therefrom has an uncertainty related to the sample size.

$$Probability = \lim_{N_o \to \infty} \frac{N}{N_o} \quad (2.3\text{-}2)$$

von Misesian probability is often called the *frequency definition of probability* although it does not have the dimensions of per unit time. The frequency results from the fact that N_o relates to a time interval.

Probability as State of Belief

As Tribus, 1969, says, all probabilities are conditional. In the example of the dice, the probabilities are conditioned on the assumption that the dice are perfect and the method of throwing has no effect on the outcome. Some writers (e.g., deMorgan, 1847) say, probability refers to the belief by a mind having uncertain knowledge. This is the interpretation of probability in the Zion-Indian Point (ZIP) and some other PSAs. Probability in this sense attempts to include all information e.g., QA that could affect the performance of a piece of equipment. Such information may be conveyed as a distribution whose height is proportional to confidence in the belief and whose width reflects uncertainty (refer to Section 2.6).

2.4 Combining Probabilities

2.4.1 Intersection or Multiplication

Consider an experiment that in N trials, events A and B occur together $N(A*B)$ times, and event B occurs $N(B)$ times. The conditional probability of A given B is equation 2.4-1 or 2.4-2. Rearranging, results in equation 2.4-3.

If A and B are independent, i.e., $P(A|B) = P(A)$ then equation 2.4-3 becomes 2.4-4 and probabilities are multiplied. This is a common approximation in PSA.

The rule for combining independent probabilities in intersection is to multiply them together (equation 2.4-5).

$$P(A|B) = \frac{\lim_{N_o \to \infty} \frac{N(A*B)}{N}}{\lim_{N_o \to \infty} \frac{N(B)}{N}} \quad (2.4\text{-}1)$$

$$P(A|B) = \frac{P(A*B)}{P(B)} \quad (2.4\text{-}2)$$

$$P(A*B) = P(A|B)*P(B) \quad (2.4\text{-}3)$$

$$P(A*B) = P(A)*P(B) \quad (2.4\text{-}4)$$

$$P(A*B*...N) = P(A)*P(B)*...P(N) \quad (2.4\text{-}5)$$

2.4.2 Union or Addition

$$P(A_1+A_2+...A_N) = \sum_{i=1}^{N} P(A_i) \quad (disjoint) \quad (2.4\text{-}6)$$

$$P(A+B) = P(A)+P(B)-P(A)*P(B) \quad (non-disjoint) \quad (2.4\text{-}7)$$

If events cannot occur at the same time, they are called *disjoint*, it is clear from the von Misesian definition of probability that equation 2.4-6 will result for this case as may be demonstrated by adding two non-disjoint probabilities (refer to the Venn diagram, Figure 2.4-1). The segments are: $P(A_1) = P(A)*P(\overline{B})$, $P(A_2) = P(A)*P(B)$, and $P(A_3) = P(\overline{A})*P(B)$. Using the definition of a complement, this becomes: $P(A_1) = (A)*[1-P(B)]$, $P(A_2) = P(A)*P(B)$, $P(A_3) = [1-P(A)]*P(B)$ which when substituted into equation 2.4-6 and rearranged gives equation 2.4-7. This result may be generalized to larger combinations by induction or by using de Morgan's theorem (the latter is easier to write as a computer program).

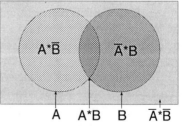

Fig. 2.4-1 Venn Diagram of Sets A and B (right and left circles respectively). Their intersection A*B is cross-hatched

The physical reason for the third term on the right of equation 2.4-7 is to correct for counting the overlap twice as seen in Figure 2.2-1. The technique of Venn diagrams is used in some PRAs to calculate mutually exclusive power states from non-mutually exclusive states.

2.4.3 M-Out-of-N-Combinations

Out of N components what is the probability of M failures? For example, if a process has N tanks what is the probability that M will fail?

Assuming that the probability of a tank failure is P, the probability of M failed tanks, if the failures are independent, is P^M, and the probability of the others not failing is $(1-P)^{N-M}$. However, this must be corrected for the number of combinations of N things taken M at a time (equation 2.4-8), where N-factorial is: $N! = N*(N-1)*(N-2)....$ The combined result is given by equation 2.4-9 which is a binomial distribution.

$$\binom{N}{M} = \frac{N!}{M!*(N-M)!} \quad (2.4\text{-}8)$$

$$P(N|M) = \binom{N}{M} P^M * (1-P)^{N-M} \quad (2.4\text{-}9)$$

As a numerical example of its use, if a plant has 4 tanks, the probability of two tanks failing is: $P(4|2) = 6*P^2$, where P is the probability of a tank failing, which is assumed to be small. Notice that this is *six-times* greater than would have been estimated if it no consideration were given for the combinatorial effect.

$$P(<M|N) = \sum_{K=1}^{M} \binom{N}{K} P^K (1-F)^{N-K} \quad (2.4\text{-}10)$$

A related problem is to find the probability of M failures *or less* out of N components. This is found by summing equation 2.4-9 for values less than M as given by equation 2.4-10 which can be used to calculate a one-sided confidence bound over a binomial distribution (Abramowitz and Stegun, p. 960).

2.5 Distributions

Distributions represent uncertainties. Often data are too sparse to provide a good estimate of the von Misesian probability. Confidence in the estimate may be found from knowing how the data are distributed. Distributions are of two types: discrete and continuous as depicted in Figure 2.5-1.

Distributions are characterized by measures of central tendency: The *median* is the value of x (e.g., crap scores) that divides the distribution into equal areas. The value of x at the peak of the curve is the *mode*. The first moment of the distribution[h] called the *mean, average, expected* or the *expectation value* of x is symbolized as *E(x)* or *<x>*. The second moment about the mean is called the *variance* or *var(x)*; its square-root is the *standard deviation* also called *sigma*. Thus median, mode and mean are measures of central tendency. Variance and standard deviation are measures of dispersion or uncertainty.

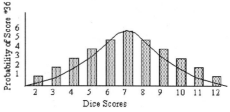

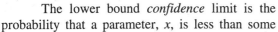

Fig. 2.5-1 The discrete distribution of probability of crap scores has impressed on it a continuous distribution with quite a different meaning

The lower bound *confidence* limit is the probability that a parameter, x, is less than some value x_o. The upper bound *confidence* limit is the probability that a parameter, x, is greater than some value x_o. Figure 2.5-1 shows that confidence may be obtained from the discrete curve by simply adding the probabilities below or above x_o for the lower or upper bound confidence respectively. If the curve is continuous it must be integrated above or below x_o. These results are normalized, by dividing the partial integral or partial sum by the full integral of the curve or complete sum.

2.5.1 Discrete Distributions

Poisson Distribution

The Poisson distribution follows naturally from the discrete binomial distribution already introduced in the craps and the *M-out-of-N* problem. As *N* becomes large, the Poisson distribution approximates the binomial distribution

The Poisson distribution for observing *M* events in time *t* is given by equation 2.5-1, where λ is the failure rate estimated as *M/t*. This model may be used if the failure rate is time dependent rather than demand dependent.

$$P_M(t) = \frac{(\lambda*t)*\exp(-\lambda*t)}{M!} \quad (2.5\text{-}1)$$

As the sample size is increased further, the Poisson goes over to the Gaussian distribution which is the name commonly used by physicists and engineers for a model of many diffusion-like physical phenomena; it is called a "normal" distribution by statisticians.

---

[h] The first moment and synonyms is the location at which the curve, if cut out, would balance on a knife-edge.

2.5.2 Continuous Distributions

2.5.2.1 Gaussian/Normal

The Gaussian/normal is distributed[i] according to equation 2.5-2, where μ is the mean, σ is the standard deviation, and x is the parameter of interest, e.g., a failure rate. By integrating over the distribution, the probability of x deviating from μ by multiples of σ are given in equations 2.5-3a-c.

For example, it might be stated that there is a 5% probability of x being more than $2*\sigma$ from the mean. This is true only if the distribution is known to be normal. However, if the distribution is unknown but not pathological, the Chebyshev inequality provides an estimate.

$$f(x) = \frac{\exp[-(x-\mu)^2/(2*\sigma^2)]}{\sigma\sqrt{(2*\pi)}} \quad (2.5\text{-}2)$$

$$P(|x-\mu| \leq \sigma) = 0.693 \quad (2.5\text{-}3a)$$
$$P(|x-\mu| \leq 2*\sigma) = 0.954 \quad (2.5\text{-}3b)$$
$$P(|x-\mu| \leq 3*\sigma) = 0.997 \quad (2.5\text{-}3c)$$
$$P(|x-\mu| \leq k*\sigma) = 1 - 1/k^2 \quad (2.5\text{-}4)$$
$$P(|x-\mu| \leq k*\sigma) = 1 - 4/(9*k^2) \quad (2.5\text{-}5)$$

$$f(x) = \frac{\exp[-(\ln(x)-\mu)^2/(2*\sigma^2)]}{x*\sigma\sqrt{(2*\pi)}} \quad (2.5\text{-}6)$$

2.5.2.2 Distribution-Free Confidence Estimators

Mean: $\alpha = \exp(\mu + \sigma^2/2)$ (2.5-7)
Variance: $\beta^2 = \alpha^2 * [\exp(\sigma^2-)1]$ (2.5-8)
Mode: $x_m = \exp(\mu + \sigma^2)$ (2.5-9)
$\sigma^2 = \ln[(\beta^2/\alpha^2) + 1]$ (2.5-10)
$\mu = \ln(\alpha) - \sigma^2/2$ (2.5-11)

Chebyshev Distribution-Free Inequality

This is a loose (conservative) bound given by equation 2.5-4. If applied to the normal for, say, $3*\sigma$, it estimates $1 - 1/9 = 0.888$, instead of 0.997, the correct result.

Gauss Distribution-Free Inequality

This, more restrictive inequality, is given by equation 2.5-5. For $3*\sigma$, it gives 0.95 which is much closer to the correct value of 0.997.

2.5.2.3 Central Limit Theorem

The Central Limit Theorem gives an a priori reason for why things tend to be normally distributed. It says: *the sum of a large number of independent random distributions having finite means and variances is normally distributed.* Furthermore, *the mean of the resulting distribution is the sum of the individual means; the combined variance is the sum of the individual variances.*

---

[i] Distributions also are called probability density functions (pdf).

2.5.2.4 Lognormal Distribution

The Reactor Safety Study extensively used the lognormal distribution (equation 2.5-6) to represent the variability in failure rates. If plotted on logarithmic graph paper, the lognormal distribution is normally distributed.

Several characteristic values of the distribution are given by equations 2.5-7 to 2.5-9 which may be solved to give equations 2.5-10 and 2.5-11.

Useful percentiles of the lognormal are given in equations 2.5-12 through 2.5-15, where κ_γ is the appropriate coefficient found in tables of the normal distribution

The error factor (EF) defined as equation 2.5-16 relates the 50th percentile to the 95th percentile (equation 2.5-17) and the 5th percentile (equatin 2.5-18).

2.5.2.5 Modified Central Limit Theorem

Since multiplication is performed by the summation of logarithms, another statement of the Central Limit Theorem is: *The multiplication (ANDing) of a large number of components having arbitrary but well-behaved distributions results in a lognormal distribution.*

2.5.2.6 Exponential Distribution

If the rate at which components are failing, dN/dt, is proportional to the number that can fail, N (equation 2.5-19), the distribution is exponential. Where λ is the proportionality constant and the negative sign indicates that the number of working components is decreasing with time. Dividing equation 2.5-19 by N_o, the initial number of components, and taking the limit as N_o becomes very large, result in the rate of change of probability being proportional to the probability (equation 2.5-20). Integration shows the exponential dependence of probability, assuming P = 1 when t = 0.

$$5th\ percentile:\ x_{05} = \exp(\mu - 1.645*\sigma) \quad (2.5\text{-}12)$$
$$50th\ percentile:\ x_{50} = \exp(\mu) = \sqrt{(x_{05}*x_{95})} \quad (2.5\text{-}13)$$
$$95th\ percentile:\ x_{95} = \exp(\mu + 1.645*\sigma) \quad (2.5\text{-}14)$$
$$\gamma\text{-}percentile:\ x_\gamma = \exp(\mu + \kappa_\gamma*\sigma) \quad (2.5\text{-}15)$$

$$EF = \sqrt{[x_{95}/x_5]} \quad (2.5\text{-}16)$$
$$x_{95} = x_{50}*EF \quad (2.5\text{-}17)$$
$$x_5 = x_{50}/EF \quad (2.5\text{-}18)$$

$$dN/dt = -\lambda*N \quad (2.5\text{-}19)$$
$$dP/dt = -\lambda*P \quad (2.5\text{-}20)$$
$$P = \exp(-\lambda*t) \quad (2.5\text{-}21)$$

2.5.2.7 Reliability Parameters

The following are commonly used terms in reliability analyses:

a) **MTTF** - the mean time to failure, is just what it says - the average time to failure. If the distribution is f(t), MTTF is given by equation 2.5-22; if f(t) is exponential, the result is equation 2.5-23.

Mathematics for Probabilistic Safety

b) **MTTR** - the mean time to repair, is just what it says. If the repair distribution is exponential with a repair rate, µ, it is given by equation 2.5-24.

c) **Reliability, R(t)** the probability of not failing from time 0 to t (equation 2.5-25), where r(t) is the probability density of success.

d) **Unreliability** is *1-R* hence equation 2.5-26, where *f(t)* is the probability density of failure.

e) **Hazard Function** the conditional probability of functioning until *t*, and failing between *t* and *t + dt* (equation 2.5-27). This is restated using conditional probability (equation 2.4-28) and written in terms of the failure rate density in equation 2.5-29.

$$MTTF = \frac{\int_0^\infty t*f(t)}{\int_0^\infty f(t)} \quad (2.5\text{-}22)$$

$$MTTF = \frac{\int_0^\infty t*\exp(-\lambda*t)}{\int_0^\infty \exp(-\lambda*t)} = 1/\lambda \quad (2.5\text{-}23)$$

$$MTTR = 1/\mu \quad (2.5\text{-}24)$$

$$R(t) = \int_0^t r(t)*dt \quad (2.5\text{-}25)$$

$$F(t) = \int_0^t f(t)*dt \quad (2.5\text{-}26)$$

The numerator of equation 2.5-29 (right) is the derivative of the denominator, hence it is the log derivative which gives equation 2.5-30. For an exponential distribution: $R(t) = \exp(-\lambda*t)$, the hazard function is just λ, and the hazard function, also called the "force of mortality," is constant. This conclusion is made without considering infant mortality and wearout.

$$h(t)*dt = P(\text{failing in dt given survival to } t) \quad (2.5\text{-}27)$$

$$h(t)*dt = \frac{P(\text{failing in dt and survival to } t)}{P(\text{survival to } t)} \quad (2.5\text{-}28)$$

$$h(t)*dt = \frac{f(t)*dt}{R(t)} = \frac{f(t)*dt}{1-F(t)} \quad (2.5\text{-}29)$$

$$h(t) = d\ln[R(t)]/dt = d\ln[F(t)]/dt \quad (2.5\text{-}30)$$

Figure 2.5-2 depicts the force of mortality as a bathtub curve for the life-death history of a component without repair. The reasons for the near universal use of the constant λ exponential distribution (which only applies to the mid-life region) are: mathematical convenience, inherent truth (equation 2.5-19), the use of repair to keep components out of the wearout region, startup testing to eliminate infant mortality, and the lack of detailed data to support a time-dependent λ.

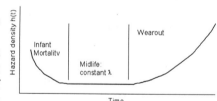

Fig. 2.5-2 Conceptual Life-Cycle of a Component

2.5.2.8 Distributions Used in WASH-1400

WASH-1400 treated the probability of failure with time as being exponentially distributed with constant λ. It treated λ itself as being lognormally distributed. There are better a priori reasons

Distributions

for the exponential distribution than there are for assuming the λs are lognormal unless the Modified Form of the Central Limit Theorem is appropriate.

2.5.3 Confidence Limits

Confidence limits are partial integrations over a probability density function. There are two special cases: failure with time and failure with demand.

2.5.3.1 Failure with Time

Suppose N identical components with an exponential distribution (constant λ) are on test; the test is terminated at T, with M failures. What is the confidence that λ is the true failure rate.

The derivation will not be provided. Suffice it to say that the failures in a time interval may be modeled using the binomial distribution. As these intervals are reduced in size, this goes over to the Poisson distribution and the MTTF is chi-square distributed according to equation 2.9-31, where $\chi^2 = 2*\lambda*N*T$ and the degrees of freedom, $f = 2(M+1)$.

Confidence is calculated as the partial integral over the chi-squared distribution, i.e., the partial integral over equation 2.5-31 which is equation 2.5-32, where $\chi^2_{\alpha,M}$ is the cumulative chi-squared distribution with percent confidence α for M failures.[j] Table 2.5-1 presents values for cumulative chi-squared.[k]

Table 2.5-1 Values of the Inverse Cumulative Chi-Squared Distribution in terms of percentage confidence with M failures

M\%	5	10	25	50	75	90	95
0	.1	.21	.58	1.4	2.8	4.6	6.0
1	.71	1.1	1.92	3.4	5.4	7.8	9.5
2	1.6	2.2	3.5	5.4	7.8	10.6	12.6
3	2.7	3.5	5.1	7.3	10.2	13.4	15.5
4	3.9	4.9	6.7	9.3	12.5	16.0	18.3
5	5.2	6.3	8.4	11.3	14.8	18.5	21.0
6	6.6	7.8	10.2	13.3	17.1	21.1	23.7
7	8.0	9.3	11.9	15.3	19.4	23.5	26.3
8	9.4	10.9	13.7	17.3	21.6	26.0	28.9
9	1.9	12.4	15.5	19.3	23.8	28.4	31.4
10	12.3	14.0	17.2	21.3	26.0	30.8	33.9
11	13.8	15.7	19.0	23.3	28.2	33.2	36.4
12	15.4	17.3	20.8	25.3	30.4	35.6	38.9
13	16.9	18.9	22.7	27.3	32.6	37.9	41.3
14	18.5	20.6	24.5	29.3	34.8	40.3	43.8
50	77.9	82.4	90.1	99.3	109.1	778.5	124.3

$$p(\chi^2) = \frac{(\chi^2)^{p-1} * \exp(\chi^2)}{2^f * \Gamma(f/2)} \quad (2.5\text{-}31)$$

$$\lambda \leq \frac{\chi^2_{\alpha,M}}{2*N*T} \quad (2.5\text{-}32)$$

It is important to note that the chi-squared estimator provides upper bounds on λ for the case of zero failures. For example, a certain type of nuclear plant may have 115 plant-years of experience using 61 control rods. If there has never been a failure of a control rod, what is λ for 50% (median) and 90% confidence?

---

[j] Note equation 2.5-31 is especially adapted for reliability purposes.

[k] Chapter 12 discusses the software provided with this book including the program Lambda that does the chi-squared and F-number calculations.

Mathematics for Probabilistic Safety

$\lambda(50\%) \le 1.4/(2*115*61) = 1.0\text{E-4/rod/years}$; $\lambda(90\%) \le 2.7/(2*115*61) = 3.3\text{E-4/ rod/years}$

2.5.3.2 Failure on Demand

The cumulative binomial distribution is given by equation 2.5-33, where M is the number of failures out of N items each having a probability of failure p. This can be worked backwards to find the implied value of p for a specified $P(M, p, N)$, which is a measure of confidence in the probability, that P is less than or equal to. This confidence, α is expressed by equation 2.5-34, $v_1 = 2*(N-M)$ $v_2 = 2*(R+1)$, and F is the F-function, where the probability is given by equation 2.5-35.

$$F(M,p,N) = \sum_{i=0}^{M} \binom{N}{i} *p*(1-p)^{N-i} \quad (2.5\text{-}33)$$

$$\alpha = 1 - \sum_{i=M+1}^{N} p^{i}*(1-p)^{N-i} = Q(F|v_1,v_2) \quad (2.5\text{-}34)$$

$$p \le \frac{M+1}{M+1+(N-M)*F} \quad (2.5\text{-}35)$$

Example: Given there have been 200,000 scram tests and scram actuations with no failure. With 90% confidence. what is the probability of failure on demand?

$v_1 = 2*200,000 = 400,000$; $v_2 = 2$. Table 26.9 of Abramowitz and Stegun (1964) gives: F= 1/2.30, hence $p \le 2.30/2\text{E}5 = 1.15\text{E-5/demand}^1$. (This calculation is easily done with the Lambda code, Chapter 12.)

2.5.4 Markov Modeling

Markov modeling is a technique for calculating system reliability as exponential transitions between various states of operability, much like atomic transitions. In addition to the use of constant transition rates, the model depends only on the initial and final states (no memory).

The concept of states, much like atomic energy levels, is integral to Markov modeling. Markov states consist of system components in various combinations of operability. If a system has only one component, and it is operable, the system is in the operating state. The failure of that component "transitions" to the failed state. Upon repair, the system "transitions" back to the operating state (Figure 2.5-3). Similarly a system of N components has $2N$ states. For example, a two component system is fully operating in state AB. A's failure takes it to state $\overline{A}B$, B's failure takes it to, both failures takes it to $\overline{AB}$.

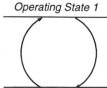

Fig. 2.5-3 Transitions between States

To illustrate the mathematics, consider a large number of two-state diesel generators (DG) with N_1 operating and N_2 failed out of a total of $N = N_1 + N_2$. The rate of change from state 1 (rate of failing) is given by equation 2.5-36, and the rate of change from state 2 (rate of succeeding) is

---

[1] For probabilities x,> 0.5, use $Q(x) = Q(1-x)$ and $F(a,b) = 1/F(b,a)$.

equation 2.5-37. Dividing by N and taking the limit as $N \to \infty$ (von Misesian Probability) gives equations 2.5-38, 39. Since, $P_1 = 1-P_2$, 2.5-40 results. Integrating equation 2.5-40 with $P_1(0) = 1$ gives equation 2.5-41. Equation 2.5-42 gives these results in more conventional notation.

After the startup transient has passed, i.e., $t \to \infty$, this becomes equation 2.5-43, where λ is the failure rate and μ is the repair rate.

Equation 2.5-43 is a definition of "availability." Since $1/\mu = $ MTTR (mean time to repair) and $1/\lambda = $ MTTF (mean time to failure) A more conventional definition is given by 2.5-44.

Unavailability = 1 availability. If $\mu > \lambda$, it is, asymptotically, found from equation 2.5-43 to be equation 2.5-45, where $\tau = 1/\mu$ is the mean time between repairs.

While not shown here, if a system has k redundancies of identical subsystems, each with failure rate λ and repair period τ, the equilibrium unavailability is given by equation 2.5-46.

$$dN_1/dt = -\lambda_{12}*N_1 + \lambda_{21}*N_2 \quad (2.5\text{-}36)$$

$$dN_2/dt = \lambda_{12}*N_1 - \lambda_{21}*N_2 \quad (2.5\text{-}37)$$

$$dp_1/dt = -\lambda_{12}*p_1 + \lambda_{21}*p_2 \quad (2.5\text{-}38)$$

$$dp_2/dt = \lambda_{12}*p_1 - \lambda_{21}*p_2 \quad (2.5\text{-}39)$$

$$p_1 = -(\lambda_{12} + \lambda_{21})*p_1 + \lambda_2 \quad (2.5\text{-}40)$$

$$p_1 = \frac{-(\lambda_{12} + \lambda_{21})*\exp[-(\lambda_{12} + \lambda_{21})*t]}{\lambda_{12} + \lambda_{21}} \quad (2.5\text{-}41)$$

$$P = \frac{\mu + \lambda*\exp[-(\lambda+\mu)*t]}{\lambda + \mu} \quad (2.5\text{-}42)$$

$$P(\infty) = \frac{\mu}{\lambda + \mu} \quad (2.5\text{-}43)$$

$$\text{Availability} = \frac{MTTF}{MTTF + MTTR} \quad (2.5\text{-}44)$$

$$Q(\infty) = \frac{\lambda}{\mu} = \lambda*\tau \quad (2.5\text{-}45)$$

$$O(\infty) = (\lambda+\tau)^k \quad (2.5\text{-}46)$$

Table 2.5-2 Mean Variance and Moment-Generating Functions for Several Distributions*				
Distribution	f(x)	E(x)	var(x)	$M_x(\theta)$
Binomial	$\binom{n}{m}*p^m*(1-p)^{n-m}$	$n*p$	$n*p*(1-p)$	$[1+p*(\exp(\theta)-1)]^n$
Poisson	$\exp(-\mu)*\mu^x/x!$	μ	μ	$\exp(\lambda)*[\exp(\theta)-1]$
Exponential	$\lambda*\exp(-\lambda*x)$	$1/\lambda$	$1/\lambda^2$	$-\lambda/(\theta-\lambda)$
Rayleigh	$K*x*\exp(-K*x^2/2)$	$\sqrt{\pi/(2*K)}$	$0.429/K$	complex
Weibull	$K*x^{m}*\exp(-K*x^{m+1}/(m+1))$	$[K/(m+1)]^{-1/m+1}*\Gamma[(m+2)/(m+1)]$	$\{[K/(m+1)]^{1-\delta}*\Gamma(\delta)-[E(x)]^2\}$; $\delta = m+3/m+1$	complex
Gamma	$x^\alpha*\exp(-x/\beta)/[\beta^{\alpha+1}*\Gamma(\alpha+1)]$	$\beta*(\alpha+1)$	$\beta^2*(\alpha+1)$	$-(1/\beta)^{\alpha+1}/(\theta-1/\beta)^{\alpha+1}$
Normal	$\exp[-(x-\mu)^2/2*\sigma^2]/(\sigma*\sqrt{(2*\pi)})$	μ	σ^2	$\exp(\mu*\theta+\theta^2*\sigma^2/2)$

* Reprinted from M. Schuman *Probabilistic Reliability: An Engineering Approach*, 2nd Edition, Copyright, Robert E. Krieger Pub. Co., 1988.

2.5.5 Summary of Functions and their Generating Functions

Table 2.5-2 provides a convenient summary of distributions, means and variances used in reliability analysis. This table also introduces a new property called the generating function ($M_x\theta$).

Generating functions are used in calculating moments of distributions for power series expansions. In general, the *nth* moment of a distribution, $f(x)$ is: $E(x^n) = \int x^n * f(x) * dx$, where the integration is over the domain of x. (If the distribution is discrete, integration is replaced by summation.)

A generating function is defined by equation 2.5-47. To illustrate it use, Table 2.5-2 gives the generating function for an exponential distribution as: $-\lambda/(\theta-\lambda)$. Each moment is obtained by successive differentiations. Equation 2.5-48 shows how to obtain the first moment. By taking the limit of higher derivatives higher moments are found.

2.6 Bayesian Methods

The Reverend Thomas Bayes, in a posthumously published paper (1763), provided a systematic framework for the introduction of prior knowledge into probability estimates (Crellin, 1972). Indeed, Bayesian methods may be viewed as nothing more than convoluting two distributions. If it were this simple, why the controversy?

The controversy (for a lucid discussion refer to Mann, Shefer and Singpurwala, 1976) between "Bayesians" and "classicists" has nothing to do with precedence, for Bayes preceded much of classical statistics. The argument hinges on: a) what prior knowledge is acceptable, and b) the treatment of probabilities as random variables themselves.

Classicists believe that probability has a precise value; uncertainty is in finding the value. Bayesians believe that probability is not precise but distributed over a range of values from heterogeneities in the database, past histories, construction tolerances, etc. This difference is subtle but changes the two approaches.

Bayes's methods aim to satisfy two needs: the concept of probability as degree of belief, and the need to use all available information in a probability estimate. Classicists reject all

$$M_x(\theta) = \int_{-\infty}^{\infty} \exp(\theta * x) * f(x) * dx \quad (2.5\text{-}47)$$

$$E(x) = \lim_{\theta \to 0} \partial[-\lambda/(\theta-\lambda)]/\partial\theta = \lim_{\theta \to 0} \lambda/(\theta-\lambda)^2 = 1/\lambda \quad (2.5\text{-}48)$$

except test information. However, a Bayesian believe that prior information from related work, laboratory tests, codes, quality control, etc. should be used to decide the state of belief in something's operability. The problem is in quantifying this prior knowledge. This section develops Bayes's equation and presents results for some distributions using conjugate functions. It closes with confidence interval estimating.

2.6.1 Bayes's Equation

Equation 2.4-2 may be considered to be composed of three variables (Tribus, 1969) as shown in equation 2.6-1, where $P(A*B|E)$ is read as the probability of A and B given E where A, B, and E are observables. (E represents the operating environment.)

$$P(A*B|E) = P(A|B*E)*P(B|E) \quad (2.6\text{-}1)$$
$$P(A*B|E) = P(B*A|E), \quad (2.6\text{-}2)$$

However, equation 2.6-2 is valid because A, B are commuting variables that lead to equation 2.6-3. Rearranging, results in one of the usual forms of the Bayes equation (equation 2.6-4). $P(A|E)$ is the prior probability of A given E. $P(B|A*E)$ is probability that is inferred from new data (call it "update"), and $P(A|B*E)$ is the posterior probability that results from updating the prior with new information. The denominator $P(B|E)$ serves the purpose of normalization.

$$P(A*B|E) = P(B*A|E) = P(A|B*E)*P(B|E) = P(B|A*E)*P(A) \quad (2.6\text{-}3)$$

$$P(A|B*E) = \frac{P(B|A*E)*P(A|E)}{P(B|E)} \quad (2.6\text{-}4)$$

From the completeness of the A_j's (the components of A) and using the multiplication rule for intersection, equation 2.6-5 is obtained. Equation 2.6-6 results from the product rule. Bayes equation may be written as the more useful equation 2.6-7, for discrete A_js, and as equation 2.6-8 for continuous A's over the domain of A. Equations 2.6-7 and 2.6-8 show that the updated probability is just the product of the prior and the new information.

$$P(B|E) = \sum_i P(B*A_i|E) \quad (2.6\text{-}5)$$

$$P(B|E) = \sum_i P(B|A_i*E)*P(A_i|E) \quad (2.6\text{-}6)$$

$$P(A|B*E) = \frac{P(B|A*E)*P(A|E)}{\sum_i P(B|A_i*E)*P(A_i|E)} \quad (2.6\text{-}7)$$

$$P(A|B*E) = \frac{P(B|A*E)*P(A|E)}{\int P(B|A_i*E)*P(A_i|E)*dA} \quad (2.6\text{-}8)$$

Now that the point has been made that all variables are conditioned by the environment, E will be dropped from the notation because it is implied.

Example of the Use of the Bayes Equation

Suppose two equal lots of bolts were supplied to an aircraft manufacturer. Fifty percent of the bolts in lot A_1 are excessively hard, and 10% in lot A_2, but the identity of the lots was lost. The problem is to determine the probability of lot identification by sampling, using equation 2.6-7 and the prior knowledge related to the fractions of bad bolts in each lot. Let B be the failed test, then the probability of B, if it were from lot A, is $P(B|A_1) = 1 - 0.5 = 0.5$, and if B is from lot A2, $P(B|A_2) = 1 - 0.9 = 0.1$. The denominator is: $P(B|A_1) P(A_1) + P(B|A_2) P(A_2) = P(B)$. Inserting values: $0.5 * 0.5 + 0.1 * 0.5 = 0.3$. Hence, the probability that lot B was selected from A_1, given the evidence of B. is: $P(A_1|B) = P(A_1)*P(B|A_1)/P(B) = 0.5 * 0.5/0.3 = 0.883$. The probability that it was selected from A_2 is: $P(A_2|B) = P(A_2) * P(B|A_2)/P(B) = 0.5 * 0.1/0.3 = 0.167$.

2.6.2 Bayes Conjugates for Including New Information

There are two special cases for which equations 2.6-7 and 2.6-8 are easily solved to fold a prior distribution with the update distribution to obtain a posterior distribution with the same form as the prior distribution. These distributions are the Bayes conjugates shown in Table 2.6-1.

This table indicates that if a beta function prior is convoluted with a binomially distributed update, the combination (the posterior) also is beta distributed. Similarly, prior information distributed as a gamma function with an exponentially distributed update gives a posterior that also is gamma distributed.

Table 2.6-1 Bayes Conjugate Self-Replication

Prior	Update	Posterior
beta →	binomial →	beta
gamma →	exponential →	gamma

For both conjugate distributions, since the posterior is the same as the prior, this process may be repeated as many times as desired. A double application is called a "two-stage Bayesian update." Bayes conjugates provide a simple, easy way to implement Bayes' equation without using the summation or integration required by equations 2.6-7 or 2.6-8 for arbitrary distributions.

2.6.3 Constant Failure Rate Model

If the failure distribution of a component is exponential, the conditional probability of observing exactly M failures in test time t given a true (but unknown) failure rate λ and a Poisson distribution, is equation 2.6-9. The continuous form of Bayes's equation is equation 2.6-10. The Bayes conjugate is the gamma prior distribution (equation 2.6-11). When equations 2.6-9 and 2.6-11 are substituted into equation 2.6-10 and the integration is performed, the posterior is given by equation 2.6-12 which is also gamma distributed as it should be.

$$p(M|\lambda*t) = \frac{(\lambda*t)^M * \exp(-\lambda*t)^M}{\Gamma(r-1)} \quad (2.6\text{-}9)$$

$$p(\lambda|M*t) = \frac{p(M|\lambda*t)*p(\lambda)}{\int_0^\infty p(M|\lambda*t)*p(\lambda)*dt} \quad (2.6\text{-}10)$$

$$p(\lambda) = \frac{\lambda*\tau)^{\phi-1} * \exp(-\lambda*t)}{\Gamma(\phi)} \quad (2.6\text{-}11)$$

$$p(\lambda|M*t) = \frac{(t+\tau)*[\lambda*(t+\tau)]^{\phi+M-1}*\exp[-\lambda*(t+\tau)]}{\Gamma(\phi+M)} \quad (2.6\text{-}12)$$

Table 2.6-2 (upper) shows that a gamma prior (equation 2.6-11) updated with exponential data produces a gamma posterior (equation 2.6-12) by adding τ to t and M to ϕ. Because the prior is derived from other than test data, τ is called pseudo-time and ϕ pseudo-failure. The mean, $E(\lambda)$ and standard deviation, σ, of the prior and posterior are given by Table 2.6-2 (lower).

Table 2.6-2 Effect of Bayes Conjugation on a Gamma prior

Prior	Update	Posterior
τ	t	$t+\tau$
M	ϕ	$\phi+M$
Statistics	Prior	Posterior
$E(\lambda)$	ϕ/τ	$(\phi+M)/(t+\tau)$
σ	$\sqrt{(\phi/\tau^2)}$	$\sqrt{(\phi+M)/(\tau+t)^2}$

2.6.3.1 Confidence Estimation for the Constant Failure Rate Model

To obtain the confidence bounds, the posterior distribution (equation 2.6-12) is integrated from zero to λ', where λ' is the upper

$$\alpha = \int_0^{\lambda'} p(\lambda|M*t)*d\lambda \quad (2.6\text{-}13)$$

$$\frac{\gamma(a,x)}{\Gamma(a)} = \frac{1}{\Gamma(a)} * \int_0^x \exp(-y)*y^{a-1} \quad (2.6\text{-}14)$$

$$\alpha = \frac{\gamma(a,x)}{\Gamma(a)} = \frac{\gamma[(\phi+M),t+\tau)]}{\Gamma(\phi+M)} \quad (2.6\text{-}15)$$

bound on the failure rate with fractional confidence α. The result is the incomplete gamma function (equation 2.6-14 from Abramowitz and Stegun (1964) - equation 6.5.2). Let $y = (t+\tau)*\lambda$, $dy = (t+\tau)*d\lambda$, $a = \phi + M$, and $x = \lambda'(t+\lambda)$ and equation 2.6-15 is obtained.

Using equation 26.4.19 from Abramowitz and Stegun, 1964 and changing to conventional notation, results in equation 2.6-16. This says that the failure rate is less than or equal to the inverse cumulative chi-squared distribution with confidence α and degrees of freedom equal to twice the number of failures including pseudo- failures divided by twice the time including psuedo-time.

$$\lambda' \leq \frac{\chi^2_{\alpha, 2*(\phi+M)}}{2*(t+\tau)} \quad (2.6\text{-}16)$$

2.6.3.2 Truncated Gamma and Step Prior

It may be decided that the gamma prior cannot be greater than a certain value x^o. This has the effect of truncating the normalizing denominator in equation 2.6-10,[m] and leads to equation 2.6-17, where $P(\chi_0^2/v)$ is the cumulative integral from 0 to χ^2 over the chi-squared density function with v degrees of freedom, α is the prescribed confidence fraction, and $\chi_0^2 = 2*\lambda^{0*}(t+\tau)$. Thus, the effect of the truncated gamma prior is to modify the confidence interval to become an effective confidence interval of $\alpha * P(\chi_0^2/v)$.

$$\lambda' \leq \frac{\chi^2_{\alpha * P(\chi_0^2/v),\, 2*(M+\phi)}}{2*(t+\tau)} \quad (2.6\text{-}17)$$

$$p(\lambda) = 1 \text{ if } \lambda < \lambda^0 \quad (2.6\text{-}18)$$
$$p(\lambda) = 0 \text{ if } \lambda > \lambda^0$$

It may be noted that if $\phi = 1$ and $\tau = 0$, the prior, equation 2.6-11, becomes $p(\lambda) = 1$. This is the flat prior indicating no prior information that leads to classical results when these parameters are inserted into equation 2.6-12.

Suppose, however, the analyst says that the prior is flat with a value of one below a certain value, λ^0 and zero above this value (equation 2.6-18). This gives equation 2.6-19 as an estimator for λ'.

$$\lambda' \leq \frac{\chi^2_{\delta * P(\chi_0^2/v),\, 2*(M+1)}}{2*\tau} \quad (2.6\text{-}19)$$

Example: Pressure Vessel Lifetime

A pressure vessel designer, on the basis of test, inspection, and experience, says that a vessel has a failure rate less than lE-6/VY. Experience has already demonstrated 9.3E5VY without failure. What is the estimated failure rate for 95% confidence?

---

[m] The derivation of these results is given in Fullwood et al., 1977, and will not be presented in detail in this book.

Mathematics for Probabilistic Safety

Using equation 2.6-19 with $\chi_o^2 = 2*\lambda^o*t = 2E-6*9.3E-5 = 1.86$, and $\nu = 2$ (no failures), and Abramowitz and Stegun (1964, table 26.7) with linear interpolation gives: $P(\chi_o^2|\nu) = 1-0.3947 = 0.6053$. For 95% confidence: $\lambda' < \chi^2_{0.575,\,2} / (2*9.3*1E5) = 1.82/((2*9.3*1E5) = 9.78E-7/VY$.

2.6.4 Failure on Demand Model

A frequently encountered problem requires estimating a failure probability based on the number of failures, M, in N tests. These updates are assumed to be binomially distributed (equation 2.4-10) as $p(r|N)$. Conjugate to the binomial distribution is the beta prior (equation 2.6-20), where f is the probability of failure.

Combining the prior with the binomial update in Bayes's equation (equation 2.6-8) for the variable range zero to one gives equation 2.6-21 which, when integrated, this gives equation 2.6-22.

$$p(f) = \frac{\Gamma(\alpha+\beta)}{\Gamma(\alpha)*\Gamma(\beta)} *f^{\alpha-1}*(1-f)^{\beta-1} \quad (2.6\text{-}20)$$

$$p(f|r*N) = \frac{p(r|N)*p(f)}{\int_0^1 p(r|N)*p(f)*df} \quad (2.6\text{-}21)$$

Comparing equations 2.6-20 and 2.6-22, the effect of the prior is to add pseudo-failures and pseudo-tests. The mean and standard deviation of this distribution are given by equations 2.6-23 and 2.6-24, respectively.

$$p(f|r*N) = \frac{\Gamma(N+\alpha+\beta)}{\Gamma(M+\alpha)*\Gamma(N-M+\beta)} *f^{M+\alpha-1}*(1-f)^{N-M+\beta-1} \quad (2.6\text{-}22)$$

$$E(f) = \frac{M+\alpha}{N+\alpha+\beta} \quad (2.6\text{-}23)$$

$$\sigma = \frac{1}{(N+\alpha+\beta)} * \frac{\sqrt{(M+\alpha)*(N-M+\beta)}}{(N+\alpha+\beta+1)} \quad (2.6\text{-}24)$$

2.6.4.1 Confidence Estimation for the Failure on Demand Model

The confidence α' that the true value of probability, f, is less than some value f' is defined as equation 2.6-25, where the probability density $p(f|r*N)$ is defined by equation 2.6-22. Equation 2.6-25 is tabulated as the incomplete beta integral $I_f^*(M, N-M)$ in Abramowitz and Stegun (1964) to give equation 2.6-26, where F (equation 2.6-27) is the variance ratio distribution function and Q is the cumulative integral over F. This is similar to the classical result (equation 2.5-33) which means that pseudo-failures, $\alpha-1$, are added to the failures, M, and pseudo-tests, $\beta-\alpha$, are added to the tests, N.

$$\alpha' = \int_0^{f'} *p(f|r)*N)*df \quad (2.6\text{-}25)$$

$$\alpha' = Q[F|2*(N+\beta-M-\alpha), 2*(M+\alpha)] \quad (2.6\text{-}26)$$

$$f = \frac{r+\alpha}{M+\alpha+(N+\beta-M-\alpha)*F} \quad (2.6\text{-}27)$$

Bayesian Methods

2.6.4.2 Truncated Flat Prior (Step Function Prior)

Suppose a designer does not know the failure probability of something, but is confident it is less than some probability f'. This statement of belief is represented by a prior such that: $p(f \leq f') = 1$, and $p(f > f') = 0$. This truncates the denominator in equation 2.6-21 from which a confidence, α' is calculated as equation 2.6-28, where I_f is the incomplete gamma function. If $M+1$ is small, equation 2.6-29 may be more convenient.

$$\alpha' = \int_0^{f'} p(f|M*N)*df = \frac{I_f(M+1,N-M+1)}{I_f^o(M+1,N-M+1)} \quad (2.6\text{-}28)$$

$$I_{f^o}(M+1,N-M+1) = 1 - \sum_{x=0}^{M} \binom{N}{x}(f^o)^x*(1-f^o)^{N-x} \quad (2.6\text{-}29)$$

2.6.4.3 Example: Scram System Failure

At one time nuclear power plant history was 1.6E5 scrams and no failure. Using all the design testing, stress calculations, reliability calculations, and engineering judgment based on experience, a designer assesses a mean failure rate $E(f) = 1E\text{-}6$/demand and a standard deviation $\sigma = 5E\text{-}7$. Equations 2.6-23 and 2.6-24, give parameters for the prior as $\alpha = 3$, and $\beta = 4E10$. Using the test observations, these engineering prior parameters, and equation 2.6-26, the 95% confidence is that the failure rate is less than 1.52E-6/demand.

2.6.5 Interpretations of Bayes Statistics

In the introduction to this section, two differences between "classical" and Bayes statistics were mentioned. One of these was the Bayes treatment of failure rate and demand probability as random variables. This subsection provides a simple illustration of a Bayes treatment for calculating the confidence interval for demand probability. The direct approach taken here uses the binomial distribution (equation 2.4-7) for the probability density function (pdf). If p is the probability of failure on demand, then the confidence α that p is less than p' is given by equation 2.6-30.

The solution is the incomplete gamma function with $\alpha = M+1$ and $\beta = N-M+1$. Equation 2.6-31 expresses this in the usual notation which also may be written in terms of the F-function as equation 2.6-32, where equation 2.6-33 gives the probability. Comparison of equation 2.7-33 with the classical results (equation 2.5-34) shows that the results are the same except that the Bayes result has $N-M+1$ in the denominator while the classical result has

$$\alpha(p<p') = \int_0^{p'} \binom{N}{M} p^M*(1-p)^{N-M}*dp \quad (2.6\text{-}30)$$

$$\alpha(p<p') = I_{p'}(M+1,N-M+1) \quad (2.6\text{-}31)$$

$$\alpha(p<p') = Q[F|2*(N-M+1),(2*M+1)] \quad (2.6\text{-}32)$$

$$p' = \frac{M+1}{M+1+(N-M+1)*F} \quad (2.6\text{-}33)$$

N-M. The methods fail to give identical results by "one" being added to the number of tests that are obtained using Bayes methods.

In conclusion, Bayesian methods are useful for incorporating prior information. If the prior is opinion - even expert opinion, the results may be controversial because another expert may have a different opinion. An alternative to the "subjective" prior was used in EPRI NP-265 in which the prior was obtained from the fault tree for the scram system. This was combined with a data distribution from nuclear power experience. It may be argued that a fault tree has subjective aspects, but the subjectivity of a fault tree is less apparent than expert opinion. If the prior information is other test data, Bayesian methods are not needed and the data from the two tests may simply be combined. Another practice uses a prior from generic data with plant-specific data updating. This also avoids problems associated with direct estimations of the prior. However, neither the use of fault trees nor generic data gives credit for the reliability obtained from design evolution, code cases, quality assurance, and the other things that affect the probability that a component/system will perform the designed function.

2.7 Uncertainty Analysis

Figure 2.5-1 illustrates the fact that probabilities are not precisely known but may be represented by a "bell-like" distribution the amplitude of which expresses the degree of belief. The probability that a system will fail is calculated by combining component probabilities as unions (addition) and intersection (multiplication) according to the system logic. Instead of point values for these probabilities, distributions are used which results in a distributed probability of system failure. This section discusses several methods for combining distributions, namely: 1) convolution, 2) moments method, 3) Taylor's series, 4) Monte Carlo, and 5) discrete probability distributions (DPD).

2.7.1 Convolution

Suppose X and Y are process components such that the failure of either one fails the train. The probaility of failing the train is the probability that one or the other or both fail, i.e., $z = x + y$ with failure rates distributed as $p_x(x)$, $p_y(y)$. Their combined distribution is expressed by the convolution integral (equation 2.7-1).

If the components are redundant, both must fail at the same time for system failure which is expressed by multiplication: $z = x*y$, the convolution is equation 2.7-2. To generalize, let $z = f(x,y)$, then y is given by equation 2.7-3, and z by equation 2.7-4, from which the convolution is equaatin 2.7-5.

$$p_z(z) = \int_{-\infty}^{\infty} p_x(x) * p_y(z-x) * dx \quad (2.7\text{-}1)$$

$$p_z(z) = \int_{-\infty}^{\infty} p_x(x) * p_y(z/x) * (1/x) * dx \quad (2.7\text{-}2)$$

$$y = f^{-1}(z,x) \quad (2.7\text{-}3)$$

$$z = [x, f^{-1}(z,x)] \quad (2.7\text{-}4)$$

$$p_z(z) = \int_{-\infty}^{\infty} p_x(x) * p_y[f^{-1}(z,x)] * \partial f^{-1}(z-x)/\partial z * dx \quad (2.7\text{-}5)$$

2.7.2 Moments Method

The moments method does not attempt to calculate the system's uncertainty distribution directly but calculates the moments of the distribution from which the distribution may be found by an expansion in its moments.

$$z = x+y \quad (2.7\text{-}6)$$
$$<z> = <x>+<y> \quad (2.7\text{-}7)$$
$$v(z) = v(x)+v(y) \quad (2.7\text{-}8)$$
$$z = x*y \quad (2.7\text{-}9)$$
$$<z> = <x>*<y> \quad (2.7\text{-}10)$$
$$v(z) = v(x)*v(y)+<x>^2*v(y)+<y>^2*v(x) \quad (2.7\text{-}11)$$

If two distributions x and y are combined additively (equation 2.7-6) the mean and variance are given by equations 2.7-7 and 2.7-8. If two distributions are combined multiplicatively (equation 2.7-9) the mean and variance are given by equations 2.7-10 and 2.7-11).

These results, presented without proof, are important and hold for any independent physical distributions. In words, they say that *the mean of a sum is the sum of the means; the mean of a product is the product of the means* for any distribution. Furthermore, the *variance of the sum is the sum of the variances, and the variance of the product is the product of the variance plus the mean of one variable squared, times the other variables' variance, plus the converse* - again for any distribution.

When the point values are average probabilities, the overall result from combining systems as combinations of sequences and redundancies is found by simply combining the mean probabilities according to the arithmetic operations.

2.7.3 Taylor's Series

Suppose a system is a function of its components (equation 2.7-12). Expand the function f about the mean values of its arguments in a multi-variable Taylor series (equation 2.7-13). The mean of Q, which is the expectation of equation 2.7-13, is equation 2.7-14, where the variance of X_j and the covariance of X_i and X_j are given by equations 2.7-15 and 2.7-16 respectively.

$$Q = f(X_1, X_2, ---X_n) \quad (2.7\text{-}12)$$

$$Q = f(<X_1>,<X_2>,---<X_n>) + \sum_{i=1}^{n}(\partial f/\partial X_i)*(X_i-<X_i>) \quad (2.7\text{-}13)$$
$$+ 1/2 * \sum_{i=1}^{n}(\partial^2 f/\partial X_i^2)*(X_i-<X_i>)^2$$
$$+ 2 * \sum_{i=1}^{n-1}\sum_{j=i+1}^{n}(\partial^2 f/\partial X_i \partial X_j)*(X_i-<X_i>)*(X_j-<X_j>)+...$$

$$Q = f(<X_1>,<X_2>,---<X_n>) + 1/2 * \sum_{i=1}^{n}(\partial^2 f/\partial X_i^2)*\mu_2(X_i) \quad (2.7\text{-}14)$$
$$+ \sum_{i=1}^{n-1}\sum_{j=i+1}^{n}(\partial^2 f/\partial X_i \partial X_j)*E(X_i-<X_i>)*(X_j-<X_j>)+...$$

The covariance is zero when the variables are uncorrelated. The variance of Q can be obtained from equation 2.7-17. $E(Q^2)$ may be found by squaring equation 2.7-13 and taking the expectation to give equation 2.7-18.

$$\mu_2(X_i) = E(X_i - \langle X_i \rangle)^2 = \text{variance of } X_i \quad (2.7\text{-}15)$$

$$E(X_i - \langle X_i \rangle)(X_j - \langle X_j \rangle) = \text{covariance of } X_i \wedge X_j \quad (2.7\text{-}16)$$

$$\mu_2(Q) = E(Q^2) - \langle Q \rangle^2 \quad (2.7\text{-}17)$$

$$\mu_2(Q) = \sum_{i=1}^{n} (\partial f/\partial X_i)^2 * \mu_2(X_i) \quad (2.7\text{-}18)$$

$$+ 2 * \sum_{i=1}^{n-1} \sum_{j=i+1}^{n} (\partial f/\partial X_j) * (\partial f/\partial X_i) * E(X_i - \langle X_i \rangle)(X_j - \langle X_j \rangle) + \ldots$$

Case 1 Union

For the special case that the system model is the sum of probabilities (equation 2.7-19) with equation 2.7-18 gives the result: equation 2.7-20. This says that the *system mean is the sum of the means of the component distributions*.

Equation 2.7-18 for the system variance is equation 2.7-21. Since the covariance of uncorrelated variables is zero, this becomes equation 2.7-23 (i.e., the *system variance is the sum of the component variances*).

$$Q = f(X_1 \ldots X_n) + \sum_{i=1}^{n} X_i \quad (2.7\text{-}19)$$

$$Q = \sum_{i=1}^{n} X_i \quad (2.7\text{-}20)$$

$$\mu_2(Q) = \sum_{i=1}^{n} \mu_2(X_i) \quad (2.7\text{-}21)$$

$$+ 2 * \sum_{i=1}^{n-1} \sum_{j=i+1}^{n} E[(X_i - \langle X_i \rangle)(X_j - \langle X_j \rangle)]$$

$$\mu_2(Q) = \sum_{i=1}^{n} \mu_2(X_i) \quad (2.7\text{-}22)$$

Case 2 Intersection

If the system distribution is the product of the component distributions (equation 2.7-23), the mean is given by equation 2.7-24 which, for uncorrelated variables becomes equation 2.7-25.

Hence, the previous result that the system mean is the product of the component means is validated.

Equation 2.7-18 gives the variance as does equation 2.7-26, which for uncorrelated variables becomes equation 2.7-27. If specialized to two

$$Q = f(X_1 \ldots X_n) + \prod_{i=1}^{n} X_i \quad (2.7\text{-}23)$$

$$\langle Q \rangle = \prod_{i=1}^{n} \langle X_i \rangle \quad (2.7\text{-}24)$$

$$+ 2 * \sum_{i=1}^{n-1} \sum_{j=i+1}^{n} \frac{\langle Q \rangle}{\langle X_i \rangle * \langle X_j \rangle} * E[(X_i - \langle X_i \rangle)(X_j - \langle X_j \rangle)]$$

$$\langle Q \rangle = \prod_{i=1}^{n} \langle X_i \rangle \quad (2.7\text{-}25)$$

$$\mu_2(Q) = \sum_{i=1}^{n} \frac{\langle Q^2 \rangle}{\langle X_i \rangle} * \mu_2(X_i) + \quad (2.7\text{-}26)$$

$$+ 2 * \sum_{i=1}^{n-1} \sum_{j=i+1}^{n} E[(X_i - \langle X_i \rangle)^2 * (X_j - \langle X_j \rangle)^2]$$

$$\mu_2(Q) = \sum_{i=1}^{n} \mu_2(X_i) + \sum_{i=1}^{n-1} \sum_{j=i+1}^{n} \mu_2(X_i) * \mu_2(X_j) \quad (2.7\text{-}27)$$

Uncertainty Analysis

components, equation 2.7-27 is the same as the equation obtained by the moments method.

2.7.4 Monte Carlo

This is a technique developed during World War II for simulating stochastic physical processes, specifically, neutron transport in atomic bomb design. Its name comes from its resemblance to gambling. Each of the random variables in a relationship is represented by a distribution (Section 2.5). A random number generator picks a number from the distribution with a probability proportional to the pdf. After physical weighting the random numbers for each of the stochastic variables, the relationship is calculated to find the value of the independent variable (top event if a fault tree) for this particular combination of dependent variables (e.g., components).

2.7.4.1 Direct Simulation

In direct simulation, the distribution of λ is simulated by selecting values distributed according to its pdf; then the experimental distribution of failure is simulated by having ones and zeros distributed according to the values of λ. From a set of trials, various combinations of zeros and ones corresponding to the component's operability state are selected and input to logical equation to determine if failure occurred. By repeating this process many times, the failure probability distribution is found. Clearly this is a very slow process. Suppose the tree has two redundant components with mean failure probabilities of 1E-2, then the system failure probability is 1E-4 which occurs only once in ten thousand samples. In a counting problem the standard deviation is estimated as $1/\sqrt{N}$ where N is the number of counts. If 10% accuracy is desired: $0.1 = 1/\sqrt{N}$ and $N=100$. But failure only occurs once in ten-thousand trials hence one-million trials are needed for 10% accuracy.

While this outlines the concept, such real time simulation generally is not practical for determining the reliability of safety systems. However, a direct simulation of safety system reliability that was performed used the electronic simulator, ERMA (Rothbart et al., 1981), in which special purpose electronics accelerated time by 1E10, so that a million years of plant life required about one hour of simulator time. Figure 2.7-1 shows the results of a simulation of the reliability of a low pressure cooling injection system for a reactor that was simulated using ERMA. It exhibits a phenomenon called "voter statistics," that is, a candidate that gets ahead, is likely to stay ahead rather than exhibit oscillations about the steady-state value. This figure

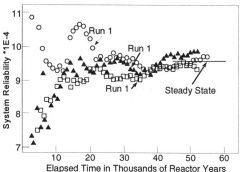

Fig. 2.7-1 ERMA simulation of a low pressure coolant injection system. The absence of oscillations in the convergence is evidence of voter statistics. Reprinted with permission of EPRI, USA

provides a caution in interpreting PSA results, namely, that a very large amount of experience is needed to reach the steady-state value--the asymptotic probability. Scatter by a factor of 2 is not surprising. These remarks are particularly appropriate in trying to draw statistical conclusions regarding the single event: TMI-2.

2.7.4.2 Simulation of Distributions[n]

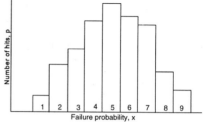

Fig. 2.7-2 *Histogram of probability distribution generated by Monte Carlo simulation. The height of each bar is the number of counts in the interval*

In the previous discussion, if all the failure rates are selected by random sampling from their distributions and these values are used directly in the exponential model with component replacement by their probability, the probability of the top event may be calculated. By sufficiently sampling of the distribution of the λs, the distribution of the top event probability is determined. The result is a histogram with each bar uncertain by the counting statistics. Figure 2.7-2 illustrates such results. To determine the amount of sampling required, suppose it is desired that the 95% wings are to be determined to 10%. This implies about 100 counts in each of the wings. But, by definition, only 5% of the area are in these locations. By proportion 200/10% = N/100%, so about 2,000 samples would be needed, fewer if there is a priori information regarding the form of the distribution; in which case the Monte Carlo process provides parameter fitting for the a priori distribution.

2.7.5 Discrete Probability Distribution (DPD)

For each component in a system, there is a distribution like Figure 2.7-2. Each bar may represent an ordered pair (x_j, P_j) where x_j is the value of the variable and P_j is the probability of the value x_{jo}.

2.7.5.1 Union (Addition)

Consider the union of only two components (equation 2.7-28), where x_1 and x_2 are each distributed as Figure 2.7-2. If there were only one bar in each histogram, the values x_1 and x_2 would be added and the probabilities multiplied as the probability of joint occurrence (equation 2.7-29).

$$z = (x_1 + x_2) \qquad (2.7\text{-}28)$$
$$z = (x_1 + x_2, P_1 * P_2) \qquad (2.7\text{-}29)$$

---

[n] This method was used in the Reactor Safety Study.

[o] In this discussion, x_j is the probability that a system's component will fail and P_j is the probability that it has the probability x_j. This confusion is avoided by just considering x_j as a variable.

Sensitivity Analysis and Importance Measures

For each distribution with a number of bars, start with the first bar of the first distribution and add successively all combinations of the bars in the second distribution, then go to the second bar of the first distribution and add all combinations of the second distribution, and so on until all combinations have been executed.

To generalize, let x_{1n} represent the nth abscissa of the first distribution and P_{1n} the corresponding ordinate of a *normalized* distribution. Then z is formed of the ordered pairs (equation 2.7-30). As an example, for x_1 (equation 2.7-31), and x_2 (equation 2.7-32 - each is a bar in the histogram), we get equation 2.7-33. It should be noted that the resulting distribution is normalized. (This may be checked by adding all the second pair numbers.)

$$z = (x_{11}+x_{21}, P_{11}*P_{21})(x_{11}+x_{21}, P_{11}*P_{21}) \quad (2.7\text{-}30)$$
$$(x_{11}+x_{21}, P_{11}*P_{2n})(x_{11}+x_{21}, P_{11}*P_{21})$$
$$\vdots$$
$$(x_{1n}+x_{21}, P_{11}*P_{21})(x_{11}+x_{21}, P_{11}*P_{21})$$

$$x_1 = (-1, 0.1)(1, 0.5)(2, 0.4) \quad (2.7\text{-}31)$$
$$x_2 = (5, 0.2)(10, 0.8) \quad (2.7\text{-}32)$$
$$z = (4, 0.02)(6, 0.1)(7, 0.08)(9, 0.08)(11, 0.4)(12, 0.32) \quad (2.7\text{-}33)$$

2.7.5.2 Intersection (Multiplication)

Suppose the components are redundant, then their probabilities, if independent, are combined by multiplication (equation 2.7-34). this is generalized by analogy with the preceding (equation 2.7-35). As a numerical example, if the distributions x_1 and x_2 in equations 2.7-30 and 2.7-31 are combined in multiplication, the results is equation 2.7-36. A bit of reflection shows that DPD is a numerical implementation of convolution integration described in Section 2.7.1.

$$z = (x_1 * x_2) \quad (2.7\text{-}34)$$

$$z = (x_{11}*x_{21}, P_{11}*P_{21})(x_{11}*x_{21}, P_{11}*P_{21}) \quad (2.7\text{-}35)$$
$$(x_{11}*x_{21}, P_{11}*P_{2n})(x_{11}*x_{21}, P_{11}*P_{21})$$
$$\vdots$$
$$(x_{1n}*x_{21}, P_{11}*P_{21})(x_{11}*x_{21}, P_{11}*P_{21})$$

2.8 Sensitivity Analysis and Importance Measures

2.8.1 Sensitivity Analysis

When a risk or reliability analysis has been performed, it is appropriate to inquire into the sensitivity of the results to uncertainties in data. One type of sensitivity analysis is the effect on system reliability that results from a small change in a component's failure probability. A problem in doing this is determining the amount of data uncertainty that is reasonable. The amount of change may be judgment or the parameter may be taken to the extreme of the data range even though such extremes may be rare.

Significance of risk contribution may be done by ordering the risk contributors from most-to-least (rank order), but because of the arbitrariness of variation of the variables, this may be meaningless. A more systematic approach is to calculate the fractional change in risk or reliability for a fractional change in a variable.

As an example, suppose risk is composed of failure probabilities $p_1, p_2, ... p_3$ i.e. $R = R(p_1, p_2, ...p_3)$. The total derivative is equation 2.8-1, where each of the partial derivatives are taken with the unincluded variables held constant. The partial derivatives are multipliers f_i that multiply the change in p_i. The larger the multiplier, the more significant that component is to the overall risk. Thus, the f_i's measure the *sensitivity* of the i-th component's probability of failure. A disadvantage of this sensitivity measure is in determining the size of the uncertainty in dp_i.

$$dR = (\partial R/\partial p_1)*dp_1 + (\partial R/\partial p_2)*dp_2 + ... + (\partial R/\partial p_n)*dp_n = \Sigma f_i*dp_i \quad (2.8\text{-}1)$$

2.8.2 Importance Measures

A number of risk importance measures have been defined for the interpretation of PSAs and for use in prioritization of operational and safety improvements. Some of these measures are similar to sensitivity defined as the total derivative (equation 2.8-1).

a) <u>Birnbaum Importance</u> - is the change in the *risk for a change in the failure probability* for a component or system (the same as f_i in equation 2.8-1). Equation 2.8-2 is the Birnbaum importance for the *i-th* component, where R (risk) is the dependent variable, p_i (probability) is the failure probability of the *i-th* component.

While Birnbaum Importance identifies systems important to safety, it does not consider the likelihood of component failure. That is, highly important but highly reliable passive components have a high Birnbaum Importance. To overcome this problem, Inspection Importance was formulated.

b) <u>Inspection Importance</u> - is *Birnbaum Importance multiplied by the probability of failure* of the component of concern equation 2.8-3). This is the same as unnormalized Fussell-Vesely Importance.

$$I_i^B = \frac{\partial R}{\partial p_i} \quad (2.8\text{-}2)$$

$$I_i^I = p_i * \partial R/\partial p_i \quad (2.8\text{-}3)$$

$$I_i^{FV} = \frac{\partial R/R}{\partial p_i/p_i} \quad (2.8\text{-}4)$$

c) <u>Fussell-Vesely Importance</u>[p] - is the *fractional change in risk for a fractional change in component failure probability* (equation 2.8-4). A problem with Fussell-Vesely importance is in determining the risk used for normalization. Should it be total risk or partial risks? If the measure is only used for prioritization, it does not matter.

---

[p] In Fussell-Vesely importance it is common to set the risk normalization to 1.

Inspection Importance is Fussell-Vesely Importance with the risk set to 1. Vesely introduced several other importance measures called "worths."

d) <u>Risk Achievement Worth Ratio (RAWR)</u> - is the *ratio of the risk with the i-th component failed to the risk with no components assumed to be failed.* This is shown in equation 2.8-5, where R_i^+ means the risk with *i-th* component failed.

$$I_i^{AR} = \frac{R_i^+}{R} \quad (2.8\text{-}5)$$

$$I_i^{AI} = R_i^+ - R \quad (2.8\text{-}6)$$

e) <u>Risk Achievement Worth Increment (RAWI)</u> - is the *incremental change in risk that result from the i-th component failed* (equation 2.8-6).

$$I_i^{RR} = \frac{R}{R_i^-} \quad (2.8\text{-}7)$$

$$I_i^{RI} = R - R_i^- \quad (2.8\text{-}8)$$

f) <u>Risk Reduction Worth Ratio (RRWR)</u> - is the *ratio of the nominal risk to the risk that results with the i-th component perfect.* This is equation 2.8-7, where R_i^- means the risk with the i-th component having a failure probability equal to zero.

g) <u>Risk Reduction Worth Increment (RRWI)</u> - is the *incremental change in risk that results from the i-th component being perfect* (equation 2.8-8).

2.8.3 Relationships between the Importance Measures

Importance measures the contributions of various components to the probability of system failure in the form of a cutset equation (Section 2.2). A cutset equation has the form of a sum of products of probability. As such it is linear in the probabilities.

The sequences involving the *i-th* component may be separated from the other sequences and this probability factored from the sum of products as shown in equation 2.8-9, where A_i is the cutset with p_i factored out. This property relates the importance measures. Substituting equation 2.8-9 into equation 2.8-2 gives equation 2.8-10. Similarly substituting equation 2.8-2 into equation 2.8-3 gives equation 2.8-11. Eliminating A between equations 2.8-10 and 2.8-11 gives the equivalence between Birnbaum and Inspection importances

$$R = \Sigma p_i * A_i + B \quad (2.8\text{-}9)$$

$$I_i^B = A_i \quad \text{Birnbaum} \quad (2.8\text{-}10)$$

$$I_i^I = p_i * A_i \quad \text{Inspection} \quad (2.8\text{-}11)$$

$$I_i^{FV} = p_i * \frac{A_i}{R} \quad \text{Fussell-Vesely} \quad (2.8\text{-}12)$$

$$I_i^{AR} = \frac{A_i + R}{R} \quad \text{Risk Achievement Worth Ratio} \quad (2.8\text{-}13)$$

$$I_i^{AI} = A_i + B - R \quad \text{Risk Achievement Worth Increment} \quad (2.8\text{-}14)$$

$$I_i^{RR} = \frac{R}{R_i^-} \quad \text{Risk Reduction Worth Ratio} \quad (2.8\text{-}15)$$

$$I_i^{RI} = R - B \quad \text{Risk Achievement Worth Increment} \quad (2.8\text{-}16)$$

(Table 2.8-1, 2). Following the same procedure produces an equivalence between Birnbaum and Fussell-Vesely importances (Table 2.8-1,3).

Achievement Worth Importance requires failing the *i-th* component by setting its probability to one: $p_i = 1$ as shown in equations 2.7-13 and 2.7-14. Substituting for A_i from 2.8-10 gives the equivalences in Table 2.8-1, 4 and 5).

The Reduction Worth Importances require that the system or component be made perfect by setting the probability to zero: $p_i = 0$ (equations 2.8-15 and equation 16). A_i and B may be eliminated to relate one importance to others (Table 2.8-1,6 and 7). This table also summarizes the importance definitions.

Table 2.8-1 Relationships between Importance Measures

#	Measure	Definition	Equivalence
1	Birnbaum	$I_i^B = \partial R/\partial p_i$	--
2	Inspection	$I_i^I = p_i * \partial R/\partial p_i$	$= p_i * I_i^B$
3	Fussell-Vesely	$I_i^{FV} = \partial R/R/\partial p_i/p_i$	$= I_i^I/R$
4	Risk Achievement Worth Ratio	$I_i^{AR} = R_i^+/R$	$= 1 + (I_i^B - I_i^I)/R$
5	Risk Achievement Worth Increment	$I_i^{AI} = R_i^+ - R$	$= I_i^B - I_i^I$
6	Risk Reduction Worth Ratio	$I_i^{RR} = R/R_i^-$	$= R/(R - I_i^I)$
7	Risk Reduction Worth Increment	$I_i^{RI} = R - R_i^-$	$= I_i^I$

2.8.4 Interpretation and Usage

The relationships between the importance measures is based on the assumption that the systems are not reconfigured in response to a component outage. If this is done, the basic definition of the importance measure is still valid but there is not such a simple relationship. Disregarding this complication, some interpretations of the importances may be made. The Birnbaum Importance is the risk that results when the i-th system has failed (i.e., it is the A_i term in Equation 2.8-9). Inspection Importance and RRWI are the risk due to accident sequences containing the i-th system. Fussell- Vesely Importance is similar except it is divided by the risk so may be interpreted as the fraction of the total risk that is in the sequences contains the Q-th system. The Risk Achievement Worth Ratio (RAWR) is the ratio of the risk with system 1 failed to the total risk and is necessarily greater than one. The Risk Achievement Worth Increment (RAWI) is the incremental risk increase if system 1 fails and the Risk Reduction Worth Ratio (RRWR) is the fraction by which the risk is reduced if system 1 were infallible.

Figures 2.8-1 and 2.8-2 from Vesely et al. (1983) show how the RAWR and RRWR measures vary for two PWR nuclear power plants (Oconee, and Sequoyah) based on their PSAs. This variation may be explained on the basis of different manufacturers of the plants (Babcock & Wilcox and Westinghouse, respectively)

It has been suggested that the rank ordering of systems is about the same regardless of the importance measure that is used. Table 2.8-2 is the result of a cluster analysis (judgement of trends) in Indian Point-3 system importances with clustering into 5 groups designated A through E. This

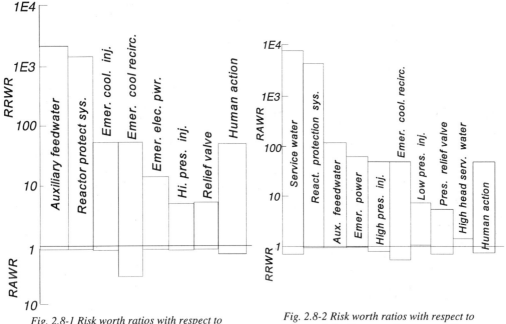

Fig. 2.8-1 Risk worth ratios with respect to core melt for Sequoyah safety systems (from Vesely et al., 1983)

Fig. 2.8-2 Risk worth ratios with respect to core melt for Oconee safety systems (from Vesely et al., 1983). See text for meaning of RRWR and RAWR

table shows that there is a significant reordering of the system ranks depending on the importance measure that is used, and depending on whether or not the importances are calculated to core melt or to public health effects.

2.9 Summary

Set theory and Venn diagram were used to show relationships between components and system, and to show the importance of the mincutset for correctly calculating system probability - one member of the risk dyad. Of the three definitions of probability, the von Misesian, limiting form of probability, is practical for a wide variety of situations, and is used repeatedly. However, the fact that it is a limit is often overlooked and the approximation may be taken to be true. Useful rules for Boolean reduction were presented for later use - especially De Morgan's theorem. Methods for combining probabilities were presented. Distributions were presented for use in estimation, for confidence calculation and for fundamental understanding. Bayes methods were presented and compared with "classical" statistics and it was found that they are different in concept but similar in results, if given the same information. Methods for estimating the uncertainty of probabilistic calculations were presented using properties of the distributions. Sensitivity calculations were presented as derivatives and a close relationship between sensitivity and common importance measures was shown.

2.10 Problems

1. The control rods in a hypothetical reactor have the pattern shown in Figure 2.10-1. Based on plant experience, it is determined that the probability of a rod failing to insert is 0.01/plant-year. Assuming that if 3 nearest-neighbor rods fail to insert, core damage will result. What is the probability of core damage from failure to scram from control rod sticking?

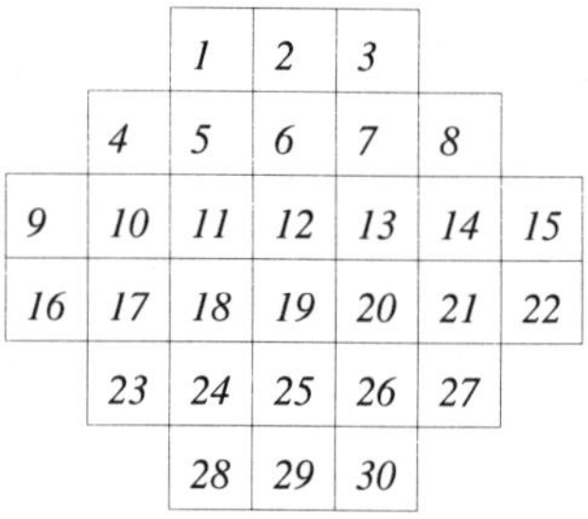

2. The Boolean equation for the probability of a chemical process system failure is: R = A*(B+C*(D+E*(B+F*G+C). Using Table 2.1-1, factor the equation into a sum of products to get the mincut representation with each of the products representing an accident sequence.

3. The probability that a system will fail is the sum of the following probabilities: 0.1, 0.2, 0.3, 0.4, and 0.5. What is the probability of failure?

	1	2	3			
	4	5	6	7	8	
9	10	11	12	13	14	15
16	17	18	19	20	21	22
	23	24	25	26	27	
		28	29	30		

Fig. 2.10-1
Control Rod Pattern

4. The probabilities for the equation in Problem 2 are: 1E-3/yr, 4E-3, 7E-4, 0.1, 0.5, 0.25, and 0.034 for the letters A through G respectively. What is the value of R? What is the probability of each of the sequences?

5. Using the information and result of Problem 4, Calculate the Birnbaum, Inspection, Fussell-Vesely, Risk Reduction Worth Ratio, Risk Reduction Worth Increment, Risk Achievement Worth Ratio, and Risk Achievement Worth Increment for each of the components A through G. Do your results agree with the equivalences in Table 2.8-1?

6. In Section 2.5.4, we found the availability of a repairable emergency generator (EG) by Markov methods. If a plant requires that two identical, independent EGs must both work for time T for success. What is the probability of this? Assume the failure rates are λ_1, λ_2, and the repair rates are μ_1, μ_2.

Chapter 3

Chemical and Nuclear Accident Analysis Methods

3.1 Guidance from the PSM Rule

3.1.1 Rule Objectives

Process safety management is a systematic approach to preventing unwanted releases of hazardous materials from affecting workers and the public.[a] This encompasses the: process technology, procedures, operational and maintenance activities, non-routine activities, emergency preparedness plans, training and other elements and procedures. The defense-in-depth safety that is incorporated into the design and operation of processes, requires evaluation to assure effectiveness. Process safety management anticipates, evaluates and mitigates chemical releases that could result from failure of process procedures/equipment.

Protection priorities are ranked according to the hazardousness of the process chemicals. The Process Safety Management Rule (PSM) aids in preventing or mitigating chemical risks. The OSHA standard is based on management's responsibility for implementing and maintaining an effective process safety management program. This OSHA standard, required by the Clean Air Act Amendments, is used in conjunction with EPA's Risk Management Plan. Full compliance comes from merging the two sets of requirements into the process safety management program to enhance the relationship with the local community.

Obviously, the risk of hazardous chemicals is reduced by a minimized inventory using just-in-time procurement. If further inventory reduction is not feasible, additional risk reduction may be achieved by dispersing the inventory to multiple site locations so a release in one location does not affect other locations of inventory.

3.1.2. Employee Involvement in Process Safety

Amendments to the Clean Air Act require employer-employees consultation to develop and implement PSM program elements and hazard assessments. Section 304 requires employee training and education on the findings of PSM incident investigations.

---

[a] Much of the following is taken from OSHA (1996).

3.1.3. Process Safety Information

Complete and accurate written documentation of chemicals properties, process technology, and process equipment is essential to the PSM program and to a process hazards analysis (PrHA). This information serves many users including the PrHA team. The needed chemical information includes: fire and explosion characteristics, reactivity hazards, safety and health hazards and the corrosion and erosion effects. Current material safety data sheet (MSDS[b]) information helps meet this requirement, but must be supplemented with process chemistry information regarding runaway reactions, and over-pressure hazards.

Process technology information includes diagrams such as Figure 3.1.3-1, and criteria for maximum inventory levels for process chemical limits beyond which the process is considered to be upset. Also included is a qualitative estimate of the consequences that could result from deviating from the limits.

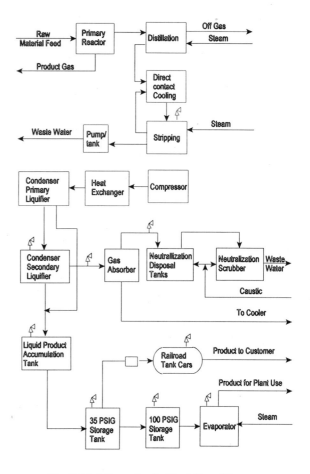

Fig. 3.1.3-1 Example of a Block Flow Diagram

A block flow diagram is simplified to show the major process equipment, interconnecting process flow lines, flow rates, stream composition, temperatures, and pressures.

Process flow diagrams (Figure 3.1.3-2) are more complex. They show all main flow streams, valves, vessels, condensers, heat exchangers, instrument locations, reflux columns, etc. Pressures

---

[b] MSDS are available from the Chemical Abstract Service of the American Chemical Society (ACS), 1155 16th St. NW Washington DC 20036, or the Internet at http//www.acs.org, and from the chemical manufacturer. A general website to these is: http://hazard.com/

Guidance from the PSM Rule

and temperatures on all feed and product lines to enhance the understanding of the process. They show materials of construction, pump capacities, compressor horsepower, and vessel design pressures and temperatures. In addition, major components of control loops are usually shown along with key utilities.

Piping and Instrument diagrams (P&IDs - not shown) may be even more appropriate for showing these details and to provide information for PSA. The P&IDs are used to describe the relationships between equipment and instrumentation and other relevant information that enhances clarity. Computer software programs, fault trees and event trees are other diagrams that provide useful information.

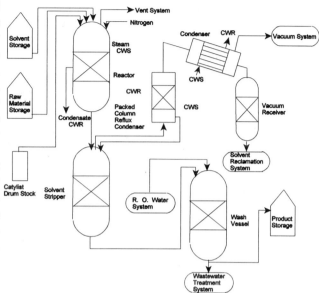

Fig. 3.1.3-2 Process Flow Diagram

This information documents the process equipment design and provides a "paper trail" to the codes and standards used in the engineering. Such codes and standards are published by such organizations as the American Society of Mechanical Engineers, American Petroleum Institute, American National Standards Institute, National Fire Protection Association, American Society for Testing and Materials, National Board of Boiler and Pressure Vessel Inspectors, National Association of Corrosion Engineers, American Society of Exchange Manufacturers Association, and model building code groups.

In addition, various engineering societies issue technical reports which impact process design. For example, the American Institute of Chemical Engineers (AIChE) publishes technical reports on topics such as two phase flow for venting devices. Referencing and using such technical reports is good engineering practice.

Existing equipment that may have been designed and constructed many years ago according to codes and standards no longer in use, requires identifying the codes and standards that were used, and the aspects of the facility and process affected. To justify that it is still suitable for the intended use may require testing, inspection, and demonstration. If the process involves departure from applicable codes and standards, the suitability of the design and construction must be documents.

3.2 Process Hazard Analysis

3.2.1 Overview

Process hazard analysis d(PrHA) is one of the most important elements of a process safety management program. A PrHA is an organized and systematic effort to identify and analyze the significance of potential hazards associated with the processing or handling of hazardous chemicals. PrHA information assists management and employees decide on safety improvements and accident consequence reductions. A PrHA is directed toward analyzing the risk of fires, explosions, releases of toxic or flammable materials. Equipment, instrumentation, utilities, human actions, and external factors are considered in the process.

Selection of a PrHA methodology requires consideration of many factors including the availability of process information such as: experience with the process, changes that have taken place, reliability, aging, maintenance, etc. If it is a new process, less reliance can be placed on experience and greater reliance must be placed on the analysis of possible accidents and accidents in similar or related processes. Size, complexity and hazard severity influences the choice of the most appropriate PrHA methodology.

Simplifying assumptions used by the PrHA team and reviewers must be understood and recorded to indicate the completeness of the PrHA, and for use in future improvements of the analysis. The PrHA team and especially the team leader must thoroughly understand the methodology that is selected

A team consists of two or more people that know the process technology, design, operating procedures, practices, alarms, emergency procedures, test and maintenance procedures, routine and non-routine tasks. They must consider: authorization and procurement of parts and supplies, safety and health standards, codes, specifications and regulations. The team leader provides management and goals to the process; the team and consultants construct and interpret the analysis.

PrHA may use different methodologies for different processes depending on the size, complexity and age of the facility by using team members that are expert for analyzing their assigned process. For example, a checklist may be used to analyze a boiler or heat exchanger, with a HAZOP for the overall process. Batch processes of similar subsystems may be analyzed with a generic PrHA that is modified according to differences in components, monomers or ingredient ratios.

Small operations, although covered by the PSM rule, may use simplified methodologies and still meet the criteria. Businesses with similar processes and equipment may pool resources and prepare a generic checklist analysis used by all members to meet the PSM rule.

If a PrHA is needed for many operations, the facility operator may prioritize the order of mitigating accidents. A preliminary hazard analysis (EPA, FEMA, and DOT, 1987) provides an overall risk perspective that can rankorders potential accidents by decreasing the number of affected individuals. Ordering may be based on maximum amounts and toxicity of the hypothetical releases, operating history of releases, age of the process and other factors that may modify the ranking. A PrHA assists in identifying improvements that give the most safety for the least cost.

3.2.2 Operating Procedures and Practices

Operating procedures specify the way tasks are to be performed, data to be recorded, operating conditions to be maintained, samples to be collected, and actions to enhance safety and health. Their accuracy, currency and utility is verified by the engineering staff and operating personnel. They include applicable safety precautions and warnings regarding: pressure limits, temperature ranges, flow rates, and the meaning and response to alarms and instruments. Procedures for startup and shutdown are included with the appropriate ranges for the process parameters. They include instructions and commands for computer process control, if used.

Training uses operating procedures and instructions to define standard operating practices (SOPs) for use by control room and operating staff. Operating procedures must be current, and validated before use in refresher training. Management of change must be coordinated and integrated into SOPs, operating personnel oriented to the change, the change is then made and the process resumed.

Upset and emergency training instructs the staff in their conduct under these conditions. It includes communications, mitigative actions, hazards of the tasks being performed, task closure, and reporting.

3.2.3. Employee Training

All employees, working with highly hazardous chemicals must understand the safety and health hazards (29CFR19l0.1200 - the Hazard Communication Standard) by knowing the properties of the chemicals with which they work, safe operating procedures, work practices, and emergency action.

Training programs identify subjects, goals, and individuals to be trained. The learning objectives define measurable goals for the training modules. Hands-on-training using a simulator is important for control room operators. Technicians for test and maintenance and chemical processes benefit from hands-on training when reinforced with theory. Training using videos or OJT is effective. The training program prepares these personnel without the risk of learning with the hazards.

Periodic evaluation of training programs is need to determine if the necessary skills, knowledge, and routines are being learned. Training evaluation is incorporated in the course design. If the employees, after training, have not learned what was expected, they must be retrained using changes that highlight their deficiencies.

3.2.4 Contractors

Chemical process contractors to do hazardous work must provide personnel trained in performing the tasks without compromising health and safety. References, safety records, experience, job skills, work methods, knowledge and certifications provide evidence of competency. These records provide audit trails to assure management's safety compliance and to aid in incident investigations.

Contract employee safety is controlled through work authorizations that applies to all workers. Work authorizations allow management to safely coordinate contract employee activities with employee activities.

3.2.5 Pre-Startup Safety

PrHA improves the safety, reliability and quality of the design and construction of a new or old process. P&IDs must be correct as constructed; operating, startup and shutdown procedures must be validated, and the operating staff must be trained before startup. Incident investigation recommendations, compliance audits or PrHA recommendations need resolution before startup.

3.2.6 Mechanical Integrity

The first line of defense is operating and maintaining the process as designed, to keep the hazardous material contained.

The second line is the controlled release of materials through venting to containment/ confinement or through filters, scrubbers, flares, or overflow tanks.

The third line includes: fixed fire protection systems (sprinklers, water sprays, or deluge systems, monitor guns, etc.), dikes, designed drainage systems, and other systems that control or mitigate hazardous releases.

Equipment used to process, store, or handle highly hazardous chemicals must be designed constructed, installed and maintained to minimize the risk of release. A systematic, scheduled, test and maintenance program is preferred over "breakdown" maintenance[c] that could compromise safety. Elements of a mechanical integrity program include: 1) identification and categorization of equipment and instrumentation, 2) documentation of manufacturer data on mean time to failure, 3) test and inspection frequencies, 4) maintenance procedures, 5) training of maintenance personnel, 6) test criteria, and 7) documentation of test and inspection results.

The first step of a mechanical integrity is the compilation and categorization of the process equipment and instrumentation. This includes pumps, pressure vessels, storage tanks, process piping, relief and vent systems, fire protection system components, emergency shutdown systems, alarms and interlocks. Mean time to failure data, and engineering judgment set the inspection and testing frequencies. Consideration is given to codes and standards such as: the National Board Inspection Code, American Society for Testing and Materials, American Petroleum Institute, National Fire Protection Association, American National Standards Institute, American Society of Mechanical Engineers and other groups.

---

[c] Breakdown maintenance means no preventive maintenance (i.e., failure is annunciated by the fact of failure).

External inspection criteria are provided by these standards for: foundation and supports, anchor bolts, concrete or steel supports, guy wires, nozzles, sprinklers pipe hangers, grounding connections, protective coatings, insulation, and external metal surfaces.

Internal inspection criteria are provided by these standards for: vessel shell, bottom and head, metallic linings, nonmetallic linings, thickness measurements, erosion, corrosion, cracking, bulges, internal equipment, baffles, sensors and erosion screens. If a corrosion rate is not known, a conservative estimate, based on corrosion rates from the codes, is used. While some of these inspections may be performed by government inspectors, management needs an inspection program to ensure the tests and inspections are conducted properly and consistently. Appropriate training for maintenance personnel ensures they understand the preventive maintenance program.

A quality assurance program ensures that construction, installation, fabrication, and inspection procedures are proper. The quality assurance program helps the first and second lines of defense. Quality assurance documentation includes: "As built", drawings, vessel certifications, verification of equipment, and materials of construction. Audits of the equipment supplier's facilities may be necessary to assure the quality of parts. Any changes in equipment must go through the management of change procedures.

3.2.7 Nonroutine Work Authorization

The hazards of management-controlled nonroutine work in the process areas must be communicated to affected individuals. The work permit prescribes the procedures that must be followed to get the permit. Work authorization procedures specify: lockout/tagout, line breaking, confined space entry, and hot work authorization through clear steps leading to job completion, closure, and return to normal.

3.2.8 Managing Change

Change in the PSM rule includes all modifications to equipment, procedures, raw materials, and processing conditions other than "replacement in kind." These changes are managed by identification and review prior to implementation. For example, any operation outside of the operating procedures' parameters requires review and approval by a written management of change.

Changes requiring management are: process technology raw materials, equipment unavailability, new equipment, new product development, catalyst and operating conditions. Equipment changes include: materials. specifications, piping, equipment, computer programs and alarms and interlocks.

Temporary changes have had dire consequences. Management must control temporary as well as permanent changes. Temporary changes must be time-limited to prevent their becoming permanent. Temporary changes are subject to the management of change that require provisions for the return to the original conditions. Documentation and review of these changes assures that safety and health considerations are incorporated in the procedures and process.

A clearance sheet tracks the changes through the management of change procedures. A typical change form may include the description and purpose of the change, the technical basis for the change, safety and health considerations, changes in the operating procedures, maintenance procedures, inspection and testing, P&IDs, electrical classification, training and communications, prestartup inspection, duration if a temporary change, approvals and authorization. Where the impact of the change is minor a checklist reviewed by an authorized person consulting with affected people may be sufficient. However, a complex or significant change requires a hazard evaluation procedure with approvals by operations, maintenance, and safety departments. Changes in documents such as P&IDs, raw materials, operating procedures, mechanical integrity programs, electrical classifications, etc., need noting so these revisions can be made permanent when the drawings and procedure manuals are updated. Copies of process changes are made accessible for operating personnel and the PrHA team.

3.2.9 Investigation of Incidents

Incident investigation is the process of identifying the underlying causes and implementing steps to prevent similar events from occurring. The incidents for which OSHA expects management awareness and investigation, are events that could have (near miss) or did result in a catastrophic release.

Employers need in-house capabilities to investigate incidents that occur in their facilities. This requires a multidisciplinary team trained in the techniques of accident investigation capable of: conducting interviews, data gathering, accident modeling, consequence estimation, documentation and report writing. Team members are selected for their training, knowledge and ability. Personnel, where the incident occurred, are interviewed for preparation of a report of the findings and recommendations for dissemination as appropriate.

3.2.10 Emergency Preparedness

Each employer must have an emergency preparedness plan in case of an unplanned release of hazardous material. This third line of defense mitigates an accident. Employees can mitigate small or minor releases, but major releases require additional resources in the form of local emergency response organizations. The emergency preparedness plans identify different levels of severity in the preparation of plans, procedures, training employees and implementing the plan.

The emergency action plan facilitates prompt evacuation of employees. Upon an alarm; personnel evacuate to safe areas, cross or up wind if possible, assisting physically impaired people as necessary. Process buildings are not suitable refuges.

Appointed, trained employees direct the emergency response to a minor release. 29CFR1910.120, the Hazardous Waste Operations and Emergency Response (HAZWOPER) standard covers plant personnel for a fire brigade, spill control, hazardous materials or first aid team. It addresses cooperative agreements between employers and state or local government emergency response organizations. Safety and health are the responsibility of employers and the on-scene incident commander and staff who are equipped and trained for emergency response. Drills, training exercises, and simulations, including the local emergency response team, are preparations for emergencies

Medium to large facilities enhance coordination and communication during emergencies with an emergency control center (ECC) that is composed of plant and local organizations and sited in a safe area for continuous occupancy throughout the emergency. It links the commander, plant management, and the local officials. The ECC contains a telephone, radio network, a backup communication network in case of power failure, the plant layout, community maps, utility drawings (fire water, emergency lighting), and reference materials (government agency notification list, company personnel phone list). It includes SARA Title III reports and material safety data sheets, emergency plans and procedures manual, a listing with the location of emergency response equipment, mutual aid information, and access to meteorological or weather data and any dispersion models.

3.2.11 Compliance Audits

A trained audit team, consisting of one or more persons audits the process safety management system. The essential elements of an audit are: planning, staffing, conducting the audit, evaluation, corrective action, follow-up, and documentation. Advanced planning regarding the format, staffing, scheduling and verification methods are essential to the success of the auditing process. It provides a checklist that details the requirements of each section of the PSM Rule, lists the audit team members, serves as the audit verification sheet to expedite and to avoid omitting any requirements, and identifies elements needing evaluation or response to correct deficiencies. This sheet is used to develop follow-up and documentation requirements.

The audit team members are selected for their experience, knowledge, training, familiarity with the processes and auditing techniques, practices, and procedures. The size of the team depends on with the size and complexity of the process. A large, complex, highly instrumented plant, may have members with expertise in: process engineering design, process chemistry instrumentation, computer controls, electrical hazards and classifications, safety and health disciplines, maintenance, emergency preparedness, warehousing, shipping, and process safety auditing.

The audit includes a review of the process safety information, inspection of the physical facilities, and interviews with all levels of plant personnel. Using the procedures and checklist, the team systematically analyzes compliance with the PSM Rule and any other relevant corporate policies. The training program is reviewed for adequacy of content, frequency and effectiveness of training. Interviews determine employee knowledge and awareness of the safety procedures, duties, rules, and emergency response assignments. The team identifies deficiencies in the application of safety and health policies, procedures, and work authorization practices to determine corrective actions.

Corrective actions are audit objectives. They address identified deficiencies, in addition to planning, follow up, and documentation. Corrective action begins with management review of the audit findings to determine appropriate actions and to establish priorities, timetables, resource allocations, requirements and responsibilities. Some corrective actions may be a simple change in procedure - subject to management of change. Complex changes require time for engineering studies and reviews. Taking no action is valid audit action; all actions require documentation.

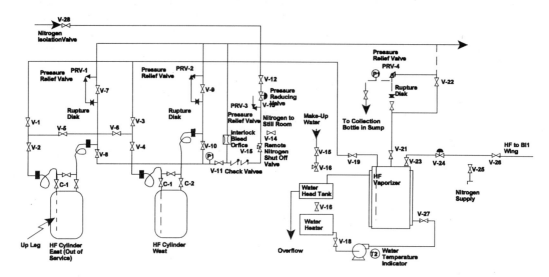

Fig. 3.3.1-1 Hydrogen Fluoride Supply System for providing HF_6 to a fluid bed reactor to produce uranium tetrafluoride for diffusive separation of U-235 from U-238. Gaseous HF is created by vaporizing anhydrous liquid HF in 850 lb amounts in shipping pressure vessels. Nitrogen at 30 psig is applied to the top of the HF cylinder that is being weighed. HF is discharged to the vaporizer which is heated to vaporization at 25 psig by a hot water blanket. Safety components are the nitrogen pressure regulator and overpressure relief valves. Overpressure protection for the vaporizer is provided by relief valves connected to the top of the vaporizer and supply cylinder. A rupture disk protects the relief valves from exposure to corrosive HF. A tee between them to a manual vent and block valve that can be opened to relieve pressure for maintenance. Vent gas is released through a plastic hose to bubble through water in a plastic bottle to absorb offgas

Each deficiency, corrective action, schedule and responsibility must be recorded and tracked with periodic status reports, engineering studies, implementation and closure report. These reports are sent to affected management and staff.

3.3 Qualitative Methods of Accident Analysis

This section describes how both hypothetical and real accidents are analyzed. These methods varying greatly in complexity and resource requirements, and multiple methods may be used in an analysis. A simple method is used for screening and prioritization followed by a more complex method for significant accident scenarios. Some methods give qualitative results; more complex methods give quantitative results in the form of estimated frequencies of accident scenarios. The process systems in Figures 3.3.1-1 and 3.3.1-2 are used in the examples.

3.3.1 Checklist

A checklist analysis (CCPS, 1992) verifies the status of a system. It is versatile, easy and applicable at any life-cycle stage of a process. It is primarily used to show compliance with standards and practices by cost-effectively identifying hazards. Checklists provide commonality for management review of hazard assessments. It may be used for controlling a process from development to decommissioning. Approvals by appropriate authorities verify each stage of a project.

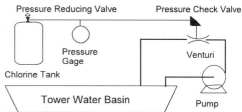

Fig. 3.3.1-2 Cooling Tower Water Chlorination Basin water is circulated by the pump through the venturi at 30 gpm. Chlorine gas, provided by a 1 ton tank, is reduced to zero pressure by the pressure reducing valve to be brought into the sub-atmospheric venturi. The pressure check valve prevents chlorine gas flow unless the pressure in the venturi is sub-atmospheric. If the first check valve fails, chlorine is vented

Description

A written list of items or procedures describe the status of a system. Checklists may vary considerably in level of detail, depending on the process. The list identifies hazards, design deficiencies, and accident scenarios associated with process equipment and operations regarding: materials, equipment, or procedures. Checklists may be used for evaluating existing processes or contemplated processes.

Developing a Checklist

A checklist analysis begins with a qualified team examining: the process description, P&IDs design procedures, and operating practices. As the checklist is developed, aspects of process design or operation that do not comply with standard industrial practices may be discovered by using detailed, process-specific checklists which be augmented by generic checkists for thoroughness.

The completeness of hazard identification would be limited to the preparer's knowledge if not for the use of multiple reviewers with diverse backgrounds. Frequently, checklists are created simply by organizing information from current relevant codes, standards, and regulations. This is insufficient, the checklist must be designed for the specific process including its accident experience. It should be a living document that is reviewed and revised from operating experience.

Table 3.3.1-1 (DOE, 1996) suggests main headings for a checklist. These are global heading starting with the process description,

Table 3.3.1-1 Possible Checklist Main Headings

- Basic Process Considerations
- Overall Considerations
- Operating Limits
- Modes of plant startup, shutdown, construction, inspection and maintenance, trigger events, and deviations of systems
- Hazardous conditions
- Ways of changing hazardous events or frequency of occurrence
- Corrective and contingency action
- Controls, safeguards, and analysis
- Documentation and responsibilities

overall considerations, operating limits, hazards and mitigation.

Table 3.3.1-2 (DOE, 1996) is an example of the use of first and second headings. The first level are the global considerations (although not as global as Table 3.3.1-1) with the second level going to the more particular subject headings. This table begins with the site and its selection and goes to the process and hazards. Then it examines the process equipment, mitigation by storage selection, protection zones, fire protection and emergency planning.

More inclusive is Table 3.3.1-3 which is appropriate at depth into the analysis. The major headings in this table address major hazardous subject areas, accident mitigation, protection and repair. For example, under the first major heading, "Storage of Raw Materials, Products, Intermediates," listed are: confinement measures, release mechanisms (valves), procedures for safe operation and limitations that must be observed for safety.

Table 3.3.1-2 Possible Headings and Subheadings

1. Site Layout	7. Civil Engineering
1.1 Site Selection	7.1 Footings
1.2 Site Layout	7.2 Foundations
2. Process Materials	7.3 Drainage Systems
2.1 Physical Properties	7.4 Roads
2.2 Chemical Properties	7.5 Buildings
2.3 Toxic Properties	7.6 Additional Items
3. Reactions, Process Conditions, and Disturbances	8. Site Zones for Fire and Hazard Protection
3.1 Reactions	8.1 Safety Considerations
3.2 Process Conditions	8.2 Zones
3.3 Disturbance Analysis	9. Fire Protection
3.4 Causes of Upsets	9.1 Introduction
3.5 Abnormal Conditions	9.2 Buildings and Plant
3.6 Critical Situations	9.3 Fire-Fighting Organization
4. Equipment	9.4 Detection and Alarm
4.1 Introduction	9.5 Classification of Fires
4.2 Design	10. General Emergency Planning
4.3 Choice of Material	10.1 Introduction
4.4 Construction	10.2 Operational Emergency Situations
4.5 Location of Equipment	10.3 Release of Liquids and Gases
5. Storage and Handling of Dangerous Substances	10.4 Fire and Explosion
5.1 Storage	10.5 Personnel Protection
5.2 Handling	10.6 Training
6. Removal of Hazardous Wastes	10.7 Communications
6.1 Introduction	10.8 Briefing and Information
6.2 Aspects of Disposal	
6.3 Reduction of Disposal	

Performing the Analysis

After preparation, a checklist may be used by less expert engineers than the preparers for inspecting the process areas to compare the process equipment and operations with the checklist either on hardcopy or laptop computer. The analyst fills in the checklist according to observations from their inspection, process documentation, and interviews with operators and their perceptions.

Documenting Results

Qualitative results of checklist analyses vary, but generally the analysis produces the answers: "yes," "no," "not applicable," or "needs more information." The checklist is included in the PrHA report to summarize the noted deficiencies. Understanding these deficiencies leads to safety improvement alternatives for consideration, and to identified hazards with suggested actions. Figures 3.3.1-4 and 3.3.1-5 present checklist analyses of the Dock 8 HF Supply and the Cooling tower chlorination respectively.

Table 3.3.1-3 Simplified Process Hazards Analysis Checklist

Storage of Raw Materials, Products, Intermediates		Personnel Protection	
Storage Tanks	Design Separations, Inerting, Materials of Construction	Protection	Barricades, Personal Protection, Shower, Escape Aids
Dikes	Capacity, Drainage	Ventilation	General, Local, Air Intakes, Flow
Emergency Valves	Remote Control - Hazardous Materials	Exposures	Other Processes, Public, Environment
Inspections, Procedures	Flash Arresters, Relief Devices	Utilities	Isolation: Air Water, Inerts, Steam
Limitations	Temperature, Time Quantity	Hazards Manual	Toxicity, Flammability, Reactivity, Corrosion, Symptoms, First Aid
Materials Handling		Environment	Sampling, Vapors, Dusts, Noise, Radiation
Pumps	Relief, Reverse Rotation, Identification, Materials of Construction, Leaks, Cavitation	Controls and Emergency Devices	
Ducts	Explosion Relief, Fire Protection, Support	Controls	Ranges, Redundancy, Fail-Safe
Conveyors, Mills	Stop Devices, Coasting, Guards	Calibration, Inspection	Frequency, Adequacy
Procedures	Spills, Leaks, Decontamination	Alarms	Adequacy, Limits, Fire, Fumes
Piping	Ratings, Codes, Cross-Connections, Materials of Construction, Corrosion, Erosion Rates	Interlocks	Tests, Bypass Procedures
Process Equipment, Facilities and Procedures		Relief Devices	Adequacy, Vent Size, Discharge, Drain, Support
Procedures	Startup, Normal, Shutdown, Emergency	Emergencies	Dump, Drown, Inhibit, Dilute
Conformance	Job Audits, Shortcuts, Suggestions	Process Isolation	Block Valves, Fire-Safe, Valves Purging, Excess Flow Valves
Loss of Utilities	Electricity, Heating, Coolant Air, Inerts, Agitation	Instruments	Air Quality, Time Lags, Reset, Windup, Materials of Construction
Vessels	Design, Materials, Codes, Access, Materials of Construction	Waste Disposal	
Identification	Vessels, Piping, Switches, Valves	Ditches	Flame Traps, Reactions, Exposures, Solids
Relief Devices	Reactors, Exchangers, Glassware	Vents	Discharge, Dispersion, Radiation, Mists
Review of Incidents	Plant, Company, Industry	Characteristics	Sludge, Residues, Fouling, Materials
Inspections, Tests	Vessels, Relief Devices, Corrosion	Sampling Facilities	
Hazards	Hang-fires, Runaways	Sample Points	Accessibility, Ventilation, Valving
Electrical	Area Classification, Conformance, Purging	Procedures	Plugging, Purging
Operating Ranges	Temperature, Pressure, Flows, Ratios, Concentrations, Densities, Levels, Time, Sequence	Samples	Containers, Storage, Disposal
Ignition Sources	Peroxides, Acetylides, Friction, Fouling, Compressors, Static Electricity, Valves, Heaters	Analysis	Procedures, Records, Feedback
Compatibility	Heating Media, Lubricants, Flushes, Packing	Maintenance	
Safety Margin	Cooling, Contamination	Decontamination	Solutions, Equipment, Procedures
		Vessel Openings	Size, Obstructions, Access
		Procedures	Vessel Entry, Welding, Lockout
		Fire Protection	
		Fixed Protection	Fire Areas, Water Demands, Distribution System, Sprinklers, Deluge, Monitors, Inspection, Testing, Procedures, Adequacy
		Extinguishers	Type, Location, Training
		Fire Walls	Adequacy, Condition, Doors, Ducts
		Drainage	Slope, Drain, Rate

Table 3.3.1-4 Checklist Analysis of Dock 8 HF Supply
Materials
☐ *Do all materials conform to specifications?*
  Yes, the cylinders, used since startup, have the same anhydrous HF specification.
☐ *Is each material receipt checked?*
  No, there have been problems with the supplier, so no check is done. Investigate consequences of receiving material other than HF. consider adding such checks on HF receipts.
☐ *Does the operating staff have access to Material Safety Data Sheets?*
  Yes, all staff are familiar with the process chemistry, including the hazards of HF.
☐ *Is fire fighting and safety equipment properly located and maintained?*
  Yes.

Equipment
☐ *Has all equipment been inspected as scheduled?*
  Yes. The maintenance personnel have inspected the equipment in the process area according to company inspection standards. Given the corrosivity of HF, inspections may have to be more frequent.
☐ *Have pressure relief valves been inspected as scheduled?*
  Yes.
☐ *Have rupture discs been inspected as scheduled?*
  Yes, none have failed; the procedure calls for inspection of rupture disc and installation after maintenance.
☐ *Are the proper maintenance materials (parts, etc.) available?*
  Yes. They include spare pigtails for the supply cylinders as well as properly rated rupture discs. Other items must be ordered.

Procedures
☐ *Are the operating procedures current?*
  Yes.
☐ *Are the operators following the operating procedures?*
  Yes. No significant violations of procedures have been noted.
☐ *Are new operating staff trained properly?*
  Yes. training includes a review of the PrHA for this process and familiarization with MSDSs.
☐ *How are communications handled at shift change?*
  If a HF cylinder needs to be changed near a shift change, the change is scheduled to be performed by either, but not both, shifts.
☐ *Is housekeeping acceptable?*
  Yes.
☐ *Are safe work permits being used?*
  Yes.

Table 3.3.1-5 Checklist of Cooling Tower Chlorination
Materials
☐ *Do all materials conform to original specifications?*
  Yes, the drums are ordered with the same chlorine specification used since startup.
☐ *Is each receipt of material checked?*
  Yes, the supplier once sent a cylinder of phosgene. Since then, a test is performed by the maintenance staff. In addition, the fusible plugs are inspected for evidence of leakage, before a cylinder is hooked up.
☐ *Does the operating staff have access to Material Safety Data Sheets?*
  Yes, all staff are familiar with the process chemistry, including the hazards of Cl_2.
☐ *Is fire fighting and safety equipment properly located and maintained?*
  Yes, this system is on a concrete building roof. Because there are no flammable materials involved in this system, if a fire occurs, there will be no special effort by fire fighting crews to concentrate on the roof area.

Equipment
☐ *Has all equipment been inspected as scheduled?*
  Yes. The maintenance personnel have inspected the equipment in the process area according to company inspection standards.
☐ *Have pressure relief valves been inspected as scheduled?*
  Yes.
☐ *Have rupture disks been inspected as scheduled?*
  Not applicable.
☐ *Are the proper maintenance materials (parts, etc.) available?*
  Yes. They include spare pigtails for the supply cylinders, as well as a rotameter and a pressure check valve. Other items must be ordered.
☐ *Is there an emergency cylinder capping kit?*
  Yes.

Procedures
☐ *Are the operating procedures current?*
  Yes.
☐ *Are the operators following the operating procedures?*
  No, it is reported that some staff do not always check the cylinder's fusible plugs for leaks. Staff should be reminded of this procedural item and its importance.
☐ *Are new operating staff trained properly?*
  Yes, training includes a review of the PrHA for this process and familiarization with MSDSs.
☐ *How are communications handled at shift change?*
  There are few open items at the end of a shift. The chlorine cylinders need to be changed only once every 45 days. If an empty chlorine cylinder needs replacement, the change is schedule during a shift.
☐ *Is housekeeping acceptable?*
  Yes.
☐ *Are safe work permits being used?*
  Yes.

Staffing and Time

Any person with knowledge of the process should be able to use a checklist. The PSM Rule team-approach requires more than one analyst to prepare the checklist and apply it to the process for review by an independent analyst. An estimate of the time required to perform a PrHA using the checklist analysis method is given in Table 3.3.1-6.

Limitations of Checklist

When a checklist is made from generic documents, it may be incomplete or erroneous. The hazards associated with the process may not be in generic information. Checklists identify hazards but not the accident scenarios that lead to the hazards.

Table 3.3.1-6 Estimated Time for a Checklist

Scope	Prep.	Eval.	Doc.
Simple (hr)	2-4	4-8	4-8
Complex (days)	1-3	3-5	2-4

3.3.2 What-If Analysis

A what-if analysis identifies hazards, hazardous situations, and accident events with undesirable consequences (CCPS, 1992). What-if analysis considers deviations from the design, construction, modification, or operating intent of a process or facility. It is applicable at any life stage of a process.

Description

"What-if" is a creative, brainstorming examination of a process or operation conducted by knowledgeable individuals asking questions. It is not as structured as, for example, HAZOP or FMEA. It requires the analysts to adapt the basic concept to the specific application.

The what-if analysis stimulates a PrHA team to ask "What-if?" Through questions, the team generates a table of possible accidents, their consequences, safety margins, and mitigation. The accidents are not ranked or evaluated.

Procedure

The PrHA team prior to meeting needs process descriptions, operating parameters, drawings, and operating procedures. For an existing plant, personnel responsible for operations, maintenance, utilities, or other services should be interviewed. The PrHA team should walk through the facility to better understand its layout, construction, and operation. Visits and interviews should be scheduled before the analysis begins. Preliminary what-if questions should be prepared to "seed" the team meetings by using old questions or making up new ones specifically for the process.

Analyzing with What-If

After setting the scope of the analysis, the analysis begins process description including safety precautions, equipment, and procedures. The meetings then focus on potential safety issues identified by the analysts through What-If questions although a concern may not be expressed as a question. For example: "*I wonder what would happen if the wrong material was delivered.*" "*A leak in Pump Y might result in flooding.*" "*Valve X's inadvertent opening could affect the instrumentation.*" Questions may address any off-normal condition related to the facility, not just component failures or process variations. The analysis usually goes from beginning to end, but it may proceed in any order that is useful, or the leader may select the order. Questions, and answers are recorded by a designated team member

Questions are organized by areas such as electrical safety, fire protection, or personnel safety and addressed by experts. In this interchange, the team asks and gets answers expressing their concerns regarding the hazard, potential consequences, engineered safety levels, and possible solutions. During the process, new what-if questions may be added. Sometimes proposed answers, developed outside of meetings, are presented to the team for endorsement or modification. For example:

"*What if the HF cylinder fails from corrosion?*" The team might respond: "*A cylinder leak would release HF to the atmosphere to the detriment of workers and eventually fail the HF feed to the vaporizer.*" The team might recommend inspecting cylinders as they are received. The meetings should be contemplative and last no longer than 4 to 6 hours per day and no more than five consecutive days. A complex process should be divided into tractable segments.

Documentation

What-if produces a table of narrative questions and answers suggesting accident scenarios, consequences, and mitigation. Table 3.3.2-1 shows a typical What-If analysis for the Dock 8. On the left in the line above the table is indicated the line/vessel that is being analyzed. To the right is the date and page numbers. The first row in the table contains the column headings beginning with the what-if question followed by the consequences, safety levels, scenario number and comments. The comments column may contain additional descriptive information or actions/ recommendations.

Table 3.3.2-1 What-If Analysis of Dock 8 HF Supply System

LINE/VESSEL: Dock 8 HF Supply System　　　　　　　　　　　　　　Date: September 29, 1997　Page __ of __

What If	Consequences	Safety Level	Scenario	Comments
... the HF cylinder corrodes through?	Cylinder leak, HF release to atmosphere, possible worker exposure via inhalation and skin, possibly fatal.	None	1	Check with supplier regarding cylinder inspection practices.
... the dock and this equipment is involved in a fire?	HF release to atmosphere via vent OR	None	2a	
	Cylinder rupture, with possible worker exposure via inhalation and skin, possibly fatal.	Relief valves, rupture discs	2b	
... the hot water jacket on the HF corrodes through?	Heat of solution, HF release via vent, possible worker exposure via inhalation and skin, possibly fatal.	None	3a	
	Possible large pipe and pipe component failures due to corrosion.		3b	
	Possible vaporizer rupture with further release and blast effects, worker injured by blast or scalded.	Relief valves, rupture discs	3c	
... moisture is introduced into the HF cylinder via the N2 supply?	Heat of solution, HF release via vent, possible worker exposure via inhalation and skin, possibly fatal.	None	4a	Prevention is procedure for monitoring N2 supply.
	HF solution attacks carbon steel, corrosion, leak or rupture, possible worker exposure via inhalation and skin, possibly fatal.		4b	

Table 3.3.2-2 What-If Analysis of the Cooling Tower Chlorination System

LINE/VESSEL: Cooling Water Chlorination System   Date: September 29, 1997 Page __ of __

What If	Consequences	Safety Level	Scenario	Comments
...the system is involved in a fire?	High pressure in chlorine cylinder, fusible plugs melt, chlorine release into fire....	Ignition control source	1	Verify the area is free of unnecessary fuel.
...the wrong material is received in the cylinder and hooked up?	Water contaminated, not sterilized.	None	2	Prevention: supplier's procedures.
...the cylinder's fusible plug prematurely fail?	Chlorine released.	None	3	Purchase and train personnel in the use of CL_2.
...the pressure check valve fails open (both pass chlorine gas)?	Built-in relief valve opens, releasing chlorine to atmosphere	None	4	
...the basin corrodes through?	Chlorinated water release.	Periodic inspection	5	
the recirculation pump fails or power is lost?	Eventually low chlorine in water, biological growth.	None	6a	
	Release of undissolved chlorine to atmosphere if pressure check valve fails.	Pressure check valve	6b	
...the chlorine cylinder is run dry and not replaced?	Eventually low chlorine to water, biological growth.	None	7	

The recommendations can be used in the report as action items for improving safety. Management reviews the results to assure the findings are considered for action. This is the standard format of a What-If analysis form. The rest of Table 3.3.2-1 is a partial analysis specific to the Dock 8 HF Supply System that is shown in Figure 3.3.1-1. Table 3.3.2-2 presents a partial What-If analysis of the cooling tower chlorination system that is shown in Figure 3.3.1-2. These partial analyses illustrate the method. For actual, complex analyses, each line or vessel would comprise a What-If table. These may be accompanied with a preliminary hazard analysis (PHA) to identify that the intrinsic hazards associated with HF are its reactivity, corrosivity, and toxicity. The N_2 supply system pressure is not considered in this example. The specific effects of loss of containment can be explicitly stated in the "loss of HF containment" scenarios. Similarly, the effects of loss of chlorine containment, including the reactivity and toxicity of chlorine, could be specified for the second example.

Staffing and Time

The PSM Rule requires that a What-If analysis be performed by a team with expertise in the process and analysis method. For a simple process, two or three people may perform the analysis, but for a complex process, a large group subdivided according to process logic into small teams is needed.

The time and cost of a What-If analysis are proportional to team size and complexity. Table 3.3.2-3 presents estimates of the time needed to perform a PrHA using the what-if analysis method.

Limitations of the What-If Analysis

The What-If analysis is a powerful PrHA method if the analysis team is experienced and well organized, but because it is a relatively unstructured approach, the results may be incomplete.

3.3.3 What-If/Checklist Analysis

What-If/Checklist analysis identifies: hazards, possible accidents, qualitatively evaluates the consequences and determines the adequacy of safety levels. It is described in CCPS (1992).

What-If/Checklist combines the creative, brainstorming features of a What-If with the systematic features of a Checklist. The What-If analysis considers accidents beyond the checklist; the checklist lends a systematic structure to the What-If analysis. A What-If/Checklist examines the potential consequences of accident scenarios at a more general level than some of the more detailed PrHA methods. It can be used for any type of process at any life cycle stage.

Procedure

The PrHA team leader assembles a qualified team to perform a What-If/Checklist analysis. If the process is large, the team is divided into subteams according to functions, physical areas, or tasks similarly to the discussion in Sections 3.3.1 and 3.3.2.

Table 3.3.2-3 Estimated Time for a What-If

Scope	Prep.	Eval.	Doc.
Simple (hr)	4-8	4-8	12-48
Complex (days)	1-3	3-5	5-15

For the checklist portion of the analysis, team leader obtains a checklist for the team which may not be as detailed as for a standard Checklist analysis. The checklist should focus on general hazardous characteristics of the process.

Covering the Gaps in What-If Questions

After the What-If questions for a process step have been developed, the previously obtained Checklist is applied. The team selects each Checklist item for accident potential and adds them to the What-If list for evaluation. The checklist is reviewed for each area or step in the process.

Evaluating the Questions

After developing questions, the PrHA team considers each to determine possible accident effects and list safety levels for prevention, mitigation, or containing the accident. The significance of each accident is determined and safety improvements to be recommended. This is repeated for each process step or area outside of team meetings for later team review

Documenting the Results

The results of a What-If/Checklist analysis are documented like the results of a What-If analysis as a table of accident scenarios, consequences, safety levels, and action items. The results may also include a completed checklist or a narrative. The PrHA team may also document the completion of the checklist to illustrate its completeness. The PSM rule requires detailed

Qualitative Methods of Accident Analysis

explanations of the analysis and recommendations for management review and transmission to those responsible for resolution.

Limitations of What-If/Checklist Analysis

Combining the What-If and Checklist analysis methods uses their positive features while compensating for their separate shortcomings. For example, a checklist is based on generic process experience and may have incomplete insights into the design,

Table 3.3.3-1 Estimated Time for a What-If/Checklist

Scope	Prep.	Eval.	Doc.
Simple (hr)	6-12	6-12	4-8
Complex (days)	1-3	4-7	5-15

Table 3.3.3-2 What-If/Checklist Analysis of Dock 8 HF Supply System

LINE/VESSEL: Dock 8 HF Supply System   Date: September 29, 1997   Page __ of __

What If	Consequences	Safety Level	Scenario	Comments
..the pressure relief valve fails closed?	Possible rupture of HF cylinder with personnel exposure to HF and blast effect, possible fatalities.	None	1	Add pressure alarm on operator console
..the operator does not valve off the empty cylinder before removing it?	HF release with personnel exposure, possible fatalities.	None	2	Review training records to make sure all staff have been trained in current procedures.

Table 3.3.3-3 What-If/Checklist Analysis of Cooling Tower Chlorination System

LINE/VESSEL: Cooling Tower Chlorination System   Date: September 29, 1997   Page __ of __

What If	Consequences	Safety Level	Scenario	Comments
..a chlorine cylinder that is not empty is removed?	If the operator does not expect it to contain chlorine, then possible Cl_2 exposure may occur via skin and inhalation.	None	1	Review training records and operating procedures to minimize the possibility of this occurring.
...the venturi is clogged with residue from the water basin?	No flow	Periodic checks of water quality	2	Review training records to make sure all staff have been trained in current procedures.
	High pressure in recirculation line with rupture and water release.	None	3	
	High pressure in recirculation line with release of Cl_2 if pressure check valve fails.	Pressure check valve	4	

procedures, and operations. The What-If part of the analysis uses the team's creativity and experience to brainstorm potential accident scenarios. Because the What-If analysis is usually not as detailed, systematic, or thorough as some more formal methods, e.g., HAZOP or FMEA, the Checklist fills in gaps in the thought process.

Staffing and Time

The number of individuals needed depends upon the complexity of the process and, to some extent, the stage at which the process is being evaluated. Normally, a PrHA using this method requires fewer people and shorter meetings than does a more structured method such as a HAZOP study. Estimates of the time needed to perform a PrHA using the what-if/ checklist analysis method are shown in Table 3.3.3-1.

Examples of What-If/Checklist Analyses

Abbreviated examples of What-If/Checklist analyses of the Dock 8 HF Supply System and the Cooling Tower Chlorination System are shown in Tables 3.3.3-2 and 3.3.3-3. The tables show the additional scenarios identified by applying the Checklist to What-If.

3.3.4 Hazard and Operability (HAZOP)

Overview of HAZOP

Hazards and Operability (HAZOP) analysis is an accident detection and prevention technique used primarily by the chemical process industry (CPI) (Lees, 1980). Even though the CPI operates in a different regulatory environment from the commercial nuclear power industry, the goals of risk reduction while maintaining productivity are similar.

HAZOP is a formal technique for eliciting insights about system behavior from a multi-disciplinary team that collectively has thorough knowledge of the plant and the physical phenomena involved in the plant (Figure 3.3.4-1).

A HAZOP team evaluates the plant through a process that involves selecting a system, applying guide words to the selected system, and identifying causes, and consequences of the postulated event. Occasionally, it is not possible to quickly resolve a postulated occurrence on the basis of the available information and expertise. In this case, the HAZOP leader may assign the best qualified team member(s), to further investigate and report the results back to the team. The team reviews the item and the proposed resolution. When the team agrees, the analysis is recorded and the item is closed out.

Composition of the HAZOP team

A key member of the HAZOP is the HAZOP leader whose duties are to:

- Select the system to be analyzed,
- Assure that all necessary disciplines are represented by the team and that the team members are familiar with the system selected for analysis,

- Provide supporting documents and diagrams for the system being investigated and the systems with which it interacts
- Assure that all systems and operating modes deemed to be important to safety are eventually addressed,
- Guide the team in addressing multiple failures,
- Resolve misunderstandings,
- Identify how to resolve technical problems, and
- Assure each HAZOP session is documented.

Responsibilities of the HAZOP team members are to:

- Provide expertise in the discipline that they represent on the team,
- Be familiar with the system being analyzed and other systems with which it interacts, and
- Resolve issues that are assigned.

The expertise of the HAZOP team depends on the system being analyzed. A typical HAZOP team should have expertise for the study in the following areas:

Instrumentation	Chemistry
Control	Stress analysis
Electrical power	Material properties
Reactor physics	Reliability specialists
Reactor systems	Human factors
Accident phenomenology	Reactor and plant operations

Individuals may provide expertise in several of the above disciplines. The size of the team should be large enough to achieve diversity of points of view, but small enough to function as a team focused on the analysis. The leader must assure all members without intimidation.

Recording the Session

The HAZOP session is recorded on a personal (laptop or desktop) computer by the person assigned to be the recorder. A program for this purpose maybe HAZSEC available from Technica.[d] This record is not a verbatim transcript; it is a recording of intermediate or final results as directed by the team leader.

HAZSEC generates two types of records. The first page is the log sheet with the time, date, revision number, team leader, and team members. This page also contains a section that describes the part of the plant design under investigation, and a statement of the design intent, i.e., the expected equipment performance under normal and accident conditions. The pages that follow repeat this

---

[d] Technica Inc., Software Products Division, 40925 County Center Dr., Suite 200, Temecula, CA 92591.

process using dates and page numbers. For each guide word[e], there are one or more causes of the postulated condition with associated statements of the consequences of this guide word operating on the equipment.

If the item is not resolved, an individual is assigned responsibility for resolving it after the session. Requests may be made for additional information.

Identification of the Operating Mode

Figure 3.3.4-1 shows that HAZOP has three looping processes. A plant generally has several operating modes (normal startup, emergency shutdown, shutdown with intention to resume soon, shutdown with intention to resume soon, and reloading) in which the design's systems may operate differently. The first loop sets the operating mode to be considered in the subsequent analyses. The mode selection is performed by the HAZOP team leader.

Process Variable Selection

Having selected the operating mode, the HAZOP leader selects the process variable to be considered for the selected mode.

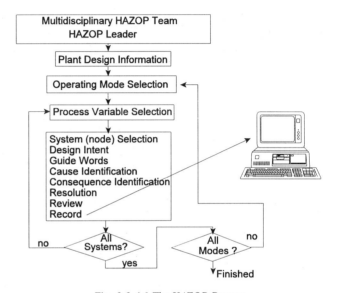

Fig. 3.3.4-1 The HAZOP Process

System (Node) Selection

The HAZOP leader selects an appropriate aspect of the plant's process systems (a process node) and associated systems that affect the selected process variable for the selected mode of plant operation. The selection may be made from the plant system classification, or it may be from the nodal analysis of the process.

Design Intent

The information assembled for the selected node includes descriptions of the hardware and physical parameters of the operating mode. For example, the information for a fluid system must include physical parameters, such as flow, pressure, temperature, temperature gradients, density, and

---

[e] A guide word (or phrase) is a perturbation on a process variable, component or system.

Qualitative Methods of Accident Analysis

chemical composition for every node[f] for which the description has meaning. This characterization of the system is called the "design intent." The first step in analyzing a node is to achieve team consensus on the design intent.

Performing a HAZOP

HAZOP focuses on study nodes, process sections, and operating steps. The number of nodes depending on the team leader and study objectives. Conservative studies consider every line and vessel. An experienced HAZOP leader may combine nodes. For example, the cooling tower water chlorination system may be divided into: a) chlorine supply to venturi, b) recirculation loop, and c) tower water basin. Alternatively, two study nodes may be used: a) recirculation loop and tower water basin, and b) chlorine supply to venturi. Or one study node for the entire process.

If the process uses a single large study node, deviations may be missed. If study nodes are small, many are needed and the HAZOP may be tedious, moreover the root cause of deviations and their potential consequences may be lost because part of the cause may be in a different node.

Each study node is examined for potentially hazardous process deviations. First, the design intent of the equipment and the process parameters is determined and recorded. Process deviations from the design are determined by associating guide words with important process parameters. Guide words for a HAZOP analysis are shown in Table 3.3.4-1; process parameters and deviations are shown in Table 3.3.4-2.

By combining elements of the first column of the guide words with elements of columns 2 or 4 of the process parameters, process deviations may be found. For example, combining the first guide word, "No" from Table 3.3.4-1 with the first process parameter, "Flow rate" from Table 3.3.4-2, the deviation "No flow rate" is found. Other deviations may be created similarly

The study team considers this deviation for possible causes (e.g., operator error causes pump blockage), the consequences of the deviation (e.g., high pressure line rupture), and mitigating safety features (e.g., pressure relief valve on pump discharge line). This consequence assumes the failure of active protection systems (e.g., relief valves, process trip signals). If the causes and consequences are significant, and the safety responses are inadequate, the team may recommend modification. In some cases, the team may identify a deviation with a realistic cause but unknown consequences (e.g., an unknown reaction product) that requires expert assistance.

The HAZOP study proceeds in a systematic manner that reduces the possibility of omission. Within a study node, all deviations associated with a given process parameter should be analyzed before the next process parameter is considered. All deviations for a study node should be analyzed before the team proceeds to the next node.

In listing the causes that lead to a guide word, engineering judgement must be used to exclude clearly incredible causes. However, it is not prudent to restrict the list to items of high probability or to items having obviously significant effects. The objective is to obtain insight into

---

[f] The term "node" means a location at which something different occurs. It may be a location at which streams converge (e.g., an orifice or a tank). However, at a region along which the physical parameters are constant, only one node would be used to represent the region.

Chemical and Nuclear Accident Analysis Methods

Table 3.3.4-1 Guide Words for HAZOP

Guide Word	Meaning	Examples
No	Negation of Intention	No forward flow when there should be. Sequential process step omitted.
More	Quantitative Increase	More of any relevant physical parameter than should be, such as more flow (rate, quantity), more pressure, higher temperature, or higher viscosity.
		Batch step allowed to proceed for too long.
Less	Quantitative Decrease	Opposite of More of.
Partial	Qualitative Decrease	System composition different from what it should be (in multicomponent stream).
As well as	Qualitative Increase	More things present than should be (extra phases, impurities).
		Transfer from more than one source or to more than one destination.
Reverse	Logical Opposite	Reverse flow.
		Sequential process steps performed in reverse order.
Other than	Complete Substitution	What may happen other than normal continuous operation (start up, normal shutdown, emergency shutdown maintenance testing, sampling).
		Transfer from wrong source or to wrong destination.

the characteristics of the system under various deviations from normal operation. It must also be noted, that the causes can originate anywhere in the system, not just in the specific node currently under consideration.

Consequence Identification

Upon completion of the list of causes, the team addresses the potential consequences from each of the listed causes to any node in the system. That is, the evaluation considers consequences arising anywhere in the system, which can effect the subject deviation in the subject node. These consequences include potential operating problems and safety concerns.

Resolution

Complex or particularly significant problems regarding a system may not be resolved immediately, and require further action. In such a case, a team member, whose expertise is most closely related to the subject, is assigned the task of obtaining the information and reporting back to the HAZOP team for close-out.

Review

Several systems (nodes) at a time may be in the process of review and resolution. In some cases, the causes and consequences for a given section, system, node, and guide word can be readily

Table 3.3.4-2 HAZOP Process Parameters and Deviations

Process Parameter	Deviation	Process Parameter	Deviation
Flow (rate)	No flow, High flow, Low flow, Reverse flow	Time	Too long, Too short, Too late, Too soon
Flow (amount)	Too much, Too little	Sequence	Omit a step, Steps reversed, Extra step
Pressure	High pressure, Low pressure	pH	High pH, Low pH
Temperature	High temperature, Low temperature	Viscosity	High viscosity, Low viscosity
Level	High level/overflow, Low level/empty	Heat value	High heat value, Low heat value
Mixing	Too much mixing, Not enough mixing, Loss of agitation, Reverse mixing	Phases	Extra phase, Phase missing
Composition	Component missing, High concentration, Low concentration	Location	Additional Source, Additional destination, Wrong source, Wrong destination
Purity	Impurities present, Catalyst deactivated/inhibited	Reaction	No reaction, Too little reaction, Too much reaction, Reaction too slow, Reaction too fast

assigned and that item closed out. In other cases, close-out may be deferred until the assigned resolution is complete. Even less frequently, items that were closed-out may be reopened in the light of new information.

These are the steps in the review that, in the end, lead to a thorough investigation of safety and operability of a plant

Record Keeping

As described, each session of the HAZOP team is recorded using a computer and special software. These records are revised whenever open items are resolved to produce the final record.

After recording the information for one process variable, another process variable is selected, as shown in Figure 3.3.4-1, and the process is repeated until all important process variables have been considered for an operating mode then another operating mode is selected and the whole process is repeated.

A HAZOP study is a systematic, tabular document of process deviations. The study gives the normal operating conditions and analysis boundary conditions for each item and lists action items for further evaluation. Tables 3.3.4-3 and 3.3.4-4 are examples of HAZOP analysis of the Dock 8 HF Supply System and the Cooling Tower Chlorination System, respectively (for a more complete study see DOE, 1973). These example show the format for HAZOP tables. A typical HAZOP study

Table 3.3.4-3 HAZOP of the Cooling Tower Chlorination System

LINE/VESSEL: Cooling Tower Chlorination System   Date: September 29, 1997 Page __ of __

Guide Word	Deviation	Cause	Consequence	Safety Level	Scenario	Action
None	No flow chlorination loop	Pump failure. Loss of electric power to pump	No chlorine flow to tower basin.. Low chlorine concentration in tower basin	Chlorination pump failure alarm	1	
		Low water level in tower basin	HF release into area; possible injuries/fatalities	None	2	No action: Unlikely event; piping protected against external impact
Less	Low flow in chlorination loop	None identified			3	
More	High flow in chlorination loop	None identified			4	Note: Pump normally runs at full speed
Reverse	No flow in chlorination loop	None identified			5	
Reverse	Backflow to HF inlet line	None			9	

report includes a brief system description, a list of drawings or equipment analyzed, the design intents, the HAZOP study tables, and a list of actions items.

Staffing and Time

Staff requirements for a HAZOP depend on the size and complexity of the process being analyzed. Time and cost are proportional to the size of the process and the experience of the study leader and team members. Table 3.3.4-5 presents estimates of the time needed to perform a PrHA using the HAZOP study method (CCPS, 1992). Study sessions should be limited to 3 consecutive days.

Limitations of HAZOP

The primary limitation of a HAZOP study is the length of time required to perform it. Because the study is designed to provide a complete analysis, study sessions can be intensive and tiring. HAZOP studies typically do not look at occupational hazards (e.g., electrical equipment, rotating equipment, hot surfaces) or chronic hazards (e.g., chronic chemical exposure, noise, heat stress). For experience with HAZOP see Swann (1995).

Qualitative Methods of Accident Analysis

Table 3.3.4-4 HAZOP of the Dock 8 HF Supply System

LINE/VESSEL: Dock 8 HF Supply System Date: September 29, 1997 Page __ of __

Guide Word	Deviation	Cause	Consequence	Safety Level	Scenario	Action
No	No flow	Valve V-19 closed HF Vaporizer inlet header plugged/ frozen	Loss of HF to B-1 process: unknown consequences	No known protection	1	Action item: Determine the level of protection available and potential consequences in B-1 wing
		Line rupture	HF release into area; possible injuries/fatalities	None	2	No action: Unlikely event; piping protected against external impact
Less	Low flow	Valve V-19 partially closed HF Vaporizer inlet header partially plugged/frozen	Insufficient HF supply to B-1 process, consequences unknown	No known protection	3	Same as Scenario 1
			Local rapid flashing, rupture disc/relief valve inadvertently open, release to stack	Stack height designed to dissipate release	4	
			Release HF into storage area; potential injuries/fatalities if occupied	Valve V-28 closed, forcing release to stack	5	Action item: Consider administrative controls or actions to ensure V-28 is closed when operating
More	High flow	None			6	
	High temp.	Fire; hot weather	Over-pressure; HF release; possible injuries/fatalities	Local temperature indication on water heating loop	7	No action; Unlikely event
	Low temp.	Cold weather	Possible plugging of lines; insufficient vaporization (see consequences of #1-5)		8	
Reverse	Backflow to HF inlet line	None			9	

Table 3.3.4-5 Estimated Time for a HAZOP			
Scope	Prep.	Eval.	Doc.
Simple (hr)	8-12	8-24	16-48
Complex (days)	2-4	5-20	10-30

3.3.5 Failure Mode and Effects Analysis

Description

FMEA examines each potential failure mode of a process to determine effects of failure on the system. A failure mode is anything that fails hardware. It may be a loss of function, unwanted function, out-of-tolerance condition, or a failure such as a leak. The significance of a failure mode depends on how the system responses to the failure.

Procedure

Three steps are identified: (1) defining the process to be analyzed, (2) doing the analysis, and (3) documentation. Defining the process and documenting the results can be performed by one person, but the PSM rule requires a team for the analysis.

Analysis Definition

The vessels, pipes, equipment, instrumentation, procedures and practices for the FMEA are identified and understood to establish the scope and level of detail. The PSM Rule requires FMEAs to be performed at the major component level - a trade-off between the time to perform the analysis and its value.

To establish boundary conditions requires:

1. Identifying systems or processes for analysis,
2. Establishing physical and analytical boundaries and interfaces,
3. Documenting the design performance of parts, processes, and systems,
4. Up-dating process equipment and functional relationship information.

Table 3.3.5-1 FMEA Consequence Severity Categories

Category	Classification	Description
Catastrophic	I	May cause death, injury, loss of system process or finances.
Critical	II	May cause severe injury, major property damage, or major system damage.
Marginal	III	May cause minor injury, major property damage, or major system damage.
Minor	IV	Does not cause injury, property damage, but may result in unscheduled maintenance or repair.

Functional system or process descriptions describe system, process and/or component behavior for each operational mode. They describe operational profiles of the components, the functions and input and outputs of each. Block diagrams assist by illustrating operations, interrelationships, and interdependencies of functional components. Interfaces should be indicated in these block diagrams.

Performing the Analysis

The FMEA is executed deliberately, and systematically to reduce the possibility of omissions. All failure modes for one component should be completed before going to the next.. The FMEA's tabular format begins with drawing references of the system boundaries and systematically evaluates the components as they appear in the process flow path using the following categories:

Failure Mode. The PrHA team lists each equipment item and interface failure mode for each operating mode being considered to find all conceivable malfunctions.

Cause(s). Root causes of the failure mode are identified to aid in preparing subsequent analytical steps and for use in hazard ranking.

Operational Mode. The operating mode indicates the operating environment for the equipment which may be used differently in different modes.

Effects. The expected effects of the failure on the overall system or process in the given operational mode are described. Usually the effect of only one failure is considered at a time, but if the successful operation of a component is critical, the FMEA may consider concurrent failures. This certainly should be done if their success depends on a common cause such as sharing the same power.

Failure Detection Method. The failure detection method should be identified, to show the likelihood of detection. If failure cannot be detected before adverse consequences, this should be noted.

Mitigation Design provisions, safety or relief devices, or operational actions that can mitigate the effects of failure should be listed.

Severity Class. Severity of the consequences of failure are expressed qualitatively as presented in Table 3.3.5-1

Remarks. Present suggested actions for reducing its likelihood or mitigating its effects and the conditions that were considered in the analysis. If the team discovers that a component's failure is not detectable, the FMEA should determine if any second failure would have bad consequences.

Documentation

An FMEA is a qualitative, systematic table of equipment, failure modes, and their effects. For each item of equipment, the failure modes and root causes for that failure are identified along with a worst-case estimate of the consequences, the method of detecting the failure and mitigation of its effects. Tables 3.3.5-2 and 3.3.5-3 present partial examples of FMEAs addressing the Cooling Tower Chlorination System, and the Dock 8 HF Supply System.

Staffing and Time

The PSM Rule requires that an FMEA be performed by a team. Multiple participants may use an FMEA prepared as blank worksheets on viewgraphs for large screen display. When the PrHA

Table 3.3.5-2 Partial FMEA for the Cooling Water Chlorination System

Date:	October 3, 1997			Page: 1/1		Preparers: RRF and JDC		
Plant:	Y-12			System: Cooling Tower Chlorination System				
Item:	Pressure check valve			References: ORNL-YYQ938596				

Failure Mode	Causes	Operating Mode	Failure Effects	Failure Detection Method	Mitigation	Severity	Remarks
Too much flow through valve	Both internal pressure valves fail open	Operation	Excessive chlorine flow to Tower Water Basin - high chlorine level to cooling water - potential for excessive corrosion in cooling water system	Rotameter Daily testing of cooling water chemistry	Relief valve on pressure check valve outlet	III	None
Too little flow through valve	One or both internal pressure valves fail closed	Operation	No/low chlorine flow to Tower Water Basin - low chlorine level in cooling water - potential for excessive biological growth in cooling water system reduction in heat transfer	Rotameter Daily testing of cooling water chemistry	Automatic temperature controllers at most heat exchangers	IV	None
Chlorine flow to environment	Internal relief valve sticks open	Operation	Potential low chlorine flow to Tower Water Basin - see above	Distinctive odor	Pressure check valve located outdoors - unlikely to accumulate significant concentration	III	Action item: Consider venting relief valve above ground level
	Both internal pressure valves fail open and relief valve opens		Chlorine released to environment - potential personnel exposure and injury				

team reaches a consensus on each item's failure modes, causes, effects, detection methods, compensating provisions, severity, and remarks these are recorded. A more effective method than viewgraph projection, may be to use a table format in a computer with large screen display and record the consensus directly in the computer program. Staff requirements for a FMEA vary with the size and complexity of the process being analyzed. The time and cost of a FMEA is proportional to the size of the process and number of components analyzed. Two to four equipment items may be analyzed in an hour assuming all necessary information is available. If processes or systems are similar, the time for preparing an FMEA is reduced. Table 3.3.5-4 presents estimates of the time needed to perform a PrHA using an FMEA (CCPS, 1992).

Limitations of Failure Mode and Effects Analysis

Human operator errors are not usually examined in a FMEA, but the effects of human error are indicated by the equipment failure mode. FMEAs rarely investigate damage or injury that could arise if the system or process operated successfully. Because FMEAs focus on single event failures, they are not efficient for identifying an exhaustive list of combinations of equipment failures that lead to accidents.

Table 3.3.5-3 Partial FMEA for the Dock 8 HF Supply System

Date: October 3, 1997				Page: 1/1		Preparers: RRF and JDC	
Plant: Y-12				System: Dock 8 HF Supply System			
Item: Pressure reducing valve V-13				References: ORNL-XWQ15396			
Failure Mode	Causes	Operating Mode	Failure Effects	Failure Detection Method	Mitigation	Severity	Remarks
Valve open too far	Internal valve failure	Operation	High N_2 pressure at HF cylinders, HF vaporizer vessel rupture - HF released to atmosphere	Local pressure indication on N_2 line	PRV-3 at V-12 outlet	II	If N_2 line relief valves lift, vaporizer relief valve should not lift.
	Operator error		High HF flow to HF vaporizer - high HF flow to B-1 wing - potential liquid HF to B-1 wing	Local pressure indication between rupture disk and PRV-4 at vaporizer	PRVs on N_2 feed lines to HF cylinders		Relief valve discharges piped to D-wing stack
	Calibration error				PRV-4 at HF vaporizer		
Valve closed too far	Internal valve failure	Operation	No N_2 pressure to HF cylinder - no HF flow to vaporizer, B-1 wing	Local pressure indication on N_2 line	None	IV	
	Operator error						
	Calibration error						
External leakage	Valve seal leakage	Operation	Waste on N_2	Audible	None	IV	
			If severe, same as "valve closed too far"	Local pressure indication on N_2 line, if severe			

3.4 Quantitative Methods of Accident Analysis

Table 3.3.5-4 Estimated Time for an FMEA

Scope	Prep.	Eval.	Doc.
Simple (hr)	2-6	8-24	8-24
Complex (days)	1-3	5-15	10-20

The preceding methods were concerned with qualitative statements like: frequent, rare, unlikely, incredible, short time process disruption, several worker injuries, possible off-site deaths etc. This section seeks ways to associate risk with estimated numbers like: average time between failures of one year, 10 estimated injuries, population exposure of 10 person-rems etc. We begin with the simplest and progress to the more complex, but first consider some underlying ideas.

The objective is to estimate, numerically, the probability that a system composed of many components will fail. The obvious question is, "Why don't you just estimate the failure rate of the system from operating experience?" There are three reasons: 1) the system may not exist, so new data are not available, 2) the injuries and fatalities from the developmental learning experience are unacceptable - the risk must be known ahead of time, and 3) by designing redundancy, the probability of the system failing can be made acceptably remote; in which case system failure data cannot be collected directly. The only practical way uses part failure statistics in a system model to estimate the system's reliability.

Chemical and Nuclear Accident Analysis Methods

A system model is a mathematical and symbolic representation of how the components of a system work together to perform the system's purpose. As described in Section 1.1, reliability analysis developed during World War II with the increasing complexity of military systems and the increasing need to make them reliable. Parts count methods combined the probabilities of successful component operation according to the system configuration. Markov models had early use, but are so arcane that their use is limited by mathematical complexity. As systems became increasing complex, Boolean logic was introduced in the 1960s for system modeling. Systems may be modeled to determine the combinations of successful component operations that result in system success. The same thing may be done to determine which failures of components will cause system failure.

Modeling failure rather than success has two advantages: 1) the number of component combinations that will cause failure is less than the number needed for success, and 2) the arithmetic is easier because failure probabilities are generally small numbers while the probability of success is nearly 1 (many 9s before the significant digits).

Boolean equations can be used to model any system; the system's reliability is calculated by factoring the equations into cutsets and substituting the probabilities for component failure. This can be done for either success or failure models. Working directly with equations is not everyone's "cup of tea" many individuals prefer graphical to mathematical methods. Thus, symbols and appearance of the methods differ but they must represent the same Boolean equation for them to be equivalent.

3.4.1 Parts Count

Perhaps the simplest way to assess the reliability of a system is to count the active parts.[g] The reliability estimate is the product of the number of parts and some nominal failure rate for the parts. In the design phase, two competing designs may be compared on the basis of the number of parts but several cautions are in order.

If designs are compared on the basis of count, then failure rates of each type of part either must be about the same, or be adjusted for the variations. For example, the parts count of vacuum-tube and solid-state television sets (using discrete components) are approximately the same, but their reliabilities are considerably different because of the better reliability of solid state components.

The other caution relates to redundant systems. Figure 3.4.1-1 shows very simple cooling systems that pump water from a tank into a cooling header. System A consists of a tank of water (T1), an open motoroperated valve (MOV), a

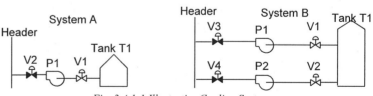

Fig. 3.4.1-1 Illustrative Cooling Systems

---

[g] Active parts, i.e., parts that produce or use energy are much less reliable than those that do not - called passive.

pump (P1) and a closed valve (V2). System B consists of 2 trains each of which are identical to System A, hence, System B is doubly redundant, i.e., either train may fail, but the system will still perform its designed purpose. System A has 3 active components and if the nominal failure rate is 1E-3/demand, the failure rate of System A is 3E-3/demand. System B has 6 active components, so a superficial estimate of the system failure rate is 6E-3/demand, but because it is doubly redundant, according to equation 2.4-4, the train probabilities, if independent, should be multiplied to give System B's failure rate as: 9E-6/demand, hence, the reliability estimation by a simple parts count is grossly conservative.

3.4.2 FMEA/FMECA

We previously encountered failure modes and effects (FMEA) and failure modes effects and criticality analysis (FMECA) as qualitative methods for accident analysis. These tabular methods for reliability analysis may be made quantitative by associating failure rates with the parts in a systems model to estimate the system reliability. FMEA/FMECA may be applied in design or operational phases (ANSI/IEEE Std 352-1975, MIL-STD-1543 and MIL-STD-1629A). Typical headings in the FMECA identify the system and component under analysis, failure modes, the effect of failure, an estimate of how critical a part is, the estimated probability of the failure, mitigators and possibly the support systems. The style and contents of a FMEA are flexible and depend upon the objectives of the analyst.

It is interesting that NASA in their review of WASH-1400 Draft (included in WASH-1400 Final Appendix II), indicated that they had discontinued the use of fault tree analysis in favor of the FMEA.

Table 3.4.2-1 Failure Modes Effects and Criticality Analysis Applied to System B of Figure 3.4.1-1					
Component	Failure mode	Effect	Criticality	Mitigation	Probability
Water storage tank T1	Sudden failure	Loss of water supply	No cooling water supply	None	1E-10/hr
MOV V1	Valve body crack	Loss of fluid to header	Reduce flow redundant train	Valve off use other train	1E-10/hr
	Motor operator inadvertent operation	Flow blockage	No flow to header	Redundant train	1E-3/demand
Pump P1	Motor failure: bearing or short	No pumping	No flow to header	Redundant train	1E-3/d
	Pump failure bearing seize				1E-3/d
MOV V3	Valve body crack	Loss of fluid to header	Reduce flow redundant train	Valve off use other train	1E-10/hr
	Motor operator fails to respond	Flow blockage	No flow to header	Redundant train	2E-3/d
	Gate disconnect from stem	Flow blockage			1E-4/d

FMEA is particularly suited for root cause analysis and is quite useful for environmental qualification and aging analysis. It is extensively used in the aerospace and nuclear power industries, but seldom used in PSAs. Possibly one reason for this is that FMEA, like parts count, is not directly suitable for redundant systems such as those that occur in nuclear power plants. Table 3.4.2-1 shows a FMEA for System B in Figure 3.4.1-1.

This FMEA/FMECA shows failure rates that are both demand and time dependent. Adding the demand failure rates gives a train failure rate of 5.1E-3/demand. The sum of the time dependent failure rates is: 3E-10/hr. A standby system such as this, does not exhibit its operability until it is actuated for which the probability is needed that the train has failed since the last use.

Valve bodies are considered to be part of the cooling envelope and may considered to be similar to piping. Piping characteristically leaks before breaking, therefore it may be expected that leakage might indicate potential fracture of the valve body. If inspected monthly, the time between inspections is 731 hours and the average time between inspections is 365 hours. If the distribution is exponential with an exponent much less than one, the probability of failure is the product of the failure rate and the average time since it was known to be operable: 3E-10* 365 = 1E-7, hence this type of failure may be ignored with respect to the other types to give a train failure rate of 5.1E-3/d. To complete the discussion of the FMEA analysis of System B with the redundant systems, the probabilities are multiplied, or (5.1E-3/d)^2 = 2.6E-5/demand. This is within a factor of 3 of the results of the parts count estimate.

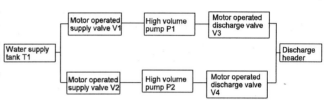

Fig. 3.4.3-1 Reliability Block Diagram of Figure 3.4.1-1

3.4.3 Reliability Block Diagram (RBD).

The preceding shows the care needed to understand the system logic and redundancies to model a system. Neither the parts count nor FMEA methods aid the analyst in conceptualizing the logical structure.

A Reliability Block Diagram (RBD - ANSI/IEEE Std 352-1975) while laid out similarly to a flow diagram, represents the components as a series of interconnecting blocks. Components are identified by the block labeling. System B in Figure 3.4.1-1 is shown as an RBD in Figure 3.4.3-1. It is customary to draw RBDs with the source on the left and the sink on the right. Even unquantified, RBDs are useful because they show the system's logic and how it can fail.

"Cutset" means one combination of component failures that fail the system. It is believed to have originated as cuts across the flow in reliability block diagrams[h]. In Figure 3.4.3-1, a cut through

---

[h] Conversation with W. Vesely 1972.

Quantitative Methods of Accident Analysis

trains must cut through V_1 or P_1 or V_3 in combination with V_2 or P_2 or V_4 making 9 combinations (3^2) with 11 cutsets for the system. Letting failure be represented by a bar over a letter (called "not"), the cutsets are: $\overline{T_1} or \overline{H_1} or \overline{V_1} and \overline{V_2} or \overline{V_1} and \overline{P_2} or \overline{V_1} and$
$\overline{V_4} or \overline{P_1} and \overline{V_2} or \overline{P_1} and \overline{P_2} or \overline{P_1} and$
$\overline{T_4} or \overline{V_3} and \overline{V_2} or \overline{V_3} and \overline{P_2} or \overline{V_2} and \overline{V}$.
A cutpath is a possible flow path from source to sink. Considering the probabilities of success (no bar) the probability of success is: $T_1, and V_1 and P_1 and$: $V_3 and H_1, or T_1, and V_2 and P_2 and V_4$ $and H_1$. Upon comparison to go from failure to success one or vice versa probability is complimented and logic is reversed: AND goes OR and vice versa, as indicated by de Morgan's theorem (Table 2.2-1).

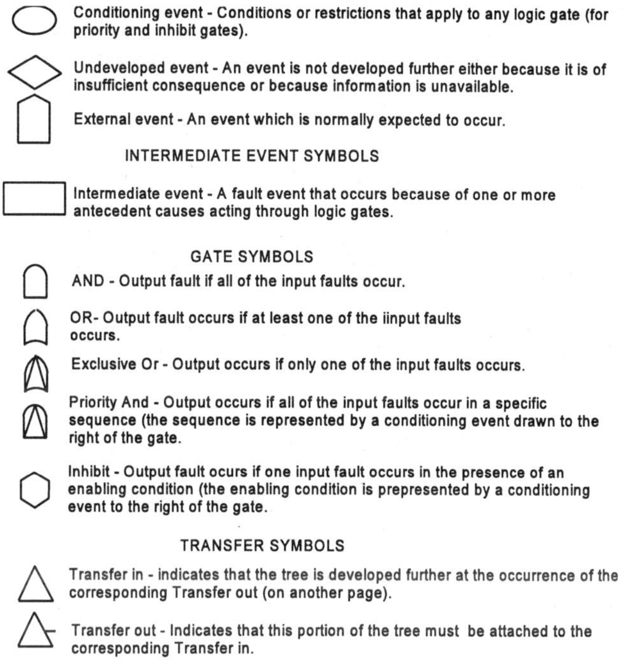

Conditioning event - Conditions or restrictions that apply to any logic gate (for priority and inhibit gates).

Undeveloped event - An event is not developed further either because it is of insufficient consequence or because information is unavailable.

External event - An event which is normally expected to occur.

INTERMEDIATE EVENT SYMBOLS

Intermediate event - A fault event that occurs because of one or more antecedent causes acting through logic gates.

GATE SYMBOLS

AND - Output fault if all of the input faults occur.

OR- Output fault occurs if at least one of the iinput faults occurs.

Exclusive Or - Output occurs if only one of the input faults occurs.

Priority And - Output occurs if all of the input faults occur in a specific sequence (the sequence is represented by a conditioning event drawn to the right of the gate.

Inhibit - Output fault ocurs if one input fault occurs in the presence of an enabling condition (the enabling condition is prepresented by a conditioning event to the right of the gate.

TRANSFER SYMBOLS

Transfer in - indicates that the tree is developed further at the occurrence of the corresponding Transfer out (on another page).

Transfer out - Indicates that this portion of the tree must be attached to the corresponding Transfer in.

Fig. 3.4.4-1 Symbols of Fault Tree Analysis

The unreliability equations for an RBD are written such that the probabilities of failure of components in a train are combined in union (added probabilistically, see Equation 2.4-6) and the probabilities of trains failing are combined in intersection (multiplied, see Equation 2.4-5). Using the failure rates presented in Table 3.4.2-1, the unreliability of the redundant trains is: (lE-3 + 2E-3 + 2.1E-3) * (lE-3 + 2E-3 + 2.1E-3) = 2.6E-5/demand as obtained before. Reliability may be obtained from the unreliability by complementing the reliability (subtracting the reliability from "one") to give 0.999974. The reliability may be calculated directly, if desired, by ANDing the probabilities of successful operation of the components and ORing the success probabilities of the trains. The disadvantage of a direct reliability calculation is the large number of "9's" involved in the probabilities of success.

RBDs are a very useful method for analyzing a simple system, but complex systems are usually analyzed by fault tree analysis which uses special symbols for the operations of union and intersection. An advantage of RBD over fault tree analysis is that the RBD logic diagram resembles a flow diagram or a piping and instrumentation diagram (P&ID) and is more easily understood without training in fault tree analysis.

3.4.4 Fault Tree Analysis

This section describes the most commonly used method for complex systems analysis - fault tree analysis. The previous section introduced cutsets as physically cutting through an RBD, here, cutsets are presented mathematically. The symbols of fault trees are introduced and a heuristic demonstration, based on von Misesian probability, shows the transformation of logic equations to probability equations. This is followed by rules for fault tree construction and a demonstration of transforming a fault to a success tree.

3.4.4.1 Fault Tree Symbols

Figure 3.4.4-1 summarizes conventional fault tree symbols.[i] The many symbols are daunting, but remember that the computer only performs AND (Boolean multiplication) and OR (Boolean addition) operations. All else are combinations of these.

Under the primary event symbols, the circle represents a component failure for which a description and failure-on-demand or failure frequency data[j] is provided. The oval symbol represents a probability or failure rate. A

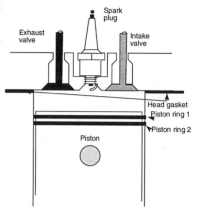

Fig. 3.4.4-2 Sketch of an Internal Combustion Engine

diamond is an event that is not developed because the preparer did not know what to do or thought it insignificant. A house-top is an expected occurrence (probability of one). Rectangles identify and explain gates.

The gate symbols are the AND gate which has a flat bottom and is true if all inputs (logic into the bottom of the gate) are true. The OR gate has a curved bottom and is true if one or more inputs are true. The Exclusive OR is true if only one input is true. Priority AND requires all inputs but in a given sequence. Inhibit acts like a gate. It is actuated by a side signal to pass the bottom signal. The transfer triangles are very useful for connecting one modular fault tree to another for size control.

3.4.4.2 From Logic to Probability

A simple example of fault tree analysis applied to an internal combustion engine (Figure 3.4.4-2) is the Figure 3.4.4-3 fault tree diagram of how the undesired event: "Low Cylinder Compression" may occur. The Boolean equation of this fault tree is in the caption of Figure 3.4.4-3. Let the occurrence of these events be represented by a *1*, non-occurrence by *0,* and consider that there may be a long history of occurrences with this engine. Several sets of occurrences (trials) are

---

[i] For an expanded discussion of the symbol meanings, refer to the Fault Tree Handbook, NUREG-0492.

[j] It is necessary to check units to be sure they make sense. Fault trees are often associated with event trees in which only the initiator has the units of frequency and the fault trees are dimensionless probability. This dimensionless is achieved by failure rates being paired with a mission time.

depicted in Table 3.4.4-1. After this history, the number of failures may be summed to find that the top event *(LCC)* occurred N_{LCC} times from the various component failures combinations (equation 3.4.4-1). Dividing equation 3.4.4-1 by the total number of trials T and letting T go to infinity, probability in the von Misesian sense results in equation 3.4.4-2. It is important to note that the symbols + and * in this equation represent the operations of union and intersection using the rules of combining probability - not ordinary algebra (Section 2.1).

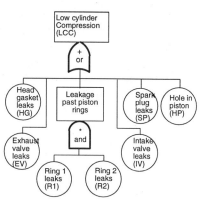

Fig. 3.4.4-3 Fault Tree of Engine

Thus, the conversion from a logical representation to a probability representation follows naturally from superposition and the von Misesian definition of probability. Simply replace the component identifier by its failure probability and combine probabilities according to the logical operations.

3.4.4.3 Cutsets

Two-state (off-on) logic being discussed and specifically, fault trees, may be represented by a Boolean equation. A fault tree (or other logical diagram) may be drawn in different ways that are equivalent if the equation of the logical diagram is the same. To show they are the same, it may be necessary to change the form of the equation by factoring and rearranging according to the rules of Boolean algebra presented in Table 2.1-1. A particularly powerful factorization of a Boolean equation is the sum of products form (equation 3.4.4-3). This is the *mincut* (minimal cutset) representation (Section 2.2) - the smallest combination of component failures that will result in the top event.

Table 3.4.4-1 Hypothetical Engine Operating Experience

Trial	LCC	HG	EV	IV	SP	P	PR1	PR2
1	1	1	0	0	0	0	0	0
2	0	0	0	0	0	0	1	0
3	0	0	0	0	0	0	0	1
4	1	0	0	0	1	0	0	0
...	...	...	...	...	...	...	...	...
T-1	1	0	1	1	0	0	0	0
T	1	0	1	1	0	0	1	1
Total	N_{LCC}	N_{HG}	N_{EV}	N_{IV}	N_{SP}	N_P	N_{PR1}	N_{PR2}

$$N_{LCC} = N_{HG} + N_{EV} + N_{IV} + N_{SP} + N_{NP} + (N_{PR1} + N_{PR2}) \quad (3.4.4\text{-}1)$$

$$P_{LCC} = P_{HG} + P_{EV} + P_{IV} + P_{SP} + P_{NP} + (P_{PR1} + P_{PR2}) \quad (3.4.4\text{-}2)$$

$$Top\ event = \sum_j \prod_i P_{ij} \quad (3.4.4\text{-}3)$$

By the definition, a minimal cutset is the collection of those events that produce the top event (no extras), although there may be many different collections. The combination is the smallest unique collection of components which if they all failed, produce the top event. If any components in the cutset do not fail then the top event will not occur - at least by that cutset.

Any fault tree consists of a finite number of minimal cutsets. One-component minimal cutsets, if there are any, represent single components whose failure alone causes the top event to occur. Two-component minimal cutsets are double failures which cause the top event to occur. N-component minimal cutsets, are the n components whose collective failure causes the top event to occur. In equation 3.4.4-3, a mincut is one of the products; whose sum constitutes all the ways of failing the mincuts.[k] For an *n*-component minimal cutset, all n components in the cutset must fail in order for the top event to occur. The minimal cutset expression for the top event can be written in the general form: $T = M_1 + M_2 + \cdots + M_k$, where T is the top event and M_i are the minimal cutsets. Each minimal cutset consists of a combination of specific component failures, hence, the general *n*-component minimal cut can be expressed as: $M_2 = X_1 * X_2 \ldots * X_n$, where $X_1, X_2, \ldots$ are basic component failures.

An example of a top event expression is: T= A+ B*C, where A, B, and C are component failures. This top event consists of a one-component minimal cutset (A) and a two-component-minimal cutset (B*C). The minimal cutsets are unique to the top event and independent of other ways the fault tree may be drawn.

Top-Down Solution

To determine the minimal cutsets of a fault tree, the tree is first translated into its equivalent Boolean equations and then either the "top-down" or "bottom-up" substitution method is used. The methods are straightforward and involve substituting and expanding Boolean expressions. Two Boolean laws, the distributive law and the law of absorption (Table 2.1-1), are used to remove superfluous items.

Consider the example fault tree (Figure 3.4.4-4); the Boolean equations taken, gate at a time, are: $T = E1*E2$, $E1 = A+E3$, $E3 = B+C$, $E2 = C+E4$, $E4=A*B$. Start with the *T*-equation, substitute, and expand until the minimal cutset expression for the top event is obtained. Substituting for *E1* and *E2* gives:
$T = (A+E3)*(C+E4) = (A*C)+(E3*C)+(E4*A)+(E3*E4)$.

Substituting for *E3*: $T = A*C +(B+C)*C + E4*A +(B+C)*E4 = A*C + B*C + C*C + E4*A + E4*B + E4*C$.

Using idempotency (*C*C=C*) gives T=A*C +B*C+C +E4*A+ E4 *B+E4*C. But $A*C +B*C + C + E4*C = C$ hence the law of absorption produces: $T = C + E4*A + E4*B$.

Finally, substituting for *E4* and applying the law of absorption twice: T=C+ (A*B)*A+ (A*B)* B =C+A*B

Thus, the fault tree of Figure 3.4.4-4 is represented in mincut form as: $T =C+A*B$ and by fault tree as Figure 3.4.4-5 which shows the remarkable simplification and reduction provided by the mincut form. This shows that the top event, *T*, consists of a single component failure, C, and a doubly-redundant failure.

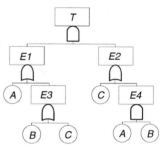

Fig. 3.4.4-4 Example Fault Tree

---

[k] The term "min" is readily understood when Boolean algebraic reduction is applied to the logic equation, because terms like A + A*C collapse to A, thereby affording considerable simplification. The reason for this is if A causes failure, then additional failures are superfluous.

Quantitative Methods of Accident Analysis

Bottom-Up Solution

The bottom-up method uses the same substitution and expansion techniques, except that now, the operation begins at the bottom of the tree and proceeds up. Equations containing only basic failures are successively substituted for higher faults. The bottom-up approach can be more laborious and time-consuming; however, the minimal cutsets are now obtained for every intermediate fault as well as the top event.

Returning to Figure 3.4.4-4: $T = E1*E2$, $E1 = A+E3$, $E3 = B+C$, $E2 = C+E4$, $E4=A*B$. Because $E4$ has only basic events, substitute it into $E2$ to obtain: $E2 = C + A B$. The minimal cutsets of $E2$ are thus C and $A*B$. $E3$ is already in reduced form having minimal cutsets B and C. Substituting $E1 = A+B+C$, hence $E1$ has three minimal cutsets A, B, and C. Finally, substituting the expressions for $E1$ and $E2$ into the equation for T, expanding and applying the absorption law gives: $T = (A+B+C)(C+A*B) = A*C+A*A*B+ B*C+B*A*B+C*C+C*A*B = A*C +A*B+ B*C +A*B +C+A*B*C = C+A*B$ - the same as was obtained previously by the top-down method.

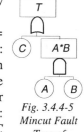

Fig. 3.4.4-5 Mincut Fault Tree of Figure 3.4.4-4

3.4.4.4 Considerations in Fault Tree Construction.

Top Event

A fault tree may either stand alone or be coupled to an event tree to quantify a nodal probability. The top event in either case is the objective of performing the analysis. If the objective is the reliability of a system under specific conditions - then that is the top event. If it is to quantify a node of an event tree the top event title is that of that particular node subject to the conditions imposed by the preceding modes.

Clearly and uniquely specify the top event to the precise requirements without over-specification which might exclude important failure modes. Even if constructed by different analysts, the top event specification should produce trees that have the same Boolean equation.

The fault tree identifies component failures that cause the top event. Systems may be required to respond in different ways to different accidents, suggesting a general top event for a general purpose fault tree that adapts to specific system configurations. This may result in ambiguity in the top event definition and difficulty in construction. It is better and easier to precisely specify the top event for one system configuration, construct its tree, and do the same for each system configuration. Such editing of trees is greatly facilitated by the computer.

An example of a multipurpose system is a reactor's low pressure coolant injection system which uses one out of four (1/4) pumps for a small LOCA, and three-out-of-four pumps for a large LOCA (3/4). The top event for the small LOCA might be: "none of four LPCI pumps fail to start within five minutes and deliver rated flow for at least two hours following a small LOCA." The large LOCA top event might be: "more than two-out-of-four LPCI pumps fail to start within five minutes and deliver rated flow for at least two hours. Such a top event specifies failure in physical and temporal terms.

System Definition.

Before the analyst begins fault tree construction, it is necessary to document and understand:

- System design,
- Causes of top event failure including substructure and components,
- Support necessary for operation,
- Environment,
- Components failure modes, and
- Test and maintenance practices.

Documentation of the system design and operating data, analytical assumptions and constraints used in the analysis in a notebook becomes part of the QA record. However, it is useful to have documents analyses and records as computer files for handling the voluminous data reporting and presenting the results of the work. It provides.

A risk assessment analyses systems at two levels. The first level defines the functions the system must perform to respond successfully to an accident. The second level identifies the hardware for the systems use. The hardware identification (in the top event statement) describes minimum system operability and system boundaries (interfaces). Experience shows that the interfaces between a frontline system and its support systems are important to the system evaluation and require a formal search to document the interactions. Such is facilitated by a failure modes and effect analysis (FMEA). Table 3.4.4-2 is an example of an interaction FMEA for the interface and support requirements for system operation.

Failure Modes and Effects Analysis (FMEA)

A failure modes and effects analysis delineates components, their interactions with each other, and the effects of their failures on their system. A key element of fault tree analysis is the identification of related fault events that can contribute to the top event. For a quantitative evaluation, the failure modes must be clearly defined and related to a numerical database. Component failure modes should be realistically and consistently postulated within the context of system operational requirements and environmental factors.

All component fault events can be described by one of three failure characteristics:

1. *Failure on Demand*. Certain components are required to start, change state, or perform a particular function at a particular time. Failure to do so is a "failure on demand."
2. *Standby Failure*. Some systems or components are normally in standby, but must operate when demanded. Because they are on standby, their failed state will be unknown unless it is caught by testing. This is "standby failure."
3. *Operational Failure*. A system or component may fail in normal operation or may start but fail to continue for the required time. This is an "operational failure."

Table 3.4.4-2 Example Format for an Auxiliary Feedwater System Interaction FMEA

Front-line System		Support system							
Div.	Comp.	System	Div.	Comp.	Failure mode	Failure effect	Detection	Diagnosis	Comments
A	MDP-1A	AC Power	A	Breaker A1131	Fail open	Concurrent failure to start or run CFSR	At pump test	Pump operability only	Treat as part of local pumping failure
B	MDP-1B	AC Power	B	Breaker A1132					
A	MDP-1A	AC Power	A	Bus E11	Low voltage	CFSR motor failure	Prompt	Control room monitors ESG EF11 voltage alarmed	Partial failure noted for future reference
B	MDP-1B	AC Power	B	Bus F12					
A	MDP-1A	HVAC	A	Rx cooler 3A	No heat removal	Pump motor burnout in 3-10 continuous service hours	Shift walk around	No warning for local faults	AC and SWS support systems of HVAC monitored but not HX
B	MDP-1B	HVAC	B	Rx cooler 3B					
A	MDP-1A	ESWP	A	Oil cooler S31	Loss of service water	Pump motor burnout in 1-3 CSH	At pump test	Local lube oil temperature gage, none in control room	ESWS header and pumps monitored but not lube oil coolers, local manual valve alignment checked in maintenance procedure xx but not periodic walk around.
B	MDP-1B	ESWP	B	Oil cooler S32					
A	MDP-1A	DC Power	A	Bus A131	Low voltage	Precludes auto or manual start, no local effect on already running pump	Prompt	Control room monitors DC bus voltage - many lamps out in control room	Effect of DC power loss on AC not Evaluated here; local motor controller latches on, needs DC to trip or close
B	MDP-1B	DC Power	B	Bus A132					

In addition, test and maintenance activities can contribute significantly to unavailability by components being out of service for test or maintenance. The amount of unavailability from this cause depends on the frequency and the duration of the test or maintenance act.

Three general types of testing that may make systems unavailability are:

1. Tests of the system control logic to ensure proper response to appropriate initiating signals.
2. System flow and operability tests of components such as pumps and valves.
3. System tests performed after discovering the unavailability of a complementary safety system, generally referred to as tests after failure.

Testing schemes generally affect complete subsystems; hence, consideration of each hardware element is unnecessary. Tests of redundant portions of a system are particularly important, and may be constrained by the technical specifications which must be reflected in the fault tree. Testing may require the reconfiguration of systems for the test, which may prevent the performance of their designed function. In this case, other members of the redundancy must be available, but may fail. Failure to restore a system after test significantly increases the risk.

Maintenance, whether routine or unscheduled contributes to system unavailability. The frequency and duration can be obtained from the maintenance procedures. Take care that outages associated with preventive maintenance are not already included in the time intervals assigned to testing.

Unscheduled maintenance activities are caused by essential equipment failures. Because these activities are not scheduled, the frequency and the mean duration of maintenance is based on statistics from experience.

Human factors, discussed in Section 4.2, enter a fault tree in the same manner as a component failure. The failure of manual actions, that prevent or mitigate an accident, are treated the same as hardware failures. The human error failure probability is conditioned by performance shaping factors imposed by stress, training and the environment.

Human errors may be dependent on the specific accident sequence displayed in the event tree, and, for that reason, may be included in the event tree. This requires the human-factors specialist to consider the context of the error in terms of stress, operator training in response to the accident, diagnostic patterns, environmental, and other performance-shaping factors.

3.4.4.5 Fault Tree Construction Rules

Fault Tree Hierarchy

Initially, a system's hierarchy is identified for subsystems, sub-subsystems and so on to the components for which data must be found. The top event specifies system failure; subsystems required for operation of the system in the mode specified are input to the top event's *OR* gate. Redundancy is represented by the redundant systems inputting an *AND* gate. This process of grouping subsystems under *OR* gates, if they can individually fail a function, or under *AND* gates if concurrent failures are necessary, is continued to the component or support system level until the tree is completed. This process grades the hierarchy from top to bottom, down the fault tree.

Trees can grow geometrically to become impractically large. Furthermore, very large trees are difficult to understand and review. The fault tree triangle linking symbol surmounts this problem by breaking a large tree into linked small trees. Subsystems that are logically identical may be represented by the same subtree but with different component identifiers for each of the subsystems represented. A fault tree may diagram a system to the component level, but including support systems in the tree would add complexity and size. This is circumvented with linking triangles to connect the fault trees of the support systems. Since the support system may support several systems, this technique minimizes the need for duplicate fault trees.

Label the Gates

Gates look alike; the are only distinguished by the gate label. Each event box is labeled to fully describe the postulated fault according to the what and when about the event. The precise statement enhances the analysis by: focusing on the event, and aiding the fault tree review.

Quantitative Methods of Accident Analysis

Complete the Gates

Complete all gate inputs. Beginning a new gate before completing a previous gate leads to confusion and incompleteness. These rules are made concrete by an example of fault tree analysis.

3.4.4.6 Example of Fault Tree Construction

Figure 3.4.4-6 presents a flow diagram of a simplified emergency cooling injection system. It is activated by a safety injection signal (SIS) which sends control signals along the dotted lines as shown. The top event selected for the fault tree

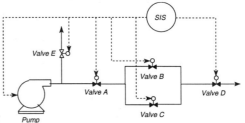

Fig. 3.4.4-6 Simple injection system

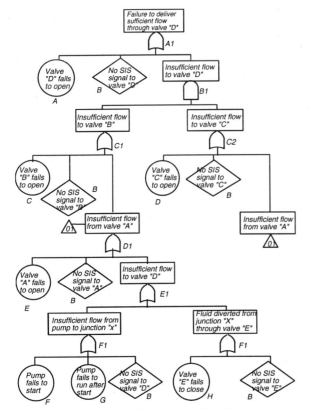

Fig. 3.4.4-7 Failure to Inject Fault Tree

A1 = A+B+B1
B1 = C1*C2
C1 = C+B+D1
C2 = D+B+D1
D1 = E+B+E1
E1 = F1+F2
F1 = F+G+B
F2 = H+B
Substituting,
E1 = (F+G+B) + (H+B)
D1 = E+B+(F+G+B)+(H+B)
C2 = D+B+E+B+(F+G+B)+(H+B)
C1 = C+B+E+B+(F+G+B1+(H+B)
B1 = [C+B+E+B+(F+G+B)+(H +B)]*[(D+B+
    E+B+(F+G+B)+(H+B)
A1 = A+B+[C+B+E+B+(F+G+B) + (H+B)]
    *[(D+B+E+B+(F+G+B)+(H+B)]
Simplifying,
A1 = A+B+(C+B+E+B+F+G+H+B)* (D+B
    + E +F+G+ H)
Multiplying,
A1= A+B+C*D+C*B+C*E+C*F+C*G+C*H
    +B*D+B*B+B*E+B*F+B*G+B*H+B*D
    +E*B+ E*E+ E*F+E*G+E*H+ F*C+F*
    B+F*E+F*G+F*H+G*D+G*B+G*E+G*
    F+G*G*H+H*D+H*B+H*E+H*F +
    H*G+H*H
Using identity: X*X=X,
A1 = A+B+C*D+C*B+C*E+C*F+ C*G+
    C*H+B*D+B+B*E+B*F+B*G+
    B*H+E*D+E*B+E+E*F+ E*G+
    E*H+F*D+F*B+F*E+F+F*G+F
    *H+G*D+G*B+G*E+G*F+G+G
    *H+H*D+H*B+H*E+ H*F +
    H*G + H
Using identity: X+(X*Y}=X,
A1 = A + B + E + F + G + H + C*D

Fig. 3.4.4-8 Boolean Reduction of "Failure to Inject" Fault Tree

Chemical and Nuclear Accident Analysis Methods

(Figure 3.4.4-7) is "Failure to Deliver Sufficient Flow through Valve D"; the analyst starts with valve D and asks how the top event could occur? Valve D could fail to open (A) or valve D may not receive a SIS (B) which, according to Figure 3.4.4-7, the analyst does not think credible, or valve D may not get flow from valve B and valve C. This process proceeds down through the system hierarchy. Notice that the labels on the gates from A to F indicate this progression. A note of caution, however; this tree is not concisely drawn as will be shown in the algebraic reduction of the tree. The progression into the hierarchy is not from complex to simple, because these systems are essentially of the same level of complexity. Figure 3.4.4-8 shows the algebraic reduction of the fault tree using the rules of Table 2.1-1.

The simplicity of the final result is the mincut representation (sum of products - Section 2.2) depicted as a fault tree in Figure 3.4.4-9. If the single double, and so on to higher redundancy components had been identified, the complex and awkward tree of Figure 3.4.4-8 would have been avoided. Some systems are so complex that this cannot be done by observation, but computer analysis will show simplicities if they exist.

3.4.4.7 Success Tree

Using de Morgan's theorem (Table 2.1-1), the fault tree in Figure 3.4.4-9 is converted into the success tree shown in Figure 3.4.4-10. Notice that the effect of the theorem is to reverse the logic: "AND" gates become "OR" gates and vice versa; failure becomes success and vice versa.

Success trees, by definition, are success oriented. Some analysts claim that a success tree is a state of mind and that a true success tree cannot be adapted from a fault tree by de Morgan's theorem but must be built from "scratch" to involve the psychology of success rather than of failure.

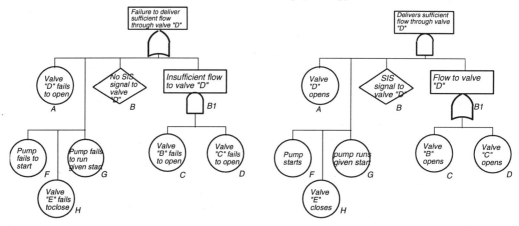

Fig. 3.4.4-9 Fault Tree from the Mincut Representation of Figure 3.4.4-8 Fault Tree

Fig. 3.4.4-10 Success Tree Transformation of Figure. 3.4.4-9 Fault Tree

3.4.4.8 Preparing Fault Trees for Evaluation

A fault tree is a graphical form of a Boolean equation, but the probability of the top event (and lesser events) can be found by substituting failure rates and probabilities for these two-state events. The graphical fault tree is prepared for computer or manual evaluation by "pruning" it of less significant events to focus on more significant events. Even pruned, the tree may be so large that it is intractable and needs division into subtrees for separate evaluations. If this is done, care must be taken to insure that no information is lost such as interconnections between subtrees.

Before quantitative evaluation, the events must be coded with an unique identifier. A systematic and orderly identification scheme minimizes the possibility of ambiguity or multiple identifier for the same event.

Some fault tree coding schemes use eight-character event identifiers; others use 16-character identifiers. A good coding scheme mnemonically identifies the system in which the component is located, the component type, the specific component identifier, and the failure mode. More complex identifiers give additional information, such as the component location that is needed for common cause searches. Graphical fault tree codes (e.g., IRRAS) supplement the coding with text describing the event.

3.4.5 Event Trees

3.4.5.1 Event Tree Construction

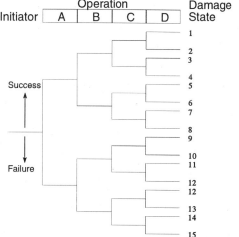

Fig. 3.4.5-1 Event Tree Scenario Depiction
The left side connects to the initiator, the right side to process damage states, the top bar gives the systems in the order that they enter the scenario. The nodes (dots) are entry points for the branching probabilities that may be determined from fault tree analysis or from a database. Success is indicated by the path going up at a node; failure is by going down.

Event trees are constructed using the following steps:

Step 1 - Initiator Selection

An event tree is a model of the process' response to an accident initiator. The initiators are found from a study of the plant's processes, its operating history, and accidents in related facilities. Collectively these initiators and their analyses define the risk envelope of the plant (Figure 1.4.2-2).

Two types of initiators are *internal* and *external*. Internal initiators result from failures within a plant or the plant's support utilities. Thus, vessel rupture, human error, cooling failure, and loss of offsite power are internal events. All others are external events: earthquakes, tornados, fires (external or internal), and floods (external or internal). Event trees can be used to analyze either type of initiator.

Step 2 - System Sequences

Given the initiator, the question is, "What operational systems or actions are involved in responding to the initiator?" The systems shown as A, B, C, and D in Figure 3.4.5-1 are generally ordered in time sequence. A support system must enter the sequence before the affected systems in order for a support system failure to fail the systems it supports. These mitigating systems are not necessarily equipment, they may be human actions.[1] Exceptions to time sequencing the events arise from the need to introduce dependencies by manipulating the branching probabilities according to preceding events.[m]

Step 3 - Branching Probabilities

The "event[n]" list, across the top of the event tree, specifies events for which the probability of failure (or success) must be specified to obtain the branching probabilities of the event tree. Events that are the failure of a complex system may require fault tree or equivalent methods to calculate the branching probability using component probabilities. In some cases, the branching probability may be obtained directly from failure rate data suitably conditioned for applicability, environment and system interactions.

Step 4 - Damage States

The right side of the event tree represents the damage state that results from that path through the event tree. Thorough knowledge of the processes, operating experience, accident history, and safety analyses (e.g., Chapter 15 of nuclear plant FSARs) are needed to determine the damage state. States do not form a continuum but cluster about specific situations, each with characteristic releases. Figure 3.4.5-1 shows 16 damage states resulting from two-state combinations of the four systems (2^{16}). This is an "unpruned" event tree. In reality, the number of branches in the tree may be reduced by suppressing physically impossible nodes or the same damage states.

Step 5 - Dependency Analysis

The success of some systems depends on other systems. An event tree is constructed by placing support systems before the supported systems. This may require iterating the event tree construction to get appropriate ordering. If the system dependency is in the fault trees, it is not reflected in the event tree and only becomes apparent when the PSA is calculated.

---

[1] Some PSAs (e.g., RSS) put human actions in the fault trees.

[m] Chapter 12 discusses the distribution software BETA for preparing event tree analysis from a work processor table. BETA allows the use of binary conditionals so the nodal probabilities in a vertical line are not necessarily equal but depend on preceding events.

[n] The event list is not necessarily hardware; it may include anything germaine to the accident sequence such as human error.

Quantitative Methods of Accident Analysis

Step 6 - Estimating the End-States

The combinations of failures and non-failed conditions define the state of the plant at the right branches. The damage associated with these plant damage states are calculated using thermal-hydraulic analyses to determine temperature profiles that are related to critical chemical reactions, explosions and high pressure. These end-states serve as initiators for breaking confinement that leads to release in the plant and aquatic and atmospheric release outside of the plant.

3.4.5.2 Event Tree Fault Tree Trade-off

Event trees and fault trees work together to model plant accidents, but the demarcation between them is the choice of the analysts. This has resulted in two styles: 1) small event tree-large fault tree (SELF), and 2) large event tree-small fault tree (LESF). SELF is the style used in WASH-1400, in NRC sponsored PSAs, and in several utility PSAs. LESF was introduced by PL&G in the Oyster Creek PSA and used subsequently by PL&G and several utilities.

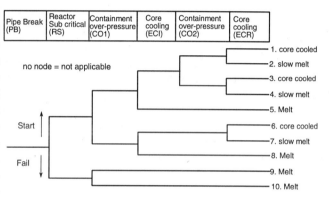

Fig. 3.4.5-2 Large LOCA as a Function Event Tree (adapted from NUREG/CR-2300)

SELF is a two-stage process going from functional event trees, to system fault trees. LESF directly models the system to include events pertinent to the accident progression.

3.4.5.4 Small Event Tree-Large Fault Tree SELF

Functional Event Trees

Function event trees are concerned with depicting functions that must happen to mitigate an initiating event. The headings of the function event tree are statements of safety functions that are required but that may fail in an accident sequence.

Function event trees are developed to represent the plant's response to each initiator. The function event tree is not an end product; it is an intermediate step that provides a baseline of information and permits a stepwise approach to sorting out the complex relationships between potential initiating events and the response of the mitigating features. They structure plant responses to accident conditions - possibly as time sequences. The transition labels of function event trees (usually along the top of the event tree) are analyzed to provide the probability of that function occurring or not occurring.

Chemical and Nuclear Accident Analysis Methods

In constructing the event tree, the analyst considers the functions that are required to prevent damage states, health consequences considering the relationships between safety functions. For example, if RCS inventory is not maintained, the heat-removal functions are depicted as failed states that may lead to core melt.

Figure 3.4.5-2 shows a typical functional event tree for a large BWR LOCA. It is ordered from left to right in the time sequence of the event, beginning with the postulated pipe break (PB). From a review of the piping, the break areas are categorized according to LOCA flow rates. The pre-break operating conditions of the reactor are defined (e.g., power level and core burnup). Initiating events for a break class are chosen from operating experience. Thus, for a large LOCA, the frequency is for the large pipe class, not necessarily the very largest pipes. Usually large, medium, and small are the classes chosen to represent the pipe break. Whether or not a guillotine break is possible must be decided from the plant design. It may be that pipe hangers prevent such a break.

The sequence of events following the LOCA, is presented in Table 3.4.5-1. In preparing the event tree, reference to the reactor's design determines the effect of the failure of the various systems. Following the pipe break, the system should scram (Figure 3.4.5-2, node 1). If scram is successful, the line following the node goes up. Successful initial steam condensation (node 2 up) protects the containment from initial overpressure. Continuing success in these events traverses the upper line of the event tree to state 1 "core cooled." Any failures cause a traversal of other paths in the event tree.

If scram fails, the core melts regardless of other successes. Scram can succeed but the core melts by failure of the core cooling. Slow melt occurs if the long-term core cooling fails. Sequences 5, 9, and 10 have missing nodes to indicate inapplicability of ECI, CO2, and ECR for the cases shown.

The branching probability at a node is determined by either fault tree analysis of the event system or by data from operating experience. Hence, it may be imagined that fault trees hang from each mode, as shown in Figure 3.4.5-3. Electric power

Table 3.4.5-1 Sequence of Events after LOCA

1. Reactor Subcritical (RS): scram termination of the fission process.
2. Containment Overpressure (COI): initial suppression of blowdown by steam condensation only.
3. Core cooling (ECI): initial removal of core heat by coolant-inventory makeup only.
4. Containment Overpressure (COR): containment temperature and pressure control by steam suppression and heat rejection.
5. Core cooling (ECR): additional heat to coolant makeup.

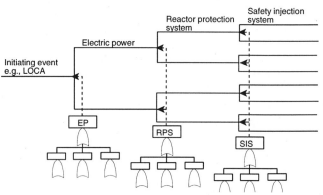

Fig. 3.4.5-3 Fault Trees at the Nodes of Event Trees to Determine the Probability. RPS is reactor protection system; SIS is safety injection system

Quantitative Methods of Accident Analysis

enters the event tree ahead of the systems that fails if it fails. The LOCA event does not use a fault tree because the frequency is obtained from operating experience.

Function event trees include primarily the engineered safety features of the plant, but other systems provide necessary support functions. For example, electric power system failure could reduce the effectiveness of the RCS heat-removal function after a transient or small LOCA. Therefore, EP should be included among the systems that perform this safety function. Support systems such as component-cooling water and electric power do not perform safety functions directly. However, they significantly contribute to the unavailability of a system or group of systems that perform safety functions. It is necessary, therefore, to identify support systems for each frontline system and include them in the system analysis.

3.4.5.5 System Event Trees

The problem with function event trees is that some functions are quite complex and must be analyzed. If a function event tree models the plant's response to an accident initiator, modeling system responses in a fault tree will not clearly exhibit the functional criteria.

A system event tree provides this display and uses the Tech Spec criteria to specify the function. Figure 3.4.5-4 shows a system event tree developed from the function event tree presented as Figure 3.4.5-2. It should be noted that the functions RS, CO1, CO2 and ECR are accomplished by systems and are thus unchanged ongoing from a function to system event tree. ECI is quite complex and may be performed by various system combinations such as 2 or 1 core spray (CS) loops, or various combinations of low pressure injection (LPCI).

Classification of accidents by safety function is the starting point for classification by mitigating system. Because of the factors listed below, classification by system usually produces more accident

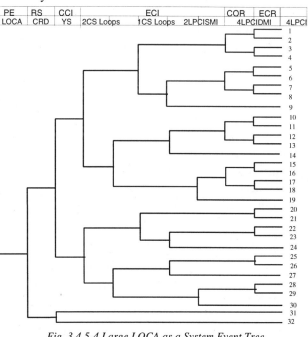

Fig. 3.4.5-4 Large LOCA as a System Event Tree (adapted from NUREG/CR-2300)

classes than does classification by safety function. The factors responsible for this are the following:
1. **Design of Systems**. Although the same set of safety functions may be required for two sets of initiating events, different systems may be employed to perform the same function because of the

115

nature of the initiating event. For example, a distinction is made between LOCAs if they require a different complement of systems for RCS inventory control.

2. ***Interactions between Initiating Events and Systems.*** Some initiating events affect either the function or the availability of mitigating systems. Therefore, the set of systems available for mitigating events will be reduced from the full complement because of interactions. An obvious example is the situation which can occur such that a loss of offsite power makes some power buses unavailable for RCS heat removal. In addition, this loss-of-power initiator affects the availability of the remaining systems, because emergency power becomes the only source of electric power.

System event trees use the information on the effects of loss of various safety functions identified in the function event trees. However, the sequences in the system event trees are likely to differ somewhat from the function event trees because system faults may fail multiple functions.

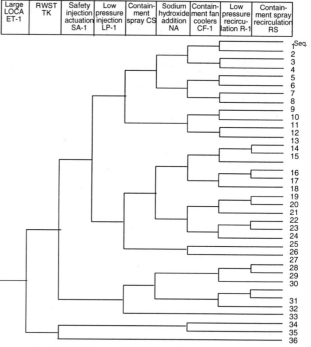

Fig. 3.4.5-5 Large LOCA Event Tree in the LESF style
(End states 26,33, 35, 36 are early core melt; 12 is late core melt; 16-18, 22,24, 27-32 are early core melt with containment spray; 25,31 are early core melt with fan-coolers operating; 13-15, 19-21 are early core melt with fans and containment sprays operating; 10 is late core melt with fans; 4 and 8 are late core melt with containment sprays, 2 and 6 are late core melt with fans and sprays; all others are no core melt)

Each system event tree has a systems heading. The exact order of the headings is not crucial to the analytical results but it is generally in the order of time and dependency. A good practice is a time sequence with the systems arrayed in the order in which they are expected to respond to an accident. Systems responding immediately (e.g., the reactor protection system) are placed first, and those responding later are listed in their time sequence of operation (e.g., high pressure injection, then high pressure recirculation). However, there may be good reasons for ordering sequences out-of-time.

Functional and hardware relationships between systems are considered in selecting the order of event tree headings. Systems that depend on the operation of other systems in order to perform their function should be listed after the other systems. For example, the decay-heat removal system may require successful operation of containment sprays, hence should be listed after containment sprays in the event tree.

Nodes in the event tree can be eliminated if the system does not:

- Affect the outcome,
- Contribute to a safety function in this context, or
- Affect the need for, or the operation of, other systems.

If any of the responses are positive, the node should be included in the event tree. Each question must be examined in the context of each potential accident sequence because of effects on other systems.

3.4.5.6 Large Event Trees-Small Fault Trees (LESF)

LESF (Figure 3.4.5-5), exemplified for the large LOCA, is compared with SELF. Event tree headings are the refueling water storage tank (RWST) a passive component, an engineered safety system (SA-1) and four elements of the containment system. Other examples of the LESF method show human error in the event tree while the criteria for system success is usually in the fault tree analysis.

Although it is not inherent in the LESF method of event tree construction, it is customary to treat support system dependencies by defining degraded states that are evaluated using separate event trees to show the degradation. That is, the nodal probabilities have one set of values for, say, full electric power and another set for a degraded state in which one bus is out. Some PSAs carry this one step further by defining "binary conditionals" (see BETA in Chapter 12). A generic set of probabilities are specified for a given tree in a given degraded state, but the probability of system failing is not the same throughout the tree and may take on different values depending on the preceding sequence of events

3.4.5.7 Comparisons of the SELF and LESF Methods

While the original event trees for the Oyster Creek PSA that introduced the LESF method were indeed large, recent implementations of LESF use event trees of comparable complexity to SELF. The primary distinction seems to that the LESF method takes great latitude as to the type of elements included in the event tree. These may be systems, components or human actions. The SELF method is restricted to functions and systems.

Practitioners of SELF explicitly include the dependencies either in the event trees or in the fault trees. Examples of the LESF method that have been examined, treat major dependencies by the definition of degraded states and reevaluate the systems and event trees for the assumed degraded state as well as for the probability of being in that state. Mathematically this is very effective but the dependency coupling is not as pictorial as the SELF method.

Either method, properly performed, provides PSA plant models that are accepted by the NRC.

3.4.5.8 Binning and Containment Event Trees [o]

A process may suffer destruction but be isolated from the public by containment. However, the containment may fail from pressure, heat and missiles - the subject of containment event trees. These analyses determine yield and ultimate-strength levels for the base structure. If the ultimate strength of a thin-shell structure is determined quite simply, the results may not be valid because of simplifying analysis that does not account for nonlinear effects associated with large plastic strains and for interactions among the various components of a complex structure.

In addition to considering the gross behavior of the structure, special consideration should be given to localized conditions, such as the following.[p]

1. Penetrations, including electrical penetrations and major openings.
2. Major discontinuities, such as the transitions from the cylindrical shell to the top head and the basemat.
3. Layout and anchorage of the reinforcement.
4. Liner walls and anchoring.
5. Interactions with surrounding structures at large deformations.

System event trees connect to the containment event tree with a large number of sequences formed by different combinations of events. The number of sequences is made tractable by grouping sequences in release categories - a process called binning. Two approaches to binning are: 1) probability screening and 2) plant-damage bins.

Approach 1 was used in the Reactor Safety Study Methodology Applications Program (Carlson et al., 1981). Trial sequences were selected using point estimates to identify those with the highest frequencies for realistic accident processes. A problem with this approach is the iteration and judgment required to determine completeness. It analyzes each dominant accident sequence and then groups, rather than grouping the release categories and then analyzing each group.

Approach 2 develops groups of system sequences called plant damage bins containing plant damage states or plant event sequence categories. The categories are identified by the characteristics of the system sequence that affect the release of radionuclides to the environment. All system sequences within a bin are assumed to have the same containment event tree because the branching probabilities are the same, and the end points are assigned to the same radionuclide release categories. The process of binning combines the PSA plant model with calculations of releases and in-plant transport. Figure 3.4.5-6 shows the containment event tree used in RSS for the PWR. The basic procedure is to order the events from most severe to least so that subsequent events are

---

[o] Adapted from NUREG/CR-2300, Section 7.4.2.

[p] Overpressure, overtemperature, hydrogen explosion, steam explosion, and core melt through are mechanisms that may fail the containment of nuclear and chemical reactors.

compounding. For example, CRVSE is the worst thing. If it occurs, overpressure, hydrogen explosion, etc., do not matter. However, if CRVSE does not occur and containment isolation (CL) is successful, a hydrogen explosion might occur or it might not, so on to the end of the tree. The convenience of using the containment event tree for examining all the combinations of compounding failure, the branch points, are given by equations 3.4.5-1a-e.

Not all containment event trees are this simple. The NSAC-60 trees are considerably more complex as are the ZIP trees made so by the addition of key time periods of interest. Although it is comparatively easy to draw these trees, quantifying the probabilities is something else, because the events are basically deterministic physical processes that might occur in the accident. As a result, the probabilities of these events are basically judgmental.

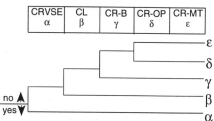

CRVSE - Containment failure from in-vessel steam explosion
CL - Containment isolation failure
CR-B Containment failure from hydrogen combustion
CR-OP - Containment failure from overpressure
CR-MT - Containment failure through basemat penetration

Fig. 3.4.5-6 Containment Event Tree (from WASH-1400)

$$\epsilon = \overline{\alpha} * \overline{\beta} * \overline{\gamma} * \overline{\delta} \quad (3.4.5\text{-}1a)$$
$$\delta = \overline{\alpha} * \overline{\beta} * \overline{\gamma} * \delta \quad (3.4.5\text{-}1b)$$
$$\gamma = \overline{\alpha} * \overline{\beta} * \gamma \quad (3.4.5\text{-}1c)$$
$$\beta = \overline{\alpha} * \beta \quad (3.4.5\text{-}1d)$$
$$\alpha = \alpha \quad (3.4.5\text{-}1e)$$

3.4.6 Alternatives to Fault Tree Analysis

System reliability can be analyzed in a number of other ways. Objections to fault tree analysis are:

- Different appearance from a process flow diagram,
- Understanding the meaning with its symbol set that is different from the symbols for flow diagrams, and
- Negative reasoning.

Alternative methods are listed in Table 3.4.6-1; three methods will be discussed in more detail: Modular Fault Trees (MFT), Go and Digraph.

MFT is a modularization of fault tree methodology to reduce the effort. Go is an old method, originally developed for the analysis of electronics, but has the advantage of appearing similar to a flow diagram or a RBD but having the ability to perform some complex operations with simple programming. Digraph was developed to be a better method for analyzing system interactions.

3.4.6.1 Modular Fault Trees (MFT)

The use of component logic models to build system fault logic has been discussed by several authors for chemical and electrical systems (Powers and Thompkins, 1974; Fussell, 1975; and Powers and Lapp, 1976). In addition, generic sabotage fault trees have been used for some time in the analysis of security concerns for nuclear power plants (NUREG/CR-0809, NUREG/CR-3121,

NUREG/CR-3098, and NUREG/CR-2635). MFT draws on this experience for logical modeling of process plants.

An early version of MFT methodology was applied in the Interim Reliability Evaluation Program (IREP) that analyzed the Calvert Cliffs and Arkansas Nuclear One, Unit 1 plants. The methodology has undergone significant evolutionary refinement based on lessons learned in IREP and other applications. Although MFT is an extension of the fault tree analysis (Section 3.4.4), it warrants a separate discussion (see NUREG/CR-3268). Objectives of MFT are:

- Improve the consistency of fault tree analyses.
- Reduce the time to develop and analyze fault trees.
- Facilitate the understanding and development of fault trees.
- Consider system interfaces and unusual system operating characteristics.
- A systematic approach to improve accuracy.
- Simplification of the review of fault trees.

Table 3.4.6-1. Summary of Other Methods

Method	Applicability	Characteristics
Phased mission analysis	Evaluation of components systems, or functions undergoing phased mission	Qualitative, quantitative, time-dependent; nonrepairable components only; assumes instantaneous transition
Digraph	Model and evaluation of components or systems	Quantitative, time-dependent multiphased inductive; complexity increases rapidly; practical only for simple systems
Markov analysis (Sec. 2.5.4)	Evaluation of components systems, or functions	Quantitative, time-dependent; modeling process complex; success oriented; has potential for modeling complete nuclear plant
GO	Identification of hazards or dependencies	Qualitative, inductive; considers only one failure at a time; simple to apply; provides orderly examination
FMEA	Identification of hazards for improving safety	Qualitative; also used for accident investigation
MORT (see Johnson, 1980)	Model of components or systems	Qualitative; used to synthesize fault trees; complexity increases rapidly
Reliability block diagram	Model and evaluation of components or systems	Quantitative
Signal flow	Model and evaluation of components or systems	Quantitative; assumes constant failure and repair rates

These are accomplished by MFT using the following:

1. A library of generic fault logic for common system configurations and components. These MFT logic models are those needed by the analyst for constructing system models.

2. Standardized procedures to aid and guide the analyst in selecting MFTs for modification to become the system models.

3. Provide an interactive computer system to prepare input data for plotting and analyzing the fault trees.

An MFT begins with the top event and determining the combination of failures that could produce it. The analyst reviews the system diagram and notes the system nodes. Each train of components between junctions is identified as a system segment. System fault logic is developed in terms of faults in the system segments through the use of a set of prescriptive rules applied at each node in the system for recurring configurations.

3.4.6.2 GO

GO is a success-oriented system analysis method using 17 operators for model construction. It looks like a flow diagram of a system or a wiring diagram. It was developed by Kaman Sciences Corporation during the 1960s for reliability analysis of electronics for the Department of Defense, but it was extensively modified for EPRI for the nuclear power industry. The first nuclear power application was a reliability analysis of a PWR containment spray system (Long, 1975); it was used for the Midland, Bellefonte and Sequoyah PSAs.

Table 3.4.6-2 Attributes of GO

- System models follow the normal process flow.
- Model elements have an almost one-to-one correspondence with physical components and elements.
- Modeling of most component and system interactions and dependencies is explicit.
- Models are compact and easy to validate.
- Model evaluations can represent both success and failure states of systems.
- Model alterations and updates are readily effected.
- Potential system faults can be identified.
- Numerical errors due are controlled by limiting the calculational detail.
- Uncertainty analysis can be performed with the same model.

Some of the features of GO (EPRI NP-3123) are given in Table 3.4.6-2. A GO model is networks GO operators to represent a system. It can be constructed from engineering drawings by replacing system elements (valves, switches, etc.) with one or more GO symbols. The GO computer code quantifies the GO model for system reliability, availability, identification of system fault sequences, and relative importance in rank of the constituent elements.

Seventeen symbols represent logical relationships between components of three basic types: (1) independent, (2) dependent, and (3) logic. Independent operators represent components requiring no input, i.e., sources such as tanks or electric power to start the modeling process. The dependent operators require at least one input in order to have an output and represent such components as valves, pumps, and relays. Logic operators combine the operators into the success logic of the system being modeled. Included are operators which represent "OR," "AND," and "ONLY IF" functions as well as operators whose logic can be defined by the users.

Each independent and dependent operator has two aspects: 1) Operation represents a component requiring actuation. For example, a motor-operated valve, is represented as a Type 6 operator since the valve needs both an input (flow) and an actuation (electric power) to pass the flow successfully. 2) The probability of successful operation for independent and dependent operators is needed. In some cases, a third probability for premature or inadvertent operation may be needed.

In addition to the 17 operators, "supertypes" may be defined for subsystems or groups of components that recur in a system. This reduces the required analysis time and adds clarity to the analysis of subsystems. A supertype is a collection of operators defined and used as a single entity.

Once the problem has been defined, the system boundaries identified, and success criteria established, the GO model may be developed. The four basic steps in the development of a GO model are:

Step 1 Defining the System - Collect the information needed to perform the analysis. Information needs include: system descriptions, schematics, P&IDs, logic diagrams, and operating procedures.
Step 2: Establishing Inputs/Outputs - Every GO model begins with at least one input and may have many interfacing inputs. The output of the model is determined by the success criteria.
Step 3: Drawing a Functional GO Chart - The system is represented by the interconnected GO symbols. Independent components are represented by triangle symbols and dependent components are represented by circles.
Step 4: Defining the Operators - Each components operation is analyzed and assigned a GO operator that most closely represents the operation of the physical component. It is identified by the Type number which is the first of two numbers within each GO symbol.

Kelley and Stillwell (1983) provide the following broad comparisons of fault tree and GO methodologies:

1. GO is suited for many PSA applications with boundary conditions well-defined by a system schematic and other design documents to quantify the components.
2. Fault trees are better suited than GO for applications requiring exhaustive investigation of the failure modes and failure mode combinations that lead to a top event. The deductive, nature of the fault trees aids in going beyond the level of the explicitly displayed components in engineering drawings. Unlike GO chart's implicit modeling of failure modes, fault trees explicitly display and catalog the faults. In summary, fault trees are best for a safety analysis that exhaustively catalogs events, and identifies primary and secondary faults and dependencies beyond those explicitly identified on a system schematic.

3.4.6.3 Digraph/Fault Graph

The fault graph method (Alesso, 1985) uses the mathematics and language of graph theory (Swany, 1981) such as "path set" (a set of nodes traveled on a path) and "reachability" (the complete set of all possible paths between any two nodes). It is also known as digraph matrix analysis (DMA) (Alesso, 1984). A related method is the logic flowgraph (Guarro and Okrent, 1984) used as an aid in fault tree construction and disturbance analysis.

This method, developed under USNRC sponsorship, was demonstrated in the analysis of system interactions in the Watts Bar plant (Sacks et al., 1983), Indian Point 3 (Alesso, 1984). It had use for sabotage analysis (Sacks, 1977; Lambert, 1977; and Sacks 1978). The digraph method (Alesso, 1983) for system modeling is similar to a GO chart, but it uses different symbols. The only operators are AND and OR gates, represented by symbols that are different from the fault tree symbols.[q] The fault graph representation of a system is converted to an adjacency matrix to indicate the connection of a node (component or gate) to the next adjacent node. The adjacency matrix is converted to a reachability matrix that gives the nodal connections in the system. These matrices are computer analyzed to give singletons (single component-caused system failure) or doubletons (pairs of component failures causing system failure).

The advantages of digraph (Alesso, 1983) are:

1. Each accident sequence is a single model, including support systems. Its partitioning by subgraphs is more rigorous than fault tree analysis; and
2. Cycles and feedback loops are analyzed by graphical methods.

It consists of four basic steps:

Step 1 - Select the combinations of systems that enter the analysis. (This is equivalent to finding accident sequences in event tree analysis.)
Step 2 - Construct a global digraph model for each accident sequence.
Step 3 - Partition digraph models into independent subdigraphs and find singleton and doubleton minimum cutsets of accident sequences.
Step 4 - Evaluate singletons and doubletons on the basis of probability and display answers.

It is necessary to convert the symbolic fault graph into a form suitable for mathematical analysis by forming an adjacency matrix to show nearest-neighbor connections and required input for an "AND" gate.

3.5 Common Cause of Failure

Overview

System models assume the independent probabilities of basic event failures. Violators of this assumed independence are called: "Systems Interactions," "Dependencies," "Common Modes," or "Common Cause Failure (CCF)" which is used here. CCF may cause deterministic, possibly delayed, failures of equipment, an increase in the random failure probability of affected equipment. The CCF may immediately affect redundant equipment with devastating effect because no time is available for mitigation. If the effect of CCF is a delayed increase in the random failure probability and known, time is available for mitigation.

---

[q] The "AND" gate is a vertical bar with inputs as arrows; the "OR" gate is a circle with arrows at angular incidence.

CCF means different things to different people. Smith and Watson (1980) define CCF as the inability of multiple components to perform when needed to cause the loss of one or more systems. Virolainen (1984) criticizes some CCF analyses for including design errors and poor quality as CCF and points out that the phenomenological methods do not address physical and statistical dependencies. Here, CCF is classed as: known deterministic coupling (KDC), known stochastic coupling (KSC), and unknown stochastic coupling (USC).

KDC has a cause and effect relationship between as the primary cause leading to secondary failures. Besides its drastic operational effects on redundant systems, the numerical effects that reduce system reliability are pronounced Equation 2.4-5 shows that the probability of failing a redundant system composed of n components is the component probability raised to the n-th power. If a common element couples the subsystems, Equation 2.4-5 is not correct and the failure rate is the failure rate of the common element. KDC is very serious because the time from primary failure to secondary failures may be too short to mitigate. The PSA Procedures Guide (NUREG/CR-2300) classifies this type as "Type 2."

KSC results from an environmental change that affects the probability of failure of the affected components. An obvious example is an increased failure rate due to a change in conditions such as fire, stresses from an earthquake, or improper maintenance practices affecting several components. NUREG/CR-2300 classifies this type of common cause as "Type 1."

USC includes mathematical procedures for deducing physical effects that are shown in data or judged for redundant systems to fail with a higher frequency than that calculated under the assumption of independent systems. The mechanism for the increase is not known or else it would be included in the system model.

USC may be modeled as a power-series expansion of non-CCF component failure rates. No a priori physical information is introduced, so the methods are ultimately dependent on the accuracy of data to support such an expansion. A fundamental problem with this method is that if the system failure rate were known such as is required for the fitting process then it would not be necessary to construct a model. In practice information on common cause coupling in systems cannot be determined directly. NUREG/CR-2300 calls this "Type 3" CCF.

3.5.1 Known Deterministic Coupling

KDC severely impacts safety but it is the CCF that is most amenable to systems analysis. References: NUREG/CR-1321, NUREG/CR-1859 and NUREG/CR-1901 present methods used by: Sandia, Lawrence Livermore, and Brookhaven National Laboratories, respectively for this purpose. These reports review previous work using: FMEA, Diversion Path Analysis, fault trees, event trees, GO, Digraph and plant walk-throughs. Buslik et al. (NUREG/CR-1901) say that single methods are insufficient; combinations of methods should be used. The most obvious method, not mentioned in the studies, is to simply assure that all the safety system dependencies have been modeled from the plant as-built drawings and inspection of the plant. Assuming the identification of KDC CCFs, they can be incorporated into the PSAs.

3.5.2 Known Stochastic Coupling

KSC increases the failure rates of components from causes such as: earthquake, fire, flooding, tornado, erroneous maintenance or mis-specifying the operating environment. Environmental qualification (EQ) of equipment, specified by Regulatory Guide 1.97, assures the operation of instruments in an accident environment. Similar qualification is required for a design-basis earthquake.

The environmental effects on the failure rate may be modeled using Arrhenius or power laws. In some cases it may be necessary to model the failure rate using the technique of overlapping stress/strength distributions (Haugen, 1972).

3.5.3 Unknown Stochastic Coupling

Waller (NUREG/CR-4314) provides a concise review of USC with 143 references and relevant papers. He quotes Evans (1975) that no new theory is needed to analyze system dependencies. The problem persists because the conditional probabilities are not known. Except for the bounding method used in WASH-1400, the other two methods presented below can be shown to be derivable from the theory of Marshall and Olkin (1967). Waller reviews methods other than the three presented here. They are not presented for absence of physical insight to the problem.

3.5.4 Modeling Known Dependencies

Modeling KDC in an Event Tree

- Direct Introduction - This involves locating, in the event tree, support systems (e.g., electric power) before the systems they affect. The event tree is constructed so that systems requiring the support systems are bypassed as failed. This is the technique in WASH-1400.
- Probability Manipulation - An alternative is to change the failure rates of the supported systems according to the state of the support system. For example, electric power may be analyzed as having a fully operating and several degraded states. The systems disabled by the power failure are found by following the buses and assigned a failure probability of 1 in the event tree evaluation. Thus, this method reconfigures the event tree by manipulating the branching probabilities.

Modeling KDC in a Fault Tree

- Direct Introduction - This method, similar to the preceding, introduces the dependency in the fault tree higher than the affected subsystems to bypass the "AND" operation of the redundancy. This is also the technique in WASH-1400. An example of this technique is found in the fault tree shown in Figure 3.4.4-9.
- Component Level Entry - The method shown in Figure 3.4.4-7 is the other method. It is more logical to enter utility and control elements with the component that is being affected

as shown in this tree. Actually, in the flow diagram (Figure 3.4.4-6), everything to the left of the redundant valves (B and C) could be considered capable of a common cause failure. An alternative to treating intersystem dependencies in the event tree is to link the system fault trees together to create a single large fault tree for the entire accident sequence. This tree may be formed from the respective system fault trees by linking them together with an AND gate. It is not necessary to physically construct the sequence logic tree to implement the fault tree linking method. An alternative is to determine the mincutsets with each system separately and to resolve the shared equipment dependence by using Boolean algebra to find the mincutsets for the sequence.

3.5.5 Geometric Mean

WASH-1400 introduced this method for unknown common mode effects. The procedure was not presented in the best light and was severely criticized by the Lewis Committee for lack of a physical basis, although the approach is not unreasonable. Basically, the procedure considered the failure rate of a system, with a common mode, to be between two bounds. The lower bound is the one with no common mode between subsystems, hence, all probabilities are treated as independent. The upper bound assumes such a strong coupling that the system failure rate is the same as the largest subsystem failure rate. The overall system failure rate including common modes was estimated to be the geometric mean of these extremes (equation 3.5.5-1). This method as superseded by the beta factor method which also lacks a physical basis.

$$\lambda_{cm} = \sqrt{(\lambda_{indep} * \lambda_{dep})} \quad (3.5.5\text{-}1)$$

Table 3.5.6-1 Generic Beta Factors for Reactor Components (from NUREG-1150)

Component	Upper bound	Mean
Reactor trip breakers	0.19	0.0792
Diesel generators	0.05	0.0208
Motor-operated valves	0.08	0.0333
Safety/relief valves (PWR)	0.07	0.0292
Safety/relief valves (BOOR)	0.22	0.0917
Safety injection pumps	0.17	0.0708
Residual heat removal pumps	0.11	0.0458
Containment spray pumps	0.05	0.0208
Auxiliary feedwater pumps	0.03	0.0125
Service water pumps	0.03	0.0125
Batteries	0.10	0.04

3.5.6 Beta Factor

This method assumes that λ, the total constant failure rate for each unit, can be expanded into independent and dependent failure contributions (equation 3.5.6-1), where λ_i is the failure rate for independent failures and λ_c is the failure rate for dependent (common cause/mode) failures. For convenience, a parameter β, is defined as the fraction of the total failure rate attributable to dependent failures, such that $\beta = \lambda_c/\lambda$. Substituting into equation 3.5.6-1, results in equation 3.5.6-2.

$$\lambda = \lambda_i + \lambda_c \quad (3.5.6\text{-}1)$$

$$\lambda_c = \frac{\beta * \lambda_i}{1-\beta} \quad (3.5.6\text{-}2)$$

Table 3.5.6-1 (from NUREG-1150) provides generic beta factors.

For systems with more than two subsystems the beta-factor model, as presented, does not distinguish between different numbers of multiple failures. This simplification can lead to conservative predictions when it is assumed that all units fail when a common cause failure occurs. Further, it may be necessary to consider dependent failures of two or three units out of a total system of n units. Note that, in general, the beta factor for the failure to continue running is not necessarily equal to the beta factor for the failure to start on demand (BD). The value of beta is obtained from the examination of the failure experience of systems for which it is developed.

Subjective assessments of the parameter values have been used when data are unavailable. The beta-factor method is most useful for analyzing dependent failures in systems with limited redundancy (two or three units). It can be applied after finding the minimal cutsets of the system or incorporated directly into the fault trees.

3.5.7 Example of the Beta-Factor Method: Emergency Electric Power

Consider a plant having two power trains A and B with diesel generators DA connected to train A and DB to train B and DS, a swing diesel that can connect to train A or B as fault tree diagramed in Figure 3.5.7-1.

Experience shows that the failure rate of one of these diesels is $\lambda_i = 8.7E-5/hr$, and they are tested every 30 days (720 hours). A study of related plant data shows that common mode contributes 10%; hence $\beta = 0.1$, and $(1+\beta) = 9.6E-5/hr$.

The failure rate of one doubly redundant system: $Q_{AB} * Q_{AC} * Q_{BC} = (½ * 9.6E-5 * 720)^2 = 0.0012$ is the average unavailability for a subsystem and 0.0036 is the average unavailability for the whole system as diagramed. If the system had been modeled without considering common mode: $Q_{AB} = Q_{AC} = Q_{BC} = (½ * 8.7 E-5 * 720)^2 = 0.00097$, hence, there is an 80% effect.

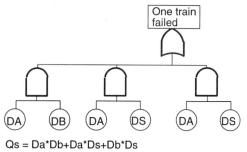

Fig. 3.5.7-1 Fault Tree Diagram of the Emergency Electric Power System

3.5.8 Common Cause Multiparameter Models.

Clearly, more parameters allow more freedom in representing common cause effects in systems of complex redundancy. NUREG-1150 describes the Multiple Greek Letter Model (MGL), the Basic Parameter (BP) model and the Binomial Failure Rate (BFR) model.

3.5.8.1 Multiple Greek Letter Model

Fleming et al. (1985) define λ as the independent failure rate and higher order effects in order of the Greek alphabet (skipping α). The conditional probability that a CCF is shared by one or more components is designated: β, by two or more components, the CCF is γ, and by three or more components, the CCF is δ. The equations for using these parameters depend on the system configuration.

3.5.8.2 Basic Parameter Model

Fleming et al. (1985) define this as similar to the model of Marshall and Olkin (1967) except that BPM is only for time-dependent failure rates. Equations 3.5.8-1a-d are for four parameters, but the method may be generalized to n components. These parameters may be related to the MGL parameters as shown in equations 3.5.8-2a-d.

3.5.8.3 Binomial Failure Rate Model

This, more physical model that visualizes failure to result from random "shocks," was specialized from the more general model of Marshall and Olkin (1967) by Vesely (1977) for sparse data for the ATWS problem. It treats these shocks as binomially distributed with parameters m and p (equation 2.4-9). The BFR model like the MGL and BPM models distinguish the number of multiple unit failures in a system with more than two units, from the Beta Factor model. NUREG/CR-2303 describes the BFR model in some detail and provides a numerical example.

$\lambda_1 =$ *failure rate for one component to fail to run* (3.5.8-1a)
$\lambda_2 =$ *failure rate for two components to fail to run* (3.5.8-1b)
$\lambda_3 =$ *failure rate for three components to fail to run* (3.5.8-1c)
$\lambda_4 =$ *failure rate for four components to fail to run.* (3.5.8-1d)

$$\lambda = \lambda_1 + 3*\lambda_2 + 3*\lambda_3 + \lambda_4 \quad (3.5.8\text{-}2a)$$

$$\beta = \frac{3*\lambda_2 + 3*\lambda_3 + \lambda_4}{\lambda_1 + 3*\lambda_2 + 3*\lambda_3 + \lambda_4} \quad (3.5.8\text{-}2b)$$

$$\gamma = \frac{3*\lambda^3 + \lambda^4}{3*\lambda_2 + 3*\lambda_3 + \lambda_4} \quad (3.5.8\text{-}2c)$$

$$\delta = \frac{\lambda^4}{3*\lambda_3 + \lambda_4} \quad (3.5.8\text{-}2d)$$

To conclude, models of USC, and particularly the multiparameter models, are no better than the data to support them. The higher order the fit, the more demands on the data with the continual temptation to use data from a dissimilar system simply because the data are available.

3.6 Computer Codes for System Reliability Analysis

Assigning the correct probabilities to the components, combining these numbers according to system logic and numerically solving the system model, only can be done practically with a computer using the codes that have been developed over the years.

The PSA Procedures Guide (NUREG/CR-2300) summarizes many of the codes, and the Fault Tree Handbook (NUREG-0492) contains an overview discussion, but these sources do not

Table 3.6-1 Computer Codes for Qualitative Analysis

Code Name/Input/	Type of gates/ restrictions	Method of generating cutsets	Fault tree truncation/ Other outputs	Other features	Computer language and availability
ALLCUTS 8-character alphanumeric names, control information, primary-event probability, fault-tree description	AND OR/ 175 primary events and 25 gates; Up to 1,000 cut sets can be generated.	Top-down successive Boolean substitution	Minimal cutsets, probability/Cutsets in specified probability range, cut set and event probability	Fault tree plotting option	IBM 360/370 CDC 7600 Fortran IV
FATRAM 8-character alphanumeric names, control information, fault-tree description	AND OR /None	Top-down successive Boolean substitution with gate coalescing option	Minimal cutsets up to specified order		CDC Cyber 74 Fortran IV. Available from EG&G Idaho Inc
FTAP 8-character alphanumeric names, control information, fault-tree description	AND OR K-of-N NOT/ None; computer memory is limiting factor; Minimal cut sets of up to 10 can be generated	Top-down, bottom up, and Nelson method prime implicants	Minimal cut sets and prime implicants	Independent subtrees found and replaced by module	IBM 360/370 CDC 6600-7600 Fortran IV. Available from Operations Research Center University of California, Berkeley
MOCUS 8-character alphanumeric names, control information, fault tree description	AND OR INHIBIT; Minimal cutsets of up to 20 can be generated	top-down successive Boolean substitution	Path sets	Cutsets can be automatically punched on cards or online data for KITT or SUPERPOCUS	IBM 360/370 CDC 7600 Fortran IV. Available from Argonne Software Center
PL-MOD 79-character alphanumeric names, control information, fault-tree description, failure data	AND OR NOT K-of-N	Bottom-up modularization and decomposition of fault tree into best modular representation	Top event probability, time dependence, characteristics, minimal cutsets, and uncertainty	Option of not generating minimal cut sets for qualifying fault tree	IBM 360/370 PL Fortran IV. Available from Argonne Software Center
PREP 8-character alphanumeric names, control information, fault-tree description	AND OR INHIBIT/ None; computer memory is limiting factor; Minimal cut-sets up to 10 can be generated	Combinatorial testing		Cutsets can be automatically punched in cards or online data for KITT or SUPERPOCUS	IBM 360/370 CDC 600 Fortran IV Available from Argonne Software Center
SETS 16-character alphanumeric names, user's program, failure data, fault tree description alphanumeric	AND OR INHIBIT PRIORITY/ Exclusive or special 2,000 primary events and 2,000 gates	Top-down Boolean substitution, but user's program can be designed for any other method	Probability of minimal cutsets, prime implicants	Automatic fault tree merging and plotting online data sets can be stored on tapes for use in other runs; independent subtrees can be obtained simplify cut set generation	CDC 7600 Fortran V. Available from Argonne Software center
SIFTA 10-character alphanumeric names, control information, failure data, fault tree description	AND OR K-of-N/ None, computer memory is limiting factor. No cutsets generated	Pattern recognition to reduce structure of tree; numerical simulation to calculate probabilities	Independent branches of tree with small probability/ New structure of tree after reduction; probability of top event	Handles trees with multiple top events, merging of fault trees possible, fault trees can be plotted	CDC 7600 Fortran V. Available from EPRI code center

Chemical and Nuclear Accident Analysis Methods

Table 3.6-1 Computer Codes for Qualitative Analysis (cont.)

Code Name/Input/Reference	Type of gates/ restrictions	Method of generating cutsets	Fault tree truncation/ Other outputs	Other features	Computer language and availability
TREEL and MICSUP / 8 character alphanumeric name control information. fault-tree description/ Pande, 1975	AND OR INHIBIT/ None computer memory is limiting factor. Minimal cutsets up to order 10 are generated	Top down successive Boolean substitution	Minimal cutsets/path sets	Can determine minimal sets of intermediate gates	CDC 6400 Fortran IV Available from Operation Research Center University of California, Berkeley
WAMCUT/ 10 character alphanumeric names control information, failure data, fault-tree description/ Erdmann, 1978	1500 primary events and 1500 gates, up to 2000 minimal cutsets of any order may be generated.	Bottom up Boolean substitution, WAMCUT II finds independent subtrees replaces them by pseudo components the uses top-down Boolean substitution.	Based on both cutset order and probability/probabilities of minimal cutsets and top event, first and second moments of minimal cutsets and top events	Not option can generate minimal cutsets of intermediate gates.	CDC7600, IBM 376 Extended Fortran IV Available from EPRI Code Center

The PSA Procedures Guide classifies the codes according to whether they perform qualitative, quantitative, uncertainty, or dependency analysis (Tables 3.6-1, 3.6-2, 3.6-3, and 3.6-4 respectively). Such classification is not rigid, and some codes in one category will perform in another. An example is SETS which will perform the most complex analyses.

3.6.1 Codes for Finding Minimal Cutsets and Tree Quantification

The codes listed in Tables 3.6-1 and 3.6-2 are useful for various purposes of system analysis, primarily fault tree analysis. The following discusses their features.

PREP, used by the Reactor Safety Study, is a preprocessor for KITT that performs the fault tree quantification for a constant repair rate at specified time intervals. PREP accepts the fault tree coding and finds the cutsets by successively failing single and double components in all combinations. Above doubles, the cutset combinations are found by an accelerated unbiased Monte Carlo process. It should be noted that qualitative cutsets are valuable for ranking systems by the order of the number of component failures necessary for system failure. Most codes provide cutsets, although they are not required for quantification. The methods for finding the cutsets have advanced considerably since PREP-KITT.

MOCUS implemented the Fussell algorithm (Fussell, 1974) for top-down solutions of the fault tree. This algorithm was used in ALLCUT and was modified to be bottom-up in WAMCUT. While cutsets are valuable for qualitative and quantitative purposes, they are not compact. They are thousands of trains (AND sequences) connected to one OR gate.

FTAP The Fault Tree Analysis Program (Willie, 1978, Section 6.4, and distribution disk) is a cutset generation code developed at the University of California, Berkeley Operation Research Center. FTAP is unique in offering the user a choice of three processing methods: top-down,

Table 3.6-2 Computer Codes for Quantitative Analysis

Code/Input/Reference	Quantitative Calculations	Importance	Other Features	Computer and Availability
FRANTIC, FRANTIC II Reduced system equation or minimal cutsets, primary event failure data/ Vesely, 1977.	Time-dependent calculation; nonrepairable, monitored, and periodically tested primary events are handled; uncertainty analysis for failure rates in conjunction with time dependent calculation	No	Can model human-error and dependent-failure contributions; FRANTIC II can handle time-dependent failure rates and incorporates effect of renewal on aging	IBM 360/370 Available from Argonne Software Center
GO GO chart and fault-tree failure data/Gately, 1968	Only time-independent calculations for gates and top event; nonrepairable or periodically tested primary events are handled	No	Cutsets for selected gates and probability truncation of cutsets up to order 4	CDC 7600 Available from EPRI Code Center
ICARUS Reduced system equation, choice of testing schemes, failure data	Average unavailability, optimal test interval, relative contributions of testing, repair, and random failures	No	Three testing schemes available: random testing, uniformly staggered testing, and nearly simultaneous testing	IBM 360/370 Available from Argonne Software Center
IMPORTANCE Minimal cutsets, primary event failure data	Top-event point-estimate probability or unavailability	Can calculate the following: Birnbaum, criticality, upgrading function, Fussell-Vesely, Barlow-Proschan steady-state Barlow-Proschan, sequential contributory	Can rank cutsets and primary events on basis of each importance measure	CDC 7600 Available from Argonne Software Center
KITT-1, KITT-2 Minimal cutsets supplied directly or by MOCUS or PREP; primary-event failure/ Vesely, 1969, and 1970	Time-dependent unavailability for primary events, minimal cut sets, and top event; failure rate, expected number of failures, and unreliability for top event and minimal cutsets	Fussell-Vesely importance calculations for primary events and minimal cutsets	KITT-2 allows each component to have unique time phases and thus failure and repair to vary from phase to phase	IBM 360/370 CDC 7600. Available from Argonne Software Center
RALLY Fault-tree description, control information, failure data	Average unavailabilities and failure frequencies for top event, time-dependent calculation possible through use of minimal cutsets uncertainty analysis possible by using minimal cutsets s normal, lognormal, Johnson, extreme value-1, Weibull, gamma, and exponential distributions are handled	code CRESSEX in RALLY performs importance calculations	Can handle up to 1500 components and 2000 gates; can determine minimal cutsets using either a simulative or analytical way	IBM 360/370. Available from Argonne Software Center
RAS Fault-tree description or minimal cutsets; failure and repair rates	Time-independent unavailability, expected number of failures, and frequency of top event	No	Phased-mission analysis possible; if fault tree is input, minimal cutsets will be calculated	CDC 7600 Available from Argonne Software Center
SUPERPOCUS Minimal cutsets, component failure data, time at which calculations are performed	Time-dependent unavailability, reliability, and expected number of failures for minimal cutsets and top event	Yes	Ranks minimal cutsets on basis of importance; can read cutsets directly from MOCUS or PREP	IBM 360/370 CDC 7600. Available from Dept. of Nuclear Engineering, University of Tennessee
WAM-BAM Fault-tree description, primary-event failure data/Leverenz, 1976.	Point unavailability for top event and intermediate gates; no time-dependent analysis possible	No	Extensive error checking possible through WAM; probability truncation of fault tree sensitivity analysis possible by using WAM-TAP preprocessor instead of WAM	CDC 7600 Available from EPRI Code Center

bottom-up, and the "Nelson" method. The top-down and bottom-up approaches are akin to the methods used in MOCUS and MICSUP, respectively. The Nelson method is a prime implicant algorithm which is applied to trees containing complement events and uses a combination of top-down and bottom-up techniques. FTAP is the only fault tree code, other than SETS, which can compute the prime implicants. FTAP uses two basic techniques to reduce the number of

Chemical and Nuclear Accident Analysis Methods

Table 3.6-3 Computer Codes for Uncertainty Analysis

Code Name/Input	Type of gates/ restrictions	Method of generating cutsets	Other features	Computer language and availability
BOUNDS Reduced system equation or minimal cutsets, primary-event failure data	Mathematical combination of uncertainties; output includes two moments of minimal cutsets and the top event	Johnson, empirical	Can handle multiple system functions with multiple data input descriptions; can fit Johnson-type distribution to the top event	IBM 360/370. Available from University of California at Los Angeles
MOCARS Minimal cutsets or reduced system equation, primary-event failure data/ Matthews, 1977.	Monte Carlo simulation	Exponential, Cauchy, Weibull, empirical, normal, lognormal, uniform	Microfilm plotting of output distribution; Kolmogorov Smirnov goodness of-fit test on output distribution is possible	CDC Cyber 76. Available from Argonne Software Center
PROSA-2 Reduced algebraic function for system representation, failure data	Monte Carlo simulation	Normal, log-normal, uniform, any distribution in the form of a histogram, truncated normal, beta	Can correlate input parameters; no sorting necessary to obtain the top-event histogram	IBM 370. Available from Argonne Software Center
SAMPLE Minimal cutsets or reduced system equation, primary-event failure data	Monte Carlo simulation Mathematical	Uniform, normal, lognormal	Used in the Reactor Safety Study; output is a probability distribution for the top event	IBM 360/370. Available from Argonne Software Center
SPASM Fault tree or reduced system equation, component failure data	Combination (similar to BOUNDS)	Lognormal	Works in conjunction with WAMCUT	CDC 7600. Available from EPRI Code Center
STADIC Reduced system II equation, primary-event failure data	Monte Carlo simulation (similar to SAMPLE)	Normal, lognormal, or log-uniform, tabular input distribution	Has a better and efficient method of sorting the probabilities obtained in each trial	PRIME, UNIVAC-1180, CDC 7600 Available from General Atomic Company

non-minimal cutsets produced and thereby increase the code's efficiency. The first technique, (bottom-up and Nelson method) is modular decomposition. The second technique (top-down and Nelson method) is the "dual algorithm" that involves transformation of a product of sums into a sum of products whose dual is then taken using a special method. Non-minimal sets that appear during construction of the dual are less than the number of such sets in the original product of sums. Other features of FTAP are: tree reduction by cutset order or probability, finding either path sets or cutsets, direct input of symmetric (k-out-of-n) gates, and considerable flexibility and user control over processing and output.

SETS is unique in that it performs symbolic manipulation of Boolean equations using the logic rules in Table 2.1-1 to achieve other forms of the equations such as minimal cutsets (Worrell, 1997). It processes equations defined by fault trees, event trees vital area analysis, common cause analysis or other Boolean forms. It has been used for non-safety-related applications such as: verification of electronic circuit designs, minimum-cost fire protection, combinatorial optimization problems with Boolean constraints, and susceptibility analysis of a facility to unauthorized access by disabling the protection system sensors.

Equations are factored before they are stored to save space and to produce a form of the equation that is useful for other kinds of processing. Results from one part of an analysis can be

Computer Codes for System Reliability Analysis

Table 3.6-4 Computer Codes for Dependent-Failure Analysis

Code Name/Input/Reference	Method of common-cause analysis	Other features	Computer language and availability
BACFIRE Cutsets, component susceptibilities and locations, and susceptibility domains (Cate, 1977).	Examines cutsets for possible common generic causes or links between all components; prints out cutsets that are common-cause candidates	Has same features as COMCAN, but allows use of multiple locations for primary events (e.g., pipes and cables)	IBM 360/370. Available from Dept. of Nuclear Engineering, University of Tennessee
COMCAN Cutsets, component susceptibilities and locations, and susceptibility domains (Burdick, 1976).	Examines cutsets for possible common generic causes or links between all components	Cutsets that are common cause candidates can be ranked by significance of common-cause failure output	IBM 360/370. Available from Argonne Software Center
COMCAN-II Fault tree, component susceptibilities and locations, and susceptibility domains	Same as COMCAN	FATRAM is used to generate cutsets before common cause analysis; other features are similar to those of COMCAN	CDC 7600. Available from Argonne Software Center
MOCUS-BACFIRE Fault tree, component susceptibilities and locations, and susceptibility domains (Cate, 1977, Fussell, 1972).	Same as BACFIRE	Similar to BACFIRE, but does not need cut-set input: cutsets are generated by MOCUS and automatically passed to BACFIRE	IBM 360/370. Available from Dept. of Nuclear Engineering, MIT
SETS Fault tree (Worrell, 1978).	Adds generic causes and links to fault tree; cutsets that include one or more generic causes are obtained and identified as common-cause candidates	Can handle large fault trees and can identify partial dependency in cutsets; attractive features of SETS as cut-set generator justify use for dependent failure analysis	CDC 7600. Available from Argonne Software Center
WAMCOM Fault tree with susceptibilities added	Uses modularization and SETS to more effectively identify cutsets that contain critical events, critical random events, and significant common cause events or to describe common cause sets for each failure	Can identify common total or partial links between fault-tree components; can handle very large fault trees	CDC 7600. Available from n EPRI Code Center

stored for later use. This has advantages for level 1 and level 2 PSAs because accident sequences can be generated one-after-the-other without having to re-solve fault trees for each sequence.

Logic Analysts version of SETS operate on: Cray X-MP, Y-MP, CDC 930, IBM 3090, all 80386 and higher personal computers, IBM RS-6000, Sun SPARC 1, DEC VAX, and HP-700 computers. WAMBAM uses the original GO algorithm of table hookup and does not provide the intermediate step of finding the minimal cutset (current versions of GO provide cutsets).

3.6.2 Truncation of a Fault Tree

Cutsets are the least compact representation of a complex plant, they may be so numerous that they are unmanageable which obscures significant risk contributors. To address this hydra-like expansion, cutsets may be truncated according to order, probability, or risk. Truncation by order is an approximation to truncation by probability as if each component has about the same probability of failure (a very gross assumption). Truncation by order and by probability are featured in most codes that calculate cutsets. A better truncation method is by risk, as provided in ALLCUTS in as much as a low probability cutoff may delete a high consequence, significant risk contributor. Truncation by risk is difficult because the consequence of a sequence may not be known when the probability is calculated. Furthermore, consequences merge with the probabilities in the event tree, not the fault tree. However, approximate consequence values can be used.

Any truncation is an approximation. WAMCUT estimates the error of truncation. WAMBAM does this also but does not display the cutsets. The GO code contains features for estimating the effects of truncation on the probability but not the risk.

3.6.3 Time Dependence

A type of time dependence that is available in most codes evaluates the exponential distribution at specified times. This is the constant failure rate - constant repair rate approximation (Section 2.5.2). This may not be realistic as indicated by Figure 2.5-2 in which the failure rate is not constant. Furthermore, Lapides (1976) shows that repair rates are not constant but in many cases appear to be lognormally distributed.

There are two reliability modeling codes used for Tech. Spec. modification to address time-dependent failure rate and repair: FRANTIC developed by BNL/NRC and SOCRATES developed by BCL/EPRI.

3.6.4 Uncertainty Analysis

Section 2.7 presented the various techniques for uncertainty analyses. Generally, these are not implemented in the fault tree codes. Exceptions are WAMCUT that calculates the first and second moments of a distribution, hence, more appropriate if the distribution is symmetrical. GO (which is not a fault tree code) contains a selectable routine for performing a DPD calculation separately from the main analysis. Most uncertainty analyses have used Monte Carlo, as done by the Reactor Safety Study using the SAMPLE-A code. This type of calculation is facilitated by the code SPASM (EPRI). Monte Carlo calculations are generally very expensive in computer time and are usually performed on components identified as significant by using a system analysis code (e.g., SETS, WAMCUT, etc.). Now personal computers reduce the cost of computing time so that Monte Carlo simulation is extensively used. The distribution disk contains the code MONTE for such uncertainty analysis

3.6.5 Importance Calculations

Component importances are used for insights into system characteristics. COMCAN-III contains within itself an option for generating some importance information. SETS does not itself generate importance information (other than the so-called structural importance), but interfaces with two publicly available companion codes (SEP and Importance) that do this. INEL developed a code (SQUIMP) operating on a VAX computer in the artificial intelligence, LISP language, to calculate importances at Indian Point 2.

Two IBM-PC interactive codes calculate importances and risk and allow the operator to investigate the effects of system and component failures on the risk and importance. The BNL code NSPKTR models the Indian Point Plants 2 and 3 using information in their PSAs. NSPKTR uses the plant model in the form of a compact Boolean equation to calculate the risk and importances of health effects to the public. This code includes both success and failed states so that the calculations are correct with any combination of failures. The Boolean representation allows the use of an untruncated plant model with de Morgan's theorem for avoiding approximating the probability calculations.

PRISIM embodies the IREP model of Arkansas 1. It includes extensive graphics of simplified flow diagrams and relevant operating history from LERs (Licensee Event Reports required by Regulatory Guide 1.16) The plant model consists of 500 cutsets truncated by probabilities determined from normal operation.

3.6.6 Processing Cutset Information

Postprocessing cutset information such as cutset acronyms and cutset searches is performed by REPORT (BNL written to aid in preparation of NUREG/CR-4207). A standard output format for the fault tree codes would facilitate postprocessing.

TEMAC (NUREG/CR-4598) by Iman and Shortencarrier is a unique code for processing cutset information in matrix form. The basic events are arrayed along one axis of the matrix (e.g., x-axis) and the initiators along the other axis with the elements of the matrix being the probabilities. The bookkeeping is performed by a separate "T" matrix that specifies how the probabilities are to be combined for calculating the top event probability specified by the cutset. For a sum of probabilities the T-matrix provides the alternating combinations of products required by equation 2.4-6 (a problem that is avoided by using de Morgan's theorem). This matrix method is both flexible and computationally convenient. It can handle complemented events such as required for event trees as well as distributions by simply inputting probabilities that the basic events take for the assumed probability distributions.

Table 3.6-5 Code Usage in Utility PSAs

Power Plant	Code
Browns Ferry	RAS, COMCAN
Oconee 3	SETS
Calvert Cliffs (IREP)	SETS
ANO-1 (IREP)	SETS
Millstone 1 (IREP)	SETS
Midland	GO & derivatives
Limerick	WAM codes
WASH-1400	Prep-Kitt
Crystal River-3 (IREP)	SETS
Shoreham	WAMCUT-II
Seabrook	SETS
Big Rock Point	WAM
Grand Gulf (RSSMAP)	SETS
Zion	RAS, Inspection

3.6.7 System Analysis Code Usage in Past PSAs

Table 3.6-5 presents code usage in the PSAs of some nuclear power plants. and Table 3.6-6 surveys codes used in major studies.

3.6.8 Logistics of Acquiring Codes

Most codes for fault tree analysis in the public domain are available from the Argonne Code Center. EPRI codes are available under special arrangements; SETS is available from Sandia National Laboratories. Advanced versions for work stations and personal computers are available from Logic Analysts, Inc., 1717 Louisiana NE, Suite 102A, Albuquerque NM 87110.

Table 3.6-6 Code Usage in PSA Studies

Study	Code
La Salle Unit 2	COMCAN-II
Accident Sequence Evaluation	COMCAN-III &SETS
Byron LCO Relaxation Study (NUREG/CR-4207)	WAMCUT
LLNL Systems Interaction Study	Digraph
Sandia Systems Interaction Study (NUREG/CR-1321)	SETS

3.7 Code Suites

Code suites are integrated collections of personal computer codes that perform different functions but work together through unified controls and data transfer.

3.7.1 SAPHIRE

The Systems Analysis Programs for Hands-on Integrated Reliability Evaluations (SAPHIRE - NUREG/CR-5964, NUREG/CR-6116, Smith, 1996),[r] developed by INEL for the NRC, is a personal computer program for creating and analyzing PSAs. SAPHIRE facilitates event tree and fault tree construction to define accident sequences using basic event failure data, to solve system and accident sequence fault trees, to quantify cutsets, and to perform uncertainty analysis of the results. Its features include text and graphics to report the results. It consists of five modules: 1) IRRAS, 2) SARA, 3) MAR-D, 4) FEP, and 5) GEM as shown in Table 3.6-7.

---

[r] SAPHIRE is available through its internet home page: http://sageftp.inel.gov/ saphire/ saphire.htm. Special arrangements and possibly fees are needed to use SAPHIRE.

Computer Codes for System Reliability Analysis

	Table 3.6-7 The Saphire Suite
IRRAS	Integrated Reliability and Risk Analysis System is an integrated PRA software tool for creating and analyzing fault trees and event trees with a personal computer using minimal cutset generation. Some general features of IRRAS are: • PC-based fault tree and event tree graphical and text editors • Cutset generation and quantification • Importance measures and uncertainty modules • Relational database with cross-referencing features
SARA	System Analysis and Risk Assessment system displays and manipulates cutsets generated by IRRAS.
MAR-D	Models and Results Database imports and exports PRA data from the SAPHIRE codes, and reads and writes output files from other PRA codes (e.g., SETS).
FEP	Fault tree, Event tree, and Piping and instrumentation diagram (P&ID) graphically develops and edits fault trees, event trees, and P&IDs.
GEM	Graphical Evaluation Module performs Accident Sequence Precursor (ASP) and event assessment analyses and automates SAPHIRE for event assessment. Such an assessment examines the impact of certain conditions on the core damage frequency according to the categories of 1) initiating events occurrence, or 2) component/system inoperability. GEM provides conditional core damage probability (CCDP) calculations for determining if an initiating event or a degraded condition adversely impacts safety. It uses SAPHIRE models and data for the ASP program, or standard SAPHIRE data however, some of the features (e.g., automatic recovery modeling) require additional information. GEM simplifies operational event analyses and automates current-value input to the analysis software. For conditions not involving an initiating event, GEM calculates a CCDP based on the modified basic event probabilities and the duration of the condition. Thus, GEM performs "what if" risk analyses for various plant configurations.

Preparation

When first running SAPHIRE[s] for Windows, after installation, the Define Constants dialog appears for assigning default values to various functions.[t] After saving the defaults the SAPHIRE window appears for selecting a family.

A family data set is a grouping of models and data for a problem (plant or system). To select a family, from the File menu choose "Select Database." The Select Family dialog appears with the name of the currently selected family in the title bar. Choose the desired family from the list by double-clicking[u] the name. A pop-up menu appears from which the following options may be chosen:

• Add a family to the database,

---

[s] SAPHIRE contains list boxes and pop-up menus. Items are selected by placing the cursor on the item and clicking the left mouse button. Multiple items are selected by holding down the control (CTRL) key and clicking the items.

[t] After first use, this dialog does not reappear, and it is necessary to select this dialog from the Utility option to change defaults.

[u] An option to double clicking is a single click and pressing "enter."

- Copy - a file (e.g. raw data and/or a MAR-D file) into the family.
- Modify - a family data record.
- Delete - a family data record from the database if no sub-families are contained.
- Family: Copy - copies database files between families. Select the family to be copied from, then choose the Family Copy button to specify an empty recipient.

"Generate" is used to change set data. Selecting this from the menu causes a dialog to appear with a list of change sets. Changes are applied to the marked change sets in the order they are marked by double clicking. To mark or unmark a change set, double click on the desired change set in the list. Other options are:

- Generate - applies the event data modifications specified by the marked change sets to the basic event data file. This must be done before any data analysis.
- Single - add, modify, or reset events in the change set. Add, Modify, and Reset are accessed from the Selected Events dialog. Add adds a changed event to the change set. Modify - modifies a changed event already in the change set. Reset - resets a changed event to its base case values.
- Class - changes event data parameters for a specified grouping of events. Class changes are identified by a "C" to the left of the affected events.

Fault Tree, Event Tree, Sequence, End State

The Fault Tree, Sequence, and End State options are:

- Solve/Gather - generates/gathers cutsets for a system or sequence/end state based on cutset generation cutoff values.
- Quantify - calculates a new minimum cutset upper bound for system, sequence, or end state cutsets using current data values (event change set and alternate cutsets).
- Uncertainty - analyzes the uncertainty of a system, sequence, or end state using either the Monte Carlo or Latin Hypercube simulation technique.
- CutSets - works with cutsets.
- Update - updates alternate cutsets based on cutset generation cutoff values using:
    - Prune - eliminates cutsets from a system or sequence that contains events that conflict in some way with one another.
    - Recover - modifies system or sequence cutsets by applying recovery action (through recovery rules).
    - Partition - defines the end states associated with each cutset in a sequence.
    - Edit- creates and edits base or alternate case cutsets.
- Display - displays the results of system, sequence, or end state analysis using:
    - CutSets - displays the cutsets generated based on current data.
    - Importance - calculates and displays three importance measures, Fussell-Vesely, Risk Reduction Ratio, Risk Increase Ratio, for each event in the system, sequence, or end state.

Uncertainty - displays distribution and confidence limits of a system, sequence, or end state for both base and current data values.
- Time Dependent - runs a time dependent analysis of system or sequence.

Fault Tree (System)
Choose this to build, edit and analyze fault tree models. The System List dialog appears listing all systems in the current family. A pop-up menu provides various functions, depending on the selected systems. With no systems selected, the pop-up menu has the following options.
- Edit Graphics - constructs the actual fault tree diagram using the graphical Fault Tree Editor (Figure 3.6-1).
- P&ID Graphics - edits the Piping & Instrumentation Diagram
- Edit Logic - modifies the logic of a system or subsystem.

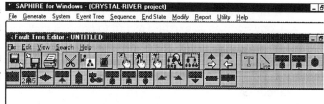

Fig. 3.6-1 The Saphire Main Menu and the Fault Tree Editor Menu. Symbols are selected and pasted into the working space

Event Tree
Choose this to build and edit-event trees. The Event Tree List dialog appears listing all event trees and subtrees in the current family. A pop-up menu provides the following options:

- Edit Graphics - constructs and edits event trees using the graphical Event Tree Editor.
- Edit Rules - defines and edits event tree linkage rules using the Linkage Rules Editor. A linkage rule is a special case, substitution or exception to the normal sequence generation.
- Edit Logic - changes the logic for an event tree using the Event Tree Logic Editor.
- Link Trees - generates sequence logic by using the event tree logic and applying the linkage rules.
- Sequence Logic - edits sequence names and end states and assigns frequencies after an event tree has been created.

Sequence
Recalculates sequence values after event failure data and/or cutsets have been modified. After choosing Sequence from the menu, the Sequence dialog appears listing all sequences in the current family. The pop-up menu lists: Solve, Quantify, Uncertainty, CutSets, Display, and Time Dependent.

End State
Provides the means to recalculate end state values after modifying event failure data and/or cutsets. After choosing End State from the menu, the End State List appears listing all end states in

the current family. A pop-up menu provides the selection of: Gather, Quantify, Uncertainty, CutSets, and Display.

Modify Database

If Modify is chosen from the menu, a drop-down menu with the available data types appears: Family, Event Trees, Systems, End States, Basic Events, Attributes, Analysis Types, Gates, Histograms, P&IDs, Change Sets, and Flag Sets. After selecting an option, a dialog containing a list of all records for the selected data type appears. The functions: Add, Copy, Modify, and Delete may be selected from another pop-up menu.

- Add - adds a new record to the database.
- Copy - copies an existing record to create a new record in the database.
- Modify - edits the selected record,
- Delete - deletes the selected record(s) from the database.
- Text - provides viewing and editing of the descriptive text associated with the specified record.
- Base Update - overwrites all base case (original) data with the alternate (current) case for the selected record(s) and analysis type.
- Clear Alternate - clears all alternate case information for the selected record(s) and analysis type.
- Sequences - modifies the sequences associated with the selected event tree.
- Transformations - replaces or adds events inside the fault tree logic.
- Remove Unused - deletes all unused basic events or gates.
- Flags - adds basic event changes to the selected flag set.

Reports

Provides information about the current family. Various reports can be generated on the following data types: Family, Attributes, Basic Events, System, Event Tree, End State, Sequence, Gate, Histogram and User Information. From the Reports Menu dialog, a specific type of report may be chosen by a single radio button for each group (data type, report type and sub type). Then choose the Process button, select data, and output destination. Enter the report title and select the type of output using Report Options radio buttons:

- File - saves the report to a file.
- Printer - prints the report.
- Print Preview - shows the report in the Report Viewer window.

Utility Options

Performs routine functions required by SAPHIRE:

- Define Constants - specify file locations, archive information, uncertainty analysis settings, cutset generation, transformations, quantification constants, and set default values for the graphical editors.
- Load and Extract - converts information to/from the generic output found in the Models and Results Database (MAR-D) file format to/from the SAPHIRE database file.
- Archive Family - compresses or expands data files for a family.
- Recover Database - restructures and re-indexes the data. The rebuild process recovers all key indexes and cross references.

3.7.2 PSAPACK

The integrated fault tree software package PSAPACK (probabilistic safety analysis package) uses a personal computer for preparing Level 1 PSAs and updating and recalculating PSAs. It was developed by the International Atomic Energy Agency (IAEA) in cooperation with member states, beginning in 1987.

The modular design integrates, as far as possible, existing software from member states, to permit the free distribution of the package. Version 1.0 of PSAPACK was used in a IAEA international training course on PSA in safety decisions (1988). Since then, PSAPACK evolved in response to users' requirements, and has been widely distributed.

PSAPACK 4.2 is an integrated fault/event tree package with capabilities for easily examining results and performing recalculations. There are two levels of processing in the package:

Level A for Operational Safety Management updates the plant status for specific components that are out of service and forecasts the effect on core melt frequency using minimal cutsets generated in Level B. Code features include:

- User friendly definition of plant status using component attributes such as system and room location,
- Different types of importance measures to represent the results, and
- Display of results in graphic and text format.

Level B for Level 1 PSA main features include:

- Fault tree editor (graphic and text);
- Fault tree analyzer (SETS code);
- Event tree construction and analysis (graphic and text);
- Accident sequence Boolean reduction and quantification;
- Reliability database (generic data, unavailability evaluation, component attributes); and
- Utilities (help functions, on-line code manual and PSA procedures guide).

The most important modules in the package are:

RDB, the reliability database module, creates a user-defined database or retrieves data from the IAEA generic reliability database. The design facilitates RDB development for: components, human actions, initiating events and the attributes of components. Component unavailabilities can be calculated from the database of reliability parameters using 10 types of predefined reliability models.

FTED, fault tree editor module, edits and manages fault trees in text format using a full screen, user-friendly, menu driven editor. FTED automatically verifies the FT logic and determines if all components are defined in the RDB. It transfers the tree structure from graphics to text format.

FTA, fault tree analyzer module, uses SETS and FTAP to reduce fault trees and generate minimal cutsets for storage as minimal cutset libraries. Cutset control uses truncation by probability or order. The user chooses the codes according to the personal computer's capabilities. The FTA module uses OR, AND, N/M, switch gates and supercomponents.

VIEW is the quantification module. All minimal cutsets are stored in the specific libraries for the fault trees, supercomponents and sequences. VIEW recalculates the point estimates. It computes and displays: the Fussel-Vesely importance, risk increase and risk reduction measures.

ETED, event tree editor module, edits and manages event trees in textual format by inputting 0s, and 1s coding in graphical format using the FEP module. Event trees can be changed from graphics to text or vice versa.

BOOL, boolean reduction module, generates minimal cutsets by combining minimal cutsets, generated by the FTA module, with a Boolean algorithm. BOOL automatically handles of success paths and truncates by probability or order.

FEP, fault tree, event tree and piping & instrumentation diagram (P&ID) editor, accesses the package of graphical tools for risk assessment. These tools include the event tree, fault tree and P&ID graphical editors. The event tree and fault tree editors are in PSAPACK; the P&ID editor is in FEP. The event tree editor is used for graphical construction and modification of event trees. The fault tree editor does the same for fault trees; the P&ID editor does the same for construction and modification of P&IDs used in a PSA.

STATUS reevaluates plant status for a specific plant configuration. The user can update the plant status according to the components that are out-of- service. It can change the data by selecting basic events according to systems, initiating events, event codes, or attributes.

CDFG, core damage frequency graphics, depict the core damage frequency for various plant configurations at various times. PSAPACK 4.2 recalculates the plant core melt frequency using the list of cutsets previously generated in Level B. Therefore, the impact of truncated cutsets cannot be determined. Moreover, special care must be taken in disabling components or systems or recalculation will erroneously not recognize that some sequences are no longer possible.

PSAPACK is free from the International Atomic Energy Agency, P.O. Box 100, A-1400 Vienna, Austria, attention of Luis Lederman. IAEA's Internet address is www.iaea.or.at/.

3.7.3 RISKMAN

RISKMAN[v] is an integrated Microsoft Windows™, personal computer software system for performing quantitative risk analysis. Used for PSAs for aerospace, nuclear power, and chemical processes, it has five main modules: Data Analysis, Systems Analysis, External Events Analysis, Event Tree Analysis, and Important Sequences. There are also modules for software system maintenance, backup, restoration, software updates, printer font, and page control. PLG has also integrated the fault tree programs: CAFTA, SETS, NRCCUT, and IRRAS into RISKMAN.

RISKMAN embodies PLG's scenario-based, engineering-approach displays and calculates: event sequence diagrams, fault trees, and event trees. It allows for system-specific data updates; and accounts for dependencies between systems.

The Event Tree Module can solve trees with up to 500 sequences with viewing, printing, and display of branch point probabilities, importance ranking and scenario frequency rankings. RISKMAN 3 does multi-branching. It does not link fault tree logic in solving sequences but uses logic rules for assigning branch point values. For the sequences, it displays the systems, operator actions, key dependencies cutsets, and basic event importances. The System Analysis module automatically adds user-specific common cause events to fault trees, performs uncertainty analysis on the system level, and builds a master frequency file for providing branching fractions for the event tree code.

The Systems Module constructs and displays fault trees using EASYFLOW which are read automatically to generate minimal cutsets that can be transferred, for solution, to SETS, CAFTA, or IRRAS and then transferred to RISKMAN for point estimates and uncertainty analysis using Monte Carlo simulations or Latin hypercube. Uncertainty analysis is performed on the systems level using a probability quantification model and using Monte Carlo simulations from unavailability distributions.

The Important Sequence Model module does sensitivity studies and importance rankings for about a thousand highest frequency sequences. The analyst zooms to the most frequent plant damage category, to the most frequent sequences in that category, to the most important top event, to the most important split fraction, and to the most important cutsets. If sensitivity analysis is needed on the model as a whole, a menu option, "CLONE a Model," makes a copy of the model, changes are made, and results compared.

RISKMAN includes extensive error checking and configuration management for quality control. Key module elements and intermediate results are time and date stamped; keystrokes of modifications leave an audit trail.

---

[v] RISKMAN is available from PLG Inc., 2260 University Drive, Newport Beach, CA 92660.

3.7.4 R&R Workstation[w]

The confluence of sharply rising Operations and Maintenance (O&M) costs, NRC requested Individual Plant Examinations (IPEs) and increased personal computer capabilities gave rise to the R&R Workstation. Its uses and maintains-current PSA models and databases for individual plants to perform O&M planning and scheduling, and uses the PSA in IPE models to identify plant design, procedure and operational vulnerabilities. The Risk and Reliability Workstation Alliance was organized by EPRI to support the R&R Workshop in order to achieve O&M cost reduction, plant productivity and safety enhancement through risk-based, user-friendly, windowed software tools (Table 3.6-8). The Alliance, initiated in 1992, includes 25 U.S. utilities and four international partners from Spain, France, Korea, and Mexico. SAIC is the prime contractor for the R&R Workstation, with participation of five other PSA vendors.

Software available for the R&R Workstation:

- Editors
    - CAFTA for Windows (Fault Tree Analysis)
    - ETA-II (Event Tree Analysis)
    - RMQS (Risk Integration)
- Analysis Tools
    - GTPROB (Exact Probability Calculations)
    - UNCERT (Uncertainty Propagation)
    - PRAQUANT (Fault Tree Linking)
    - SS-INT (Support State Analysis)
    - BROWSER (Fault Tree Inspection)
    - ONE4ALL (Integrate PSA into Single Top)
    - RECOVER (Automated Rule-Based Cutset Recovery)
    - EQUIMP (Generate Importances of Grouped Events)
    - RDUPDATE (Update PSA Databases)
    - FTCOMP (Compare Fault Trees)
- O&M Support
    - EOOS Monitor (Online Risk Monitor)
    - Rapid/Tag (Automated Tagging System)
    - PEAT/CARP (Bayesian Updates, Trending, etc.)
    - PRIOR (Combinatory Ranking of Importance)
- Fire Risk Analysis
    - FIVE (Fire Induced Vulnerability Evaluation Methodology)
    - FRANC (Fire PSA Methodology)
- Others

---

[w] For more information contact: Frank J. Rahn, Electric Power Research Institute, 3412 Hillview Ave., Palo Alto, CA 94304, (415) 855-2037.

Computer Codes for System Reliability Analysis

Table 3.6-9 Applications of the R&R Workstation

Application	Software
Online Maintenance/Maintenance Rule	The Equipment Out-of-Service (EOOS) provides quantitative and qualitative status information that meets the needs of: operators, schedulers, and system engineers
Fire Protection	FIVE and FRANC support quantitative and qualitative fire risk assessment. FRANC also supports the analysis of alternative safe shutdown paths during an Appendix R review. System train availability is exhibited on a system status panel. Using this a strategy can be defined to upgrade Thermo-Lag and identify risk significant areas.
Risk-Based Regulation	EQUIMP calculates component importance measures based on groups of basic events. RDUPDATE performs sensitivity on entire classes of events. These and other tools produce results to support a risk-based argument for code relief.
Technical Specification Optimization	EOOS can be adapted to show changes in tech spec status, "What-if analyses," show the effects on plant risk of removing trains of equipment from service.
Issue Resolution, JCOs, Plant Modifications	EOOS performs sensitivity analyses to determine the significance of, e.g., the removal of offsite power or decay heat removal equipment.
PSA Model Development	Tools in the CAFTA Workstation include an event tree developer, fault tree editor, quantification tools, cutset editor, and automated sequence editor.

- RBDA (Reliability Block Diagram Analysis)

3.7.5 WinNUPRA, NUCAP+, and SAFETY MONITOR

WinNUPRA

NUS/SCIENTECH's WinNUPRA is a Windows version of NUPRA that:

- Builds, solves, plots and manipulates all Level 1 tasks;
- Supports Living PSAs;
- Solves fault trees linked into event trees;
- Merges system and functional equations into accident sequence cutsets;
- Performs importance, sensitivity, and uncertainty calculations;
- Analyzes shutdown and full-power operations for internal and external events;
- Evaluates the risk of current or hypothetical configurations;
- Analyzes up to 5,000 gates;
- Performs a batch solution of PSA functional equations, sequence equations and point estimates using the "Big Red Button" providing QA by recording analysis steps,
- Performs spatial event analysis using subset data files associated with equipment location in the plant;

- Models plant normal, off-normal and shutdown states.

NUCAP+

NUCAP+ prepares Level 2 PSAs for nuclear power plants by organizing and automating the numerical, graphical, and results aspects of Level 2 analytical work using Level 1 accident sequences and frequencies data. Containment degradation is modeled with a containment event tree (CET); results may be extended to Level 3. It features Boolean logic for sorting and binning operations; data must be provided by the analyst. NUCAP+ aggregates Level 1 core melt sequences using a grouping logic diagram with a CET accident model. Each path through the CET is calculated to determine the endpoint probabilities. The end points are grouped by source term categories; the frequencies are the sum of the products of the CET sequence probabilities and the Level 1 plant damage state frequencies. These source term frequencies are measures of risk.

A CET is developed for each plant damage state to describes the possible ways whereby a severe accident can proceed in the containment. Each heading in the tree represents an event in the accident progression that significantly affects subsequent events and the characteristics of the fission product source terms. Each CET event is associated with a "decomposition event tree" (DET) which is a secondary event tree related to a particular CET event to provide details and dependencies. DETs indicate the subjective "degree-of-belief" to turn off and on branches according to the plant's condition.

Safety Monitor

Safety Monitor™ is an interactive computer program (Stamm, 1996) that performs real-time assessments of configuration-specific plant accident risk. Originally used at Southern California Edison's San Onofre station in 1994, further development was sponsored by three nuclear utilities and EPRI, to include shutdown operation and expanded user features. It will be enhanced to calculate large early release frequency (a Level 2 risk).

Compatible with the R&R Workstation, it assesses plant accident risk by calculating the plant's PRA model, using the current plant configuration, with high-speed algorithms and logic model optimization. It can accommodate "swing" components and alternate train alignments. While intended primarily for operating personnel, experience showed that maintenance personnel are the primary users for preparing long-term maintenance schedules and evaluating the risk of maintenance strategies. The Safety Monitor includes features that address requirements of the NRC's Maintenance Rule[x] including the effects of maintenance on risk and the collecting of component-level availability data. Other features are the tabulation of train and system unavailability, tracking component unavailability against preestablished criteria, and additional risk profile reports.

The main screen displays a 90 days history of the plant's risk profile in terms of: CDF, release, or boiling. Movement of a slider positions the time interval that is desired. It has two

---

[x] 10CFR50.65, the "Maintenance Rule" requires nuclear plant licensees in the U.S. to monitor the condition of structures, systems, or components (SSC) against licensee goals to assure the SSC are fulfilling their intended function.

"thermometers" that display the real-time risk level for CDF, or release using redlines for warning levels. Plant personnel may use buttons to view: a list of currently important components, a ranked list of the risk improvement by restoring of out-of-service components, and quantitative contingency advice on out-of-service components. Changes in plant configuration are entered from other screens by: importing a schedule file, data file update using automated communication features, or manually specifying changes such as component removal or restoration.

3.8 Summary

The very practical guidance from the PSM rule began this chapter by discussing process safety information in flow diagrams and block diagrams. The strong kinship between chemical and nuclear engineering results in overlap in PSA techniques. After all, it is no accident that a reactor is called a reactor. Methods of process hazard analysis are similar; nuclear hazard analysis tends to be more formal and mathematical but this depends on system complexity and practical experience. The qualitative methods presented are: checklist, what-if, what-if/checklist, HAZOP, FMEA; the quantitative methods are: parts count, FMEA/FMECA, RBD, fault tree, event tree, and miscellaneous methods. Common causes of failure are discussed - how to analyze them and how to quantify them. Computer codes aids for reliability analysis (fault tree analysis codes) were discussed from the early codes to the code suites that perform the modeling by providing a platform for the analyst to prepare the fault tree, provide data for quantification, analyze the tree and perform importance and sensitivity analysis.

3.9 Problems

1a. Transform the event tree of Figure 3.4.5-2 into several fault trees for each plant damage state noting that end points 1, 3, and 6; 2, 4 and 7; 5, 8, 9, and 10 lead to the same damage state.

1b. Find the cutsets of each fault tree.

1c. Assuming the following failure rates: PB = 3.1E-2, RS = 1.0E-5, CO1 = 2.3E-2, ECI = 3.2E-3, CO2 = 0.1, ECR = 2.5E-2, calculate the Birnbaum, Inspection and Risk Achievement Worth Ratio for each system including the initiator.

2a. Construct RBD of the simple injection system shown in Figure 3.4.4-6.

2b. Discuss the relative merits of these models and the fault tree model of the same system.

2c. Assuming the following failure probabilities: pump = 0.01, SIS = 1.5E-3, valves A through E = 4.3E-3, evaluate each of your models to determine the probability of this system functioning on demand. Calculate the Birnbaum, Inspection and Risk Achievement Worth Ratio for each of the components using whichever model you prefer. (Once you get the cutsets, it does not matter which model was used as long as it is a correct system model.)

2d. Rank order the systems for-each importance measure.

2e. Why doesn't each measure result in the same ordering? Discuss which is the "best measure.

3a. In the ZIP, members of a redundant train are treated as identical and for this reason correlated. In such case the means and variances do not propagate as independent distributions and recourse is made to Equations 2.7-24 and 2.7-26. For the "V" sequence (release bypassing containment), the mean failure rate of an MOV if 1.4E-8/hr with a variance of 5.3E-15/hr*hr. For two doubly redundant valves, what is the mean failure rate and variance?

3b. What is the beta factor under this assumption of the valves being identical?

3c. Discuss your opinion of the validity of the identical assumption and suggest experimental tests of the validity of such strong coupling.

4a. Prepare a FMECA for Valve A (a motor-operated valve - MOV) in Figure 3.4.4-6.

4b. How do you handle the redundancy between Valve B and Valve C in a FMECA? Note this problem requires some understanding of MOVS and how they can fail. Most MOVs in the U.S. are made by the Limitorque Corp.

5a. Prepare a fault tree of a dual emergency power system, considering that the compressed air used for starting the diesels is a separate system which may be a common-cause failure to start the diesels.

5b. Using λ_1 and λ_2 for the failure rates and μ_1 and μ_2 for the repair rates, what is the frequency of both systems working? One or the other? None?

5c. Compare these results with the Markov model prepared in Section 2.5.4. Under what conditions do the two methods agree?

6. The "gedanken" experiment in Section 3.4.4.2 results in equation 3.4.4-2 suggesting the operations of summation and multiplication in the algebraic sense - not as probabilities. Since this is a simulation why are not the results correct until given the probability interpretation? Hint: refer to the Venn Diagram discussion (Section 2.2).

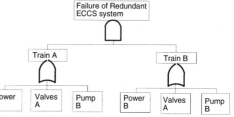

Fig. 3.7-1 Fault tree of a doubly redundant ECCS

7. Taking a safety subsystem out for test or repair during plant operation can be done only if the system of which it is a part is redundant. Figure 3.7-l shows the system as a fault tree whose risk increases during AOT or ST because if either train A or B is out of service, the top gate no longer multiplies two small probabilities (e.g., 0.01 * 0.01 = 0.0001); but the probability of system failure is that of one train alone (e.g., 1*0.01 = 0.01). This figure also shows that if Power A and Power B are from the same bus, then the failure of power in that bus will bring down the redundant system with a probability that is the probability of this bus failing.

The probability of the redundant system failing is the product of the single system time-dependent unavailability which includes approximately linearity between test and repair. There are two ways of performing the surveillance tests on these two systems: sequential and staggered testing.

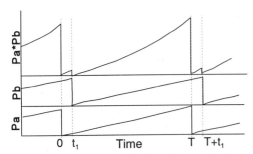

Fig. 3.7-2 Time-sequential testing. A second train is tested soon after a previous test. The combined probability rises parabolically with a small rise between tests

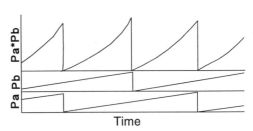

Fig. 3.7-3 Staggered testing. Train B is tested halfway through the Train A test interval

Sequential Testing - In this procedure, when a test of one train is completed and restored, the other train is tested. This is illustrated in Figure 3.7-2 with the combined effects presented as the product of the train probabilities.

Staggered Testing - Another surveillance test strategy is staggered testing in which the second test is performed halfway through the Train A test interval. This is illustrated in Figure 3.7-3. Because of the parabolic nature of the system failure rate, staggered testing considerably reduces both the average probability of system failure and the highest probability.

7a. Develop a relationship that shows the improvement that may be achieved by staggered testing.

7b. Assume that the failure rate of system A or B is 0.1/yr, test interval, T, is monthly and t_1 is one day for sequential testing, what is the numerical ratio of the risks of the two schemes?

Chapter 4

Failure Rates, Incidents and Human Factors Data

Systems analyses are like formulas, they have little usefulness until the variables are assigned probabilistic numbers from nuclear or chemical data bases. These data concern the probability of failing: vessels, pipes, valves, instruments and controls. The primary difference between chemical and nuclear data is that the former may operate in a more chemically active environment, while the later operate in radiation. This chapter addresses both, but most of the data were gathered for nuclear systems. It covers: 1) failure rate databases, 2) incident databases, 3) how to prepare failure rates from incidents, and 4) human factors for nuclear and chemical analyses.

4.1 Databases

4.1.1 Background

With the Industrial Revolution, life became more complex but it was not until World War II that reliability engineering was needed to keep the complex airplanes, tanks, vehicles and ships operating. Of particular concern was the reliability of radar. Prior to this time equipment was known qualitatively to be reliable or unreliable. To quantify reliability requires collecting statistics on part failures in order to calculate the mean time to failure and the mean time to repair. Since then, NASA and the military has included reliability specifications in procurements thereby sustaining the collection and evaluation of data build statistical accuracy although it adds to the cost.

Table 4.1-1 overviews U.S. reliability databases. Initially, FARADA was the program that collected reliability data from their suppliers in the 1960s for NASA and the military services. It is now called GIDEP (Government Industry Data Exchange Program) operated by the U.S. Navy at China Lake, California. Non-suppliers (class B) may voluntarily share the body of data by participating and contributing data to the program.

4.1.2 Some Reliability Data Compilations

BNLDATA

The distribution disk with this book includes a folder entitled "BNLDATA" which contains the file "bnlgener.xls," a spreadsheet in EXCEL format. It is a collection of failure rate data drawn from many sources. The file "refern.txt" contains the references to the table's data. There are 1,311 data entries; many cite different estimates by different organizations for the failure rate of the same

Failure rates, Incidents, and Human Factors Data

Table 4.1-1 Partial History of Reliability Data Collection			
Time	Industry	Data type/use	Progress
1900-1930	Auto	Maintenance frequency	Systematic collection to get the frequency
1930-1940	Commercial air	Engine failures vs number of engines	Collection to address the tradeoff
1940-1950	Military	Selected part failure frequency to decrease "weak link" problems	Collection for system analysis
1950-1960	Military/space (Titan)	Martin reliability data generic part failure rates, failure modes in lifetimes (1959)	Exponential rates, modes correlated to type, and use of 1E-6/hr units
1960-1965	Military space	Mil HDBK 217-1961 US Navy	Extensive electronics data
		RADC Notebook 1961 USAF	Surveys and rates tied to failure mode
		MRF failure rates 1962 (ARINC, COLLINS, Boeing, etc.)	Creation of comprehensive data collection systems 66-1, TAMS, 3M
1965-1970	Civil space	NASA and contractor work related to Apollo development and unmanned S/C	PRACA detailed failure mode investigation, and Alert System, GIDEP
	Commercial	Commercial carrier failure rate data and alert based on 2nd-order statistics	Tracking of reliability performance measures in operation and directly relating performance to relaxed regulatory requirements MSG
1965-1975	Commercial Power	Failure frequency collection	Systematic data collection from nuclear plants reports
		Reporting for Reg. Guide 1.16 (later LER)	Comprehensive application of hybrid Delphi method to database development
		Human error	Comprehensive component failure rate development of data collection approach
		Unit performance reporting	Comprehensive user run component failure data systems in United States
1975-1982	Commercial Power	Safety equipment failure rate calculation from plant reports - NPRD	Systematic data collection from nuclear plants reports
		Failure rates for generic E&E equipment - IEEE 500 (1977)	Comprehensive application of hybrid Delphi method to database development
		Pump, valve, diesel failure rates from plant reports - IPRD, EPRI	Comprehensive component failure data collection
		Utility data systems	Comprehensive user component failure data systems in U.S. and Europe
1982-1997	Commercial Nuclear Power	NUREG/CR-4550	Data used for NUREG-1150 and most of the IPE
	Chemical Processing	CCPS, 1989	A collection of process and non-process data from various for use in chemical safety

type of component. These data include hardware used in the chemical and nuclear industries: sensors, valves, instruments, pumps, pipes, vessels, etc.

GIDEP

The GIDEP Reliability-maintainability Data Bank (RMDB) has failure rates, failure modes, replacement rates, mean time between failure (MTBF) and mean time to repair (MTTR) on components, equipment, subsystems and systems. The RMDB includes field experience data, laboratory accelerated life test data, reliability and maintainability demonstration test results. The

RMDB is supplemented by microfilmed reports containing failure rates/modes and analyzed data. Reports relating to the theory and procedural techniques to obtain and analyze reliability-maintainability data are cataloged, abstracted, microfilmed and computer listed in the R-M Data Summaries.

The RMDB consists of two mayor sections: computer data storage and retrieval system, and backup microfilm data bank file. The computer data storage and retrieval system is used for the standard reliability and maintainability data listing, special calculations, and searches. This data bank system has been established to facilitate remote terminal access compatible with the GIDEP remote terminal programs. The microfilm data bank file is used for storage and distribution of supplier's documents, failure analysis curves, description of methods used in the collection of the data, and additional background information too extensive to include in the computer data bank.

The RMDB may be queried directly by qualified participants. Data needed urgently may be obtained by an Urgent Data Request (UDR) sent to the GIDEP Operations Center.

The RMDB lists 1,106 major subject categories for which there are subcategories. The subjects relate to space, aviation and military equipment.

RADC

The Rome Air Development Command (RADC - Rome NY) provides the MIL HDBK 217 series of detailed electronics information. Early reports in this series provided failure rates for electronic components. The development of integrated circuits resulted in the approach of providing parameters for mathematical models of transistors and integrated circuits. RADC also publishes "Nonelectronic Parts Reliability Data" covering the failure rates of components ranging from batteries to valves.

Chemical Process Safety

CCPS, 1989b, Process Equipment Reliability Data (Table 4.1-1) is a compilation of chemical and nuclear data. It assesses failure rates for 75 types of chemical process equipment. A taxonomic classification is established and data such as the mean, median, upper and lower (95% and 5%) values, source of information, failure by time and failure by demands are presented.

Reactor Safety Study

Appendix III, of WASH-1400 presents a database from 52 references that were used in the study. It includes raw data, notes on test and maintenance time and frequency, human-reliability estimates, aircraft-crash probabilities, frequency of initiating events, and information on common-cause failures. Using this information, it assesses the range for each failure rate.

IEEE 500

IEEE Std 500-1984 (IEEE, 1984) contains failure rate and out of service, repair and restoration times for electrical electronic, and sensing component, and mechanical equipment. It is a considerable improvement over IEEE STD 500-1977. The reported values are the consensus of over 200 experts. Each expert submitted a low, recommended, and a high value for the failure rate under normal conditions and a maximum value that would be applicable under all conditions (including abnormal ones). The consensus is expressed as the geometric mean.

Data Summaries of Licensee Event Reports at U.S. Nuclear Power Plants

Incidents described in LERs (Regulatory Guide 1.16) can be analyzed to obtain failure rate data albeit of accuracy limited by the number of incidents and appropriateness in the compilation. Six reports summarizing different systems are presented in Table 4.1-2. Component failures are reported for individual plants, by reactor vendor, by failure mode, and for all plants considered together. This information provides failure rates, failures on demand, and some information on repair times. The estimates are based on population, demands, and exposure time. The statistical analysis includes estimated information together with actual plant data.

Table 4.1-2 Data Collections
1. Diesel Generators (NUREG/CR-1362)
2. Pumps (NUREG/CR-1201).
3. Valves (NUREG/CR-1363).
4. Selected Instrumentation and Control Components (NUREG/CR-1740; EG&G-EA-5388).
5. Primary Containment Penetrations (NUREG/CR-1730).
6. Control Rods and Drive Mechanisms (NUREG/CR-1331).

In-Plant Reliability Data System

IPRDS prepares nuclear plant data under the auspices of the ANSI/Failure and Incidents Reports Review (FIRR) Data Subcommittee. Data collection teams visit plants to extract data from in-plant maintenance records for a reliability database. Because of the amount of work involved, the scope is limited to a few plants. NUREG/CR-3154 assesses valve reliability for a BWR and PWR; NUREG/CR-2886 is a similar pump study. Repair time data are collected as a part of the IPRDS.

Nuclear Plant Reliability Data System (NPRDS), Institute of Nuclear Plant Operations

The NPRDS collects failure data on safety-related systems and components. At present, more than 60 plants are reporting data. The data are compiled and disseminated in periodic reports to the participants of the program and other potential users. In addition, special searches of the database may be requested by the participants and others, or the users can access the data through their computer terminals. Typical information provided by NPRDS is presented in Table 4.1-3. Contact INPO, 1100 Cir. 75 PKWY, Ste 1500, Atlanta, GA 30339, (404) 953-5443. The main disadvantage is the dependence of the NPRDS on voluntary reporting. As the value of the information is more widely recognized, greater participation may be achieved.

Table 4.1-3 NPDRS Contents
1. Plant operating mode (i.e., operating, standby, and shutdown),
2. Calculated in-service hours of the system,
3. Outage times,
4. Number of failures per million in-service hours,
5. Number of applicable tests,
6. Number of actuations for standby equipment, and
7. Component failure modes and effects.

National Electric Reliability Council (NARC)

On January 1, 1979, the Edison Electric Institute (EEI) transferred to NARC the responsibility for operating its equipment-availability data system--the prime utility-industry source for the collection, processing, analysis, and reporting of information on power plant outages and overall performance. The Unit Year Summary computer program produces a report for each individual unit, including statistics for the latest year and cumulative statistics for the life of the unit. In addition, the Equipment Availability Task Force produces an annual report on equipment availability for a 10-year period. Finally, the EEI has established a procedure for processing special requests for the analysis of reliability data.

NUCLARR

The Nuclear Computerized Library for Assessing Reactor Reliability is a PC-based risk analysis software package, data manual, and user's guide developed for the NRC to support safety analysis of nuclear power plant operations (Gertman, 1994). It guides the locating of data records, aggregates values, graphs data, and copies data to a file, or a report. Data references are provided to query the database and determine the applicability. The information includes performance shaping factors (PSFs): type of plant, description of the error, and documentation. Currently, the data base contains over 2,500 individual data points; half are human failure rates, the other half are failure rates for mechanical and electrical components found in nuclear power plants. The data base may become more and more useful as the data store size increases. Examples of data sources for human failure rates are current plant-specific PRAs, databases from individual plants, scientific reports, and simulator studies. Data from U.S. utilities, special PRAs such as those conducted for interfacing system loss-of-coolant accidents (ISLOCAs), and data obtained from German utilities are incorporated in the data bank. George Mason University devised an algorithm that allows human error probabilities (HEPs) identified in other industries, to be matched to the data categories and action verbs in the NUCLARR classification scheme.

Fig. 4.1-4 Sample of NUCLARR Component Data, Gilbert, 1990

Component	Failure* Mode	Error Rate	Upper Bound	Number of Records
Power electronics	FTO	4.5E-3 /d	2.5E-2 /d	48
Diesel generator	FTO	5.2E-3 /d	1.9E-2 /d	172
Pipe	LKG/RPTR	8.0E-8 /hr	4.4E-7 /hr	42
Channels	FTO	1.0E-5 /hr	3.6E-5 /hr	242
Check valves	FTO	1 AK-4 /d	1.8E-3 /d	71
		6.6E-7 /hr	3.7E-6 /hr	
Relief valves	FTO	1.6E-3 /d	1.4E-2 /d	80
		4.9E-8 /hr	2.8E-7 /hr	
Motor-driven pumps	FTO	4.1E-4 /d	3.7E-3 /d	235
		4.5E-6 /hr	7.2E-5 /hr	
Turbine-driven pumps	FTO	9.4E-3 /d	3.4E-2 /d	36
		2.4E-5 /hr	4.0E-4/hr	
Circuit breakers	FTO	3.5E-4 /d	1.9E-3 /d	19
Switches	FTO	7.3E-7 /d	5.4E-5 /d	6
		4.1 E-7/hr	3.6E-6 /hr	
Control rod drives	FTO	3.2E-5 /hr	1.0E-4 /hr	3
Tanks	EXT.LKG/ RPTR	4.2E-7 /hr	2.3E-6 /hr	2

* FTO- failure to operate, LKG/RPTR - leakage or rupture, EXT.LKG/RPTR - external leakage or rupture.

Failure rates, Incidents, and Human Factors Data

Table 4.1-5 NUREG 1150 Nuclear Plant Reliability data			
Item	PRA/other studies	Assessed mean	Error factor
Transient caused by loss of DC bus	5E-4 to 6E-2/y	5E-3/y	3
Transient caused by loss of AC bus	9E-4 to 6E-2/y	5E-3/y	3
Transient caused by loss of offsite power	0.1 to 0.3/y	0.1/y	3
Transient caused by other than loss of power conversion system	3.7 to 7.1/y	4.77/y BWR FW 0.56/y BWR no FW 6.85/y PWR	3
Transient caused by loss of power conversion system	1.8 to 5.2/y	1.56/y BWR 1.41/y PWR	3 3
Small small LOCA (loss of coolant accident) 50 to 500 gpm	2E-2/y BWR 2E-2/y PWR	3E-2/y BWR 2E-2/y PWR	3 3
Small LOCA	1.4E-3/y BWR 1E-3 to 9.8E-4/y PWR	3E-3/y BWR 1E-3/y PWR	3 3
Intermediate LOCA	6.7-4/y BWR 4E-3 to 9.8E-4/y PWR	3E-4/y BWR 1E-3/y PWR	3 3
Large LOCA	2.1-3/y BWR 1E-4/y PWR	1E-4/y BWR 5E-4/y PWR	3 3
Inadvertent Open Relief Valves	0.21/y BWR ~0/y PWR	1.4E-1/y BWR ~0/y PWR	3
Solenoid operated valve — fails to operate	3E-4 to 2E-2/D	1E-3/D	3
plugs	2E-5 to 1E-4/D	4E-5/D	3
unavailable - test and maintenance	6E-5 to 6E-3/D	2E-4/D	10
Hydraulic operated valve — fails to operate	3E-4 to 2E-2/D	1E-3/D	3
plugs	2E-5 to 1E-4/D	4E-5/D	3
unavailable - test and maintenance	6E-5 to 6E-3/D	2E-4/D	10
Explosive operated valve — fails to operate	1E-3 to 9E-3/D	3E-3/D	10
plugs	2E-5 to 1E-4/D	4E-5/D	3
unavailable - test and maintenance	6E-5 to 6E-3/D	2E-4/D	10
Manual operated valve — plugs	2E-5 to 1E-4/D	4E-5/D	3
unavailable - test and maintenance	6E-5 to 6E-3/D	2E-4/D	10
Check valve — fails to open	1E-4 to 6E-3/D	1E-4/D	3
fails to close		1E-3/D	3
Power operated relief valve & safety relief valve:			
demanded to open: Westinghouse	0.1 to 0.2/D	0.1/D	1
Babcock and Wilcox	0.01 to 0.02/D	0.1/D	1
Combustion Engineering	0.07 to 1.0/D	0.1 - 0.01/D SRV only	1
fails to close: General Electric		1E-5/D	3
fails to reclose: BWRs and PWRs	0.1 to 0.003	3E-2/D	10
Electric motor driven pumps:			
fails to start	5E-4 to 1E-2/D	3E-3/D	10
fails to run	1E-6 to 1E-3/hr	3E-5/hr	10
unavailable from test and maintenance	1E-4 to 1E-2/D	2E-3/D	10
Turbine driven pumps:			
fails to start	5E-3 to 9E-2/D	3E-3/D	10
fails to run	8E-6 to 1E-3/hr	5E-3/hr	10
unavailable from test and maintenance	6E-3 to 4E-2/D	1E-2/D	10

Table 4.1-5 NUREG 1150 Nuclear Plant Reliability Data (cont.)

Item	PRA/other studies	Assessed mean	Error factor
Diesel driven pumps:			
fails to start	4E-3 to 1E-2/D	1E-3/D	10
fails to run	2E-5 to 1E-3/hr	8E-4/hr	10
unavailable from test and maintenance		1E-2/D	30
Heat exchanger			
blockage		5.7E-6/hr	10
Rupture (leakage)		3E-6/hr	10
Emergency diesel Generator			
fails to start	8E-3 to 1E-1/D	3E-2/D	3
fails to run	2E-4 to 3E-3/hr	2E-3/hr	10
unavailable from test and maintenance	~0 to 4E-2/D	6E-3/D	10
Offsite power failure (not initiator)	1E-4 to 3.3E-3/D	2E-4/D	
Hardware failure probability			
battery	2E-5 to 1E-3/D	4E-4/D	3
bus	"	9E-5/D	5
charger	"	4E-4/D	3
inverter	"	4E-2/D	3
Test and maintenance unavailability			
battery	~0 to 1E-3/D	1E-3/D	10
bus		6E-5/D	10
charger		3E-4/D	10
inverter		1E-3/D	10
Battery depletion time			
Westinghouse and CE plants	2 to >12 hr	7 hr	
B&W plants	"	5 hr	
GE plants	"	7 hr	

NUREG 1150

NUREG-1150 and its companion document, NUREG/CR-4550, are considered to be the NRCs acme of PSAs for nuclear power plants. Many IPE analyses used this data, shown as Table 4.1-5. The sources for this table are drawn from the WASH-1400, RSSMAP, IREP, Zion, Limerick, Big Rock Point, NUREG/CR-1363, NUREG/CR-2989, NUREG/CR-3226, NUREG/CR-0666, and NUREG/ CR-1032.

Miscellaneous References

The frequency of anticipated transients is addressed by EPRI NP2330, 1982 to give information on the type and frequency of initiating events that lead to reactor scram.

Loss of offsite power at nuclear power plants is addressed in EPRI NP-2301, 1982 giving data on the frequency of offsite power loss and subsequent recovery at nuclear power plants. Data analysis includes point estimate frequency with confidence limits, assuming a constant rate of occurrence. Recovery time is analyzed with a lognormal distribution for the time to recover.

EPRI NP-2433 addresses diesel generator reliability at nuclear power plants. The sources include plant records, utility records, and LERs. The report gives frequency of failure to start, failure to continue running, and mean repair times.

Failure rates, Incidents, and Human Factors Data

Table 4.2-1 Internet Websites Relevant to Chemical Safety	
American Industrial Hygiene Association (AIHA)	http://www.aiha.org/
American Plastics Council	http://www.plasticsresource.com
American Chemical Society	http://www.acs.org
American Conference of Government Industrial Hygienist (ACGIH)	http://www.acgih.org
American Institute of Chemical Engineers (AIChE)	http://www.aiche.org/
Chemical Manufacturers Association	http://www.cmahq.com
American Society of Safety Engineers	http://www.asse.org
Royal Society of Chemistry	http://chemistry.rsc.org/rsc
Synthetic Organic Chemical Manufacturers Association (SOCMA)	http://www.socma.com
DOE Operating Experience Analysis and Feedback	http://dewey.tis.eh.doe.gov/web/oeaf
OSHWEB	http://turva.me.tut.fi:80/~oshweb/
US FDA/CFSAN, Chemistry Information via the Internet	http://vm.cfsan.fda.gov/~dms/chemist.html
Molecule of the Month	http://www.bris.ac.uk/Depts/Chemistry/MOTM/motm.htm
Strategic Environmental Health and Safey Management (SEM)	http://www.estd.battelle.org/sehsm/
Laboratory Chemical Safety Summaries	http://www.hhmi.org/science/labsafe/lcss
Operating Experience Analysis	http://dewey.tis.eh.doe.gov/web/oeaf/oe_analysis.html
Chemfinder	http://chemfinder.camsoft.com
What's New	http://dewey.tis.eh.doe.gov/web/chem_safety/whatsnew.html
Requirements and Guidelines	http://dewey.tis.eh.doe.gov/web/chem_safety/reqs_info.html
DOE Documents	http://dewey.tis.eh.doe.gov/web/chem_safety/doe_doc.html
Chemical Occurrences	http://www.dne.bnl.gov/etd/csc
Lessons Learned	http://dewey.tis.eh.doe.gov/web/chem_safety/lessons_learned.html
Human Factors	http://dewey.tis.eh.doe.gov/web/chem_safety/human_factors/index.html
Chemical Safety Networking	http://dewey.tis.eh.doe.gov/web/chem_safety/chem_saf_network.html
Molecule of the Month	http://dewey.tis.eh.doe.gov/web/chem_safety/mom.html
Risk Management	http://dewey.tis.eh.doe.gov/web/chem_safety/rmq/index.html
Chemical Safety Tools	http://dewey.tis.eh.doe.gov/web/chem_safety/chem_saf_tools.html
Chemical Information Links	http://dewey.tis.eh.doe.gov/web/chem_safety/other_links.html
Questions & Feedback	http://dewey.tis.eh.doe.gov/web/chem_safety/qandf.html

4.2 Incident Reports

Nuclear Power Licensee Event Reports

According to 10 CFR 50.73, the holder of an operating license for a nuclear power plant (the licensee) must submit an LER for a reportable event, within 30 days after discovery. An event is reportable regardless of the plant mode, power level, structure, system, or component that initiated the event. In addition the licensee must report the completion of any nuclear plant shutdown required by the plant's Technical Specifications; or any operation or condition prohibited by the plant's Technical Specifications, or any deviation from the plant's Technical Specifications. LERs are available on the Internet at http://www.nrc.gov/NRR/DAILY/97mmdddr.htm, where mm is the month in two digits and dd is the day in two digits. They may be obtained in paper copy from the NRC for years not covered by the Internet.

U.S. Department of Energy Chemical Safety Program

The DOE Office of Environmental Safety and Health distributes monthly summaries based on data retreived from the DOE Occurrence Reporting and Processing System (ORPS) to share chemical safety concerns throughout the industry to alert operators of similar processes. In addition quarterly and annual reviews feature lessons learned from ORPS by trending analyses of information from the monthly summaries. Analyses include evaluations of occurrences, causes, corrective actions, and lessons-learned to identify common or recurring chemical safety concerns. Occurrences with significant actual or potential consequences are identified and analyzed in depth to communicate information useful for resolving chemical safety problems. The trending analysis incorporated in the quarterly reviews provides the basis for the Chemical Hazard Events indicator used in the reports entitled "DOE Performance Indicators for Environment, Safety and Health."

The DOE Chemical Safety Internet site at "http://dewey.tis.eh.doe.gov/web/chem_safety/" has the following hypertext buttons: 1) What's New, 2) Requirements and Guidelines, 3) DOE Documents, 4) Chemical Occurrences 5) Lessons Learned, 6) Human Factors, 7) Chemical Safety Networking, 8) Molecule of the Month, 9) Risk Management Quarterly, 10) Chemical Safety Tools, 11) Other Chemical Information Links, 12) PRF Response Group, and 13) Questions & Feedback. Some Internet sites that are relevant to chemical safety are presented in Table 4.2-1; addresses related to the environment are given in Table 4.2-2.

An example of DOE chemical incidence reporting is:

Table 4.2-2 Some Environmental Internet Addresses

Environmental Compliance Assistance Center	http://www.hazmat.frcc.cccoes.edu
National Wildlife Federation Campus Ecology	http://www.nfw.org/nwf/campus
National Resources Defense Council	http://www.nrdc.org/nrdc
Sierra Club	http://www.nfw.sierraclub.org
Greenpeace	http://wwwgreenpeace.org
Green Money Online	http://www.greenmoney.com
John Denver's Windstar Foundation	http://www.nfw.wstar.org
Environmental Defense Fund	http://www.edf.org
Environlink Network	http://www.envirolink.org
The Earth Times	http://www.earthtimes.org
U.S. Environmental Protection Agency	http://www.epa.gov
The Environmental News Network	http://www.enn.com
Global Warming Org. part of National Consumer Coalition	http://www.globalwarming.org

Chemical Storage Tank Explosion
   (W): (RL--PHMC-PFP-1997-0023)
   On May 14, 1997, at Hanford, an explosion occurred at the Plutonium Reclamation Facility (PRF), part of the Plutonium Finishing Plant (PFP) in the 200 West Area. The explosion occurred in Room 40 where bulk chemicals were mixed to support the now-discontinued plutonium recovery process. Information from air samples inside the facility and air monitors in the surrounding area show no signs of a radioactive release. No employees were injured, although several (8-10) were sent to a local hospital and some complained of metallic tastes in their mouths. A Type B Investigation is underway at Hanford; updates/results of this investigation will be posted on the Chemical Safety Homepage and will be summarized in an upcoming Quarterly Review. Also note that ES&H Safety Alert 97-1 (DOE/EH-0554), "Chemical Explosion at Hanford was issued on May 22.

Investigators believe evaporation caused two substances inside the chemical tank to react and explode. Investigators believe the explosion was caused by a volatile mix of hydroxylamine nitrates and nitric acid. In 1993, a solution of nitric acid and hydroxylamine nitrates - heavily diluted by water - was put in the tank. The solution had a volume of 185 to 200 gallons. At that volume, the water kept the nitric acid and hydroxylamine nitrates in a safe, stable condition, but over time water evaporated out of the solution, and the vapor escaped through tiny openings in the tank. The volume of liquid shrank to less than 40 gallons - leading to much greater concentrations of hydroxylamine nitrates and nitric acid. The greater concentrations heated up the chemicals, causing them to react.

In an earlier, related occurrence involving the same chemicals, on December 28, 1996, at Savannah River (SR--WSRC-FCAN-1996-0030), a sump high-level alarm activated. Investigation determined that the source of liquid level increase was a warm canyon sump receipt vessel. Indications were that the vessel eructed resulting in some spillage of solution to the sump. Chemicals (hydroxylamine nitrate [HAN] and nitric acid) were added to the receipt tank in preparation of sump receipts to ensure criticality safety. Eructation of receipt tank contents occurred as a result of the auto-catalytic decomposition of HAN in the presence of strong nitric acid. Over a period of time, several sump receipts were made to the tank. The acidity and HAN concentrations in the sump receipt tank increased through the required chemical additions. At these conditions (including high temperature), the reaction initiated and proceeded to completion resulting in 250 gallons of solution being eructed to the sump.

4.3 Database Preparation

Commonly, there are components that are not in any database of failure rates, or the data do not apply for the environment or test and maintenance at your plant. In addition, site specific data may be needed for regulatory purposes or for making the plant run safer and better. For both cases there is a need for calculating failure rate data from incident data, and the mechanics of database preparation and processing.

4.3.1 How to Estimate Failure Rates

Practical probability is the limit of two ratios (Section 2.2). The numerator is the number of cases of failure of the type of interest (N); the denominator, the normalizing term is the time duration over which the failures occurred or the total number of challenges to the system. The former has the units of per time and may be larger than 1, hence it cannot be probability which must be less than 1. The latter is a dimensionless number that must be less than 1 and can be treated as probability.

The numerator is a random normally distributed variable whose precision may be estimated as $\sqrt{(N)}$; the percent of its error is $\sqrt{(N)}/N = 1/\sqrt{(N)}$. For example, if a certain type of component has had 100 failures, there is a 10% error in the estimated failure rate if there is no uncertainty in the denominator. Estimating the error bounds by this method has two weaknesses: 1) the approximate mathematics, and the case of no failures, for which the estimated probability is zero which is absurd. A better way is to use the chi-squared estimator (equation 2.5.3.1) for failure per time or the F-number estimator (equation 2.5.3.2) for failure per demand. (See Lambda Chapter 12.).

4.3.2 Making a Database

The work of database preparation consists of: gathering the data, preparing and entering it into a computer program for data manipulation, assuring validity and removing redundant reporting. The data sources are plant records and operating procedures. Table 4.3-1 lists parameters, data required and possible sources of the data. How these sources can be used to extract needed information is briefly explained below.

4.3.3 Test Reports and Procedures

Test reports and procedures provide data on failures, demands, and operating time for periodically tested components. These reports for components or systems describe the test procedure and the checklist used for the test. For example, an emergency diesel-generator operating test procedure requires starting the diesel and running it - typically for an hour. The test report indicates any failure to meet requirements. Another example is a test of an emergency cooling system's response to a test signal's actuation of motor-operated valves and starting and running pumps to provide the emergency cooling.

The mode of failure in a test is examined carefully before the failure is included in the database. In the diesel-generator example, unsatisfactory performance may have been reported because of a trip on a low oil pressure signal, high oil temperature, or both. Because these trips are disabled by a loss of cooling signal, they should be disregarded in calculating the diesel reliability in a loss of cooling sequence. If the trip were bypassed the diesel would have failed, and it is counted as a failure. Similarly, the test would fail from lack of compressed air even though it was not the fault of the diesel.

If test records are not readily available, test procedures specify the number of test demands or operating test durations. For this, the number of demands or the operating time of a single test is multiplied by the frequency of the test and time being considered. The validity of this, assumes that tests were conducted at the prescribed frequency. Some tests may be more frequent than specified; plant personnel interviews will reveal the true state.

In doing this, a failure count obtained from different sources may not indicate clearly which failures occurred during a test. This over-counting results in a conservative estimate. In addition to test demands, the number of operational demands must be obtained which may be difficult to extract from available data sources.

Table 4.3-1 Sources of Data

Parameter	Data requirements	Potential sources
1. Probability of failure on demand	a) Number of failures	Periodic test reports, maintenance reports,
	b) Number of demands	Control room log, periodic test reports, periodic test procedures, operating procedures, controlroom log
2. Standby failure data	a) Number of failures	See 1a above
	b) Time in standby	Control room log
3. Operating failure rate	a) Number of failures	See 1a above
	b) Time in operation	Control room log, periodic test reports, periodic test procedures
4. Repair-time distribution parameters	Repair times	Maintenance reports, control room log
5. Unavailability due to maintenance and testing	Frequency and length of test and maintenance	Maintenance reports, control room log, periodic test procedures
6. Recovery	Length of time to recover	Maintenance reports, control-room log
7. Human errors	Number of errors	Maintenance reports, control room log, periodic test procedures, operating procedures

Failure rates, Incidents, and Human Factors Data

4.3.4 Maintenance Reports

Reports of maintenance on components are potential sources of data on failures, repair times and unavailability due to maintenance. Table 4.3-2 shows their typical contents. The report may indicate maintenance because of failure to operate partially or completely. Such an event is added to the count of component failures.

Table 4.3-2 Contents of Maintenance Reports
Plant identification number for the component undergoing maintenance and a description of the component
Description of the reason for maintenance
Description of the work performed
Indication of the time required for the work or the duration of the component's unavailability

Information may be given about the failure mode and time spent in repair after the failure was discovered. Careful interpretation is needed, because the repair time may be a fraction of the time that the component was unavailable between failure detection and the repair completion. Additionally, the repair time is in person-hours, which means that the actual repair time depends on the size of the work crew; treating it as a one person crew leads to a conservative estimate of repair time. The out-of-service time for the component is the interval from the date the failure was discovered until the date of restoration to service.

Preventive maintenance reports provide information on a component's unavailability. They show the dates of removal from and restoration to service. The frequency can be found from the number of preventive-maintenance reports in a time interval.

Unfortunately, maintenance reports do not always present all the information indicated in Table 4.3-2. Descriptions of component unavailability or work performed may be unclear, requiring engineering judgment regarding whether a component was made unavailable by maintenance or maintenance was required because of component failure.

4.3.5 Operating Procedures

Operating procedures can provide the estimated demands on components in normal operation in addition to the demands from periodic testing. The number of test demands is estimated by multiplying the number of demands on a component in performing a procedure by the number of times the procedure is executed during a time of interest. The number of times the procedure was performed during plant startup or shutdown is readily obtainable, but procedures during operation only are available from the control room log.

4.3.6 Control Room Log

Gaps in a component's data compiled from test and maintenance records can be filled from the control room log of events at the plant which records the demands made on components from which operating-time data may be found. The log notes the use of various operating procedures from which demand information is obtained. The log records periods when certain components and systems are out of service, and is often more accurate than maintenance reports. A problem with obtaining component data from the controlroom log is the events are chronological, without system, or event classification which requires much searching to find entries for the database.

4.3.7 Information Flow in Plant Data Collecting

Figure 4.3-1 shows the types of information flow for processing plant. In addition to counting the number of failures for each type of component, the number of components of each type in service at the plant is needed as well as time in service, different operational modes, and the demands.

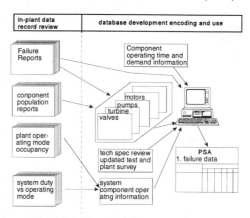

Fig. 4.3-1 Plant Data Processing for PSA

4.3.8 Data Sources Used in Past PSAs

Table 4.3-3 from Joksimovich et al. (1983) presents the data sources used in preparing the Big Rock Point and Zion nuclear power plant PSAs. It is seen that both PSAs used plant records extensively.

4.4 Human Reliability Analysis

Humans control all chemical and nuclear processes, and to some extent all accidents result from human error, if not directly in the accident then in the process design and in the process' inadequate design to prevent human error. Some automatic systems such used in nuclear power reactors because the response time required is too short for human decisions. Even in these, human error can contribute to failure by inhibiting the systems.

4.4.1 Human Error in System Accidents

Human error may be the major cause of death, injury and property loss in the CPI. It impacts quality, production, and profitability. Garrison (1989), shows that human errors (on-site errors) account for $563 million, the second highest cause. If off-site errors are included (e.g., Flixborough which was an engineering error), human error dominates. Further analyses by Garrison for 1985-1990, shows human error in more than $2 billion of CPI property damage. Uehara and Hasegawa (1986), who studied fire accidents in the Japanese CPI during 1968 - 1980 found that 120 accidents (45%) were attributed to human error. This becomes 58%, if human error in improper design and materials are included. Table 4.4-1, from CCPS (1994), indicates the central importance of human error in CPI safety.

Table 4.3-3 Reliability Data Sources for Big Rock Point and Zion PSAs

Source	No. of failures	No. of demands or service time	Test interval	No. of mainte-nances	No. of dependent failures	Failure rate	Failures/ demand	Repair time
Big Rock Point								
WASH-1400						x	x	
IEEE-500						x	x	
NUREG/CR-1205						x	x	
NUREG/CR-1363						x	x	
Operating logs	counted	counted	counted	counted		calculated	calculated	calculated
Maintenance reports	counted			counted		calculated	calculated	counted
Test surveillance reports	counted	counted	counted				calculated	
Event reports	counted				x	calculated	calculated	
Technical Spec			x					x
ZION								
WASH-1400				x		x	x	review
NPRDS				x			x	review
NUREG/CR-1205 (pumps)	counted	estimated				x	x	
NUREG/CR-1362 (diesel generators)	counted	estimated				x	x	
NUREG/CR-1363 (valves)	counted	estimated				x	x	
IEEE-500						x	x	
Technical Specifications			x	estimated				estimate
Testing Procedures		calculat	counted			calculated	calculated	
Testing Records		calculat	counted			calculated	calculated	
LERs	counted				review	calculated	calculated	
Deviation Reports	counted				review	calculated	calculated	
Operating Records		calculat				calculated	calculated	
Out of service log	counted			counted		calculated	calculated	counted

In addition to these formal studies of human error in the CPI, almost all the major accident investigations in recent years, for example, Texas City, Piper Alpha, Phillips 66, Feyzin, Mexico City, have shown human error as a significant causal factors in design, operations, maintenance or the management of the process. Figures 4.4-1 and 4.4-2 show the effects of human error on nuclear plant operation.

Organizational Factors

Organizational factors create preconditions for errors. At the operational level, plant and corporate management inadvertently support conditions for errors. The safety culture of the

Human Reliability Analysis

organization influences safety. Attributing blame may result in lack of motivation, in coverup, and in deliberate unsafe behavior. Participation between management and the workforce impacts positively on the safety culture. Clear policies ensure good procedures; training reduces error likelihood.

Organizational and plant design policies, directed by senior management; plant and corporate management, and implemented by line management affect conditions that influence error. While suitable policies adopted by senior management may be frustrated if not supported by the line management.

At the next organizational level are factors directly causing error: 1) job characteristics such as complexity, time stress, noise, lighting, environment, or mental requirements, and 2) individual factors such as personality, and team performance. These, collectively, are called performance-influencing factors, or PIFs.

Next, product production activities include human interactions with hardware, physical operations such as opening and closing valves, charging reactors and carrying out repairs. These are labor intensive tasks such as batch processing, or in modern, highly automated plants, they are "cognitive" tasks such as problem solving, diagnosis, and decisionmaking in process and production optimization. In all facilities, humans are involved in test, maintenance and repair.

At the lowest level are defenses against foreseeable hazards. These include engineered system features such as emergency shutdown systems, relief valves, burst disks, valves or trips. In addition to hardware systems are human systems such as: 1) emergency response, 2) administrative controls, such as work permits and training for accident mitigation, and 3) operators acting as process control elements.

Table 4.4-1 Studies of Human Error in the CPI Magnitude of the Human Error Problem

Study	Results
Garrison (1989)	Human error accounted for $563 million of major chemical accidents up to 1984
Joshchek (1981)	80-90% of all accidents in the CPI due to human error
Rasmussen (1989)	Study of 190 accidents in CPI facility: •insufficient knowledge 32% •design errors 30% •procedure errors 23% •personnel errors 15%
Butikofer (1986)	Accidents in petrochemical industry: •equipment and design failures 41% •personnel and maintenance failures 41% •inadequate procedures 11% •inadequate inspection 5% •other 2%
Uehara and Hoosegow (1986)	Human error accounted for 58% of the fire accidents in refineries due to: •improper management 12% •improper design 12% •improper materials 10% •misoperation 11% •improper inspection 19% •improper repair 9% •other errors 27%
Oil Insurance Association Report on Boiler Safety (1971)	Human error accounted for 73% and 67% of total damage for boiler start-up and on-line explosions, respectively.

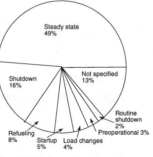

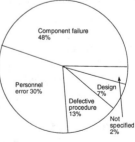

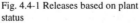

Fig. 4.4-1 Releases based on plant status

Fig. 4.4-2 Releases based on cause code

Modeling Humans

Human behavior is responsive and adaptive to what is happening. This is very difficult to model as a fault tree, or equivalent model. Humans are a programmable control element that is beyond the capability of system models. An additional problem occurs when trying to model human response to insufficient, contradictory, and ambiguous information in an accident. In such cases, the human performs extremely well in solving the risk-benefit problem. In cognitive tasks, human-reliability is a function of the time available, and depends on training and procedures. If the actions are determined and practiced, human action may be rapid and reliable (e.g., the decision time in baseball is a fraction of a second). In a PSA, human performance is modeled as occurring under normal operating conditions, or during or subsequent to accidents. Errors also may occur in maintenance, calibration, or testing.

Error Causation

Error conditions at various levels of the organization result from inappropriate policies, or ineffective implementation of good policies by line management. Poor practices may produce "latent failures" that are not actuated until in combination with other conditions or failure that produce an accident.

Accident defenses may be regarded as a series of barriers (engineered safety systems, safety procedures, emergency training, etc.). As barriers fail, incipient failures become real. Inappropriate management policies create inadequate PIFs, which give rise to opportunities for error, when initiated by local triggers or unusual conditions.

4.4.2 Lack of Human Error Considerations

Human performance problems are a threat to CPI safety, that has been neglected. Reasons for this neglect is the belief among engineers and managers that human error is inevitable and unpredictable. It is neither of these, but arises from failure of designs to encompass human weaknesses while using human strengths.

Systematic consideration of human error is neglected because of the belief that computerization of processes will make the human unnecessary. Experience shows numerous accidents in computer controlled plants. Human involvement in critical areas of maintenance and plant modification, continues even in the most automated processes.

Human error has been cited for deficiencies in the overall plant management. Accident attribution to a single error made by a fallible process worker is convenient, but the worker may be the last link in a chain made vulnerable by poor management. Human error is neglected in the CPI for lack of knowledge of its significance for safety, reliability, and quality, even though methodologies are available to address them in a systematic, scientific manner.

Quality and Safety

The CPI would benefit from the application of human factors principles to improve safety, quality, and productivity. These arise from applying quality management to get at the underlying causes of errors rather than after-the-fact blame or punishment. Crosby (1984) advocates error cause removal programs; Deming (1986) and Juran (1979) emphasize the central importance of uniform human performance to achieve quality. Internationally, there is increasing interest in the relationship between quality and safety (see, e.g., Whiston and Eddershaw, 1989; Dumas, 1987). Quality and safety failures are related to the same types of human errors and to the same underlying causes. Whether a particular error has a safety or quality consequence depends on when or where it occurs. An investment in error reduction yields cost effective safety and quality improvements. Another reason for investing in error reduction is to conform with regulatory standards. Error reduction yields regulatory relief.

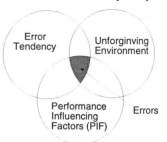

Fig. 4.4-1 Venn Diagram of Error Cause

There is increasing interest in human factors issues in the CPI. Kletz (1991), Lorenzo (1990) and Mill (1992) address human error in the CPI, Kletz (1994) addresses human factors through case studies.

Traditional and Systems Approach to Human Error

The traditional approach to human error assumes that errors are primarily the result of inadequate knowledge or motivation. The systems approach views error and accident causation by the way human errors are assessed and the preventative measures adopted. The systems approach considers PIFs (e.g., poor design, training, and procedures) as direct causes of errors, and organization and management as creating them. This is illustrated in Figure 4.4-2, a Venn diagram of the cause of errors. The upper left circle represents error tendencies intrinsic to error-prone such as a limited capability for processing information, reliance on rules to handle commonly occurring situations, and variability in performing unfamiliar actions. The upper right circle represents an unforgiving environment, and the lower, centered circle, represents the error-inducing environment that results from negative performance-influencing factors. Where the three circles combine, error results. For example, limited information processing capabilities may lead to overload when combined with distractions and PIFs such as management influences - the result is error.

The size of the overlap of the three circles represents the likelihood of error. However, given appropriate conditions, recovery from an error may be possible. Recovery may arise from error correction, before its consequences (accidents, product loss, degraded quality) occur, or the system may be insensitive to individual human errors. Thus, the shaded area in the center may not occur. The organization may reduce the likelihood of error by designing the system to reduce the mismatch between the demands of the job and the capabilities of the worker. This may be done by: improving PIFs to reduce the demands, or improving human capability through better job design, training, procedures, and team organization.

Errors may be reduced by reducing the demands placed on personnel. These demands are craft skills (breaking flanges, welding pipe work, etc.) mental skills (diagnosing problems, interpreting trends) and sensory skills (e.g., responding to changes in process information). Errors may be reduced by improving human capabilities through improved design of jobs and tasks, or by constituting teams of personnel whose combined resources meet the requirements. If resources exceed demands, the organization is "excellent" in the terminology of Peters and Waterman (1982).

Excess resources can contribute to the Total Quality Management continuous improvement process. Spare capacity provides resilience to unusual demands. Increasing resources does not necessarily mean more people, but it may be achieved by better design of jobs, equipment, or procedures.

4.4.3 Cases Involving Human Error

The following are some cases (Oil Insurance Association, 1971) in which human error is involved as one of several errors in a chain.

4.4.3.1 Maintenance Failure

An exothermic batch process was cooled by water flowing through a cooling jacket. The pump, circulating the cooling water, failed and the reaction went out of control causing a violent explosion. The low flow alarm was inoperable. The circulating pump bearing which had not been lubricated during maintenance, seized and the pump failed. Human error was reported, in the incident report, as the cause. Available maintenance procedures had not been used. The maintenance technician was disciplined and a directive was issued to use the maintenance and procedures.

This is the traditional view of human error. There were several reasons why the maintenance procedures were not followed:

1) The highly technical language used by the pump manufacturer,
2) Obscure format made it difficult to find the information,
3) Hard cover binding unsuitable for plant conditions,
4) The procedures were not current,
5) A culture of only novices using procedures,
6) No technicians participation in procedures development, hence, no sense of ownership,
7) Training was normally carried out "on the job," and
8) No confirmation of competence.

Chaos resulted from many pumps being repaired without an effective scheduling policy or policies for procedures preparation, or training. Maintenance on pump bearings that had been overlooked previously was caught before restoring the pumps to service, but not reported for lack of an incident reporting systems. Management failed to restore the low-flow alarm.

Pump maintenance steps were performed from memory without a checklist or independent checking. Two types of pumps were being maintained; the other type did not require bearing maintenance. This mix up of procedures is called a "stereotype error."

Independent checking would have avoided this accident; checklists should have been in use; the low flow alarm should have been operative; the design was faulty for lacking redundancy in the cooling water supply and flow.

4.4.3.2 Stress and Reconfiguration

Failure of combustion air supply shut down one boiler in a boiler house with eight large boilers attended by two men. They lost control and the lacking of proper combustion control equipment caused six more of the boilers to shut down. With low instrument air pressure, low steam pressure, constantly alarming boiler panels, the blocking valves and attempts to get the boilers back on line, one boiler exploded. Contributing factors were a purge interlock system manual operation of the individual burner valves, and the inability to charge the fuel gas header until completion of a purge time-out.

The manual individual burner valves on the boiler that exploded were not closed when the boiler shut down. After the purging fuel gas was admitted to the header from remote manual controls in the control room and into the firebox. Low fuel gas pressure tripped the master safety valve after each attempt to pressure the fuel header. Three attempts were made to purge the boiler and on each of these occasions fuel gas was dumped into the furnace through the open manual burner gas valves. On the third attempt the severe explosion occurred.

4.4.3.3 Poor Display and Accessibility

The pump feeding an oil stream to the tubes of a furnace failed. A worker closed the oil valve and intended to open a steam valve to purge the furnace tubes of oil, but opened the wrong valve, cutting off water flow to the furnace resulting in overheating and tubes collapse. The worker knew which was the right valve but due to poor labeling opened the wrong one.

This is a typical case of blaming human error but investigation showed deficiencies:

1) Access to the steam valve was poor and the right valve was difficult to see,
2) No flow through the furnace coils was not indicated in the control room, and
3) No low-flow alarm or trip on the furnace.

This case is an example of "system-induced error." The poor design of the information display and poor steam valve accessibility created the conditions for failure.

4.4.3.4 Poor Instrument Location - Lack of Temperature Confirmation

A chemical reactor, being started, was filled with the reaction mixture from another reactor which was already on line. The panel operator increased the flow of fresh feed while watching an eye level temperature recorder. He intended to start cooling water flow to the reactor when the temperature began to rise, but did not because the temperature recorder was faulty, thus a runaway reaction.

He did not notice the rise in temperature indicated on a six-point temperature recorder at a about three feet above the floor.

4.4.3.5 Lack of Service Line Labeling

A mechanic was instructed to fit a steam supply at a gauge pressure of 200 psi to a process line in order to clear a blockage. By mistake, a steam supply at a gauge pressure of 40 psi was connected. Neither supply was labeled; the 40 psi supply was not fitted with a check valve resulting in backflow of process material into the steam supply line. Later the steam supply caught fire when used to disperse a small leak.

4.4.3.6 Incorrect Transportation Container Label

Tank cars were used for shipping both nitrogen and oxygen. Before filling the tank cars with oxygen, filling connections were changed and hinged boards on both sides of the tanker were folded down to read "oxygen" instead of "nitrogen." A tank car was fitted with nitrogen connections and labeled "nitrogen." Vibration caused one of the hinged boards to fall down, to read "oxygen." The filling station staff therefore changed the connections and put oxygen in it. The tank car, labeled "nitrogen" on the other side was used to fill nitrogen tank trucks for a customer who off-loaded oxygen into his plant, thinking it was nitrogen. Fortunately, the customer noticed the weigh scale indicated that the tanker weighed three tons more than usual. A check then showed that the plant nitrogen system contained 30% oxygen.

4.4.3.7 Mistaken Calibration

A workman, pressure testing pipe work with a hand-operated hydraulic pump, told his foreman that he could not get the gauge reading above 200 psi. The foreman told him to pump harder. He did so bursting the pipeline. The gauge was calibrated in atmospheres and not psi. The abbreviation "atm." in small letters was not understood.

4.4.3.8 Confusing Calibration

A worker was told to control the temperature of a chemical reaction at 60° C; he adjusted the setpoint of the temperature controller to be 60. The scale indicated 0-100% of a temperature range of 0-200° C, so the set point was really 120° C. A runaway reaction resulted which overpressured the vessel. Discharged liquid and injured the worker.

4.4.3.9 False Assumptions due to Rule-Based Operation

A worker noticing that the level in a tank was falling faster than usual, reported that the level gauge was out of order and asked an instrument mechanic to check it. It was afternoon before it could be checked and found to be OK. Then the worker found a leaking drain valve. Ten tons of material had been lost. Rule-based instructions said "If the level in tank decreases rapidly then level gauge is faulty." It should have instructed: "check loss of material" because the important thing is the loss of material to the environment and the monetary loss. A bad gauge is a secondary effect.

4.4.3.10 Bad Indication from Erroneous Wiring

Following some modifications to a pump, an operator pressed the stop button on the control panel and saw that the "pump running" light went out hence closed a remotely operated valve in the pump delivery line. Several hours later the high-temperature alarm on the pump sounded; soon afterward there was an explosion in the pump.
The operator thought the alarm was faulty because he stopped the pump and saw the running light go out. When the pump was modified, an error was made in the wiring such that pressing the stop button did not stop the pump but switched off the "pump running" light. The deadheading pump overheated, and the material in it exploded.

4.4.3.11 Misleading Indication from Design Error

When an ethylene oxide plant tripped, a light on the panel told the operator that the oxygen valve had closed. Because the plant was going to be restarted immediately, he did not close the hand-operated isolation valve, but relied on the automatic valves. Before the plant could be restarted an explosion occurred. The oxygen valve had not closed and oxygen continued to enter the plant
The oxygen valve was closed by venting the air supply to the valve diaphragm by means of a solenoid valve. The light on the panel merely said that the solenoid had been deenergized not, as the operator assumed, that the oxygen valve had closed. Even though the solenoid deenergized the oxygen valve, flow continued because: 1) the solenoid valve did not open, 2) the air was not vented, and 3) the trip valve did not close. In fact, the air was not vented. The vent line on the air supply was choked by a wasp's nest. Although this example, primarily illustrates a wrong assumption, a second factor was the inadequate indication of the state of the oxygen valve by the panel light.

4.4.3.12 Permit Confusion

A permit was issued to remove a pump for overhaul. The pump was deenergized, removed, and the open ends blanked. Next morning the maintenance foreman signed the permit to show that removing the pump was complete. The morning shift lead operator glanced at the permit. Seeing that the job was complete, he asked the electrician to replace the fuses. The electrician replaced them and signed the permit to show this. By this time the afternoon shift lead operator had come on duty. He went out to check the pump and found that it was not there.

The job on the permit was to remove the pump for overhaul. Some permits are to remove a pump, overhaul and replace it, but this permit was for removal only. When the maintenance foreman signed the permit showing that the removal was complete. The lead operator, did not read the permit thoroughly and assumed that the overhaul was complete.

4.4.3.13 Over-Reliance on Automatic Control

A small tank was filled every day with sufficient raw material to last until the following day by watching the tank level and stopping the filling pump when the tank was 90% full. This worked for several years until the worker overfilled the tank. A high level trip was installed to turn off the pump automatically if the level exceeded 90%. To the surprise of engineering staff the tank overflowed again after about a year because the worker decided to rely on the trip and stopped watching the level. The supervisor and foreman knew this, but were pleased that the worker's time was being used more productively. The engineers had assumed the worker would still be monitoring the fill level, and a highly reliable shutoff mechanism was not needed. The trip was being used as a process controller.

4.4.3.14 Ineffective Work Organization I

A manhole cover was removed from a reactor to put in extra catalyst. After cover removal, it was found that the necessary manpower was not available until the next day. The supervisor decided to replace the manhole cover and regenerate the catalyst overnight. By this time it was evening and the maintenance foreman had gone home and left the work permit in his office, which was locked. The reactor was therefore boxed up and catalyst regeneration carried out with the permit still in force. The next day a mechanic, armed with the work permit, proceeded to remove the manhole cover again, and doing so was drenched with process liquid.

4.4.3.15 Ineffective Work Organization II

A pump was being dismantled for repair when the casing was removed, hot oil, above its autoignition temperature caught fire killing three men and destroying the plant. Examination showed that the pump suction valve was open and the pump drain valve was shut. The pump had been awaiting repair for several days; a work permit was issued, but the foreman who issued the permit should have checked that the pump suction and delivery valves were shut and the drain valve open. He claimed that he did, if so, someone closed the drain valve and opened the suction valve. When the valves were closed, they were not tagged. A worker might have opened the suction valve and shut the drain valve so that the pump could be put on line quickly if required. A complicating factor was that the maintenance team originally intended to work only on the pump bearings. When they found that they had to open up the pump, they told the process team, but no further checks of the isolations were carried out.

4.4.3.16 Failure to Explicitly Allocate Responsibility

A flare stack was used to dispose of surplus fuel gas, from the gas holder by a booster through valves B and C. Valve C was normally left open because valve B was more accessible. One day the worker responsible for the gas holder saw that the gas pressure had started to fall. He got gas from another unit, but a half hour later the gas holder was sucked in. Another flare stack at a different plant had to be taken out of service for repair. A worker at this plant therefore locked open valves A and B so that he could use the "gas holder flare stack." He had done this before, though not recently, and some changes had been made since he last used the flare stack. He did not realize that this action would result in the gas holder emptying itself through valves C and B. He told three other men what he was going to do but he did not tell the gas holder worker because he did not know that this man needed to know.

4.4.3.17 Organizational Failures I

A leak of ethylene from a bad joint on a high pressure plant was ignited by an unknown cause and exploded, killing four men and causing extensive damage.

Poor joints and the consequent leaks had been tolerated, for a long time before the explosion, by elimination of ignition sources to prevent leaks from igniting. The plant was part of a large corporation in which individual divisions were autonomous in technical matters. The other plants in the corporation had believed that leaks of flammable gas could ignite. Experience had taught them the impossibility of eliminating all sources of ignition, and therefore strenuous efforts had been made to prevent leaks. Unfortunately the managers of the ethylene plant had little technical contact with the other plants. Handling flammable gases at high pressure was, they believed, a specialized technology and little could be learned from those who handled them at low pressure. After the accident, a corporation-wide policy of leak elimination was instituted.

4.4.3.18 Organizational Failures II

Traces of water were removed from a flammable solvent in two vessels containing a drying agent. While one vessel was on-line, the other was emptied by blowing nitrogen into it before regeneration. The changeover valves were operated electrically with their control gear located in a Division 2 area because non-sparking control gear could not be obtained. It was housed in a metal cabinet purged with nitrogen to prevent ingress of flammable gas from the surrounding atmosphere. If the nitrogen pressure fell below a preset value (about 0.5-inch water gauge), a switch isolated the power supply. Despite these precautions an explosion occurred in the metal cabinet, injuring the engineer starting up the unit.

The nitrogen supply used to purge the metal cabinet was also used to blow out the dryers. When the nitrogen supply pressure fell occasionally, solvent from the dryers entered through leaking valves into the nitrogen supply line, and into the metal cabinet. Low nitrogen pressure allowed air diffusion into the cabinet.

Because the nitrogen pressure was unreliable, it was difficult to maintain a pressure of 0.5 inch water gauge in the metal cabinet. Workers complained that the safety switch kept isolating the electricity supply, so an electrician reduced the setpoint first to 0.25-inch and then to zero, thus effectively bypassing the switch. The setpoint could not be seen unless the cover of the switch was removed and the electrician told no one what he had done. The workers thought he was a good electrician who had prevented spurious trips. Solvent and air leaked into the cabinet, as already described, and the next time the electricity supply was switched there was an explosion.

Explosion causes were: the contamination of the nitrogen, the leaky cabinet, and the lack of any procedure for authorizing, recording, and checking changes in trip settings. A design error was from not realizing that the nitrogen supply was unreliable and liable to contamination..

Familiarization
· Information gathering
· Plant visit
· Review of procedures and information from system analysis

↓

Qualitative assessment
· Determine performance requirements
· Evaluate performance situation
· Specify performance objectives
· Identify potential human errors
· Model human performance

↓

Quantitative assessment
· Determine probabilities of human errors
· Identify factors and interactions affecting human performance
· Quantify effects of factors and interactions
· Account for probabilities of recovery from errors
· Calculate human-errors contribution to probability of system failure

↓

Incorporation
· Perform sensitivity analysis
· Input results to system analysis

Fig. 4.5-1 Phases of Human Reliability Analysis (from PRA Procedures Guide)

It is difficult to maintain pressure in thin metal boxes, had a HAZOP had been conducted, these facts, well known to the operating staff, should have been made known to the designers. It should have been known that that compressed air could have been used instead of nitrogen to prevent diffusion into the cabinet. The control cabinet did not have to be in a Division 2 area.

4.5 Incorporating Human Reliability Into a PSA

Incorporating human reliability is a function of education, training, ergonomics, stress, and physical condition. Incorporating this accurately into PSA is difficult. References for doing this are Gertmann (1994) and Dougherty (1988) as well as the many technical reports.

4.5.1 Human Reliability Analysis Models

The PRA procedures guide, NUREG/ CR-2300, partitions human reliability analysis (HRA) into four phases (Figure 4.5-1). The familiarization phase, evaluates a sequence of events to identify human actions that directly affect critical process components. From plant visits and review, this part of HRA identifies plant-specific factors that affect human performance such as good or bad procedures used in the sequence under consideration. The familiarization phase notes items overlooked during systems evaluation.

Table 4.5-1 HRA Analysis Methods

Acronym	Title	Reference
AIPA	Accident Initiation and Progression Analysis	Swain, 1989
ASEP	Accident Sequence Evaluation Procedure	NUREG/CR-4772, 1987
ATHEA	A Technique for Human Error Analysis	NUREG/CR-6350
CADA	Critical Action and Decision Approach	Gull, 1990
CBDTM	Cause Based Decision Tree Method	Singh et al., 1993
CES	Cognitive Environment Simulation	NUREG/CR-5213, 1990
CM	Confusion Matrix	Potash et al., 1981
CREAM	Cognitive Reliability and Error Analysis Method	Hollnagel, 1993
DNE/EE	Direct Numerical Estimation/Expert Estimation	NUREG/CR-3688
HAZOP	HAZard and Operability Study	Kletz, 1974
HCR	Human Cognitive Reliability model	EPRI RP-2170-3, 1984
HEART	Human Error rate Assessment and Reduction Technique	Williams, 1988
HRMS	Human Reliability Management System	Kirwan, 1992
INTENT	Human error rate assessment for INTENTion-based errors 1	Gertman et al., 1992
MAPPS	Maintenance Personnel Performance Simulation model	NUREG/CR-3626, 1984
MD	Murphy Diagrams	Pew et al., 1981
MORT	Management Oversight and Risk Tree analysis	Johnson, 1973
MSFM	Multiple-Sequential Failure Model	NUREG/CR-2211, 1981
OAT	Operator Action Tree system	Wreathall, 1982
ORCA	Operator Reliability Calculation and Assessment	Dougherty, 1990
PC	Paired Comparisons	Swain, 1989
PHECA	Potential Human Error Cause Analysis	Kirwan, 1992
SAINT	Systems Analysis of Integrated Networks of Tasks	Swain, 1989
SHARP, SHARP1	Systematic Human Action Reliability Procedure	EPRI NP-3583, 1984 Wakefield, 1992
SLIM	Success Likelihood Index Methodology	NUREG/CR-3518, 1984
SRM	Sandia Recovery Model	NUREG/CR-4834, 1987
STAHR	Socio-Technical Approach to assessing Human Reliability	Phillips et al., 1983
TALENT	Task Analysis Linked EvaluatioN Technique	NUREG/CR-5534, 1991
THERP	Technique for Human Error Rate Prediction	NUREG/CR-1278, 1983
TRC	Time-Reliability Correlation	Dougherty and Fragola, 1988

For example, if performance of a noncritical element

subsequently affects performance of a critical element, it must be considered, even though that task in itself is not important to the reliability of the system.

The next phase, Qualitative Assessment, in general uses procedural talk-throughs and task analysis to model human reliability. It may be performed in conjunction with the control room survey, or during interviews with operations personnel, or talk-throughs of the procedures in question. The analyst questions the operator on points of procedure until the analyst's understanding of the task is thorough. Performance specifics such as time requirements, personnel assignments, skill-of-the-craft requirements, alerting cues, and recovery factors are identified. (The talk-through can also be performed for activities not defined by a specific plant procedure, but the effort required of the human reliability analyst for such an analysis is greatly increased.)

At this point, a task analysis is performed. A "task" is defined as an activity or performance that the operator sees as a unit either because of its performance characteristics or because that activity is required as whole to achieve some part of the system goal. Only the tasks that are relevant to the system safety are considered. A task analysis involves decomposition of each task into individual units of behavior. Unusually, this analysis tabulates information about each specific human action. The format of such a table is not rigid - any style that allows easy retrieval of the information can be used. The format reflects the level of detail and the type of task analysis performed. The analysis yields qualitative or quantitative information.

Errors of omission and commission are identified for every human action appearing in the task-analysis table. A human action (or lack thereof) constitutes an error only if it has at least the potential for reducing the probability of some desired event or condition. The existence of this potential should be identified with the system analysts. For every human action appearing in the task-analysis table, errors of omission and commission should be pinpointed.

The HRA evaluates errors that may affect the probabilities of system success and failure but do not appear in the task analysis. Some of these can be disregarded by assuming that a certain condition does or does not exist. For example, for a post-maintenance test, the concern is for

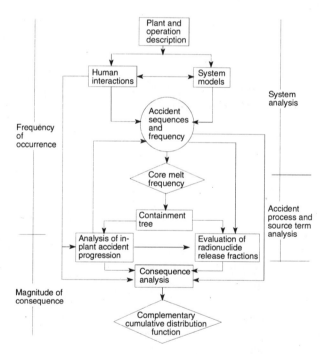

Fig. 4.5-2 General incorporation of HRA into the PSA process (from EPRI NP3583). Reprinted with permission from the Electric Power Research Institute U.S.A.

174

the conduct of the test under the assumption that the supervisor ordered the test. In determining which of these assumptions may be made, great care is taken.

At this point in the HRA, an applicable model of human action is chosen. Thirty-one methods are listed in Table 4.5-1, some of which are briefly reviewed. Swain (1989) is a thorough review of some of these methods, as is Dougherty (1988).

Phase three of a typical HRA begins with developing human error probabilities that can be applied to the selected model. In some cases, a set of nominal human errors can be derived from plant data, however, due to the sparseness and low confidence of these data industry generic information may be used. Chapter 20 of NUREG/CR-1278 includes a typical set of such data.

The human error probabilities estimated for a given task can now be modified to reflect the actual performance situation. For example, if the labeling scheme at a particular plant is very poor, the probability should be increased towards an upper bound. If the tagging control system at a plant is particularly good, the probability for certain errors should be decreased toward a lower bound.

Some of the performance-shaping factors (PSFs) affect a whole task or the whole procedure, whereas others affect certain types of errors, regardless of the tasks in which they occur. Still other PSFs have an overriding influence on the probability of all types of error in all conditions.

In any given situation, there may be different levels of dependence between an operator's performance on one task and on another because of the characteristics of the tasks themselves, or because of the manner in which the operator was cued to perform the tasks. Dependence levels between the performances of two (or more) operators also may differ. The analyses should account for dependency in human-error probabilities. In addition, each sequence may have a set of human recovery actions that if successfully performed will terminate or reduce the consequences of the sequence. This information, coupled with a knowledge of the system success criteria leads to the development of human success and failure probabilities which are input to the quantification of the fault trees or event trees. With this last step, the HRA is integrated into the PSA, and Phase 4 is complete.

Since this text is not solely devoted to HRA, the above process is by necessity simplified. A more defined interaction is shown in Figure 4.5-2. Here HRA interacts with the development of the system models and feeds into not only the accident sequences but also the physical analysis of the inplant and explant accident progress.

4.5.2 Quantifying Human Error Probabilities (HEPs)

Numerous methods have been used in PSAs for estimating HEPs and other methods are emerging, several of which were developed and evaluated by the NRC. This section describes very briefly the methods presented in Table 4.5.1.

Table 4.5-2 Requirements for Confusion Matrix
• List of the possible misdiagnoses and subsequent operator actions under consideration which could give rise to system interactions from operator diagnostic errors,
• Confusion matrix developed for each diagnosis and operator action,
• List of the misdiagnoses which do not cause system interaction, and
• Reason for selecting diagnostic errors.

Four general classes of HRA methods are: (1) expert judgment, (2) performance process simulation, (3) performance data analysis, and (4) dependency calculations. These classes are encompassed in the ten methods; many of which contain multiple classes of the methods. No attempt is made to classify them according to the methods.

4.5.2.1 Confusion Matrix

The confusion matrix (NSAC-60) is a method that identifies potential operator errors stemming from incorrect diagnosis of an event. It can be used to identify the potential for an operator to conclude that a small LOCA has occurred, when it is actually a steam line break. This provides a method for identifying a wrong operator response to an off-normal plant condition. It is particularly useful in Step 5 of the SHARP procedures. Documentation requirements are presented in Table 4.5-2.

4.5.2.2 Direct Numerical Estimation

Direct numerical estimation (NUREG/CR-3688) is a structured expert group judgment technique (i.e., psychological scaling). A list of human errors is given to a group of experts who assign their best estimate of the probability of error with associated uncertainty bounds. The error rates and uncertainty bounds are combined as a geometric mean. Documentation requirements are presented in Table 4.5-3.

Table 4.5-3 Requirements for Direct Numerical Estimation

- Human errors considered by the expert group,
- Composition of the expert group,
- Description of the training and background of the facilitator instructing the expert group,
- Individual data, and
- Final probability estimates and associated uncertainty bounds.

Table 4.5-4 Requirements for LER-HEP

- List of all human errors analyzed using this method,
- Source of field data used (e.g., LER file, I&E reports, etc.),
- Assumptions used to group reported errors to represent the error being analyzed,
- Analyses used to determine the number of opportunities for each type of error to occur,
- Opportunities, and resultant rates for each error analyzed, and
- Any modifications of the rates to estimate the HEPs.

Table 4.5-5 Requirements for MAPPS

- List of the maintenance tasks analyzed using this method,
- List of the subtasks used for each error analyzed,
- Input variables specified for each error analyzed and how they were selected,
- Assumptions used in analyzing the output of the method and how final probability values for use in the overall risk assessment were determined, and
- Uses for the output of the method to assist in any other part of the risk assessment.

4.5.2.3 LER-HEP Methodology

The LER-HEP method (NUREG/ CR-3519) is a means of analyzing field data to estimate HEPs. It considers available data on specific human errors in similar to those being considered in the risk assessment. The application of the method in NUREG/CR-3519 is to the Licensee Event Report (LER) file so it is called the LER-HEP method. For each error analyzed by this method, an error rate is derived by dividing the number of similar errors by the estimated number of opportunities for that specific type of error. This ratio is the HEP for use in the risk assessment. Requirements for LER-HEP are given in Table 4.5-4.

4.5.2.4 Maintenance Personnel Performance Simulation (MAPPS) Model

MAPPS (NUREG/CR-3634) is a task-oriented, computer-based model for simulating actual processes in maintenance activities. It includes environmental, motivational, task, and organizational variables which affect human performance and it yields information such as predicted errors, personnel requirements, maintenance stress and fatigue, performance time, and required ability levels for any corrective or preventive maintenance actions. Documentation needs are given in Table 4.5-5.

4.5.2.5 Multiple Sequential Failure (MSF) Model

The MSF model (NUREG/CR-3837) is used principally to determine the level of dependence between safety systems introduced by maintenance, testing, and calibration activities. It is a mathematical model which modifies the independent failure probability of any single component by considering that a component with which it is redundant has already failed. This allows the conditional failure probabilities of redundant components to be calculated to determine the overall system failure probability. Documentation requirements are given in Table 4.5-6.

4.5.2.6 Operator Action Tree (OAT)

The Operator Action Tree (OAT) is a time reliability correlation method (NUREG/ CR-3010) specifically designed to estimate HEPs for actions requiring decision making on the part of the plant operator. It employs two steps: (1) the development of an operator action tree which reflects the specific actions that may be taken by an operator faced with a particular decision, and (2) a time

---

Table 4.5-6 Requirements for MSF

- List of all systems analyzed using this method,
- Sources of independent failure probabilities for the components involved,
- Dependency data that were analyzed,
- The dependency factors used in each calculation and how they were determined,
- System failure probability for each system analyzed, and the specific conditional failure probabilities for each component in the system.

Table 4.5-7 Requirements for OAT

- List of all human errors analyzed using this method,
- Operator action tree used to develop the errors, and the specific time reliability curves used and how they were developed,
- Critical considerations in determining the time available for each action, and
- List of any weighting factors modifying the HEPs used in the risk assessment.

Table 4.5-8 Requirements for Paired Comparisons

- List of all human errors analyzed using this method,
- Expert tables generated during the pairwise comparisons,
- Final ranking of the human errors considered,
- List of the specific anchor points, their sources, and scaling factors used to determine probabilities for other errors analyzed,
- Composition of the expert group used, and
- Description of the training and background of the facilitator instructing the expert group.

reliability correlation which relates the time available for action to the likelihood of an error. Documentation requirements are given in Table 4.5-7.

4.5.2.7 Paired Comparisons

The paired comparisons method (NUREG/CR-3688) is a structured expert judgment method in which human errors are compared in pairs. By combining the judgments of the group of experts, the errors are arranged in order of likelihood of occurrence. If assessed probabilities are available for several of the human errors considered, they can be used as "anchor points" to assess the probabilities on the list. Documentation requirements are given in Table 4.5-8.

4.5.2.8 Success Likelihood Index Method/Multi-Attribute Utility Decomposition (SLIM-MAUD)

SLIM-MAUD (NUREG/CR-3518) is a structured expert judgment method that is facilitated by a computer terminal to classify, rate, and weight HEPs. It involves a specific process by which errors are put into separate classes depending on their degree of similarity. The errors in each class are then analyzed according to the factors that are perceived by the expert group to affect performance. The expert group weights these factors to arrive at an index for each error analyzed which is then converted into an HEP for the risk assessment. Documentation requirements are given in Table 4.5-9.

4.5.2.9 Technique for Human Error Rate Prediction (THERP)

THERP (NUREG/CR-1278), is used to estimate HEPs for a risk assessment. It provides error probabilities for generic tasks and describes the process used to modify these rates depending on the specific performance shaping factors (PSFs) involved in the task

Table 4.5-9 Requirements for SLIM-MAUD
• List of human errors analyzed using this method,
• Description of the composition of the expert group used, including the size of the group and the backgrounds of the individuals involved,
• Errors analyzed according to the final classifications assigned by the group,
• List of the factors considered for each error and the relative weight of each factor assigned, and
• Final index values and the calculated HEP for each error considered.

Table 4.5-10 Requirements for THERP
Documentation for Human Errors
• List of all human errors analyzed using this method,
• Human action trees used and all input used in constructing those trees (e.g., procedures),
• List of the generic human error probabilities used to determine a base error rate for each human error considered, and
• List, for each error of all PSFs considered and the considerations that went into scaling the generic probability to represent each error.
Documentation of Dependency Calculations
• List of all systems which were analyzed using this method,
• Event trees used to identify complete and partial failure combinations,
• High, medium, or low dependence level assigned to each system and the resultant conditional failure probabilities,
• Reasons for selecting high, medium, or low dependence for each system, and
• Results of any sensitivity analyses conducted during the analysis.

being analyzed. THERP also provides a method for estimating dependencies introduced into redundant systems by human activities. Documentation requirements are given in Table 4.5-10.

4.5.3 Human Factors Data

Human error contributed to about 50% of the accident sequences in the RSS but none of the human error data came from the nuclear power industry. Furthermore, very high failure rates (0.5 to 0.1/action) were predicted but are not supported by the plant operating experience. Since the publication of WASH-1400, human factors data have improved considerably.

Fragola and Collins (1985) emphasize that a HEP model must satisfy system interface requirements and be developed verifiably and as accurately as possible into the larger plant risk model. To this end, they classify information as procedural and cognitive. HEPs, based on plant experience, for procedural data, are available from ORNL /TM-9066; similar information based on judgment is available from NUREG/CR-1278, and NUREG/CR-2554. Plant data for cognitive actions are provided in Fullwood and Gilbert (1976) based on reporting under Reg. Guide 1.16 (before LERs), Fullwood (1978), for time-dependent response based on plant data from the ANSI N666 committee. Cognitive data from plants, simulators, and judgment are available in NUREG/ CR-3010, and Fragola (1983). Plant and simulator

Table 4.5-11 Sample of NUCLARR Human Error Probability Data (NUREG/CR-4639)[*]

Component	Operator	Action	Mean HEP	UCB	LCB	Error Type
Circuit breaker	AO	Opens/ closes	1.5E-2 (4)	3.8E-2	2.9E-3	OM
	AO	Opens/ closes	6.0E-3 (7)	2.0E-2	4.5E-4	COMM
Control rod drive mechanism	CRO	Operates	2.7E-4 (10)	1.0E-3	1.0E-5	OM
Electrical equipment	CRO	Starts/ stops	1.8E-3 (5)	5.6E-3	2.2E-4	OM
Electrical equipment	MT	Calibrate	1.7E-1 (2)	3.1E-1	7.7E-2	OM
Electric motor, ac	CRO	Operates	8.7E-3 (2)	1.5E-2	4.3E-3	OM
	CRO	Operates	4.0E-3 (10)	1.5E-2	1.5E-4	COMM
Generator	AO	Operates	2.7E-2 (10)	1.0E-1	1.0E-3	OM
Pressure control instrumentation	MT	Calibrate	2.7E-4 (10)	1.0E-3	1.0E-5	OM
Pumps	CRO	Operates	5.5E-2 (4)	1.5E-1	1.0E-2	OM
	CRO	Operates	2.6E-2 (5)	8.2E-2	3.2E-3	COMM
Switch	MT	Tests	1.8E-2 (3)	4.0E-2	5.4E-3	OM
Valves	AO	Opens/closes	1.3E-1 (3)	3.0E-1	3.3E-2	OM
	AO	Opens/closes	2.8E-3 (5)	1.4E-2	5.5E-4	COMM
Vessel/tanks	AO	Inspects	6.7E-5	2.5E-5	2.5E-6	COMM

* AO - auxiliary operator, CRO - control room operator, MT - maintenance technician, UCB - upper confidence bound, LCB - lower confidence bound, OM - error of omission, COMM - error of commission, and numbers in parenthesis are error factors.

Table 4.5-12 Estimated Human Error

#	Potential Errors	HEP	EF
1	Select wrong control on a panel from an		
	a. Identified by labels only	0.003	3
	b. Arranged in well-delineated	0.001	3
	c. Part of a well-defined mimic layout	0.0005	10
2	Turn rotary control in wrong direction		
	a. When there is no violation of	0.0005	10
	b. When design violates a strong	0.05	5
	c. When design violates a strong	0.5	5
3	Turn a two-position switch in wrong	2a-c/5	same
4	Set a rotary control to an incorrect setting	0.001	10
5	Failure to complete change of state of	0.003	3
	a. Densely grouped and identified by	0.005	3
	b. In which the PSFs are more favorable	0.003	3
6	Improperly mate a connector	0.003	3

* The HEPs are for errors of commission and do not include errors in selecting the control.

Failure rates, Incidents, and Human Factors Data

cognitive data are available in NUREG/CR-3092 and simulator data in Peterson et al. (1981), NUREG/CR-2534. Stimulus response measurements in a control room can be found in Ablitt (1969). Table 4.5-11 provides HEP data from NUCLARR; Table 4.5-12 gives examples of HEPs from NUREG/CR-1278 (Swain and Guttman), the most commonly used reference for HEPs.

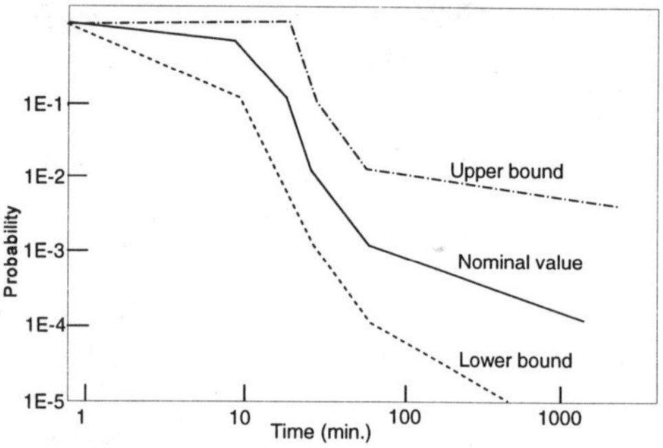

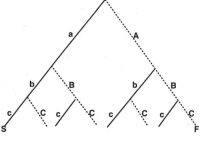

Fig. 4.5-3 Failure vs Problem-Solving Time (NUREG/CR-2815 based on NUREG/CR-3010)

4.5.4 Example of Human Error Analysis:

Operator Actuation of Automatic Depressurization

In some BWR transient scenarios, the high pressure injection systems are postulated to fail. To make use of the low pressure injection system, it is necessary to depressurize the reactor coolant system, a function performed by the automatic depressurization system (ADS). In the scenario considered, ADS actuation is manual because the signals for automatic initiation of the system are not present.

Initially, assume that failure occurs if the water level drops below the top of the core. This occurs 30 minutes after the initiating event, if at least one of the injection systems did not operate successfully.

The thinking interval is the difference between 30 minutes (the total available time) and the sum of a) the time required for the clues to become available to the operator,

Fig. 4.5-4 HRA Event Tree.
Here A, B, and C are the first, second, and third tasks performed.
Solid lines are success, broken lines are errors

and b) the time required for his actions to take effect (the time required for ADS to reduce the pressure and for LPI to begin to inject). Assume that 8 minutes are required for the clues to materialize; this is the interval during which it becomes clear that no water is being injected. Furthermore, assume that 5 minutes are required for ADS and for LPI to initiate. This leaves a thinking interval of 17 minutes [30-8-5]. For this thinking interval, Figure 4.5-3 gives a failure probability of approximately 0.15. Possible modifications to this reasoning may be taken into account. For some transients, the 30-minute time interval might be judged too long, while for others it may be too short. If the definition of the top event is modified to be "uncovery of more than X% of the core," rather than "uncovery of the top of the core," the thinking interval may be lengthened from 17 to, say, 22 minutes where the probability is 0.081. If the top event allows 35 minutes, cues are available after 3 minutes. The time required for action to be effective is again 5 minutes, the thinking interval is [35-3-5]) = 28, and the corresponding failure probability is approximately 0.015. Further examples are given in Wreathall (1982) and Hall et al. (1982) (NUREG/CR-3010).

4.5.5 HRA Event Tree (NUREG/CR-1278)

A probabilistic statement of the likelihood of human-error events presents each error in the task analysis as the right limb in a binary branch of the HRA event tree. These binary branches form the chronological limbs of the HRA event tree, with the first potential error starting from the highest point on the tree. (Figure 4.5-4). Any given task appears as a two-limb branch; the left limb represents the probability of success; the right limb represents the probability of failure. Each binary fork is attached to a branch of the preceding fork and is conditioned by the success or failure represented by that branch. Thus, every fork, represents conditional probability. Each limb of the HRA event tree is described or labeled, in shorthand. Capital letters (A) represent failure; lower case letters (a) represent success. The same convention applies to Greek letters, which represent non-human error events, such as equipment failures. The letters S and F are exceptions to this rule in that they represent system success and failure respectively. In practice, the limbs may be labeled with a short description of the error to eliminate the need for a legend. The labeling format is unimportant: the critical task in developing HRA event trees is the definition of the events themselves and their translation to the trees.

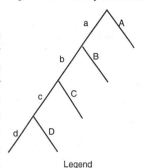

Legend
A - Control room operator omits ordering the following tasks
B - Operator omits verifying the position of MU-13
C - Operator omits verifying the opening of the DH valves.
D - Operator omits isolating the DH pumping rooom

Fig. 4.5-5 HRA Event Tree for Actions Outside the Control Room

A PSA analyst is usually interested in determining the probability of error for a task or the probability that a set of tasks will be performed incorrectly. For the first case, no HRA event tree is needed unless performance on that task is affected by other factors whose probabilities should be diagramed. A description of the task and knowledge of the performance-shaping factors are sufficient to determine the probability of a single human error.

For the second case, the probability of all tasks being performed without error, a complete success path through the HRA event tree is followed. Once an error has been made on any task, a criterion for system failure has been met. Given such a failure, no further analysis along that limb is necessary at this point. In effect, the probabilities of event, success that follow a failure and still end in a system success-probability constitute recovery factors and should be analyzed later, if at all. Thus, as shown in Figure 4.5-5, there are HRA event trees that are developed along the complete success path only. This does not mean that this is the only possible combination of events; it means only that the initial analysis goes no further once a system failure criterion has been met.

The development of the HRA event tree is one of the most critical parts of the quantification of human error probabilities. If the task analysis lists the possible human error events in the order of their potential occurrence, the transfer of this information to the HRA event tree is facilitated. Each potential error and success is represented as a binary branch on the HRA event tree, with subsequent errors and successes following directly from the immediately preceding ones. Care should be taken not to omit the errors that are not included in the task analysis table but might affect the probabilities listed in the table. For example, administrative control errors that affect a task being performed may not appear in the task analysis table but must be included in the HRA event tree.

Failure rates, Incidents, and Human Factors Data

Table 4.5-13 Comparison of Human Interaction Evaluations in Utility and NRC PSA Studies

Utility Sponsored	Zion	Big Rock Point	Limerick
Human reliability data	NUREG/CR-1278 was supplemented by judgment of system analysts and plant personnel.	Human error probabilities were developed from NUREG/CR-12;8, human action time windows from system analysis; and some recovery times from analysis of plant specific experience.	Data sources were WASH-1400 HEPs, Fullwood and Gilbert assessment of US power reactor exp., NUREG/CR-1278, and selected aerospace data.
Methods of defining actions	Important human actions were defined jointly by the analysts and utility personnel.	Discussions with plant personnel, plant visits review of operating, maintenance and test procedures.	Important human actions were defined by plant visits, talk through events, previous experience as reflected in WASH-1400, and review of fault tree analysis information.
Techniques for quantifying	Techniques used were basically from NUREG/CR-1278 or PRA Procedures Guide, Chapter 4.	(1) THERP (1980 version NUREG/CR-1278 with appropriate factors for a tree was used in procedure defined events. (2) A time dependent modeling approach was used to quantify operator actions during a sequence.	THERP methods were used as described in WASH-1400.
Considerations of dependencies between human and human system interactions	Human-human or team dependencies were patterned after NUREG/CR-1278. Human system dependencies were treated by developing recovery distributions based upon the time required to repair various failure modes of the machine (e.g., diesel generator repair).	Human-human dependencies were modeled as NUREG/CR-1278. Human system dependencies were modeled on a case I by case as analyzed in the time dependent models.	Dependencies were evaluated as recommended in WASH-1400. Stress factors of WASH-1400 were utilized in time dependent sequences. Human errors from NUREG/CR-1278 were assumed to be independent of accident sequences. Human-system interactions were assessed from a review of data.
NRC Sponsored	WASH-1400	Grand Gulf (BWR)	Arkansas 1 (PWR)
Human reliability data	Developed by a team of psychologists led by A. Swain.	Used WASH-1400 data.	Data from NUREG/CR-1278 was used.
Methods of defining actions	The following considerations were used to define human actions: operating procedure review, interviews with plant personnel, design of controls, observation of control room, test, maintenance tasks at plant study of written materials and photographs, and review of training.	WASH-1400 used as guidance in which operating procedure review, interviews with plant personnel, design of controls, observation of rf control room, test, maintenance tasks at plant study of written materials and photographs, and review of training.	Actions were defined as a result of plant visits, discussions with system analysts, and review of emergency procedures with plant personnel.
Techniques for quantifying	Human actions were decomposed into constituent acts for which data could be assigned, and a probability tree diagram composed to determine the success and failure probabilities.	Quantified the same as WASH-1400 where human actions were decomposed into constituent acts for which data could be assigned, and a probability tree diagram composed to determine the success and failure probabilities.	THERP modeling was used as described in NUREG/CR-1278 for testing and maintenance. Pessimistic bounding assumptions were employed to model operator actions during an accident sequence.
Considerations of dependencies between human and human system interactions	An upper and lower bounding technique was used to model human coupling. Human-system interactions were incorporated in the tree structure.	No explicit task for study of human dependencies was included. The results were numerically the same as WASH-1400.	Human dependencies were addressed using the dependency models in NUREG/CR-1278.

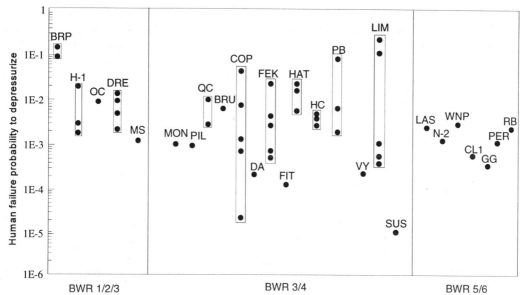

Fig. 4.5-6 Example of Spread in HEPs from IPE for BWRs "Failure to Depressurize"
Acronyms are: BRP - Big Rock Point, BRU - Brunswick 1&2, CL1 - Clinton 1&2, COP - Cooper, DRE - Dresden 2&3, DA - Duane Arnold, FER - Fermi 2, FIT - Fitzpatrick, GG- Grand Gulf, HAT - Hatch 1&2, HC - Hope Creek 1&2, LAS - LaSalle 1&2, LIM - Limerick 1&2, MON - Monticello, MS - Millstone, N-1 - Nine Mile Point, OC - Oyster Creek, PB - Peach Bottom 2&3, PER - Perry, PIL - Pilgrim 1&2, QC - Quad Cities 1&2, RB - River Bend, SUS - Susquehanna, VY - Vermont Yankee, 1&2, WNP - WNP2

4.5.6 Comparison of Human Factors in PSAs

The treatment of human factors in both NRC and utility-sponsored PSAs is presented in Table 4.5.13 (adapted from Joksimovich et al., 1983, with permission of EPRI).

Individual Plant Examinations (IPE) (NUREG-1560) were performed by the U.S. utilities using their staff and utilities. Results from the IPEs indicate that human error can significantly increase or decrease the CDF. Certain human actions are consistently important for BWRs and PWRs.

The only human actions important in more than 50% of the BWR IPEs are manual depressurization, containment venting, initiation of standby liquid control (SLC), and system alignment for decay heat removal. In PWRs, only switch over to recirculation, feed-and-bleed, and the actions associated with depressurization and cooldown are important in more than 50% of the IPEs.

Many factors influence the importance of human actions in the IPEs such as: plant characteristics, modeling details, sequence specific attributes, dependencies, HRA method and the performance shaping factors (PSFs) and the biases of both the analysts performing the HRA and

Failure rates, Incidents, and Human Factors Data

plant personnel providing information and judgments. HEPs cause the risk accident sequences and human actions to be unrealistically reported. The variability of HEPs for several of the more important human actions was examined across IPEs as shown in Figure 4.5-6. This shows how variable the HEPs are for depressurization of BWRs; the variability for PWRs is similar and covers several orders of magnitude. The reasons for the variability in some HEPs was unexplainable, but most of the variability resulted from the assumed time for operator response to the initiator and the sequence modeling related influences, dependencies, and plant-specific system characteristics.

4.7 Summary

This chapter discussed the importance of human reliability in process accidents. It reviewed the databases that are presently available. It presented Internet addresses from which the most current data may be obtained. It discussed how human errors have caused or affected plant accidents. The SHARP method was presented for integrating HRA into a PSA. Next, it introduced ten methods for assessing and modeling human error mostly using expert judgment. Statistical human error data sources were presented with a preponderance coming from the nuclear power industry. The chapter concluded with the presentation of two of the more popular methods. First an example of an assessment of human error for operator actuation of the ADS, and second as an alternative a discussion of the HRA event tree approach. Both of these methods are widely used in current PSAs. A integrated approach applying each of the methods as needed is presented from Luckas, et al. (1986). Here the analysis concentrated on an anticipated transient without scram (ATWS) sequence with the closure of the main steam isolation valves. With this background, we can now attack the problem of reliability modeling of complex systems in the next chapter.

4.8 Problems

1. Table 4.1-5 shows in these records, there has been 3 control rod drive failures. Assume 100 plants in the U.S. with an average of 30 control rods/plant and 10.7 years of experience in this database. Estimate, the mode, 90% and 10% confidence limits for the failure rate.

2. Prepare a reliability data acquisition plan for your or a hypothetical plant. Detail the sources of information, method of recording, analysis methods, and interfaces with management, maintenance, and operations.

3. Prepare an HRA event tree for driving from your home to work or school. Use this tree to estimate your annual error rate. Compare this with your actual rate. Explain disagreements.

4. Bar-code readers (like those in grocery stores) are to identify equipment in a plant. Suggest a way that this may be used to check the temperature, acid type and quantity, and aluminum quantity for acid dissolution of aluminum.

Chapter 5

External Events

"External events" are accident initiators that do not fit well into the central PSA structure used for "internal events." Some "external events" such as fire due to ignition of electrical wires, or flood from a ruptured service water pipe occur inside the plant. Others, such as earthquakes and tornados, occur outside of the plant. Either may cause failures in a plant like internal events. External initiators may cause multiple failures of independent equipment thereby preventing the action of presumably redundant protection systems. For example, severe offsite flooding may flood the pump room and disable cooling systems. An earthquake may impede evacuation of the nearby populace. These multiple effects must be considered in the analysis of the effects of external events.

Plants are built to withstand a wide variety of accidents, e.g., the design basis earthquake (DBE), the design basis tornado (DBT), the design basis flood (DBF). Their analysis may be simplified by conservative bounding calculations. Consider aircraft impact, if the site is not near a major airport, simple calculations, using appropriate data from NUREG-0533, often suffice to show that severe accidents due to aircraft impact are so rare that the risk is negligible. Ground transportation accidents, like those resulting from nearby industrial sites, military bases, or pipeline ruptures my be treated in a similar fashion. Section 10.3.1 of the PSA Procedures Guide (NUREG/CR-2300) screening dismisses external events if the maximum hazard is less than that for which the plant was designed. Table 5-1, adapted from the PSA Procedures Guide, lists external events with the aim of completeness, not necessarily their risk significance. Process plant PSAs consider dominant external events to be: seismic, internal fire, and internal flood. These externalities are the subject of this chapter and are among the leading risk contributors. They are also some of the most uncertain because of modeling complexity, common cause failures, and sparse data.

5.1 Seismic Events

5.1.1 Overview

WASH-1400 did not analyze seismic events, but if they had, the risk they assessed would have been larger. Earthquake analysis was part of the FSAR for licensing a plant and had been practiced for some time before the RSS. This analysis included: earthquake frequencies from historical records, ground-coupling models to estimate the intensity and frequency spectrum of the acceleration, and spring-mass models to estimate the forces on plant components under the

Table 5-1 Natural and Man-caused Risks

Event	Comment	Event	Comment
Avalanche	Excluded by siting considerations	Flooding	Plant specific; requires detailed study
Coastal erosion	Considered in the effects of external flooding	Lightning	Considered in plant design and siting
Drought	Excluded by assuming that there are multiple sources of the ultimate heat sink or that the ultimate heat sink is not affected by drought (e.g., cooling tower with adequately sized basin)	Low lake Internal or river water	Ultimate heat sink is designed for at least 30 days of operation
External flooding	Site specific; requires detailed study	Low winter temperature	Thermal stresses and embrittlement are covered by design codes and standards for plant design; also see "ice."
Extreme winds and tornadoes	Site specific; requires detailed study	Meteorite	All sites are equally susceptible
Fire	Plant specific; requires detailed study	Pipeline accident	Site specific; requires detailed study
Fog	Possible cause of plant impact by surface vehicles or aircraft	Release of chemicals	Plant specific; requires detailed study
Forest fire	Fire cannot propagate to the site because the site is cleared; plant design and fire-protection provisions a adequate to mitigate the effects	Seiche	Included under external flooding
Frost	Subset of ice and snow	Seismic	Site specific; requires detailed study
Hail	Subset of missile analysis	Snow	Plant designed for high loading; snow melt causing river flooding is included under external flooding
High water	Included under external flooding	Tsunami	Included under external flooding and seismic events
High summer temperature	Ultimate heat sink is designed for at least 30 days of operation	Toxic gas	Site specific; requires detailed study
Hurricane	Included under external flooding; wind forces are covered under extreme winds and tornadoes	Turbine-generated missile	Plant specific; requires detailed study
Ice	Loss of heat sink considered in plant design	Volcanic activity	Excluded by siting analysis
Industrial or military accident	Site specific; requires detailed study		

assumption that everything works as designed. PSA, on the other hand, considers pre-existing failures and failures from the seismic stress. In fact, an earthquake is a giant common cause failure.

The first study to enter this realm was conducted in 1977 under NRC request as part of the licensing process for Pacific Gas and Electric's Diablo Canyon Plant. As specified by the NRC, the study was to consider an earthquake of Richter Magnitude 7.5 M occurring on the Hosgri Fault. Although PG&E and its consultants believed, on the basis of independent investigations, that the assumed magnitude of the Hosgri event and the resulting assumed ground acceleration (0.759) at the site are unreasonably severe, the analysis was undertaken using procedures diagramed in Figure 5.1-1.

Since this study, the same basic method has been applied to other power plants such as Oyster Creek, Zion, Indian Point, and Oconee. The NRC has funded methods development in a

Seismic Events

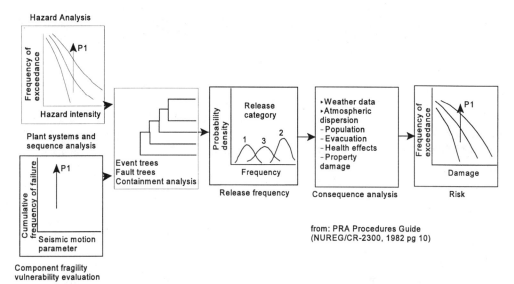

Fig. 5.1-1 Procedures for Analyzing External Events

Table 5.1-1 Modified Mercali Scale (Wood-Neumann Scale) Perceived Intensity
I. Not felt except by very few, favorably situated.
II. Felt only on upper floors, by a few people at rest. Swinging of some suspended objects.
III. Quite noticeable indoors, especially on upper floors, but many people fail to recognize it as an earthquake. Standing automobiles may sway. Feels like passing truck.
IV. Felt indoors by many during day, outdoors by few. If at night, awakens some. Dishes, windows, and doors rattle, walls creak. Standing cars may rock noticeably. Sensation like heavy truck striking building.
V. Felt by nearly all. Many wakened. Some fragile objects broken and unstable objects overturned. A little cracked plaster. Trees and poles notably disturbed. Pendulum clocks may stop.
VI. Felt by all. Many run outdoors. Slight damage. Heavy furniture moved. Some fallen plaster.
VII. Everyone runs outdoors. Slight damage to moderately well-built structures, but considerable damage to those built poorly. Some chimneys broken. Noticed by automobile drivers.
VIII. Damage slight in well-built structures considerable in ordinary substantial buildings with some collapse. Great damage in poor structures. Panels thrown out of frame structures. Chimneys, monuments, factory stacks thrown down. Heavy furniture overturned. Some sand and mud ejected, wells disturbed. Automobile drivers disturbed.
IX. Damage considerable even in well-designed buildings. Frame structures thrown out of plumb. Substantial buildings greatly damaged, shifted off foundations, partial collapse. Conspicuous ground cracks, buried pipes broken.
X. Some well-built wooden structures destroyed. Most masonry and frame structures destroyed, with their foundations. Rails bent, ground cracked. Landslides on steep slopes and river banks. Water slopped over from tanks and rivers.
XI. Few, if any, masonry structures left standing. Bridges destroyed. Underground pipes completely out of service. Rails bent greatly. Broad cracks in ground and earth slumps and landslides in soft ground. Damage total. Waves left in ground surface and lines of sight disturbed. Objects thrown upward into the air.

Source: J. Gilluly et al., *Principles of Geology*, W.H. Freeman, San Francisco, 1954, p. 456.

program at Lawrence Livermore Laboratories called Seismic Safety Margins Research Program (SSMRP, Smith et al., 1981). The most recent guide, although not probabilistic, is Eagling (1996). Reference should also be given to the PSA Procedures Guide (NUREG/CR-2300).

The process of seismic risk analysis involves several well-defined tasks: a) determination of the Richter magnitude-frequency distribution, b) estimation of the ground shaking at the reactor building by ground-coupling models, c) a damped spring-mass model of the plant to determine the stresses on the equipment in addition to the normal stress, d) fragility curves to determine the stress levels that will cause system failure, e) systems analysis to determine system failures from component failures, and f) determination of the consequences to the public of such an accident. The logical progression just presented may not be rigorously followed; systems analysis may be used to identify critical components to focus the analysis on them for simplification. Some sources of seismic information are: DOC (1969) and DOC (1973).

5.1.2 Richter Magnitude - Frequency of Occurrence Distribution

Earthquake measurements, essentially, began in the 1930s with the development of the Richter Scale relating seismographic measurements to the energy released. Before this, a perception scale called the Modified Mercali was used (Table 5.1-1). The Modified Mercali Intensity (MMI) is a qualitative estimate of ground acceleration by witnesses. Table 5.1-2, for comparison, is the Richter Scale that relates energy release to magnitude. This table includes the qualitative effects of ground coupling for comparison with the MMI scale.

Many independent researchers worked to develop reliable methods for estimating recurrence intervals for earthquakes (e.g., Algermissen and Perkins, 1976 and Perkins et al. 1980. These methods include deterministic extrapolation of short-term or long-term frequency magnitude data, application of the theory of extreme values to frequency magnitude data (Gumbel, 1958), application of Bayes theory to expert opinion on recurrence intervals, deterministic and/or probabilistic extrapolation of recurrence intervals, deterministic and/or probabilistic extrapolation of strain-rate data, and utilization of geologic data from fault-zone investigations. Unfortunately

Table 5.1-2 Richter Magnitudes, Energy, Effects and Frequencies

Characteristic effects of shallow shocks in populated areas	Approximate Richter magnitude	Number of earthquakes per year	Energy (megawatt-hours)
Felt but not recorded	2.0-3.4	800,000	1 - 2500
Felt by some	3.5 .4.2	30,000	0.4 - 21E4
Felt by many	4.3-4.8	4,800	4 - 75E5
Felt by all	4.9-5.4	1,400	1 - 15E7
Slight damage to buildings	5.5-6.1	500	0.3 - 7.5E9
Considerable damage to buildings	6.2-6.9	100	1.4 64E10
Serious damage, rails bent	7.0-7.3	15	
Great damage	7.4	4	1.1E14
Damage nearly total	8.0	0.0-0.2	2.8E14

Adapted from *Earth*, 4th edition by Frank Press and Raymond Siever. Copyright 1974, 1978, 1982, 1986 W.H. Freeman and Company. Reprinted with permission.

consensus among the various methods is lacking; the uncertainties are great, and the results differ widely.

Bloom (1981) approached this situation as a three stage process. In the first stage, various earthquake databases which contained information on origin and size of earthquakes were examined for their completeness and level of world coverage. The most complete database with world-wide coverage is the National Geophysical and Solar-Terrestrial Data Center (NGSDC). This database was analyzed for completeness (Bloom and Erdmann, 1979) to show a biases in the data. Large events are reported more completely than small events. There is time bias due to improvements in record keeping and in the increase of recording stations allowing the collection of data throughout the world. Another bias results from the shift from Modified Mercali to Richter scales for defining earthquake intensity.

The second stage was a statistical analysis of the data available from the NGSDC database and several other smaller databases to extract the frequency-magnitude (f-M) relations for earthquakes originating in widely different regions of the earth. Previous studies indicated that each region was different. However, the f-M distribution after correction for magnitude and time bias is applicable worldwide.

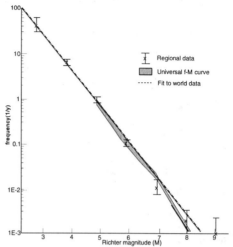

Fig. 5.1-2 Comparison of the world frequency-magnitude shape with data from the CMMG. The xs are California data. The shaded region is the universal f-M shape normalized to the data. The dashed curve is the fit to world data (equations 5.1-1 and 5.1-2.

The f-M distributions were obtained from sets of earthquake events originating within eight widely separated regions on the earth, and were compared to the worldwide f-M distribution. The regions have geological diversity and areas greater than 4E4 mi². To within the estimated error, the shape of total world data agrees with similar plots of data subsets taken from the eight separate regions of the earth (Bloom and Erdmann, 1980).

Figure 5.1-2 shows the world f-M shape obtained from fitting (dashed line) to equation 5.1-1 which with the fitting parameter becomes equation 5.1-2, the usual relationship of Gutenberg and Richter (1942). The world f-M distribution rolls-off at high intensity possibly due to finite lengths of faults that can release energy in an earthquake. The data for California are shown in the figure for comparison.

This f-M distribution estimates a lower risk from earthquakes because the high intensity earthquakes that exceed the DBA and result in plant damage are in the roll-off region of the curve. Unfortunately, this roll-off at high intensities is not universally accepted by the seismological community (PSA Procedures Guide 11-14).

$$\ln(f) = a + b*M \quad (5.1\text{-}1)$$
$$\ln(f_w) = 4.13 - 0.844*M \,(4.5 < M < 7.5) \quad (5.1\text{-}2)$$

External Events

5.1.3 Ground Coupling with Attenuation

The energy of the earthquake is ground-coupled and attenuated according to equation 5.1-3, where a is the maximum ground acceleration (cm/sec^2), at the location of interest, M is the magnitude of the earthquake (Richter or local magnitude), R is the distance (km) to the energy center or the causative fault, and b_1, b_2, and b_3 are coefficients that are determined from strong-motion data. For example, Donovan and Borstein, 1977, reported: b_1 = 2.154E6 R$^{-2.1}$, b_2 = 0.046 + 0.445*log$_{10}$(R), and b_3 = 2.515 - 0.4* log$_{10}$(R).

$$a = b_1 * \exp[b_2 * M * (R+25) - b_3] \quad (5.1\text{-}3)$$
$$I_s = I_o + 2.35 - 0.00316 * R \quad (R<2km) \quad (5.1\text{-}4)$$
$$\log_{10}(a_{pi}) = 0.25 * I_s + 0.25 \quad (5.1\text{-}5)$$

For the eastern and midwestern United States, where most recorded earthquake data are in MM intensity units, there are two approaches for determining the attenuation of ground motion. Method 1 is an intensity-attenuation relationship appropriate for the region selected. An example is given by Gupta (1976) for the Central United States (equation 5.1-4), where I_s is the site intensity in MM units and I_o is the epicentral intensity in MM units. The site intensity I_s is converted to the instrumental peak ground acceleration a_{pi} using an equation like equation 5.1-5.

The other method uses computer codes that predict ground acceleration:

- SRA (Seismic Risk Analysis), developed by C.A. Cornell at the Massachusetts Institute of Technology, 1975.
- EQRISK, a FORTRAN code developed by McGuire, 1976. It is available from the National Information Service for Earthquake Engineering, University of California, Berkeley.
- FRISK, a code for seismic risk analysis using faults as earthquake sources, developed by McGuire, 1978.
- Seismic Risk Analysis Program by C.P. Mortgat, the John A. Blume Earthquake Engineering Center, Stanford University, Stanford, California, 1978.
- HAZARD, developed at the Lawrence Livermore National Laboratory, Livermore, California, 1980.

5.1.4 Shaking Model of the Plant

Failure frequencies of structures, equipment, and piping are related to their acceleration which is related to the ground-motion of the plant's foundation (e.g., the peak ground acceleration). For PSA, it is useful to present the seismic hazard at the site as a family of hazard curves with different nonexceedence-probability levels (Figure 5.1-3). By selecting various values of the peak ground acceleration, the acceleration and forces on the plant components may be obtained as described in the following.

Seismic Events

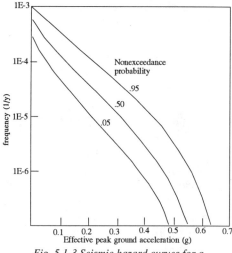

Fig. 5.1-3 Seismic hazard curves for a hypothetical site (NUREG/CR-2300d)

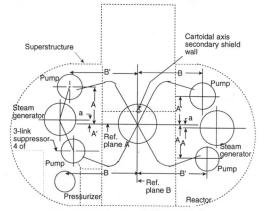

Fig. 5.1-4 Plan view of nuclear steam supply system

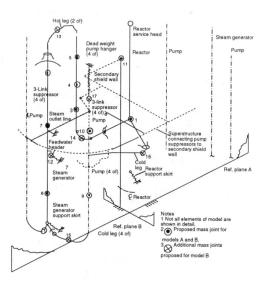

Fig. 5.1-5 Nodal diagram of a nuclear steam supply system

A typical plan view of a 4-loop PWR is shown in Figure 5.1-4. At first glance, the arrangement appears to be symmetrical about reference planes A and B, but the symmetry about plane B is less than it is about plane A.

The four primary coolant pumps are connected to the secondary shield wall by three-link snubbers designed to be flexible under static applied loads (thus, allowing thermal expansion) but become stiff under dynamic loads that might occur during an earthquake. Accordingly, the system is coupled to the wall under seismic loading.

Figure 5.1-5 shows a perspective view of one loop of the system. Note that in this system, kinematic constraints in the form of pipe supports, anchors, and hangers are minimal. Only the pumps are laterally braced by the suppressors. The pumps in this system are supported by hangers from above.

The initial problem is one of idealizing the piping system as a mathematical model that will reveal the significant modes of the structure. The primary piping system, including its components, can be adequately represented by three-dimensional beam elements. These have straight or curved centroidal axes, uniform or

191

nonuniform cross section, connections at structural joints, mass (translational as well as rotational) concentrations at selected structural "mass joints."

Although the concern is primarily for the response of the piping system, the possibility of dynamic coupling with the containment structure should not be neglected. A concern is whether or not the secondary shield wall will withstand the dynamic interaction between the walls and the pump. This is answered by examining the mode shapes if there were no coupling between the walls and the pump.

The natural frequency, ω_p, associated with the mode shape that exhibits a large displacement of the pump is compared with the fundamental frequency, ω_o, of the wall. If ω_p is much less than ω_o, then the dynamic interaction between the wall and the loop may be neglected, but the kinematic constraint on the pump imposed by the lateral bracing is retained. If ω_p nearly equals ω_o, the wall and steam supply systems are dynamically coupled. In which case it may be sufficient to model the wall as a one-mass system such that the fundamental frequency, ω_o is retained. The mathematical model of the piping systems should be capable of revealing the response to the anticipated ground motion (dominantly translational). The mathematics necessary to analyze the damped spring mass system become quite formidable, and the reader is referred to Berkowitz (1969).

The Seismic Safety Margins Research Program developed a computer code called SMACS (Seismic Methodology Analysis Chain with Statistics) for calculating the seismic responses of structures, systems, and components. This code links the seismic input as ensembles of acceleration time histories with the calculations of the soil-structure interactions, the responses of major structures, and the responses of subsystems. Since uses a multi-support approach to perform the time-history response calculations for piping subsystems, the correlations between component responses can be handled explicitly. SMACS is an example of the codes that are available for calculating seismic response for PSA purposes.

5.1.5 Fragility Curves

The fragility is the failure probability of a component conditioned on parameters, such as stress, moment, or spectral acceleration.

A fragility curve for a component is calculated by knowing the conditions that will fail it. These conditions are calculated, deterministically by a structure analyst. The PSA specialist determines the variability of the conditions which together gives the probability of failure vs acceleration forces for given operating conditions. A separate fragility curve is needed for each mode that must to be considered.

For example, Table 5.1-3. shows that a motor-operated valve can fail in several ways.

Table 5.1-3 Motor-Operated Valve Failure Modes

- Failure of power or controls to the valve (generally related to the seismic capacity of the cable trays, control room, and emergency power). These failure modes are analyzed as failures of separate systems linked to the equipment since they are not related to the specific piece of equipment (i.e., a motor-operated valve) and are common to all active equipment.
- Failure of the motor.
- Valve binding due to distortion.
- Rupture of the pressure boundary.

Analysis may show one mode of failure to be most likely in which case further analysis concentrates on that mode thereby considerably reducing the scope of the analysis. If there are several equally likely modes, they must all be analyzed.

The PSA Procedures Guide presents two methods for determining the fragility curves. The second method, outlined here was used in the Oyster Creek PSA.

Component fragility is expressed as the conditional probability of failure for a given peak ground acceleration. Data on seismically induced fragilities are generally not available for equipment and structures and must be developed from analysis supplemented with engineering judgment and limited test data. In view of this, maximum use is made of the licensing response-analysis that is performed to show that the plant will be safe under a DBE.

The component fragility for a particular failure mode is expressed in terms of the ground-acceleration A. The fragility is therefore the frequency at which the random variable A is less than or equal to a critical value, a. The ground-acceleration capacity is, in turn, modeled as equation 5.1-6, where $\overline{A}$ is the median ground-acceleration, $\epsilon_{A,R}$ is a random variable (with unit median) representing the inherent randomness about A, and $\epsilon_{A,U}$ is a random variable (with unit median) representing the uncertainty in the median value.

$$A = \overline{A} * \epsilon_{A,R} * \epsilon_{A,U} \quad (5.1\text{-}6)$$

It is assumed that both $\epsilon_{A,R}$ and $\epsilon_{A,U}$ are lognormally distributed with logarithmic standard deviations of β_{AR} and β_{AU}, respectively. The advantages of this formulation are:

- The entire fragility curve and its uncertainty can be expressed by only three parameters: $\overline{A}$, $\beta_{A,R}$, and $\beta_{A,U}$. With the limited data available on component fragility, it is necessary to estimate these parameters rather than the entire shape of the fragility curve and its uncertainty.
- The product form of equation 5.1-6, and its assumed lognormal-distribution make the fragility computations mathematically tractable.

The justification for the use of the lognormal is the modified Central Limit Theorem (Section 2.5.2.5). However, if the lognormal distribution is used for estimating the very low failure frequencies associated with the tails of the distribution, this approach is conservative because the low-frequency tails of the lognormal distribution generally extend farther from the median than the actual structural resistance or response data can extend.

Using equation 5.1-6 and the assumed lognormal-distribution, the fragility (i.e., the frequency of failure, f') at any nonexceedence probability level Q is given by equation 5.1-7, where $Q = P(f < f'/a)$ is the probability that the conditional frequency f is less than f' for a peak ground acceleration a. The quantity Φ is the Gaussian cumulative distribution function, and Φ^{-1} is its inverse. The nonexceedence-probability level Q is used for displaying the fragility curves. Subsequent computations are made easier by

$$f' = \Phi\left[\frac{\ln(a/A) + \beta_{A,U} * \Phi^{-1}(Q)}{\beta_{A,U}}\right] \quad (5.1\text{-}7)$$

treating the probability distribution as a histogram in values q_i associated with different values of the failure frequency f. A family of fragility curves, each with an associated probability q_i, is thereby developed.

For example, suppose a component has fragility parameters: $A = 0.739$, $\beta_{AR} = 0.30$, and $\beta_{AU} = 0.28$; then, from equation 5.1-7, the conditional failure frequency that is not exceeded with a 95% confidence at a ground acceleration of 0.59 is found to be 0.60. At 90% nonexceedence probability, the conditional failure frequency for a ground acceleration of 0.59 is approximately 0.52.

In estimating the fragility parameters, it is convenient to work in terms of an intermediate random variable known as the factor of safety F. This is defined as the ratio of the ground-acceleration capacity A to the safe shutdown earthquake (SSE) acceleration used in plant design.

The selection of components for fragility analysis is an iterative process between the system and structural analyst, with the system analyst providing the first iteration as was done for Diablo Canyon. An aid in selection may be those components previously identified as major contributors in seismic risk. For a typical nuclear plant, this list may include about 100 to 300 components, depending on the detail employed in the plant-system and sequence analysis. Some studies (Smith et al., 1981; Commonwealth Edison Company, 1981) have grouped the equipment in generic categories. The structural analyst develops the fragility curves for significant failure modes for each of these structures, systems, and equipment and may modify the list after reviewing plant design criteria, stress reports, and equipment-quantification reports and performing a walk-through inspection of the plant. By ordering the components by the acceleration necessary to cause failure, the list may be reduced even further to perhaps 10 major components.

5.1.6 System Analysis

The frequencies of plant damage and public consequence are calculated using plant logic combined with component fragilities. Event and fault trees are constructed to identify the accident sequences and the damage that may result from an earthquake. In performing a plant-system and accident-sequence analysis, the major differences between seismic and internal events analysis are given in Table 5.1-4

The first step-in plant-system and accident-sequence analysis is the identification of earthquake-induced initiating events. This is done by reviewing the internal analysis initiating events to identify initiating events relevant to seismic risk. For example, Table 5.1-5 shows the initiating events that were used in the Seismic Safety Margins Research Program for a PWR plant (Smith et al., 1981)

Table 5.1-4 Differences between Seismic and Internal Events Analysis

• Initiating events,
• Greater likelihood of multiple failures of safety systems requiring a more detailed event-tree development, and
• Greater dependencies between component failures as result of the correlation between component responses and between capacities.

The conditional frequency of each initiating event is calculated for different levels of acceleration (e.g., 0 to 0.159, 0.15 to 0.309, 0.30 to 0.459). The joint frequency distribution of responses at different critical locations in the piping is calculated. The convolution of the frequency distribution with the fragilities yields the conditional frequency of the initiating event. In preparing the list of the initiating events, it is necessary to consider the possibility of multiple initiating events rather than single, as is usually done for internal initiators.

The construction of event trees for earthquake-induced initiating events follows the methods described in Section 4.3. From these event trees, plant damage sequences are identified. Each of these may be followed by a containment analysis which establishes the release sequence. Figure 5.1-6 shows an event tree for a large LOCA in a PWR (Smith et al., 1981); it contains 28 core-melt sequences; each sequence can lead to a release through the potential containment-failure modes designated α, β, γ, δ, and ϵ.

Table 5.1-5 Initiating Events used in the SSMRP

- Reactor-vessel rupture,
- Large LOCA (rupture of a pipe larger than 6 inches in diameter or equivalent,
- Medium LOCA (rupture of a pipe 3 to 6 inches in diameter or the equivalent,
- Small LOCA (rupture of a pipe 1.5 to 3.0 inches in diameter or equivalent,
- Small-small LOCA (rupture of a pipe 0.5 to 1.5 inches in diameter or equivalent),
- Transient with the power-conversion system (PCS) operable, and
- Transient with the PCS inoperable.

5.2 Fires

5.2.1 Introduction

Considering the relatively small, controlled amount of combustibles in an LWR, it is surprising that fires at nuclear power plants should be important. The first incident to attract attention was the fire in the San Onofre cable trays (1969). This was followed by spontaneous combustion of uncured polyurethane foam in the cable seals at Peach Bottom 1 (1971). The incident at Browns Ferry in 1975 was similar, except that a candle ignited the polyurethane foam. These events showed the effectiveness of fire as an initiator of multiple system failures - some of which had been regarded as independent and redundant. Although there is little agreement

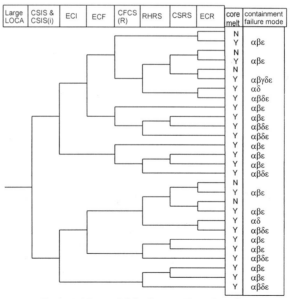

Fig. 5.1-6 Large LOCA Seismic Event Tree for a PWR

as to how close any came to causing a serious accident, it is clear that the impact could have been extensive. PSAs performed after the Browns Ferry fire have considered fires in estimating the risk.

The frequency of fire-induced core melt, calculated by averaging the observed frequency of the Browns Ferry type of fire over the experience of U.S. commercial nuclear power plants, was found to be 1E-5 per reactor-year, or about 20% of the total core-melt probability estimated in the Reactor Safety Study. Kazarians and Apostolakis (1978) performed the same type of calculations under different assumptions and concluded that the frequency of core melt could be higher by a factor of 10.

A fairly detailed risk analysis of fires was in the Clinch River Breeder Reactor (CRBR) Risk Assessment Study, 1977. In this study, FMEA was used to identify important fire locations for a wide variety of combustibles, including cables, oil, and sodium. The resulting estimate of the frequency of fire-induced core melt, 5E-7 per reactor-year, is substantially below the estimates discussed above.

It should be pointed out that the San Onofre fire occurred during construction, at which time many more combustibles are present than in normal operation. Plant operators maintain strict control over combustibles as specified by 10 CFR 50 Appendix R.

5.2.2 Procedures for Fire Analysis

The same basic procedures are common to the CRBR PSA (1977); Apostolokis and Kazarians (1980); and Gallucci (1980). They consist of the following steps:

1. Identify critical locations where fires can cause an initiating event that could fail redundant engineered safety functions, or disable redundant and diverse safety-related equipment.
2. Calculate the frequency of fires in these identified areas.
3. Calculate fire propagation considering the effects of detection and suppression.
4. Assess the effects of initiating fires and subsequent fire growth on the initiating events, such as LOCAs and transients.
5. Assess the effects of these fire scenarios on accident sequences in event trees for fire-induced initiating events.
6. Estimate the frequencies of various fire-initiated plant damage states.

These six major steps require consideration of: a) the occurrence frequency of fires, b) the physical effects of fires, and c) the response of the plant. Crucial is the plant response as it is affected by components damaged by fire, and by components unavailable for other reasons (e.g., random failures, maintenance, and fire-fighting activities).

Ruger et al. (1985) presents limitations of the methodology as it has been implemented.

5.2.3 Screening Analysis

The simplest type of screening analysis uses engineering judgment. Analysts survey the plant and identify areas where a fire would have safety significance. Although this method may identify the critical areas, it is not guaranteed to be complete since dependent failures, spatial interactions, and the possibility of zone-to-zone propagation may not be fully addressed. Fire protection reviews (Gallucci, 1980) employ such a screening approach. Because there are many rooms that contain some safety equipment, many locations may be identified, but in most plants only a small number of locations contribute significantly to the fire risk. Less critical locations may be eliminated by FMEAs.

The plant internal PSA can be used to identify critical equipment that could be damaged by fire. This form of screening was employed in the fire-risk portions of ZIP. At each location considered, the loss of all the equipment in the zone is postulated regardless of the size or position of the fire in the zone. If this does not show the occurrence of an initiating event (LOCA or transient) or if the safety functions are not damage to required for safe shutdown, the location is eliminated from consideration. If the location is found to be critical, it is considered further for detailed fire growth and fire suppression analyses.

5.2.4 Fire Frequencies

It is not sufficient to determine critical zones; the likelihood of fire in the zone is also important. For example, the fire-risk analysis for the Limerick Generating Station (Philadelphia Electric Co, 1983) included a systematic ranking of each fire zone by its contribution to the fire-induced core-melt frequency. After establishing which zones could have a significant fire that would both cause an initiating event and adversely affect the performance of mitigating systems, the frequency of such fires was determined. Data were used to estimate the frequency of fires in plant locations such as the reactor enclosure and the control structure (Philadelphia Electric, 1983). Then the frequencies of fires for the individual fire zones were calculated by partitioning them according to their location based on the ratio of the weight of combustible material at the location to the total weight of combustible material.

Table 5.2-1 Statistical Evidence of Fires in Light Water Nuclear Power Plants* (<May 1978)

Area	Number	Years
Control room	1	288.5
Cable-spreading room	2	301.3
Diesel-generator room	10	543.0
Containment	5	337
Turbine building	9	295.3
Auxiliary	10	303.3

* From Apolostakis and Kazarians, 1980, with permission.

Considerable effort has been spent in evaluating the fire events cataloged in licensee event reports and data from the American Nuclear Insurers (Hockenbury and Yeater, 1980; Fleming et al., 1979). The nature and the frequency of fires at nuclear power plants change dramatically between construction, preoperational testing, and plant operation, hence, data should be appropriate to the phase being evaluated.

A number of issues arise in using the available data to estimate the rates of location-dependent fire occurrence. These include the possible reduction in the frequency of fires due to increased awareness. Apostolakis and Kazarians (1980) use the data of Table 5.2-1 and Bayesian analysis to obtain the results in Table 5.2-2 using conjugate priors (Section 2.6.2). Since the data of Table 5.2-1 are binomially distributed, a gamma prior is used, with α and β being the parameters of the gamma prior as presented inspection 2.6.3.2. For example, in the cable- spreading room fromTable 5.2-2, the values of α and β (0.182 and 0.96) yield a mean frequency of 0.21, while the posterior distribution α and β (2.182 and 302.26) yields a mean frequency of 0.0072.

Table 5.2-2 Distribution of the Frequency of Fires[a]

Area	Gamma para.		Frequency of fires/room-year			
	α	β	5%	mode	95%	mean
Control room						
Prior	0.182	0..96	5E-8	0.0105	1.0	0.21
Posterior	2.182	289.46	3.1E-4	0.003	0.012	0.0041
Cable-room						
Prior	0.182	0.96	5E-8	0.015	1.0	0.21
posterior	543.29	302.26	1.4E-8	0.0062	0.017	0.0072
Diesel-room						
Prior	0.32	0.29	2.1E-4	0.30	5.0	1.11
Posterior	10.32	542.29	1.1E-2	0.018	0.03	0.019
Containment						
Prior	0.32	0.29	2.1E-4	0.30	5.0	1.11
Posterior	5.32	295.59	6.2E-3	0.014	0.028	0.106
Turbine bldg						
Prior	0.32	0.29	2.1E-4	0.30	5.0	1.11
Posterior	9.32	295.59	1.7E-2	0.014	0.05	0.32
Auxiliary bldg						
Prior	0.32	0.29	2.1E-4	0.30	5.0	1.11
Posterior	10.32	303.59	1.9E-2	0.033	0.053	0.034

[a] Apostolakis and Kazarians, 1980 with permission.

5.2.5 Fire Growth Modeling

Three different approaches have been used for fire propagation. The first uses a statistical model from experience (Fleming, 1979). The second uses a multistage event tree model (Gallucci, 1980), and the third employs deterministic physical models (Siu, 1980; Siu and Apostolakis, 1982). Because the behavior and effects of fire propagation depend on the geometry of the combustibles and the surroundings, the physical modeling approach was used later in fire-risk analyses. The deterministic model contains methodology explicitly incorporating the physics of enclosure fires using the computer code COMPBRN (Siu, 1982).

Three primary assumptions underlie the methodology are:

- The large size of nuclear power plant enclosures and the relatively small amounts of readily ignitable fuel in those enclosures make flash combustion unlikely; therefore, the concentration is on the fire growth phase.
- The growth of fires is treated with simple models.
- Fire growth and suppression are treated separately as independent processes.

The COMPBRN code uses simple physical models to calculate the heat transferred from a fire to its surroundings, the time to ignite or damage affected materials, and the rate of fire growth. The deterministic modeling is a synthesis of simplified, quasi-steady models called the zone approach. There are other computer codes (Mitler and Emmons, 1981; Quintiere, 1977; Tatem, 1982; Zukowski and Kubota, 1980; and Delichatsios, 1982) for analyzing fires in enclosures that also use the unit approach. Of particular interest is DACFIR (MacArthur, 1982), a code for modeling fire growth in an aircraft cabin from seat to seat - similar a fire spreading from cable tray to cable tray as analyzed in COMBRN.

Generally these codes/models are limited in ability to incorporate all of the aspects of fire while still maintaining a simple physical description of how enclosure fires develop. This requires balancing mathematical detail against physical realism.

The zone model has the following basic limitations: 1) complex enclosure geometries cannot be addressed, 2) forced ventilation cannot be realistically modeled using simple unit models, 3) burning other combustibles remote from the initiating source are not modeled, and 4) suppression activities are not included.

5.2.6 Fragility Curves

Using fire models, locations of equipment, heat transfer calculations, and environmental qualifications of the equipment, it is possible to estimate the time to failure. Fragility curves that relate fire durations and equipment damage while considering the probability of fire suppression are produced to relate to the overall PSA. These fragility curves and their use is similar the methods used for seismic analysis.

5.2.7 Systems Analysis

The fire systems analysis is similar to seismic systems analysis (Section 5.1.6). However, caution is needed regarding the inclusion of operator actions, with their large uncertainties, in fire risk analysis. Operators can substantially influence accident scenarios by extinguishing fires, manually operating equipment, repairing or improvising equipment. Negatively, faulty information may cause detrimental actions. Past analyses used crude models of operator actions with large uncertainties. Human action was responsible both for causing the Browns Ferry fire and for preventing it from going to core melt.

The quantification of accident sequences involving fires follows the general methodology for event trees and fault trees. Special attention must be paid to intersystem dependencies introduced by fire. Although early analysis based on simple system reliability models indicated low accident probability, more recent estimates employing sophisticated plant and system-level models give higher risk. These estimates tend to be dominated by the effects of interactions that increase the probability of successive failures in an accident chain. Reactor operating experience indicates that successive multiple failures are more likely to result from human or physical interaction than random events.

It is unclear whether previously published fire risk analyses have adequately treated dependent failures and systems interactions. Examples of either experienced or postulated system interactions that have been missed include unrelated systems that share common locations and the attendant spatially related physical interactions arising from fire. Incomplete enumeration of causes of failure and cavalier assumptions of independence can lead to underestimation of accident frequencies by many orders of magnitude.

5.3 Flood

5.3.1 Overview of Internal Flooding

A characteristic of "external" events is their propensity for causing multiple failures. This is simplified somewhat for internal flooding by assuming equipment fails when it is submerged, hence, failure is determined by the water level and the location of equipment.

In this sense flooding fragility is practically a step function. Some fuzzing of the step due to spray occurs as water reaches rotating machinery, but for the most part, electrical shorts cause the equipment to fail as the water rises to the circuitry. Using a cross-section of the plant, critical water levels are identified from the elevations of critical components found in the internal events analysis. The event trees/fault trees for internal states are examined for common mode failure of the equipment as it is immersed. Thus, the extent of the common cause failures continues to grow as the water depth increases to the maximum possible for the water sources and discharges available. The analysis of internal flooding consists of the following:

- Identify the sources of flooding water
- Determine the probability of release and flow rate of the flooding water
- Determine the probability of failure of flooded protection systems
- Identify the equipment failures as a function of water depth
- Enter failure probabilities (ones) in the system model to determine the plant damage states as a function of water depth
- Combine the above with the internal events analysis to obtain plant risk and propagate the uncertainties

5.3.2 Internal Flooding Incidents

In June 1972, at Quad Cities Unit 1, a rupture in the circulating-water system caused the rapid flooding of a room containing a number of pumps for different systems. The equipment damaged

Table 5.3-1 Turbine Building Flooding in U.S. Nuclear Power Plants

Date of occurrence	Plant	Affected safety component	Spill rate	Remarks
\multicolumn{5}{c}{Source: Service water}				
June 1985	Surry 2	Service-water valve		Pump developed seal leak
October 1977	Surry 2	Service-water valves of all redundant trains		Personnel forgot to close valves that were opened for maintenance
October 1978	E. Hatch 1	Service-water valve		Valve body blew out during repair
October 1979	Dresden 2	Diesel-generator control cabinet		Fire-water leak
\multicolumn{5}{c}{Source: Condenser Circulating Water}				
January 1979	Crystal River 3		Large	Solenoid valve failed open and led to flooding
April 1977	Three Mile Island		Large	Circulating-water pump casing split 360°
October 1976	Oconee	Emergency feedwater pumps	Large	Pneumatic isolation valve opened when condenser manhole was open and spilled lake water into turbine building
October 1978	Surry 2	Service-water valves	Small	Intentionally flooded during maintenance
June 1972	Quad Cities 1	Many redundant and diverse safety related components	Very large	Valve closed inadvertently, and water hammer ruptured expansion joint

Data from the start of commercial nuclear power to July 1981 from Verna, 1982.

by the flood included four service-water pumps for residual heat removal, two diesel-generator cooling-water pumps, four condensate booster pumps, and three condensate-transfer pumps. In addition, the floor-drain sump pumps, the hypochlorite system analyzer, and the condensate pressure gauges were damaged. Although the reactor was not damaged, this flood was impressive in terms of the failure of multiple components and systems. Modifications were made at Quad Cities to enhance the physical separation of safety-related pumps to prevent a recurrence of a flood in the same room. As a result of the NRC follow-up, various modifications were also made at 10 other plants to enhance protection against the flood-induced loss of safety functions (Verna, 1981).

Table 5.3-2 Flooding Frequencies for Turbine and Auxiliary Buildings

Location	Severity	Flooding frequency (1/yr)			
		5%	Median	95%	Mean
Auxiliary building	Small	2.0E-6	3.4E-4	1.0E-2	3.1E-3
	Moderate	1.6E-4	7.4E-3	3.1E-2	1.5E-2
	Large	1.0E-6	4.5E-4	2.0E-2	6.3E-3
	Moderate & large	2.5E-5	3.3E-3	1.6E-2	1.6E-2
Turbine building service water source	Moderate to large	2.9E-7	1.4E-3	2.5E-2	4.9E-3
Circulating water source	Moderate to large	2.2E-3	1.2E-2	1.3E-1	2.8E-2

External Events

A similar flood occurred at Three Mile Island Unit 1 in April, 1977, caused by a leak in the circulating-water system casing of one of the circulating-water pumps. However, because of the plant's layout, damage was confined to the six circulating-water pumps and did not affect any other systems (Verna, 1981).

These and other incidents summarized in Table 5.3-1, are taken from the PSA Procedures Guide (NUREG/CR-2300). The evidence of this experience is that flooding is a credible event. Table 5.3-2 (ibid.) presents the frequency of occurrence using the following categories:

<u>Small</u> - Hundreds of gallons (e.g., valve pit flooding, flooding of an instrument, or flooding with a component).
<u>Moderate</u> - Thousands of gallons (e.g., a few feet of water on the floor of a typical pump room).
<u>Large</u> - Tens of thousands of gallons [e.g., a few feet of water in large rooms, very deep water (more than 10 feet) in a typical pump room].
<u>Very large</u> - Hundreds of thousands of gallons (e.g., floods involving circulating-water or service-water piping).

Of course, such generic data only give an idea of the importance of the problem. Actual plant-specific data must be developed, using plant models.

5.3.3 Internal Flood Modeling

Location and the elevation are important in flooding analysis. Codes such as COMCAN III or SETS for dependent failure analysis listed in Table 3.6-4 may be used for locating components that would be affected by flooding. These codes serve this purpose with a complex component identifier for location and elevation.

A flooding analysis for major plant systems uses a qualitative fault tree that takes into account the elevation of system components. This procedure, illustrated in Figure 5.3-1, uses a fault tree captioned for the top event, "core melt due to internal flood." The fault tree is developed under the assumption that a flood causes a transient, small, or large LOCA, or causes the failure of

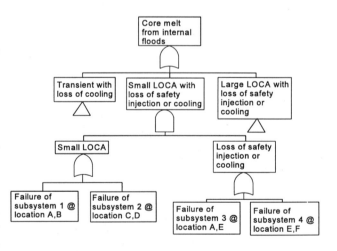

Fig. 5.3.1 Fault Tree Identification of Flood-Critical Equipment; the elevation information enters in the subtrees

a system or subsystem for mitigating the initiating event, or a combination thereof. The tree is developed only to the level of major subsystems (e.g., high-pressure-injection train A) to resolve the effect of losing groups of components in specific locations. However, support systems whose failure has a significant impact on the availability of the main safety-related systems must be included.

The simplest way to rank locations is inversely to the size of the minimal cutsets that result from postulating the flood. In this case, location A ranks above the rest, since flooding at this location produces the top event (common to both trains). If system unavailabilities from causes independent of the flood are known, a better ranking can be made using the conditional frequency of core melt given each failed location. After determining the important flood-critical locations, the analyst lists the major sources of water at the plant: circulating water, condensate, feedwater, service water, component cooling water, makeup and purification, spent-fuel pool, reactor-coolant, safety injection, and decay-heat-removal systems. A qualitative evaluation of each system and flooding source identifies and selects systems for quantification. An FMEA (Section 3.3.5) especially designed for this is useful.

If the PSA were constructed in the ZIP format, these common cause effects would be incorporated by changing the event tree branching probabilities as a function of water depth as the flood fails the various systems.

IPE Flood Findings

The contribution from internal flooding was found to vary from negligible to 40% of the CDF. The CDF from flooding ranges from negligible to 7E-5/y. It is not a major contributor to the CDF for most plants, but for two of the IPEs (3 units) it contributes about 40% of the total CDF. The highest CDF from internal flooding for the group is 7E-5/ry at H.B. Robinson and Surry 1&2. This constitutes 20% of the total CDF for H.B. Robinson and 40% for Surry 1&2 associated with breaks in SW or CCW lines that cannot be isolated by closing valves. These breaks lead to flooding of the turbine building, switchgear room, and auxiliary building.

The Surry IPE indicates that the high CDF from internal flooding is attributable to the relatively small size of the auxiliary building and the location of safety related equipment in areas in that building are sensitive to flooding. One IPE reports internal flooding as negligible but a number is not reported, hence, the lowest CDF is not known. The lowest reported CDF is 2E-6/y for Summer; the average is 2E-5/y.

5.3.4 External Flooding

River flooding has been studied by many authors, and it is generally believed to have an extreme value distribution (Gumbel, 1958). Wall (1974) considered this problem with regard to nuclear power plants. His approach was the statistical analysis of river-discharge data by means of various curve-fitting techniques. Different types of distributions were fitted to the same 44 years of river discharge data, using maximum-likelihood estimators: log-Pearson type III, lognormal, and extreme value distributions. Wall noted that excessive extrapolation of these curves (Figure 5.3-2) beyond the range of the data requires an estimate of the frequency of floods approaching the magnitude of the PMF (probable maximum flood) for the site in question. The consequences of the PMF were evaluated in terms of the warning time, damage to offsite-power supplies, and the role of watertight barriers. Wall concluded that the risk of a serious reactor accident due to rising water levels is negligible and proposed that the design-basis flood be redefined as that having a frequency of exceedance of 5E-4 over the next 50 years, or 1E-6 per reactor-year.

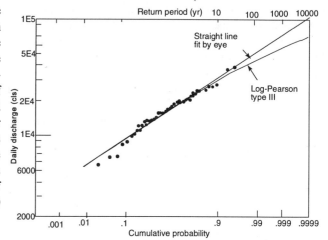

Fig. 5.3-2 Flood data plotted lognormally, from Wall, 1974

5.4 Summary

This chapter overviews the techniques for incorporating external events into a PSA. The discussion was primarily aimed at nuclear power plants but is equally applicable to chemical process plants. The types of external events discussed were: earthquakes, fires and floods. Notably absent were severe winds and tornados. Tornados are analyzed as missiles impacting the structures and causing common-cause failures of systems (EPRI NP-768). Missile propagation and the resulting damage is a specialized subject usually solved with computer codes.

The analysis methods are similar for all external events: probability of the external event, probability of failures, effects of failures on safety systems, and estimating the effects of failures for the workers, public and environment.

Chapter 6

Analyzing Nuclear Reactor Safety Systems

A PSA requires an intimate understanding of a facility or process before beginning. It has wide applicability, but its basic development was for nuclear power safety. The reasons it came into fruition with nuclear power were: to understand the insurance liability, little operating experience, and ample resources from government and industry. This chapter discusses current U.S. light water, reactors,[a] and advanced light water reactors (the parallel chapter for chemical processes follows). It goes on to discuss how PSA is applied to reactors with a demonstration fault tree analysis of an emergency diesel-electric power system using the FTAP code that is provided.

6.1 Nuclear Power Reactors

The Fission Reaction

When a neutron enters the nucleus of a fissionable atom (U-235 when the reactor core is new, but Pu-239 after operation) several neutrons come out of the atomic debris. These neutrons also can produce fission and more neutrons. Given a suitable geometry, a chain reaction may be sustained. The effectiveness of neutrons in causing fission is greatly enhanced by slowing them down (moderating) in a low atomic mass material such as water which also may be the coolant. Nuclear fission results in the prompt release of kinetic energy followed by a slow release of a smaller amount of energy as the unstable fission fragments decay.

The basic requirements of a reactor are: 1) fissionable material in a geometry that inhibits the escape of neutrons, 2) a high likelihood that neutron capture causes fission, 3) control of the neutron production to prevent a runaway reaction, and 4) removal of the heat generated in operation and after shutdown. The inability to completely turnoff the heat evolution when the chain reaction stops is a safety problem that distinguishes a nuclear reactor from a fossil-fuel burning power plant.

---

[a] Many types of reactors were investigated in the 1950s and 1960s such as: homogenous, molten salt, liquid metal, liquid metal breeder, thorium cycle, gas cooled. Heavy water cooling and graphite-reflected reactors were used to produce plutonium. Presently, it seems that the slightly enriched, light water cooled and reflected reactor is becoming the standard throughout the world. If uranium supplies become short, the liquid metal breeder may have a resurgence.

6.1.1 U.S. Light Water Reactors

Nuclear electric power, in the U.S. and most of the world, is produced by light water reactors (LWR), i.e., nuclear reactors using ordinary water for cooling and moderation. Because water absorbs neutrons it is necessary to enrich U-235 to about 3% of the total uranium before a sustained reaction can occur. As the reaction continues, the fissionable material gradually depletes until refueling is needed. The fuel is in the form of small cylindrical ceramic uranium dioxide pellets. Many of these are inserted in a thin-wall sealed zirconium tube to form a fuel rod. Many rods are assembled to form a fuel element which are arranged to approximate a cylinder with space between elements for cooling water to flow and for neutron-absorbing control rods to regulate the reactivity and to stop the nuclear chain reaction.

LWRs are of two designs: Pressurized Water Reactors (PWR) and Boiling Water Reactors (BWR). PWRs were developed for submarine propulsion, and adapted to electric power production, in the U.S., by Westinghouse (W), Combustion Engineering (CE) and Babcock and Wilcox (B&W). The BWR was developed and manufactured by General Electric (GE) specifically to produce nuclear electric power. The information presented here is primarily taken from the generic safety analyses prepared by these companies (RESSAR, GESSAR, CESSAR and BOWSAR). Additional references are Lish (1972) and Lewis (1977).

Although the Westinghouse's PWR Shippingport reactor was the first LWR to generate electricity in the U.S., GE's BWR Dresden 1 reactor followed within a year. Operating power reactors range from 600 to 1,200 MWe (million watts of electric power). Since the thermodynamic efficiency is ~33%, the thermal heat production is 1,800 to 3,600 MWt. Both types of reactor operate at about the same temperature (~600°F).

The BWR operates at a pressure such that cooling water boils as it passes through the core; the steam is dried, exits containment and drives the turbine-generator to produce electricity. This direct cycle is like a fossil-burning power plant.

The PWR operates at a pressure such that cooling water does not boil as it passes through the core. It goes to a heat exchanger (steam generator) to form steam that leaves containment, and drives a turbine generator to produce electricity.

Nuclear power plants have many elements in common regardless of manufacturer. Electric power is conducted to and from the plants by the same transmission lines and electrical equipment. The generator voltage is about 20 to 25kV and must be transformed into 100s of kilovolts for long distance transmission. For use in the plant, it must be reduced to 7 kV for high power, to 480V for medium power and 120V for low power. Redundant emergency electric power is generally provided by diesel generators but gas turbines and hydroelectric sources are also used. Instrumentation, and some control power, is supplied from banks of continuously charged storage using battery-inverters to provide "uninterruptable" AC power.

The plant, including reactor, is cooled by service water (SW) flowing to a heat sink such as a cooling tower, cooling pond, river or sea water. The Closed Cooling Water (CCW) loop transfers heat to the SW using a heat exchanger so that CCW water containing radioactivity is not exposed to the environment

A containment structure encloses the PWR or BWR reactors. The BWR containment is in

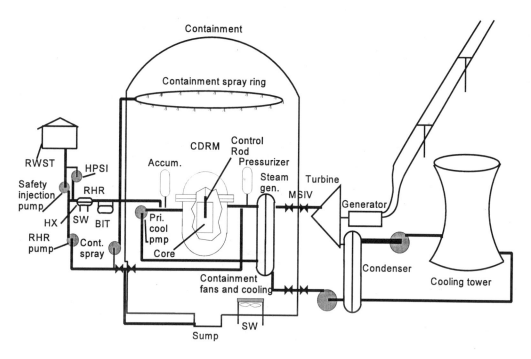

Fig. 6.1-1 Sketch of a Pressurized Water Reactor (PWR)
Heat is generated by the core, reactivity is controlled by neutron-absorbing control rods, water is circulated through the core by primary pumps to the steam generator. In a LOCA, the pressurizers inject water as the pressure drops, steam in the containment is condensed by the containment spray and containment cooling that is connected to the SW. The MSIV (main steam isolation valves) and other double valves isolate the containment. The refueling water storage tank (RWST) and pumping water drained into the sump supplies water for the high pressure safety injection (HPSI), the safety injection pump and the residual heat removal system (RHR). In normal operation, steam driving the turbine is condensed and returned by the feedwater pump to the steam generator. The turbine is driven by steam to power the generator and send electricity to the network. Water flows through the secondary side of the condenser through the cooling tower heat sink

another, secondary containment, building.

An auxiliary building houses support equipment for the reactor that can be located outside of containment.[b] The turbine building houses: turbine, electric generator, and related electrical equipment. The control room building contains the controls for operating the plant or multiple plants in case of the latter, multiple control panels are located in the same large room. Controls rooms are air conditioned and are capable of being isolated to assure a habitable environment for controlling the reactor in spite of possible hazardous chemical and smoke environments outside of the control room. Control rooms have breathing apparatus available. If the control room must be evacuated, a remotely located shutdown panel is available.

---

[b] The limited access to the containment building either to facilitate isolation or for vapor or hydrogen explosion suppression makes it desirable to locate as much equipment outside of containment as possible.

6.1.2 Pressurized Water Reactors

Figure 6.1-1 illustrates a PWR. The rectangle in the reactor represents the core through which water serving as moderator, coolant, and reflector flows to the primary side of the steam generator and back to the main cooling pump. This figure shows only one primary loop, for simplicity, but PWRs typically have 3 or 4 loops depending on the power capability. The pressure in the primary loop is controlled to be ~2,250 psi by the pressurizer by alternately heating its water to steam or by condensing its steam with a water spray. Safety relief valves (not shown) on the pressurizer prevent potentially damaging excessive pressure. About 50 to 100 control rods, depending on the reactor, control the reactivity using a neutron absorber such as boron.

A scram causes the control rods to drop into the core, absorb neutrons and stop the chain reaction. Some rods perform both controlling and scram functions. The control rods are raised to increase the neutron flux (and power) or lowered to reduce it by magnetic jacks (W and CE) or a magnetic "clamshell" screw (B&W). The chemical volume and control system (CVCS - not shown) controls the water quality, removes radioactivity, and varies the reactivity by controlling the amount of a boron compound that is dissolved in the water - called a "poison." Thus, a PWR controls reactivity two ways: by the amount of poison in the water and by moving the control rods.

Steam from the steam generator exits containment through a main steam isolation valve (MSIV) to drive the turbine. When most of the energy in the steam has been removed by the turbine, the steam is condensed to liquid and returned to the steam generator by the feedwater pump. The heat removed by the condenser goes to the heat sink.

Manifold barriers confine the radioactivity to the: 1) ceramic fuel pellet; 2) clad; 3) cooling water, as demonstrated by the TMI-2 accident; 4) primary cooling loop; 5) containment; and 6) separation from the public by siting. Further protection is provided by engineered safety systems: pressurizers, depressurization, low pressure injection, high pressure injection and residual heat removal systems.

Emergency Cooling

If a small break occurs in the primary cooling boundary, emergency equipment cannot inject cooling water against the normal reactor pressure. When the pressure drops to about 1,800 psi, the High Pressure Safety Injection (HPSI) pump begins injecting cooling water. When the pressure drops to around 600 to 800 psi, gas pressure in the accumulator exceeds the primary pressure and water in the accumulators is discharged into the primary. As the pressure drops lower, the high-volume low-pressure, safety injection pumps discharge either containment sump water or water from the Refueling Water Storage Tank (RWST) into the reactor vessel.

TABLE 6.1-1. List of PWR Transient Initiating Events (from NUREG-2300)

1. Loss of RCS flow (one loop)
2. Uncontrolled rod withdrawal
3. Problems with control-rod drive mechanism and/or rod drop
4. Leakage from control rods
5. Leakage in primary system
6. Low pressurizer pressure
7. Pressurizer leakage
8. High pressurizer pressure
9. Inadvertent safety injection signal
10. Containment pressure problems
11. CVCS malfunction - boron dilution
12. Pressure, temperature, power imbalance-rod-position error
13. Startup of inactive coolant pump
14. Total loss of RCS flow
15. Loss or reduction in feedwater flow (one loop)
16. Total loss of feedwater flow (all loops)
17. Full or partial closure of MSIV (one loop)
18. Closure of all MSIVs
19. Increase in feedwater flow (one loop)
20.. Increase in feedwater flow (all loops)
21. Feedwater flow instability - operator error
22. Feedwater flow instability from mechanical causes
23. Loss of condensate pumps (one loop)
24. Loss of condensate pumps (all loops)
25. Loss of condenser vacuum
26. Steam-generator leakage
27. Condenser leakage
28. Miscellaneous leakage in secondary system
29. Sudden opening of steam relief valves
30. Loss of circulating water
31. Loss of component cooling
32. Loss of service-water system
33. Turbine trip, throttle valve closure, EHC problems
34. Generator trip or generator-caused faults
35. Loss of all offsite power
36. Pressurizer spray failure
37. Loss of power to necessary plant systems
38. Spurious trips - cause unknown
39. Automatic trip - no transient condition
40. Manual trip - no transient condition
41. Fire within plant

Table 6.1-2 Typical Frontline Systems for PWRs

LOCA		Transients	
Frontline system	Function	Frontline system	Function
Render reactor subcritical	Reactor protection system	Render reactor subcritical	Reactor protection system Chemical volume and control High pressure injection system
Remove core decay heat	High pressure injection system Low pressure injection system High pressure recirculation system Core flood tanks Auxiliary feedwater system Power conversion system	Remove core decay heat	Auxiliary feedwater system Power conversion system High pressure injection system power-operated relief valves
Prevent containment overpressure	Reactor building spray injection system Reactor building spray recirculation system Reactor building fan coolers Ice condensers	Prevent containment overpressure	Containment spray injection system, containment spray recirculation system, containment fan cooling system, ice condenser
Scrub radioactive materials	Reactor building spray injection system Reactor building spray recirculation system Ice condensers	Scrub radioactive materials	Containment spray injection system, containment spray recirculation system, ice condenser

Table 6.1-3 ANO-1 Frontline vs Support and Support vs Support Dependencies
(All requirements for diesel generators assume loss of station power.)

Support Systems → Frontline systems ↓	Offsite AC power	Diesel AC generators	125VDC power	Engineered safeguards actuation system	Emergency feedwater initiation and control system	Service water system	Instrument air system	Integrated control system	Intermediate cooling system	AC switchgear room cooling	DC switchgear room cooling	High pressure pump room	Low pressure spray pump	Non-nuclear instrumentation power
Reactor protection systems														
Core flood systems														
High pressure injection recirculation	X	X	X	X		X				X	X	X		
Low pressure injection decay heat removal	X	X	X	X		X				X	X		X	
Reactor building spray injection recirculation	X	X	X	X		X				X	X		X	
Reactor building cooling system	X	X	X	X		X				X	X			
Power conversion system	X	X	X	X		X				X	X			X
Emergency feedwater system	X	X	X	X		X				X	X			
Pressurizer safety relief valves														
Support systems														
Offsite AC power														
Diesel AC generators			X	X		X				X	X			
125 VDC power	X	X						X			X			
Engineered safeguards actuation system	X	X	X							X				
Emergency feedwater start/control system	X	X	X							X	X			
Service water system	X	X	X	X						X	X			
Instrument air system	X		X			X		X	X					
Integrated control system	X	X	X							X	X			
Intermediate cooling system	X		X	X		X				X	X			
AC switchgear room cooling	X	X				X								
DC switchgear room cooling	X	X				X				X				
High pressure pump room cooling	X	X		X		X				X				
Low pressure spray pump room cooling	X	X		X		X				X				
Non-nuclear instrumentation power	X	X	X								X			

In an accident, the primary loop is isolated so normal heat removal paths are not available, but the residual heat removal (RHR) system discharges heat outside of containment through its heat exchanger. It may inject additional boron, depending on the design, using the Boron Injection Tank (BIT) to prevent the reactivity from rising as the temperature drops. The RHR system is used in normal shutdown to remove the residual heat from the decay of the fission fragments. If this were a large pipe break the pressure would drop rapidly and HPSI would not actuate. In case of a transient, heat is removed by the condenser and the RHR. Steam released by such accidents could fail the containment by heat and pressure. To prevent this, containment spray condenses the steam using water pumped from the sump through a heat exchanger (not shown) and sprays it from upper containment. The containment atmosphere is also cooled by flowing air through an air-water heat exchanger called the Containment Fans and Coolers.

Frontline and Support Systems

Nuclear power plant systems may be classified as "Frontline" and "Support" according to their service in an accident. Frontline systems are the engineered safety systems that deal directly with an accident. Support systems support the frontline systems. Accident initiators are broadly grouped as loss of cooling accidents (LOCAs) or transients. In a LOCA, water cooling the reactor is lost by failure of the cooling envelope. These are typically classified as small-small (SSLOCA), small (SLOCA), medium (MLOCA) and large (LLOCA).

- SSLOCA ranges in pipe break size up to 3 inches; it is mitigated by high pressure injection from typically 1 of 3 pumps
- SLOCA encompasses pipe break sizes in the range of 1 to 8 inches; it is mitigated by high pressure injection from 2 out of 3 pumps and 2 out of 3 accumulators
- MLOCA range from 6 to 18 inches; they are mitigated by 2 out of 3 accumulators and 1 out of 2 low pressure pumps
- LLOCA encompasses the largest pipes in the plant; it is mitigated by the accumulators and 1 out of 3 low pressure - high volume pumps

A transient, is a passing event which may upset the reactor operation but does not physically damage the primary cooling envelope. Table 6.1-1 lists PWR transient initiating events that have been used in PRA preparation. Typical frontline systems that mitigate LOCAs and transients for a PWR are presented in Table 6.1-2. The frontline systems must be supported by support systems; interactions between both are presented in Table 6.1-3 for ANO-1 (Arkansas Nuclear Unit 1).

6.1.3 Boiling Water Reactors

Figure 6.1-2 is a simplified illustration of a BWR. The pressure of the moderator-cooling water at about half the pressure in a PWR forms steam as it flows upward through the core. Steam passes through a moisture separator (shown as vertical lines just above the core) exits the containment through Main Steam Isolation Valves (MSIV) drives the turbine and generates electricity. After the steam is cooled by the turbine, it is condensed, and pumped back to the reactor by the feedwater pump.

BWRs do not operate with dissolved boron like a PWR but use pure, demineralized water with a continuous water quality control system. The reactivity is controlled by the large number of control rods (>100) containing burnable neutron poisons, and by varying the flow rate through the reactor for normal, fine control. Two recirculation loops using variable speed recirculation pumps inject water into the jet pumps inside of the reactor vessel to increase the flow rate by several times over that in the recirculation loops. The steam bubble formation reduces the moderator density and

Analyzing Nuclear Reactor Safety Systems

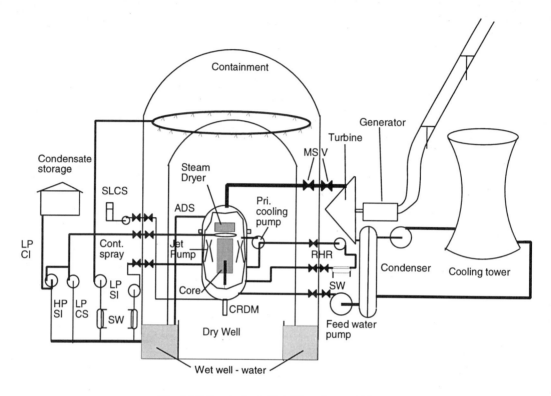

Fig. 6.1-2 Sketch of a Boiling Water Reactor(BWR)
The core heats water to steam which is dried, sent to the turbine that powers the generator sending electricity to the network.. Steam depleted of its energy is condensed and returned to the reactor vessel by feedwater pumps. Water is circulated in the reactor vessel by jet pumps which are powered by water circulated by the primary cooling pumps. The control rod drive mechanism (CRDM) is located in the bottom (away from the steam). In case of a LOCA, the automatic depressurization system discharges into the wet well to reduce the pressure so the high pressure safety injection (HPSI), low pressure safety injection (LPSI), and the low pressure cooling system can operate. The standby liquid control system injects borated water to help shutdown the reactor

consequently the reactivity. Higher flow through the core reduces the bubble density and increases the reactivity. If a rapid shutdown (scram) is required, the control rods are inserted in from the bottom by hydraulic drives. Each control rod drive is separate and is actuated by deenergizing a solenoid valve to release high pressure accumulator water to drive the pistons and insert the rods. In case of scram accumulator failure, reactor pressure will insert the rods.

This system is backed-up by a Standby Liquid Control System (SLCS) that injects sodium pentaborate into the moderator using a positive displacement pump (shown as a piston pump). Steam that originates in the core of a BWR, unlike the primary coolant in a PWR, exits the containment. The closing of the MSIVs isolates the radioactivity from the environment but when this is done, normal heat removal is not possible. The Residual Heat Removal (RHR) system

Table 6.1-4 List of BWR Transient Initiating Events (adapted from NUREG-2300)	
1. Electric load rejection	2. Electric load rejection with turbine bypass valve failure
3. Turbine trip	4. Turbine trip with turbine bypass valve failure
5. Main steam isolation valve (MSIV) closure	6. Inadvertent closure of one MSIV
7. Partial MSIV closure	8. Loss of normal condenser vacuum
9. Pressure regulator fails open	10. Pressure regulator fails closed
11. Inadvertent opening of a safety/relief valve (stuck)	12. Turbine bypass fails open
13. Turbine bypass or control valves cause increase in pressure	14. Recirculation control failure increasing flow
15. Recirculation control failure decreasing flow	16. Trip of one recirculation pump
17. Trip of all recirculation pumps	18. Abnormal startup of idle recirculation pump
19. Recirculation pump seizure	20. Feedwater increasing flow at power
21. Loss of feedwater heater	22. Loss of all feedwater flow
23. Trip of one feedwater pump (or condensate pump)	24. Feedwater low flow
25. Low feedwater flow during startup or shutdown	26. High feedwater flow during startup or shutdown
27. Rod withdrawal at power	28. High flux due to rod withdrawal at startup
29. Inadvertent insertion of control rod or rods	30. Detected fault in reactor protection system
31. Loss of offsite power	32. Loss of auxiliary power (loss of auxiliary transformer)
33. Inadvertent startup of HPCI/HPCS	34. Scram due to plant occurrences
35. Spurious trip via instrumentation, RPS fault	36. Manual scram; no out of tolerance condition

provides another path for removing the residual heat. Initially in an accident, the reactor will be near operating pressure and the High Pressure Core Spray (HPSI) is the only way to inject cooling water but not at a high flow rate. The HPSI takes water from the suppression pool or from the condensate storage tank. An alternative is to trip the Automatic Depressurization System (ADS) and release the steam by way of downcomers below the water level in the suppression pool. This condenses the steam but heats the suppression pool water which is cooled by circulation through heat exchangers (a function of the RHR not shown in the Figure). As the pressure drops, the Low Pressure Core Spray begins and pumps water from the suppression pool through a heat exchanger and sprays the reactor core. During an accident, steam accumulates in the containment, hence, another mode of the RHR is to circulate suppression pool water through heat exchangers and spray the inside of the containment with cooling water. Table 6.1-4 lists BWR transient initiating events; Table 6.1-5 lists BWR frontline systems.

6.1.4 Advanced Light Water Reactors

Sections 6.1.2 and 6.1.3 describe first generation reactors designed in the 1950s. These designs began with requirements for producing power - auxiliary systems were added as needed. This incremental approach led to: complexity, high initial and operating costs, regulatory complexity, and safety concerns, especially following the TMI-2 accident.

The first generation of power reactors was designed by the vendors and offered to the

Table 6.1-5 Typical BWR Frontline Systems

LOCA		Transients	
Frontline system	Function	Frontline system	Function
Render reactor subcritical	Reactor protection system	Render reactor subcritical	Reactor protection system Standby liquid control system
Remove core decay heat	Main feedwater system Low pressure coolant injection system Low pressure core spray system Automatic pressure relief system High pressure coolant injection system Reactor core isolation system	Remove core decay heat	Power conversion system High pressure core spray system High pressure coolant injection system Low pressure core spray system Low pressure coolant injection system Reactor core isolation cooling system Feedwater coolant injection Standby coolant supply system Isolation condensers Control rod drive system Condensate pumps
Prevent containment overpressure	Suppression pool Residual heat removal system Containment spray system	Prevent reactor coolant overpressure	Safety relief valves Power conversion system Isolation condenser
		Prevent containment overpressure	Residual heat removal system Shutdown cooling system Containment spray system
Scrub radioactive materials	Suppression pool Containment spray system	Scrub radioactive materials	Suppression pool Containment spray system

utilities. With approximately 1,000 reactor-years of experience, the utilities, led by EPRI reversed this process by specifying the attributes for the next generation of plants. Requirements are given in Table 6.1-6. By inference, this table criticizes the first generation reactors.

6.1.4.1 Westinghouse AP-600

The development of the Advanced Passive 600 MWe reactor began in 1985 with the EPRI Small Plant Study, and continued under DOE support to result in an NRC certifiable design by the end of 1994 (McIntyre, 1992 and Vijuk, 1988). The design incorporates the experience and knowledge of Westinghouse, Bechtel Power Corporation, Burns and Roe, Avondale Industries, CBI Services, MK-Ferguson, and Southern Electric.

Table 6.1-6 Attributes for Second-Generation Reactors

Core damage frequency < 1E-5/reactor-year,
Severe events with a frequency of >1E-6/reactor-year to result in <25R (0.25 sieverts) a half-mile from the reactor
Greater than 87% availability averaged on the plant lifetime
Two-year refueling capability
Spurious scrams less than one per year
Sixty year lifetime
Fifty-four month construction time
Less than 2500 ft^3/year shipable rad waste
Less than 100 man-rem/year average lifetime exposure

The innovative features, such as: passive, gravity-driven ECCS, simplified RCS, digital I&C, optimized plant arrangement, and modular construction methods, characterize the AP600 design. These changes result in fewer systems, equipment, operations, inspections, maintenance, and QA requirements (Table 6.1-7). Features of the AP-600 are as follows:

Nuclear Power Reactors

NSSS

The RCS illustrated in Figure 6.1-3 has two cold legs and one hot-leg for each of the two loops. Two canned-motor pumps are welded to the bottom of steam generator. A large pressurizer is attached to one of the hot legs.

The low-power-density, low enrichment reactor core uses soluble boron and burnable poisons for shutdown and fuel burnup reactivity control. Low worth grey rods provide load following. A heavy uranium flywheel extends the pump coastdown to allow for emergency action during loss-of-flow transients.

The I&C uses unified, distributed, digital microprocessor-based technology with electrical and fiber-optic data links. Integrated I&C includes: plant protection, nuclear and balance-of-plant control, operational display, alarms, accident monitoring, plant computer control and emergency control boards. Other equipment, such as radiation and loose parts monitoring, flux mapping, and failed fuel detection, share a common plant parameter data base with the integrated systems via the monitor bus. Human factors guided the I&C and control room design to enhance operability and to decrease the probability of operator error. CRTs and qualified plasma displays provide the operator with information for normal and emergency operation of the plant. An advanced alarm system categorizes, prioritizes, and displays alarm messages and suppresses minor, "nuisance," alarms.

Passive Safety Systems

The AP600 passive safety systems use natural forces: gravity, natural circulation, and compressed gas. Simple, mostly fail-safe, valves align the passive safety systems upon

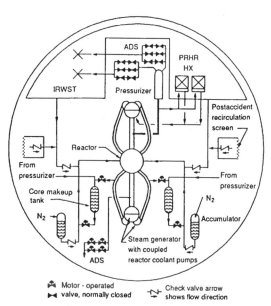

Fig. 6.1-3 The In-Containment Passive Safety Injection System: IRWST, PRHR HX and the ADS

Table 6.1-7 Comparison Showing Reductions in AP-600 Compared with a Comparable First-Generation Plant

Plant features	Reference 600-MW(e)	AP600	% reduce
Pumps - Safety - Nonnuc. safety	25 188	0 139	100 26
HVAC - Fans - Filter units	52 16	27 7	48 56
Valves - NSSS - BOP > 2 in.	512 2,041	215 1,530	58 25
Pipe - NSSS - BOP>2 in.	44,300 ft 97,000 ft	11,042 ft. 67,000 ft.	75 31
Evaporators	2	0	100
Diesel generators	2 (safety grade)	1 (non-nuc. safety)	50
Building volume - Seismic (inc. containment) - Nonseismic	9.4E6 ft 6.2E6 ft	4.6E6 ft 6.1E ft	51 2

automatic actuation. The passive safety systems do not require the large network of safety support systems needed in typical nuclear plants, e.g., AC power, HVAC, cooling water systems, and seismic structures. Safety-grade emergency diesel generators and their support systems: air start, fuel storage tanks, transfer pumps, air intake/exhaust systems, essential service water system and associated safety cooling towers are eliminated.

The AP600 passive safety system includes subsystems for: safety injection, residual heat removal, containment cooling, and control room habitability under emergency conditions. Several of these aspects are in existing nuclear plants such as: accumulators, isolation condensers as natural-circulation closed loop heat removal systems (in early BWRs), automatic depressurization systems (ADS - in BWRs) and spargers (in BWRs).

The passive safety injection system (PSIS; see Figure 6.1-3) performs three major functions: residual heat removal, reactor coolant makeup for inventory control, and safety injection. Calculations show that the PSIS prevents core damage for: breaks as large as 8 in., vessel injection lines, and 700° F margin to the maximum peak clad temperature limit for a double-ended rupture of a main reactor coolant pipe. The passive residual heat removal heat exchanger (PRHR HX) protects the plant against transients that upset the normal steam generator feedwater and steam systems. The PRHRHX consists of two elevated 100% capacity banks of tubes located in the in-containment refueling water storage tank (IRWST). These tubes, connected to the RCS in a natural circulation loop, remove decay heat under accident conditions. To prevent heat loss during normal operation, flow is blocked by two 10-in. air-operated valves. The IRWST water volume is sufficient to absorb decay heat for about 2 hours before it boils. Steam, from IRWST boiling, condenses on the inside of the steel containment vessel and drains back into the IRWST (Figure 6.1-4). The PSIS uses three sources of water to maintain core cooling: core makeup tanks (CMTs), accumulators, and the IRWST that supply through two nozzles in the reactor vessel. Passive reactor coolant makeup accommodates small leaks following transients or whenever the normal makeup system is unavailable. Gravity provides this cooling by providing water from two borated core makeup tanks (CMTs) at any RCS pressure. These tanks, that take full RCS pressure, are located above the RCS loop piping. On low-low reactor level, flow begins when the CMT discharge isolation valves open, reactor trips, and reactor coolant pumps trip. For large breaks (LOCAs up to the largest loop pipe) initial safety injection is provided by two CMTs and two accumulators. Each CMT accumulator pair have separate injection lines.

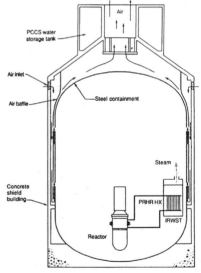

Fig. 6.1-4 AP500 Passively Cooled containment

Like current PWRs, accumulators provide the initial makeup flow in case of a LOCA. Each 1,700 ft^3 accumulator is pressurized with 300 ft^3 of nitrogen at 700 psi to force open check valves and inject into the RCS when pressure falls below this pressure. The IRWST provides long-term gravity injection water through check valves when the RCS is depressurized. The ADS uses four valves for a relatively slow, controlled depressurization. During a LOCA the IRWST provides injection for at least 10 hours. When the IRWST empties, the containment water level will exceed the RCS level, hence, it flows into the RCS where core heat turns it into steam which condenses on the containment wall and flows into the RCS again.

Westinghouse PSAs show that the AP600 has significantly reduced the core damage frequency from that of other PWRs. Accident consequences are reduced to make it technically feasible to eliminate the emergency planning zone. The AP600 containment penetrations are located so that any leakage is into the auxiliary building. The containment leak rate is so low that, if credit is taken for settling of particulate matter in the auxiliary building, there is no need for any mechanical systems for radionuclide reduction, thus no containment spray is needed for the AP600.

6.1.4.2 Combustion Engineering System 80+

In 1985, ABB Combustion Engineering Nuclear Power (ABB-CENP) and Duke Engineering and Services, Inc. (DE&S) under the aegis of EPRI, collaborated on the design of the next generation of nuclear power plants. With DOE support, the team prepared documents for NRC licensing.

Key features of the 1,300 MWe System 80+ (Bagnal, 1992) are: 15% improvement in thermal margin, slower response to upset conditions, no fuel damage up to a 6-in. break, and the ability to withstand an 8-hour station blackout. System 80+ integrates the nuclear island, turbine island, and balance of plant.

The safety improvements use redundancy and diversity to prevent and mitigate accidents. The safety injection system (SIS) and emergency feedwater system (EFWS) are dedicated four train systems. Containment spray and safety injection pumps take water from the in-containment water storage tank (IRWST), thus, eliminating the need to switch from an external source and provide a semi-closed system with continuous recirculation. Emergency core coolant flows directly into the reactor vessel to provide a simpler, more reliable system that avoids the need for orificing and valve adjustments or the potential for valve misalignment inherent in cold-leg injection schemes. The containment spray system (CSS) and shutdown cooling system (SCS) are integrated, and the pumps are interchangeable, thus improving backup and reliability. Shutdown cooling maintains pipe integrity even if it is accidentally exposed to the full primary system pressure; thus precluding the interfacing LOCA problem. Cavitating venturis minimize excess emergency feedwater flow to a steam generator with broken feed or steam lines. A spherical steel containment provides 75% more space on the operating floor than a typical cylindrical containment of equal volume. Steam generators may be removed in one piece. Safety systems located in the lower subsphere result in shortened pipe and cables to minimizes fire, flood, and sabotage concerns. A cylindrical, concrete shield building provides the added protection of dual containment.

System 80+ is designed to withstand an accident causing appreciable core damage with the large, 200 ft.-diameter, 3.4 million ft^3 free volume containment. Other features are: a reactor cavity that ensures coolability and retention of molten core debris, a passive cavity flooding system, and 39 hydrogen ignitors that operate independently of site power. A safety depressurization system (SDS) prevents containment failure from core-melt ejection heating. The reactor cavity configuration is designed to prevent debris transport and to provide coolability; it incorporates an exit area greater than the area around the vessel, a collection volume twice the core volume, and a floor area/power ratio greater than 0.02 m^2/MW(t). The flooding system uses passive gravity flow from the IRWST to the cavity via a holdup volume. The containment design prevents hydrogen buildup by natural circulation and can passively accommodate a metal-water reaction of up to 75% of the core metal without exceeding a hydrogen concentration of 13% by volume. The reactor cavity design prevents debris transport and provides core coolability. A dedicated SDS provides an alternative decay heat removal path through primary feed and bleed to rapidly decrease pressure and keep the core covered even when all feedwater is cut off. The SDS provides safety grade backup to the pressurizer spray for cooldown and steam generator tube rupture. PSA results show

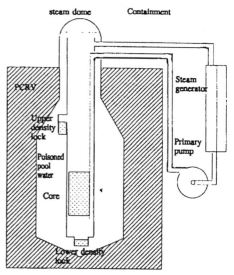

Fig. 6.1-5 Sketch of the PIUS Principle

reduction in core damage frequency by 100 compared with the preceding System 80.

6.1.4.3 ABB PIUS

The Process Inherent Ultimate Safety reactor, conceived by K. Hannerz (1983) is a 600 MWe pressurized water reactor with no control rods, and no active ECCS. It uses a prestressed concrete reactor vessel (PCRV) in a pressure suppression containment with circulation provided by "wet" motor variable-speed pumps. The primary loop consists of a flow assembly, containing the reactor, which is connected by pipes to four steam generators cooled by four variable speed pumps (one loop is shown in the Figure 6.1-5 sketch). The flow structure is immersed in a large (1E6 gal.), highly borated-water reactor pool. The pre-stressed concrete reactor vessel is capped by a removable steam dome for refueling. The slightly borated primary water rises into the pressure dome to interface with steam in the top of the dome. An external steam generator controls the reactor pressure (1,300 psi). The nuclear reactivity is controlled by the boron concentration in the primary loop that is regulated by an external chemical volume control system.

No mechanical barrier prevents entry of highly borated pool water into the primary loop through the density locks. Such entry is prevented by balancing the natural convective flow through the core with the primary pump flow. If the convective flow and the pump flow do not balance, such as occurs in an upset condition, flow through the pool flow begins. This highly borated water reduces the reactivity which further imbalances the flow to increase the loop flow - passive scram. There are three types of scram: active, manual and passive. Both active and manual methods scram by tripping power to a dedicated pump that unbalances the flows to the passively scram the reactor. Once the pool loop is activated, natural circulation flows through the lower density lock, the core, the upper density lock, the pool, and back to the lower density lock. This loop continues to circulate and transfer the residual heat to the pool water. The pool water is cooled by redundant active and passive cooling systems (NUREG/CR-6111).

6.1.4.4 General Electric ABWR

The 1,356 MWe Advanced Boiling Water Reactor was jointly developed by General Electric, Hitachi, and Toshiba and BWR suppliers based on world experience with the previous BWRs. Tokyo Electric Power operates two ABWRs as units 6 and 7 of the Kashiwazaki-Kariwa Nuclear Power Station. Features of the ABWR are (Wilkins, 1992):

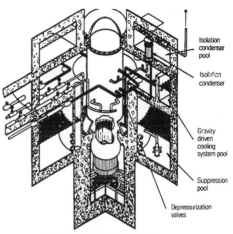

Fig. 6.1-6 SWBR LOCA Response
The suppression pool absorbs the blowdown energy, the reactor vessel is depressurized by the depressurization valves and then flooded by the gravity driven cooling system (GDCS). The isolation Condensers removes decay heat. No containment flooding is needed for most accidents

- No recirculation lines or pumps. They are eliminated by recirculation within the RPV using bottom-located electric pumps (Figure 6.1-2).
- Large pipe nozzles below the core are eliminated by eliminating the recirculation lines. This reduces ECCS requirements as the fuel is covered in spite of a full spectrum of postulated LOCAs with a single failure.
- A flatter flux profile provides wider safety margins than previous designs.
- Fine-motion of control rods is provided by hydraulic and electric drives.
- There are three redundant and independent divisions of ECCS and containment heat removal. The high pressure configuration consists of two motor-driven High Pressure Core Flooders, each with its independent sparger over the core. The Reactor Core Isolation Cooling system is safety grade to provide the dual function of high pressure ECCS flow following a LOCA, and reactor cooling inventory control for isolation transients. RCIC is steam-turbine powered in case of a LOOP. The lower pressure ECCS uses three RHR pumps to provide post-LOCA cooling, low pressure flooding and core makeup. The ADS reduces the RPV pressure to allow higher volume, lower pressure pumps to operate. The RHR provides cooling during normal shutdown and pressure-suppression pool and containment cooling during a LOCA.
- PSA estimates the risk to be 10% less than BWR 5 or 6.

6.1.4.5 General Electric SBWR

The 600 MWe Simplified Boiling Water Reactor was designed by an international team consisting of EPRI, General Electric, Bechtel, Burns and Roe, Foster Wheeler, Southern Company, Massachusetts Institute of Technology, University of California (Berkeley), other U.S. utilities,

Ansaldo, ENEL (Italy), Hatachi, Toshiba (Japan), KEMA, and Nucon (Netherlands). The design was selected and supported by the DOE. Key SBWR features satisfy design objectives consistent with today's needs (McCandless, 1989):

- No recirculation lines or pumps - the SBWR goes beyond the ABWR by eliminating all recirculation lines (Figure 6.1-6).
- It has a large reactor vessel surrounded by a passive pressure-suppression containment system that includes large water pools that inject by gravity to ensure water covers the core.
- The passive containment cooling system can remove residual heat from the containment for three days without operator action.
- Safety-grade equipment such as pumping systems and diesel generators and associated support equipment and electric power distribution is eliminated.
- The improved turbine design has a 52 in. last stage bucket to heat the feedwater.
- The advanced control room uses fiber optics, microprocessors, and digital monitoring and control. This includes self-testing, automatic calibration, user interactive front panels, full multiplex, standardization of the man-machine interface, common circuit cards, and wide-range flux monitors to eliminate range switching on startup.
- Like all BWRs, there are no steam generators.
- All of the safety-grade equipment is in one area.
- The pressure vessel is 79 ft high with an upper diameter of 23 ft and lower diameter of 20 ft. This height is key to establishing natural circulation core flow by providing a "chimney" in the space between the top of the core and the steam separator assembly. This large top diameter increases the water inventory above the core (no accumulators needed), and the smaller lower diameter reduces the volume of water needed to be replaced to provide core cooling.
- The large water inventory above the core provides a long time to respond to a feedwater flow interruption or a loss-of-cool accident.
- The large water inventory provides the reduced SBWR pressurization rate following rapid isolation of the reactor from the normal heat sink. This reduces the requirements for pressure relief.
- SBWR power density is a low 42 kW/liter. BWRs are typically ~50 kW/liter and about half the power density of PWRs. Low power density means more thermal and hydrodynamic stability margins. Vessel embrittlement, which has not been a problem for BWRs, is even less concern for the SBWR.
- 24-month operating cycles reduce shutdown risk and increase plant output.
- Adjustable speed motor driven feedwater pumps and high-capacity control rod drive pumps with backup power improve the safety margin by improving correct operator response to non-routine events.
- Passive safety grade systems such as gravity-driven emergency core cooling systems and simple condensing heat exchangers provide residual core heat removal and improve the plant's ability to handle transients and accidents. Several non-safety grade motor-driven systems backup the passive systems.

- No diesel generators are used.
- Only the reactor building is Seismic Category I.
- In case of a LOCA, the core will not uncover or the fuel heat up because of:
  - Eliminating all large nozzles from the lower region of the RPV.
  - Providing a large inventory of water in the RPV region above the core.
  - Depressurizing to near ambient conditions.
  - Flooding the reactor with low pressure gravity-driven flow from the elevated pools located in the containment.
  - Providing sufficient water in the GDCS and suppression pools to flood the containment to at least one meter above the top of the active core.
- Integrated testing of the SBWR emergency core cooling is provided.
- Decay heat is removed by isolation condensers.
- Containment heat is removed by isolation condensers.
- There are only four conventional pneumatic/ spring- actuated safety relief valves,
- Reactor depressurization (for GDCS operation) is provided by six squib depressurization valves.
- The risk is less than the ABWR.

6.2 TMI-2 and Chernobyl Accidents

The preceding overviewed the operation and engineered safety features of current and advanced LWRs. Before preceding to describe how PSA is performed on nuclear power plants, two accidents are described that have profoundly affected the industry

6.2.1 The TMI-2 Accident

6.2.1.1 Description

On March 28, 1979, the accident began which greatly affected the use of nuclear power in the U.S. and throughout the world, although the effects did not exceed regulations.

At the Three-Mile Island-2 (TMI-2) 880 MWe plant designed and constructed by Babcock and Wilcox on the Susquehanna

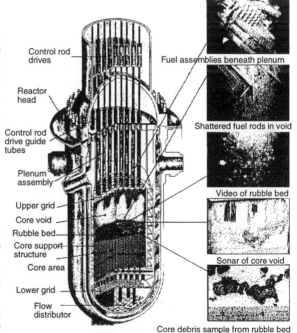

Fig. 6.2-1 The Damaged TMI-2 Core (Courtesy of GPU Nuclear)

River about 10 miles SE of Harrisburg, Pennsylvania, the PORV (pilot operated relief valve) on a pressurizer was leaking in excess of allowed limits. This caused ambiguous temperature readings in the drain piping that later disguised a stuck open relief valve. Two valves were closed following a recent servicing of the auxiliary feedwater system - in violation of Tech. Specs. A resin blockage caused a trip of the condensate pumps, followed by a trip of the main feedwater pumps. But the closed valves blocked the auxiliary feedwater flow to the steam generator, with subsequent overheating that tripped the pressure relief valve and scrammed the reactor. When the pressure subsided, the pressure relief valve failed to close, but this was not known to the operator because of the ambiguous temperature readings. Not understanding what was happening, the operator turned off the emergency cooling, which was not restored until nearly four hours later. By then, the damage had been done. Figure 6.2-1 shows a cross section of the accident. Adam (1984) estimated that 8 to 16 tons of fuel debris were dispersed outside the immediate core area. Some was on the bottom of the reactor vessel; some was throughout the primary cooling system including steam generators, pressurizer, pumps, and piping.

6.2.1.2 Causes of the TMI-2 Accident?

Accidents do not occur spontaneously - warnings are called "precursors". Precursors may be accidents or near-accidents from related experience or they may be warnings preceding an accident although the time between warnings and an accident may be too short for response to the warnings. Table 6.2-1 lists six precursors to this accident, but precursor 5, the failure of the relief valves to reclose thereby constituting a small LOCA is the direct cause. This accident initiator was not recognized in the context in which it occurred to prevent the accident although there had been prior experience at similar plants. If the precursors had been acted upon, the accident might have been prevented.

Table 6.2-1 Precursors to the TMI-2 Accident

1.	Leaky PORV on the pressurizer caused ambiguous readings of the drain piping,
2.	Two valves were closed in the auxiliary feedwater system (a violation of the Tech. Specs. for the operation of the plant),
3.	Resin blockage caused trip of the condensate pump and main feedwater pumps,
4.	Steam generator overheated and tripped its pressure relief,
5.	Pressure relief valves failed to reclose, and
6.	Operator error in failing to diagnose the accident and in tripping the automatic ECCS operation.

6.2.2 The Chernobyl Accident

6.2.2.1 Accident Overview

On the evening of April 27, 1986, a radiation monitoring station in Finland indicated several times the normal background from something unusual in the Soviet Union. A rain shower had brought down fission products. The next morning, other monitoring stations also reported high radiation readings. Measurements in Sweden made it clear that the radioactivity was not from a nuclear explosion, but from a nuclear reactor plant release in the Soviet Union. At 9 P.M., April 28th, the Soviet Union announced an accident at the Chernobyl nuclear power station in the Ukraine (67 hours after the accident). By August 1986, 31 people had died from fire, radiation and other causes. The estimated number of cancers over the following 70 years is 160, using the linear hypothesis.

6.2.2.2 Reactor Description

Chernobyl Unit 4 was an RBMK reactor; it was graphite moderated, and cooled by boiling ordinary water. The reactor core was made of uranium fuel in pressure tubes about 3-1/2 inches in diameter. Cooling water flowing through pressure tubes was converted to steam to drive two turbine-generators. Each of the 1,661 pressure tubes, containing 36 fuel rods, was made of low neutron absorption zircalloy with its good high temperature properties. The core was surrounded by a massive graphite (hence, combustible) moderator. The reactor had an ECCS. Scram was actuated by de-energizing 211 magnetic clutch winches each of which suspended 211 control rods by cables. The reactor was refueled on-line using a large refueling machine.

The partial containment system had many rooms; in one room, with a pressure capability of about 26 psi, was the reactor core. The steam drums were in two rooms; four main recirculation pumps were in each of two rooms.

Figure 6.2-2 shows its operation. A mixture of 14% steam and 86% water from the pressure tubes went to steam drums, used to separate steam from liquid water with steam on top and liquid on the bottom. Steam drives the turbine, leaving at reduced temperature and pressure, condensing in the condenser and combining with liquid from the steam drier as feedwater for recycle to the reactor.

The reactor had a loop for each half of the core. Each circuit had two steam drums and two 500 MW(e) turbines. Water was circulated in each of the two circuits by four pumps in parallel (8 total), though usually each circuit had three in use and one on standby. The pumps supplied a complex of pipes under the reactor that fed water to the separate pressure tubes.

The 211 control rods were moved in and out of the core by winches driven by electric motors. Power and neutron flux distribution were measured by in-core self-powered ion chambers, which were inaccurate at lower power. At low power, ion chambers in the graphite reflector were used.

Below the Chernobyl reactor were water pools meant to capture and condense any steam released from a pipe break or any other failure in the containment rooms. A system of relief valves and ducts led from the containment rooms to these suppression pools. RBMK's were built in pairs;

Analyzing Nuclear Reactor Safety Systems

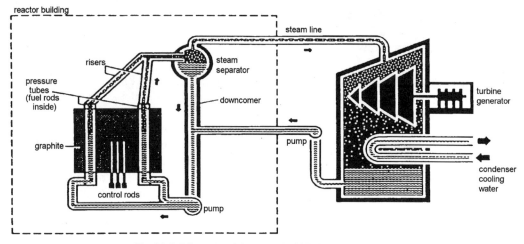

Fig.6.2-2 Schematic of the RBMK-1000 (from Kouts, 1986)

Another pair was Units 3 and 4 opposite the main building with the four turbines of the two reactors in the middle.

6.2.2.3 A Positive Void Coefficient and the Accident

All RBMK reactors have positive void coefficients: which means that increasing the boiling rate increases the steam fraction in the core which increases reactivity causing more steam void which causes more reactivity and so on. Competing factors provide stability, but startup, shutdown and maneuvering below about 600 MWt are unstable, hence, there is a rule prohibiting extended operation below 700 Mwt.

The accident resulted from a routine safety test of some electrical control equipment at the start of a normal reactor shutdown for routine maintenance. The test was to determine the ability to continue to draw electrical power from a turbine generator during the first minute of coast-down following a station blackout. In a blackout, the reactor automatically scrams and diesel generators start to assume load (about 1 minute required).

It had been found that the turbine generator output voltage fell more rapidly than desirable. A new control circuit had been added to compensate for the voltage reduction, and the purpose of the test was to determine how well the new circuit worked. The staff focused attention on the test and did not pay attention to the reactor when its power dropped from its normal level of 3,200 MWt to 700 MWt - the beginning of the unstable region. One of the two turbine-generators was switched to provide power to four main circulation pumps, thus simulating the generator's load during deceleration. Steam flow to the turbine was stopped while the reactor was scrammed. This approved test had been performed previously without problems.

At 0100, April 25, a slow power reduction for the scheduled shutdown began. At 1300 and twelve hours later, the power was 1,600 MWt, when the grid controller asked that the plant continue to supply power for the grid. Power reduction was stopped and turbine 8, scheduled for the test, continued to supply power for the next twelve hours to the first loop. Three other main circulation pumps feeding the second core loop were transferred from turbine 7 to the external transmission line and turbine 7 was stopped with the steam supply going to turbine 8. The pumps on standby, received power from the grid or turbine 8.

An hour later, in preparation for the test, the ECCS was blocked for unknown reasons, in violation of regulations. This had no effect on the accident but indicates something about plant safety although the operators were reputed to be the best.

At 2300, the grid controller released the reactor from supplying power and power reduction resumed. After 90 minutes as the power approached 700 MWt, the minimum for the test, reactor control was switched from "local" to "global." Local measures the flux using self-powered detectors; global uses ion chambers. The operator neglected to signal the control system to hold the power level steady when this transfer was made. The power level decreased rapidly to 30 MWt before the operator could move control rods to less than the minimum of 30 required for this unstable region.

At 0100, the two idle recirculation pumps were turned on causing a greater than normal circulation flow with the pumps verging on cavitation. The four pumps powered by the grid were on to run the test a second time if it failed the first time and to continue core cooling when the others stop.

At 0119:10, the operator began to increase the rate of feedwater return to reduce the recirculation flow to increase the water level in the steam drums. At 0119:45, the reduced inlet water stopped water from boiling in the core. The absence of the steam voids reduced the reactivity, and control rods were withdrawn, such that only 6 to 8 rods were in the reactor, rather than the required 30. Then, to avoid reactor trip from steam drum or feedwater signals, their scram circuits were locked out (a safety regulation violation).

At 0121:55, the operator began feedwater flow reduction from its four times the equilibrium rate; at 0122:10 boiling began again in the core with dramatic effects on reactivity. Rod bank AR1 began to drive into the reactor, reaching 90% insertion in 20 seconds. At 0122:25, rod bank AR3 began to enter the reactor again; the feedwater flow was about 2/3 the equilibrium rate for 200 MWt. Rod bank AR 1 responded violently for nearly a minute as the operator tried to establish the right setting.

At 0123:04 the planned test began with closure of the turbine stop valve for turbine 8. Before this, the operator committed his sixth and worst violation by blocking the scram circuit that would have scrammed the reactor on a turbine trip. This prevented a scram that would have avoided the accident. With the turbine stop valve closed, the core inlet flow rate diminished; the heat transfer rate fell, and the steam void fraction increased to add reactivity. Closing the turbine stop valve caused a pressure surge in the steam drums which reduced reactivity so one bank of safety rods started out of the reactor (0123:10) and then immediately reversed as the steam void fraction increased (0123:21). A second bank of rods moved in for about five seconds (0123:26), as the automatic power controller sought to compensate for the rising reactivity from the slow increase of steam void fraction in the core. Six seconds later, the third bank inserted. Eleven seconds later, all three banks were fully inserted, but the reactor power level had risen far above 3,200 MWt in less than three seconds. The high power and short reactor period warning signals sounded. The operator pressed the scram button, but further control rod insertion, if it were possible, would not help. In one more second the reactor power was at a level estimated to be above 300,000 MWt; the reactor was destroyed in another second.

Outside observers reported two explosions in several seconds. The first explosion sent projectiles, which may have been removable roof blocks, or refueling caps from the pressure tubes, through the roof of the reactor building. Fuel disintegrated into fine particles into the coolant, causing an almost instantaneous pressure surge which expelled refueling caps, destroyed piping below the core; and burst coolant channel pressure tubes. The second explosion occurred when steam was released at primary system pressure into the vault containing the reactor core to lift the 1,000 tonne cover plate of this vault, and shear off the tops of all the pressure tubes. Lifting this cover pulled any remaining control rods from the reactor core. This exposed all fuel channels and the entire core structure to the environment.

The explosions expelled hot fuel and graphite, ignited asphalt roofing material, and started a fierce fire. Fire departments from Pripyat and Chernobyl, fought the fire until 0500, over three hours after the accident. The firemen were hampered by high radiation fields, by lack of lifting equipment to access the roofs, and by inadequate protective clothing. Indeed, fatalities were all among fire-fighters and operating personnel of the reactor. All during this period, with fires raging on roofs, and high radiation fields everywhere, Unit 3 continued operating to supply power for the firefighting. Unit 3 was shut down when the fire was out. Units 1 and 2 continued to operate for another 20 hours.

Local reactor management personnel informed Moscow of the problem, but insisted that they could control the situation. The inhabitants of nearby Pripyat were told to remain indoors with their windows closed. Potassium iodide tablets were distributed, door-to-door, by young Communist League members. Next day, with rising radiation levels, the inhabitants were told to evacuate. Over the next few days, all 135,000 people in a radius of 30 km evacuated in thousands of buses commandeered from as far away as Kiev. Tens of thousands cattle were evacuated in trucks. Radionuclides were emitted for nine days; 10% of the graphite burned. Soviet emergency teams tried to put out the fire and stop the emission of radioisotopes by dropping 40 tonnes of boron carbide, 800 tonnes of dolomite (forms CO_2 when heated) 1,800 tonnes of clay and sand, and 2,400 tonnes of lead. More radioactivity may have been added to the environment during the second five-days than in the first five-days due to debris heating that had been insulated by the materials dropped on the core. On the fifth day, the emergency team began to pump liquid nitrogen into the space under the reactor, in part to cool the debris, but also to put out the fire with its inert atmosphere.

Table 6.2-2 Summary of the Chernobyl Accident
1. The unstable reactor used slow-acting safety systems,
2. Operators committed six safety violations to cause the accident.
3. Containment may have helped protect the public and environment, although the energetics of the explosions might have failed any containment built to U.S. standards.
4. The cost of the accident is estimated by the Soviets to be $3 billion. Hence the financial effects are very severe.
5. Chernobyl may represent the upper limit that is possible in a nuclear power plant accident.

6.2.2.4 The Chernobyl Accident Consequences

About 50 megacuries of activity were emitted, including that blown out in the first explosions and during the fire. It is estimated that all of the noble gases were released, about 10 to 20% of volatile iodine, bromine, cesium, and tellurium, and 3 to 5% of the fractions of other material. These amounts were in an aerosol release of fine particles generated at the time of initial fuel damage which is the reason why the Soviets believe the fuel did not melt. Heat from the release carried the cloud high into the atmosphere to be dispersed by the winds aloft. Initially, it drifted to the northwest, where it was detected by Finnish and Swedish monitoring stations. Subsequently it went to Poland, Czechoslovakia, Austria, Switzerland, Italy, the Balkans, and then the rest of Western Europe. Other wind changes bore the activity over other parts of the Soviet Union, the Orient, and the United States,

In terms of health effects, none of the evacuees from the 30 km radius evacuation zone displayed any symptoms of radiation sickness. Their collective dose from external exposure was estimated to be 1.5 million person-rem. The number of cancers over the next 70 years from this exposure, was estimated using the conservative linear model to be 160 which should be compared with 27,000 cancers the evacuees will get from natural causes over 70 years. Thus, the long-term effect of the accident will be undetectable. Table 6.6-2 summarizes the accident.

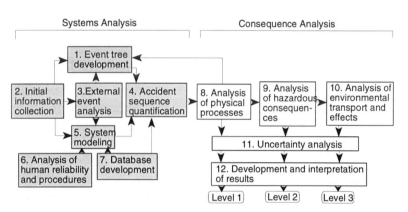

Fig. 6.3-1 Overview of the PSA process (Adapted from NUREG/CR-3200)

6.3 Preparing a Nuclear Power Plant PSA

A PSA begins with the initial plant and accident definition and splits into probability and consequence paths interacting with each other to come together as risk. Such a process has been diagramed in many references at various levels of complexity. Figure 6.3-l is a diagrams the activities as indicated in the PRA Procedures Guide (NUREG/CR-2300). The event tree has a central role of connecting accident initiators to consequences (core melt or plant damage states) through a transformation involving systems reliability.

6.3.1 Overview of the Probabilistic Safety Process Using Event Trees

The process begins with initial system and accident definition for which accident the probabilities and consequences must be determined to give the risk (Figure 6.3-1). Item 1, the event tree is central to PSA because it diagrams the accident scenarios to connect accident initiators to consequences.

To prepare an event tree, initial information (2) is collected, and external events (3) are analyzed. Steps 2 and 3 feed into the preparation of system models that also receive information from the analysis of human factors (6), and the database (7). The event trees (1), system modeling (5), and data base (7) provide the information for the accident sequence quantification (4) that gives the accident frequency (probability) estimate. The accident consequences are determined by analysis of the physical processes (8) to give the hazards (9) and how the hazards reach the public (10). In any complex analysis, uncertainties are analyzed (11), and the results interpreted (12). Three levels of complexity have been defined for PSA. Level 1 analysis is to the level of determining the physical damage of the accident (e.g., core melt, chemical explosion), level 2 proceeds to the release of hazards, and level 3 goes to long and short term effects on the public.[c]

While event trees[d] are not so common in chemical plant PSA they are extensively used to analyze nuclear accidents, but before beginning the analysis, preliminaries that are needed are: plant familiarity and initiator selection

6.3.2 Preparing for the PSA

6.3.2.1 Plant and Accident Familiarity

A thorough understanding of the plant is necessary before the analysis begins. The importance of the familiarization depends on the composition of the PSA team. If it consists primarily of experienced plant personnel, the team needs familiarity with PSA, if it is primarily a PSA team, the team needs familiarity with the plant.

Accident familiarity is gained by studying plant incidents using the databases described in Chapter 4 and the plant's operating experience.

6.3.2.2 PSA Management

A PSA for a large process is a large and expensive project that requires the support of higher management to provide the resources for its completion and to use the results. The PSA manager

---

[c] Worker risk is not usually included in nuclear power plant PSAs. If it were, it would be conducted similarly to chemical plant PSAs.

[d] Event trees are adaptations of "decision trees" used for analyzing the risk of financial decisions.

should be in the plant management to provide the PSA-management interface and be a PSA generalist to supervise the work of the PSA specialists. The PSA manager may choose an assistant technical manager to guide technical aspects and to integrate the results. The technical manager supervises team leaders having specialized disciplines as shown in Figure 6.3-2.

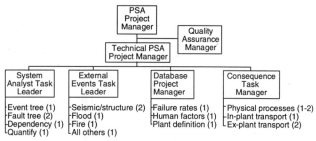

Fig. 6.3-2 Possible PSA Organization Chart (NUREG/CR-2300). The numbers in parenthesis are typical membership numbers; some individuals may do several tasks

6.3.2.3 Schedule and Manpower

A PSA consists of many tasks and subtasks; some tasks are performed in parallel; others are in series. The experience of personnel, availability of data, computer programs, and manpower determine the PSA schedule. Other considerations are the age of the plant, its operational status, availability of documentation, peculiarities of the design, and the availability of the analyses of similar plants.

Table 6.3-1 compares resource estimates from NUREG/CR-2300 (PRA Procedures Guide) with actual expenditures (Joksimovich, 1983). In this comparison of analyses, Grand Gulf is a level 1 PSA; Arkansas Nuclear 1 (ANO-1) is an IREP level 2 PSA; Big Rock Point (BRP) is a level 3 PSA with limited treatment of external events but thorough in consideration of environmental effects on equipment. The Zion PSA thoroughly treats both internal and external events as reflected by the resource expenditure. The total expenditure estimates

Table 6.3-1 Comparison of Estimated and Actual PSA Resources (person-months)

Task	NR-2300	BRP	Zion	Lim.	GGulf	ANO-1
Initial information collection	1-2	3	21-24	10	1	3
Event tree development	29-38	2	105-120	10	3	3
Systems modeling		18		30	3	50
Human reliability and procedure analysis	2-3	4		10		5
Data development	5-6	12		5		1
Accident sequence quantification	9-12	10		10	3	20
Uncertainty analysis	3-4	2		5		4
Physical processes	15-137	6	210-240	20	8	1
Radionuclide release and transport	5-20	7		10	2	
Environmental transport and consequences	3-4	2	42-48	20		
External events	3-4		42-48			
Development and interpretation of results	1-2	6	5% of total effort	10	1	6
Total effort	51-86, 78-285, 30-295	120	420-480	144	24	96
Start and end dates		4/80-3/81	12/79-8/81	5/80-3/81	7/5-10/81	10/80-8/82

in NR-2300 are for level 1, level 2, and level 3 PSAs respectively.[e]

The minimum time needed to complete a level 1 PSA is about one year. To allocate only 12 months from start of funding to final report is very ambitious, requiring complete manpower availability at the beginning of the effort and a large writing and editorial effort at the end. This is not the best use of either personnel or funds. The technical quality may suffer in such an accelerated project, depending upon the team leadership, completeness, and consistency from the beginning. But if a recent PSA for a similar plant is available and the objectives allow the use of many of the system models and most of the failure data, then a schedule extending over less than a year is feasible.

It is more common for a complete PSA to take 16 to 24 months with several months for preparation, review, and revision of the final report. The final report for a level 3 nuclear plant PSA, includes an analysis of external events, in several large volumes. Completeness and consistency in such a large document requires several months of team leadership and selected analysts. Given these resources, it may be possible to complete the technical analyses for a Level 1 PSA in a year or less, but the final report will take several more months to prepare.

6.3.2.4 Information Requirements

The information needs for a PSA are very extensive and encompass nearly all that is known about the plant. Table 6.3-2 lists, as typical, the sources used in preparing the Zion PRA.

6.3.2.5 Quality Assurance and Peer Review

The organization chart shown in Figure 6.3-2 shows the Quality Assurance Manager high in the organization, to achieve independence of QA as required by 10CFR50 Appendix B.

A PSA is of high technical quality if it accurately and completely portrays reality. Such is very difficult to measure, therefore, a PSA is considered to be of high technical quality if it has the following:

- PSA methods are used correctly and appropriately;
- Information about the plant is complete and current;
- Methods, assumptions, and judgments are clearly set forth;
- Databases are appropriate and use all available sources;
- Risk calculations are scrutable and traceable from the database through the final results; and
- Documentation is clear and complete.

PSA quality is difficult to achieve, but several steps enhance the quality. Initial program

---

[e] A Level 1 PSA is a core-melt analysis that may or may not include external events; a Level 2 PSA encompasses Level 1 and extends to the physical processes of the accident and their effect on containment; a Level 3 PSA encompasses Levels 1 and 2 and extends to public risk.

Preparing a Nuclear Power Plant PSA

planning sets the quality of the PSA through: planning, selection of methods, internal review, documentation, and computer codes.

Role of Review

Generally, PSAs are reviewed by the: 1) study team, 2) plant operating personnel, 3) peers, and 4) management. The study team consisting of the technical manager and the internal peer review group examines it for the accuracy of the characterizing the plant by models and accuracy and clarity of results. Operating personnel examine it for the accuracy of interpreting the plant and operating practices. Peers, independent of the study, concentrate on the appropriateness of methods, information sources, judgments, assumptions, perspective, scope, and its suitability for meeting objectives. These reviews advise the management review for judging its use in corporate policies.

6.3.2.6 Selecting Accident Initiators

The accident initiators specify the event trees. A systematic methodology is needed to assure that the list includes all significant accidents. Systems analysis methods presented in Chapter 3 are of little help in preparing this list because accidents develop through the physics of the process.

Completeness cannot be achieved in a mathematical sense but it is believed that these methods, singly and in combination, approach practical completeness in the sense that that which is left out is insignificant. Some of the methods are: Precedence, Preliminary Hazards Analysis (PHA), Leak Path Analysis, Heat Imbalance, Independent Review and

TABLE 6.3-2 Zion Principal Information Sources (from Joksimovich. 1983)

1	Final Safety Analysis Report (including all supplements, appendixes, and NRC questions and responses).
2	General arrangement drawings.
3	Piping and instrumentation drawings.
4	Electrical schematic and one-line diagrams (AC and DC).
5	Instrumentation and control diagrams on risk-related systems.
6	System descriptions.
7	Reactor operator training notes, lesson plans, or manuals.
8	Indices to all procedures.
9	All annual/semiannual operating reports.
10	Maintenance/Periodic test schedule.
11	Plant emergency procedures.
12	Maintenance/Periodic test procedures (risk-related systems only).
13	Technical specifications.
14	Zion specific equipment and system failure history.
15	Zion specific initiating event frequency history.
16	Annual meteorological data from designated regional stations.
17	Population density data surrounding the plant site.
18	Topographic maps for the region of influence surrounding the site
19	Evacuation plans indicating planned routes from the plant site
20	All submittals to the NRC concerning geology and seismic design basis for both structures and equipment.
21	Seismic analysis of the containment, reactor building, and auxiliary building (including input and response).
22	Representative design calculations showing load combinations and code compliance of a typical building element such as a shear wall in the auxiliary of containment building.
23	Structural design drawings for typical critical structures such as the lower elevation of the auxiliary building, reactor vessel, containment, reactor building and substructure, and concrete..
24	Seismic design basis for the reactor vessel (including attachments, pressure boundary, and CRD and internals).
25	Seismic design basis for the main coolant loop piping and pumps, and for typical category 1 piping, e.g., the auxiliary feedwater line.
26	Method of seismic analysis and any limiting results or typical structural calculations for seismically qualified equipment. a) Typical large exterior tank and three small interior tanks. b) Three different pumps. c) Valves.
27	Method of seismic and qualification and results for typical electrical cabinets and switchgear.
28	Reactor vessel and internals- materials, masses.
29	Core design and performance details.
30	Primary coolant system component parameters
31	Containment design details - basic structure, major contents (heat structures), internal safety systems performance data, special features, reactor cavity/sump details, layout elevations and floor plans, materials specifications, design limits, etc.
32	Key service buildings - basic structures and major contents, layout elevations and floor plans, ventilation details (filters), design limits, etc.
33	Blowdown data curves, existing inputs for ECC performance and DBA analyses.
34	Containment performance analyses - existing data inputs for DBA analyses

Operating staff review.

Copying other PSAs

One way to prepare a PSA is to find one that was prepared for a plant similar to yours and copy it suitably modified for your plant, precursors and relevant PSA information.

Caution should be observed in copying another analysis. No previous analysis would have anticipated the TMI-2 accident. Other near misses were the Davis-Besse loss of feedwater incident of 1985, the Salem breaker problem of 1983, the Brown's Ferry Fire of 1975 and the scram problem at Brown's Ferry.

Preliminary Hazards Analysis (PHA)

PHA is required for new systems that are designed and construction for the Department of Defense (MIL-STD-882A); it is useful for initial PSA of any system (Lambert, 1975). A PHA is a tabular listing of possible accident initiators, the parts of the plant that could be affected and possible effects of an accident. Because a PHA is performed before a PSA,

Table 6.3.3 Qualitative Categorizations from MIL-STD-882A

Probability		Consequences	
I.D.	Name	I.D.	Name
A	frequent	I	catastrophic
B	probable	II	critical
C	occasional	III	marginal
D	remote	IV	negligible
E	negligible		
F	impossible		

Table 6.3-4 Sample Preliminary Hazards Analysis

Initiator	Cause	Effect	Location	Mitigation	Risk
Excess reactivity	Failure of automatic rod control	Heat imbalance temperature increase	Control rod drive	Operator action, interlocks	IIIE
	Failure of automatic solution poison control		Chemical volume control system	Operator action, limit to rate of poison introduction	IIIC
Reduction in cooling	Loss of offsite power	Temperature increase from reduced heat removal	General	Automatic scram, turbine trip, coolant switchover	IIIA
	Loss of onsite power	Loss of power for cooling	Cooling system	Emergency power, shutdown mode	IIIC
	Loss of cooling pump	Loss of cooling	Cooling system	Pump redundancy, operator action, reduce power	IIIC

the probabilities and consequences of accidents are known only qualitatively. Table 6.3-3, adapted from MIL-STD-882A, presents a possible classification for the sample PHA presented in Table 6.3-4. The initiators list all energy sources in the plant that could initiate an accident. Accident consequences are developed by identifying hazardous materials, and confining mechanisms. The radiological hazards in a nuclear plant, arise from the core's radioactivity, in a nuclear process plant, the radioactivity is in many locations as are the hazards in a chemical plant. Both nuclear and chemical plants are subject to fires, explosions and release of toxic materials.

Leak Path Analysis

Gauss's theorem, in mathematics, says a change in a volume's density must be accompanied by flow through the boundary. Leak path analysis is a qualitative interpretation of Gauss's theorem.

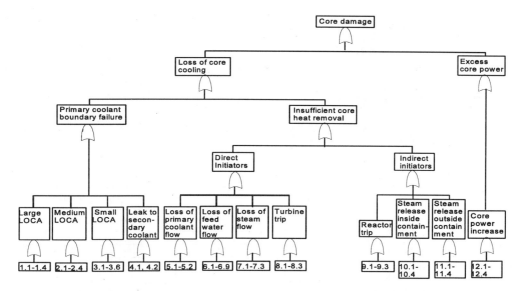

Fig. 6.3-3 Core Damage Master Logic Diagrams for Indian Point 2 and 3 Nuclear Power Stations. Numbers in boxes are locations in the report where addressed. (From their PSA, reproduced with permission of Consolidated Edison Company of New York Inc. and Power Authority of New York.)

It is suited to the defense-in-depth philosophy.

It is illustrated in Section 3.4.4 by tracing the paths for leaking engine compression and applied to fault tree construction for the FFTF reactor (Fullwood and Erdmann, 1974). The method involves writing Boolean equations for all paths whereby hazardous material may be released. It is primarily useful for enumerating release paths, but not for what started the release; It was used to enumerate the possible paths for stealing nuclear bomb material from a facility.

Heat Imbalance

Core damage can result most likely from heat imbalance. Figure 6.3-3 is an example from the Indian Point PRA that uses heat imbalance to approach completeness. This diagram shows that core damage may result from either a loss of cooling or excess power (or both). The direct causes of insufficient heat removal may be loss of: flow, makeup water, steam flow, or heat extraction by the turbine. Indirect causes are reactor trip or steam line break inside or outside of containment. Causes of excess power production are rod withdrawal, boron removal, and cold water injection.

This figure presents only the so-called internal events. Not included are the "V" sequences (valve rupture that bypasses containment), earthquake, fire, flood, tornado and air crash.

6.3.2.7 Grouping and Quantification of Accident Initiators

It is not feasible to construct event trees for 30 or 40 initiators, thus some sort of grouping and consolidation is needed. LOCAs are grouped according to the systems required to respond to

Table 6.3-5 Initiating Event Summaries

Initiating event	Event tree grouping	Freq. estimate	Error factor	WASH-1400	Plant specific data	EPRI NP-801	Other
Zion (PWR 65 categories of initiating events reduced to 16 composite groupings -frequency estimates are median.)							
SG tube rupture		1.5E-2	5.2		X	X	X
Loss of primary flow		3.4E-1	1.8		X	X	X
Loss of feed flow		5	1.2		X	X	X
Loss of one MSIV		2.3E-1	2.3		X	X	X
Turbine trip (general)		3.7	1.3		X	X	X
Turbine trip (LOOP)		3.8E-2	4.4		X	X	X
Turbine trip (loss of SW)		3.4E-4	10		X		X
Spurious safety injection		6.0E-1	6.0		X		X
Reactor trip (general)		3.7	1.3		X		X
Reactor trip (loss of CCW)		3.4E-4	10	X		X	
Loss of steam (inside containment)		3.4E-4	10		X		X
Loss of steam (outside containment)		3.4-4	10		X		X
Core power increase		1.7E-2	3.6		X		X
Large LOCA		3.4E-4	10		X		X
Medium LOCA		3.4E-4	10		X		X
Small LOCA		3.1E-2	2.4		X		X
Arkansas Nuclear One-1 (PWR frequency estimates are point values)							
Loss of offsite power		3.2E-1			X	X	
Loss of SW		2.6E-3					X
Loss of PCS	Loss of offsite power	1.			X	X	
Loss of AC Bus A5		3.5E-2			X	X	
Loss of AC Bus A2		3.5E-2			X	X	
Loss of DC Bus D01		1.8E-2			X	X	
Loss of DC Bus D02		1.8E-2			X	X	
Other transients		7.1			X	X	
LOCA (0.38"<D<1.2")		2.0E-2					X
LOCA (1.26"<D<1.66")		3.1E-4		X			
LOCA (1.66"<D<4")		3.8E-4		X			
LOCA (4" <D<10")		1.6E-4		X			
LOCA (10<D<13.5")		1.2E-5		X			
LOCA (13.5" <D)		7.5E-6		X			

Table 6.3-5 Initiating Event Summaries (cont.)

Initiating event	Event tree grouping	Freq. estimate	Error factor	WASH-1400	Plant specific data	EPRI NP-801	Other
Limerick (BWR)							
Closure of all MSIVs	MSIV closure	1.0				X	X
Turbine trip w/o bypass		1.0E-2				X	X
Loss of condenser	MSIV closure	6.75E-2				X	X
Partial closure of MSIVs	Turbine trip	2.0E-2				X	X
Turbine trip/ bypass		1.3				X	X
Recirculation problem		2.5E-1				X	X
Pressure regulator failure		6.7E-1				X	X
Inadvert. bypass open		0.0				X	X
Rod withdraw./insert.		1.0E-1				X	X
Feedwater distrib.		6.8E-1				X	X
Electric load reject		7.5E-1				X	X
Loss of offsite power		5.3E-2		X			
Inadvert. open. relief valve		6.0E-2				X	X
Loss of feedwater		7.0E-1				X	X
Small LOCA		1.0E-2				X	X
Medium LOCA		2.0E-3				X	X
Large LOCA		4.0E-4				X	X
Grand Gulf (BWR)							
Loss of offsite power		2.0E-1		X			
Other transients		7.0E+0		X			
Small LOCA		1.4E-3		X			
Large LOCA		1.0E-4		X			
Big Rock Point (BWR)							
Turbine trip	Turbine trip	1.4	10	X			
Loss of condenser vacuum		6.0E-2	11	X			
Spurious closure of MSIV		6.0E-2	10	X			
Loss of feedwater		1.6E-1	10	X			
Spurious opening of turbine bypass valve		1.0E-1	10		X		
Feedwater controller failures decreasing flow or power	Loss of FW	1.0E-1	10				X
Loss of main condenser ATWS	ATWS	6.3E-6					X
Spurious closure of both recirc. line valves		1.3E-1	10				X
Loss of offsite power ATWS		4.6E-6					X
Misc. scram ATWS		2.0E-6				X	
Fire in cable penetration area (inside containment) which could affect safety system cables	Fire	1.8E-3	3.6^e			X	
Fire in station power room which could affect all safety system cables		3.3E-3	1.9^e			X	
Fire in control room which affects all cooling systems		1.0E-4	3^e			X	
Fire in cable spreading room (outside containment) which could affect all safety system cables		9.0E-4	3.4			X	

the initiating event. Since the power conversion system (PCS) provides the means for heat removal (the condenser) and water supply to the core (the feedwater systems), transients are often divided into groups depending on whether the PCS is available or not. Transients, with offsite electrical power lost, are given special consideration. Comprehensive lists like those in Tables 6.1-1 and 6.1-2 are complex and transient initiators chosen from generic lists such as these must be examined for applicability to the specific plant in question. If the plant is currently in operation, its history should be examined for transient event information to supplement the generic data. The grouping of transients should take into account the nature and extent of the specific mitigating systems. For most transients, the flow of coolant from the core is uninterrupted so that emergency injection is required only to replace coolant that may be released by the safety relief valves. If one of these valves sticks open, the transient is transformed into a small LOCA. In some PSAs, this type of accident has been treated by special transient event trees, while in others they are grouped with the LOCAs of the appropriate size.

Table 6.3-6 Inspection Importance Ranking with Health Effects for Indian Point 3 Accident Initiators

	Description	Importance
1	Small LOCA	8.0E-5
2	Large LOCA	1.7E-5
3	Medium LOCA	1.3E-5
4	Fire total	1.3E-5
	Switchgear room 34%	
	Aux. feedwater pump room 26%	
	Other zones 17%	
	Cable spreading room 13%	
	Electrical tunnel 10%	
5	Turbine trip loss of power	3.9E-5
6	Seismic	3.6E-6
7	Steam generator tube rupture	1.6E-6
8	Tornado	1.3E-6
9	Loss of main feedwater	6.0E-7
10	Interfacing LOCA	5.7E-7
11	Turbine trip	4.3E-7
12	Reactor trip	3.6E-7
13	Steam break inside of containment	2.6E-7
14	Steam break outside of containment	2.6E-7
15	Turbine trip loss of service water	5.0E-8
16	Reactor trip loss of comp. cool.	3.1E-8
17	Loss of RCS flow	2.7E-8
18	Loss of MSIV	1.2E-8

Examples of initiating events considered in five PRAs are provided by Joksimovich et al. (1983) and presented as Table 6.3-5. The occurrence frequencies vary from a high of 3.7/yr for turbine and reactor trip at Zion to a low of 1E-6/yr for a large steam line break outside of containment at Big Rock Point. Another low frequency is 2E-6/yr for ATWS from the loss of one feedwater pump, also at Big Rock Point. Surprisingly, these are less than a large LOCA (1E-5/yr) at the same plant. Except for Big Rock Point, this table provides no information on externalities.

Primary emphasis is often placed on the frequently occurring initiators, but they may not present the largest public health risk. To gain a perception of the relative importance of various initiators, Table 6.3-6 from Taylor et al. (1986) shows an ordering of the accident initiators for Indian Point 3 by Inspection Importance (Section 2.8.2) which considers both frequency of occurrence and health effects based on of one or more latent fatalities.

Analyzing Nuclear Reactor Safety Systems

6.3.3 PSA Construction

Step 1 - Initiator List Preparation

The initiators are separated into two classes: those for which the event tree/fault tree analysis is appropriate and those for which it is not. The former are called *internal initiators* and the latter, *external initiators* (externalities).[f] If dependencies are accounted for by modifying the branching probabilities of the event tree, both internal and external initiators can be accounted for in the same event tree.

Step 2 - System Sequences

Given the initiator, the question is, "What operational systems or actions are involved in responding to the initiator?" The systems shown as A, B, C, and D in Figure 6.3-4 may be ordered in the time sequence that they occur even if dependencies are accounted for in interspersing support systems or by manipulating the branching probabilities. This allows a support system to fail the systems it supports. Mitigators may be mitigating systems, but not a necessarily. Human actions may be included in the event trees as mitigators, in contrast to RSS-type procedures where human actions enter in the fault trees.

Step 3 - Branching Probabilities

The systems list, across the top of the event tree, specifies the systems that must be analyzed to obtain the branching probabilities of the event tree. For complex reliable systems, fault tree or equivalent analysis may be needed to obtain system probability from component probabilities. For less reliable systems, the branching probability may be obtained from plant records with cautions regarding system interactions.

Step 4 - Damage States

Determining which accident sequences lead to which states requires a thorough knowledge of plant and process operations, and previous safety analyses of the plant such as, for nuclear plants, in Chapter 15 of their FSAR. These states do not form a continuum but cluster about specific situations, each with characteristic releases. The maximum number of damage states for a two-branch event trees is 2^S where S is the number of systems along the top of the event tree. For example, if there are 10 systems there are 2^{10} = 1,024 end-states. This is true for an "unpruned" event tree, but, in reality, simpler trees result from nodes being bypassed for physical reasons. An additional simplification results

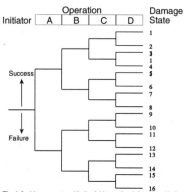

Fig. 6.3-4 *Organizing the Analysis via Event Tree*

---

[f] The characterization: internal and external, refers more to the analytical procedure than to the plant configuration.

from many event sequences leading to the same plant damage state.

Step 5 - Dependency Analysis
A dependency analysis determines which systems depend on other systems. This may require iterating the event tree construction to put a support system before the systems affected.

Step 6 - Analysis of Physical Processes
Given the damage states, the analysis flows much as shown in Figure 6.3-1, depending on the process. For a nuclear power plant, thermal-hydraulic analyses determine the spatial temperature of the damaged core, and consequently the ability of the core to retain radioactive materials. Analysis of the physical processes reveals the amounts of hazardous materials that may be released.

Step 7 - In-Plant Transport
Thermo-diffusion calculations analyze the migration of hazardous material from compartment to compartment to release in containment. These calculations use physico-chemical parameters to predict the retention of hazardous materials by filtration, deposition on cold surfaces and other retention processes in the operation. Containment event trees aid in determining the amount, duration and types of hazardous material that leaves the containment.

Step 8 - Ex-plant Transport
The materials leaving containment are source terms for offsite convective-diffusion transport calculations. Codes such as CRAC-2 calculate atmospheric diffusion with different probabilities of meteorological conditions to estimate the radiological health effects and costs.

Step 9 - Integrating the Analysis
The analysis consists of many calculations of frequency and consequences of the various accident sequences. In a matrix analysis, four large matrices assemble the risk from the internal initiators, but in a non-matrix analysis the separate pieces must be assembled to which the risk from external initiators must be added. It is at this point in the assembly process that the project QA in the form of good record-keeping aids in putting the pieces together such as the branching probabilities into the event trees and calculated probabilities of each plant damage state. Iteration on this assembly process back to Step 1 may be necessary to assure consistency. The consequence calculations, health effects and monetary costs of each damage state compose the plant risk.

Upon completion of Step 9, an uncertainty analysis is performed to determine the confidence that can be given to the results. This is done at the end of the analysis to be sure that the process has settled down and to avoid needless repetition of the uncertainty propagation. Most of the effort in uncertainty analysis has been applied to accident probabilities, but there are large uncertainties in the consequence calculations. The uncertainty calculations are performed by combining the frequency distributions of the underlying data. These are propagated through the fault trees to provide the branching frequency distributions in the event trees. The branching frequency distributions must then be combined according to the event tree logic to obtain the distributions of the plant damage state frequencies which must be combined with the distributions from external events.

Analyzing Nuclear Reactor Safety Systems

This PSA construction process provides results but tends to obscure the interpretation of what led to these results. Sensitivity analyses are basically studies of the effect on risk due to small or large changes in system/component reliability or operability (see Section 2.8).

Step 10 - Presentation of Results

Having completed the risk analyses, computed the uncertainties, and identified critical systems by importance measures (which also identifies valuable systems improvements having low costs), the PSA results must be presented. An executive summary compares the risk of operations that were analyzed with the risks of similar operations. It identifies and explains the main contributors to the risk to people untrained in PSA and statistical methods. Figure 6.3-5 shows two pie-charts that show the risk contributions of various initiators for PWRs and BWRs. A chart similar to one of these would be an effective way of showing the risk contributions in simplified form.

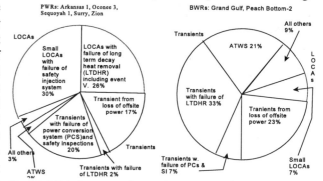

Fig. 6.3-5 Relative core melt contributions. Reprinted with permission of the Electric Power Research Institute, USA

The next level of presentation is a technical summary that gives details of the risks including the system's importance measures systems, effects of data changes, and assumptions that are critical to the conclusions. It details the conduct of the analysis - especially the treatment of controversial points. The last level of presentation includes all of the details including a roadmap to the analysis so a peer can trace the calculations and repeat them for verification.

6.4 Example: Analyzing an Emergency Electric Power System

The preceding detailed how to perform a PSA on a plant or system. However it is one thing to describe how to do something and another thing to actually do it. Fault tree or equivalent analysis is key to PSA. Small logical structures may be evaluated by hand using the principles of Chapter 2 but at some point computer support is needed. Included with this book is a distribution disk for a DOS-type computer. The files on the disk are described in Chapter 12, however here we will "walk-through" an analysis of an emergency electric power system. This would be easy using a code

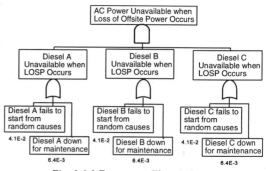

Fig. 6.4-1 Emergency Electric Power

Example: Analyzing an Emergency Electric Power System

suite such as SAPHIRE but the steps in performing the analysis would be obscured by the "canned" code. Practically all computer codes must be purchased or are restricted in usage. The exception is FTAP[g] and associated codes. With the set of codes collectively called "FTAPSUIT" we will input a fault tree of the emergency electric power system, find the cutsets, determine the probability of the top event, find the importances, and perform uncertainty analysis using Monte Carlo methods.

6.4.1 The Problem: Emergency Electric Power

Figure 6.4-1 shows a fault tree of a representative emergency electric power system. It consists of three diesel engine-electrical generator combinations. When offsite power is lost, compressed air is released to power the starter motors to start the diesel engines. When these reach the regulated speed, the generators are excited to provide 4.16 kV to their buses. The fault tree shows that all three generator sets are required, but a generator may fail randomly (4.1E-2/d) or be down for maintenance (6.4E-3/d). This is the problem to be analyzed.

6.4.2 Overview of FTAPSUIT

The suite consists of the following FORTRAN programs:

1. FTAP - Give it a fault tree or equivalent as linked Boolean equations, and it will find the cutsets (Section 2.2), but it does not numerically evaluate the probabilities.
2. PREPROCESSOR - Modifies the FTAP punch file output for common cause and dependent analyses (conditional probabilities), removes complemented events, and corrects for mutually exclusive events.
3. BUILD - Accepts a file from FTAP or PREPROCESSOR to calculate the quantitative results.
4. IMPORTANCE - Accepts the output from BUILD and calculates importance measures (Section 2.8) and provides input for the uncertainty analysis.
5. MONTE - Performs a Monte Carlo uncertainty analysis using the uncertainties in the data to estimate the uncertainty in the calculation of the system and subsystem failure probabilities.

These programs were written for a main frame computer using card input. To make the set of codes easier to use, a DOS.bat code, "FTAPSUIT" was written to provide a menu for calling FTAP and codes provided with it, and for calling an auxiliary code, "FTAPLUS" a Pascal program that organizes the procedures and permits the user to provide input in comma-delimited format. This organizes the "deck" input and assures proper column placement of the data.

Figure 6.4-2 shows the sequence of actions. Selections from FTAPSUIT are in the left

---

[g] Dr. Howard Lambert of Lawrence Livermore National Laboratory and FTA Associates, 3728 Brunell Dr., Oakland CA, 94602, e-mail lambert8@llnl.gov, provided the public domain codes FTAP (fault tree analysis program) and companions. He authored some of the codes, was associated with their development, teaches and consults for PSA.

column; the right column uses three selections from FTAPLUS.

Although the suite of programs can be run from the floppy disk, for reasons of speed and problem size, running from a hard disk is recommended. To do this, in DOS, type MD "directory name," enter, and cd\directory name, enter. Type copy a:*.* and the disk will be copied to that directory.[h] Type FTAPSUIT and the screen shown in Figure 6.4-3 is presented.

First, system information must be input. From FTAPSUIT, type 1 and enter. This starts FTAPLUS with its menu as shown in Figure 6.4-4.

Select 1 and type a title for the problem; the default title is "Station Blackout from Loss of Offsite Power," unless you want to use this enter your title.

Select 2 to enter a less than 7 character file name. This name is used throughout the suite with the extender being changed as shown in Figure 6.4-2. (In this Figure, F. stands for file name.) It is advisable to keep the name short because it is requested for the FORTRAN programs. The default designator is "DG" (diesel generator).

Select 3 to input the linked logic sequences using a comma delimiter, no comma at the end, and enter to indicate the end of the line. Type # and enter to mark the end of input. For example, the test logic is: BLACKOUT,*,LOSP*NO-AC; NO-AC,*,A,B,C; A,+,DG-A,DG-A-M; B,+,DG-B,DG-B-M; C,+,DG-C,DG-C-M;#; where the semicolon designates enter. Notice these equations are linked with BLACKOUT being the top event. This completes the preparation of FTAP input; selection 6 exits FTAPlus to return to the FTAPSUIT menu (Figure 6.4-3). Figure 6.4-2 shows file FN.FI was created. This is an ASCII file which can be edited, if needed, with a text editor.

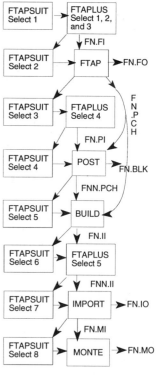

Fig. 6.4-2 Procedure for Using FTAPSUIT

The next step is to run FTAP using selection 2 from the FTAPSUIT menu (Figure 6.4-3). It asks you to enter the file name that you assigned. **Do not give an extender**. A "successful completion" sign indicates proper FTAP execution. At this point, the cutsets have been found and can be seen by reading file FN.FO with a text editor. While this qualitative analysis is useful, usually

---

[h] In Windows 3.1, go to the file manager click on the hard drive, click on file, click on "create directory" type FTAPSUIT, and copy the "A" drive to this directory. In Windows 95 go to the explorer, click on hard drive, right click and choose "make folder" and copy disk to it.

Example: Analyzing an Emergency Electric Power System

```
Master Program for FTA Associates
Codes

1. Prepare Input for FTAP
2. Run FTAP
3. Prepare Input for POSTPROCESSOR
4. Run POSTPROCESSOR
5. Run BUILD
6. Prepare Input for IMPORTANCE
7. Run IMPORTANCE
8. Run MONTE
H. Help File
Q. Quit

Input designation of Task, Press ENTER
```

Fig. 6.4-3 Menu for FTAPSUIT

```
FTA Associates FTAPlus Main Menu
====================================

1: Provide Title for FTAP
2: Provide File Name
3: Input FTAP Tree Logic and Save
4: Input Data for POSTPROCESSOR and Save
5: Input FT Data for IMPORTANCE and Save
6: Leave the Analysis Session

Select the Number Identifying Your Job
```

Fig. 6.4-4 FTAPlus Menu

a quantitative analysis is required. Figure 6.4-2 shows that the qualitative analysis can be used asis by using BUILD via file FN.PCH or POSTPROCESS to modify the operation and indicate conditional probabilities.

The postprocessor reads the punch file from FTAP and generates an output file in a format which can be read by BUILD. It can execute commands not found in FTAP. Features of the postprocessor include:

- Conducting common cause and dependent event analysis
- Dropping complemented events and performing the subsequent minimization
- Generating block files (i.e., a set of Boolean equations) for subsystems
- Eliminating min cut sets with mutually exclusive events.

```
FTA Associates PostProcessor Input
====================================

1: Eliminate Complemented Events
     (NOCMPL)
2: Print MinCuts in FTAP Format (PRTDNF)
3: Same as 2 and Boole Equa. for Top Event
     (FRMBLK)
4: SORT
5: Mutually Exclusive Input (MEX)
6: Substitute CutSets for Dependencies
     (SUB)
7: Exit Menu

Select the Number Identifying Your Job
```

Fig. 6.4-5 PostProcessor Menu

Selection 3 from the FTAPSUIT menu (Figure 6.4-3) runs FTAPlus to prepare input for the postprocessor (Figure 6.4-5). To print the contents of the block file, the postprocessor input must contain either the command PRTDNF or FRMBLK (selections 2 and 3 respectively). The command PRTDNF (print equation in disjunctive normal form) allows you to display the contents of a punch file in readable form. FRMBLK (form block) is required to generate an FTAP input file. The

241

command NOCMPL drops complemented events (i.e., assumes that they are true) and performs the subsequent minimization. The command SORT sorts cutsets according to order. It is necessary to sort the punch file to run IMPORTANCE if cutsets are not listed according to order in the punch file. (This occurs when probabilistically culling in FTAP using the IMPORT instruction.)

MEX and SUB are optional commands. MEX command must precede the SUB command. The MEX command sets to false, basic events that are pairwise mutually exclusive. For the SUB command, the user identifies sets of basic events within cutsets (original sets) that are to be replaced by another set of basic events (substituted sets). The SUB command accounts for basic events within cutsets that have either a common cause or statistical dependency.

After selecting the MEX command (selection 5) the comma delimited input consists of groups of basic events that are pairwise mutually exclusive. To extend the group to the next and subsequent lines, the first field must be blank. Up to 100 names per group are allowed with a new group starting when a non-blank character is encountered in the first field. The number of pairs that are mutually exclusive is: $n!/[(n-2)! *2!]$, where n is the number of basic events in the group. An example for the plant blackout fault tree is: DG-A-M,DG-B-M,DG-C-M; (where ";" means enter).

Selection 6 (Figure 6.4-5) executes the SUB command. The comma-delimited input alternates between the cutset to be replaced and the cutset replacing it. For example for the plant blackout problem, the input is: DG-A,DG-B,DG-C;DG-A,DG-B/A,DG-C/A&B;DG-A,DG-B;DG-A,DG-B/A;DG-A,DG-C;DG-A,DG-C/A;DG-B,DG-C;DG-B,DG-C/B. The actual input to BUILD alternates lines with * and ** to identify lines to be replaced with replacing lines, but this bookkeeping is done automatically by FTAPlus. The original set of basic events and the substituted set cannot exceed 99 basic events. The SUB command completes the postprocessor input. Select 7 to exit the postprocessor menu and 6 to exit FTAPlus for the FTAPSUIT menu.

Select 4 from this menu and POST (postprocessor) begins by requesting the data input file name (e.g., dg.pi - prepared by FTAPlus), the punch file name (e.g., dg.pch), the block file name (e.g. dg.blk) and the name to be assigned to the new punch file (e.g., dgn.pch). Two outputs are produced: FN.BLK and FNN.PCH.

Select 5 from the FTAPSUIT menu and BUILD starts by requesting the punch file name (e.g., dgn.pch or dg.pch depending on whether you want to run the original fault tree or the postprocessed fault tree. It requests the name to be given the importance input file (e.g., dg.ii) to select the type of input. Option 5 is pure probability, option 1 is failure rate. BUILD generates file FN.II for editing.

Selection 6 goes to FTAPlus from which menu 5 is selected, to provide the information needed for IMPORT. This program presents event names as given in FTAP and POST and requests failure rates or probabilities (e-format), uncertainty (f-format), and an event description. Each time this set is given, it presents the next event name until all events are specified. Then it returns to the main FTAPlus menu to be exited by selecting 6 to go to the FTAPSUIT menu.

Selection 7 causes IMPORT (importance) to start by requesting the name of the input file FN.II (e.g., dgn.ii). The importance output is contained in the FN.IO file. The output needed by MONTE is FN.MI.

Selection 8 from the FTAPSUIT menu runs MONTE (Monte Carlo). It requests the name of the input file, FN.MI (e.g., dgn.mi). The Monte Carlo analysis is contained in file FN.MO.

For examples of the output files refer to Section 7.4 for the example of analyzing a fault tree for a chemical process vessel rupture.

6.5 Summary

Previously, discussions have been on the theory and practices of PSA, this chapter focuses on the analysis of nuclear power plant systems. It began by discussing the first generation of light water cooled and moderated nuclear power: BWRs and PWRs. The major systems were presented with emphasis on safety systems. Safety concerns with the first generation of LWRs are addressed in the design of the advanced reactors of which a few have been constructed in Japan. Advanced reactors that are discussed are the Westinghouse AP-600, ABB System 80+, ABB's PIUS, and General Electric's ABWR and SBWR. The two major accidents of nuclear power plant are described: TMI-2 and Chernobyl the latter of which caused a few radiation deaths and many more deaths from conventional causes. The next section discussed the preparation of nuclear PSAs: management and organization of a PSA team, preliminary information needs, identifying the initiators, event tree focus of the analysis, database preparation, construction of event trees and fault trees, determination of the plant damage states, and the consequences of an accident. The chapter concludes with the example: analyzing an emergency electric power system using FTAP and supporting codes that are provided on a distribution disk with this book. FTAP and supporting codes were written for main frame computers. To facilitate the bookkeeping on a PC, FTAPSUIT and FTAPlus are provided and the reader is walked through the analysis of the emergency electric power system to obtain point probabilities, confidence limits, and importances.

6.6 Problems

1. A critical assembly is a split bed on which fissionable material used to mock up up a separated reactor core that is stacked half on each half. One half is on roller guides so that the two halves may be quickly pulled apart if the neutron multiplication gets too high. Use the Preliminary Hazards Analysis method described in section 3.2.1 to identify the possible accidents that may occur and the qualitative probabilities and consequences. List the initiators in a matrix to systematically investigate the whole process. Don't forget human error.

2. Removing decay heat has been the "Achilles Heel" of nuclear power. The designs shown in this section use active methods to remove the heat. (a) Sketch and discuss a design that removes the heat passively. (b) It would seem that the energy in the decay heat could be used for its own removal. Sketch and discuss a design that uses this property to remove the decay heat.

3. When Hamlet said to Horatio, "there are many things twixt heaven and earth not conceived of in your philosophy," was he complete in the scope of his coverage? If so can this approach be used for completeness in identifying all the possible accidents?

4. In the WASH-1400 analyses of nuclear power accidents, it was calculated that it is possible to overpressure and rupture the containment. Discuss whether this is better or worse than a pressure relief that releases radioactivity but prevents the pressure from exceeding the rupture

point. Spent fuel reprocessing plants are designed to release filtered and processed gases. Typically the final filters are large containers of sand or HEPA (high efficiency particulate air) filters. Discuss filtered containment with consideration for possible failure modes of the filter.

5. Early BWRs used an isolation condenser, although such is not specific to the direct cycle. This device removes decay heat by steam flow through a heat exchanger the other side of which is water vented to the atmosphere. Discuss the relative merits of such a boiler.

# Chapter 7

Analyzing Chemical Process Safety Systems

As the previous chapter discussed nuclear power reactor operation and how to perform a PSA on it, this chapter attempts to apply a similar framework to chemical processing. The problem is the diversity of chemical processing that blurs the focus. This chapter begins by showing that accidents in the chemical process industry cost lives and dollars. Descriptions of deadly chemical accidents are presented to show the chain of sequences that were involved to suggest how their PSA may be structured. Background on selected hazardous chemical process is presented followed by descriptions of how their PSA have structured. The chapter concludes by applying FTAPSUIT to a pressure vessel rupture analysis.

7.1 Chemical Process Accidents

7.1.1 The Nature of Process Accidents

Accidents follow three steps as illustrated by the following chemical plant accident: A worker, walking across a high walkway in a process plant, stumbles and falls towards the edge, grabs a nearby valve stem, but the valve stem breaks and flammable liquid discharges forming a vapor that is ignited by a truck. The explosion and fire spreads for six days to destroy the plant with a loss of $4,161,000. The lesson: minor accidents can lead to large consequences.

The three steps are:

Table 7.1-1 Safety Enhancement

Step	Action
Initiation	Use: Grounding and Bonding, Inerting, Electrical Explosion-Proof, Guardrails and Guards, Maintenance Procedures, Hot-Work Permits, Human Factors Design, Standard Process Designs, and Awareness of Hazards.
Propagation	Minimize Hazardous Materials, and Flammable Quantities, Maximize: Use of Non-flammable, Equipment Spacing, and Install: Check and Emergency Shutoffs.
Termination	Have: Firefighting Equipment and Procedures, Relief and Sprinkler Systems.

- Initiation - the event that starts the accident.
- Propagation - the event or events that maintain or expand the accident.
- Termination - the event(s) that stop the accident or diminish it in size.

Initiation was the worker tripping; propagation was the valve breaking, followed by explosion and fire; termination was the burning of all combustibles. Safety engineering tries to eliminate initiators and replace propagation steps by termination events. Table 7.1-1 from Crowl and Louvar (1990) presents some ways to achieve this goal. While accidents are prevented by eliminating initiators, it is more practical to attack all steps.

Nature of Process Accidents

J&H Marsh and McLendon (M&M, 1997), every five years, reviews and analyzes the 100 largest property damage losses in the hydrocarbon- chemical industries that occurred over the previous 30 years. Most of the losses involved fires or explosions, flood, windstorm, and pressure rupture losses.

Figure 7.1-1 shows an increasing trend in the cost of these accidents, including inflation. The five-year period from 1992-1996 had three times more losses than the period 1967-1971. The period 1987-1991 had the largest total loss and also the largest number of losses (31), which is about five times more loss than the period 1967- 71.

The total cost of the 100 losses is $7.52 billion in 1997 dollars, The largest loss was $252,500,000, from a vapor cloud explosion at a gas processing plant at Cactus, Mexico. The average loss was $75,800,000.

Figure 7.1-2 shows the losses

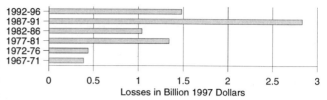

Fig. 7.1-1 Increasing Cost of Process Accidents per 4 Year Period from M&M, 1997

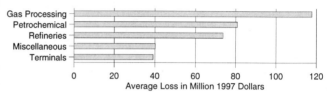

Fig. 7.1-2 Average Loss by Type of Plant

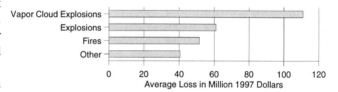

Fig. 7.1-3 Average Loss by Loss Mechanism (from M&M, 1997)

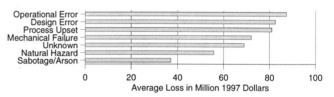

Fig. 7.1-4 Average Loss by Loss Cause (from M&M, 1997)

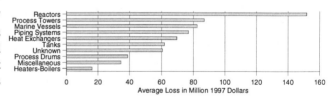

Fig. 7.1-5 Average Loss by Type of Equipment (from M&M, 1997)

by type of plant: Refineries, Petrochemical Plants, Gas Processing Plants, Terminals, and Miscellaneous, which includes: an industrial resin coatings plant, a paint manufacturing plant, a synthetic fuels plant, a natural gas transmission compressor station, and a crude oil pipeline pump station.

Single losses were most costly at refineries than in any other type of plant at an average loss of $73,700,000. The second most costly accidents occurred at petrochemical plants with an average loss of $80,800,000. Although only 7% of the losses occurred at gas processing plants, the highest average loss was $117,800,000.

Figure 7.1-4 ranks fires third, but it is the second most frequent type of loss. The most devastating losses involved delayed ignition of vapor clouds, which account for the highest average loss of $110,900,000. Vapor cloud explosions accounted for the greatest percent of large losses, however, if losses of lesser magnitude were considered, the percent of losses involving fires would exceed that of vapor cloud explosions.

The most frequent loss mechanism (43%) was mechanical failure of equipment which was second in terms of average loss at $72,100,000 (Figure 7.1-4). Operational error was the second most frequent cause of loss (21%), but it had the highest average loss ($87,400,000).

The type of equipment most frequently involved in accidents was Piping Systems (33%) at an average loss of $76,900,000 (Figure 7.1-5). The second most frequently involved type of equipment was tanks (15%) with an average loss of $61,900,000. While Reactors accounted for only 10% of the losses, but had the highest average loss of $151,800,000.

This information shows the large costs of accidents in the process industries. The next section shows the deadly effects that process accidents can have and commends us to reducing their frequency and effect through the use of PSA.

7.1.2 Some Deadly and Severe Chemical Accidents

7.1.2.1 Five Hundred and Seventy Six Die in Texas City Disaster[a]

The port, along with much of Texas City, was ripped apart on April 16, 1947 after a cargo ship, the S.S. Grandcamp, exploded during a shipboard fire. Fueled by 2,300 tons of ammonium nitrate fertilizer (the same potentially explosive material used in the Oklahoma City bombing) the tremendous blast triggered a series of explosions and fires that killed 576 people, injured 4,000 more and damaged every building in Texas City.

On April 11, 1947, the Grandcamp arrived at Texas City, loaded with 16 cases of small arms ammunition, 59,000 bales of fisal binder twine, 380 bales of cotton, 9,334 bags of shelled peanuts, some oil field, refrigeration and farm machinery. The Grandcamp docked at Warehouse Pier O, next to Monsanto Chemical Co. plant. Over the next few days, it was loaded with the rest of its cargo. The ammonium nitrate fertilizer was in 100-pound, 6-ply paper bags. (Although Alfred Nobel had included ammonium nitrate in his original formula for TNT, few people - even experts - recognized the potential danger of the granulated, fertilizer-grade variety.) The U.S. Department of Agriculture didn't consider the wax-covered granules hazardous when packed in paper bags or wooden containers, as long as the fertilizer was segregated from other explosives.

The Grandcamp was loaded with: 2,300 tons of ammonium nitrate fertilizer, 29,000 bales of sisal twine, 9,334 bags of shelled peanuts, automobiles, trucks, and agricultural machinery, oil,

---

[a] From Texas City Disaster website, http://rf.rexl.boise.id.us/v/research.pl/rfrexlboise/ 029277723X/av/880107/key1d/qpremiums

tobacco, 16 cases of small arms ammunition, and 380 bales of cotton.

Shortly after 8 AM, a longshoremen smelled smoke. No one knows what started the fire, but it was speculated that it was caused by a discarded cigarette on the ship in disregard of "No Smoking" signs posted on the wharf. The fire didn't cause much alarm; there were no warnings about the explosive potential of ammonium nitrate - the dockyard workers thought it was inert. The longshoreman, who spotted the smoke, alerted his coworkers who moved several fertilizer bags and found some flames between the cargo and the hull. They tossed a jug of drinking water on the fire, and sprayed it with a fire extinguisher. The flames just got worse. Some of the crew members started hauling boxes of the small arms ammunition out of hold No. 5. The men fighting the fire in the No. 4 hold called for a hose line, but before they could use it, the ship's captain, Charles de Guillebon, intervened. Water would ruin the cargo, he said. The captain instructed his men to close the hatch to the No. 4 hold, cover it with a wet, heavy tarpaulin and activate the ship's steam smothering system in an attempt to suffocate the flames. The steam and decomposing fertilizer, however, combined to create combustible gas. The hatch cover blew off. Thick columns of orange and brown smoke billowed into the air. De Guillebon gave the order to abandon ship. Someone called the Texas City Fire Department; the first firefighters arrived at 8:40 AM; the Grandcamp was out of control. All four engines of the Texas City Fire Department were at the dock, along with several ambulances and 27 of the city's 50 volunteer firefighters. Streams of water from their fire hoses vaporized as soon as they hit the deck of the burning ship. Flames shot out of the hatch. The inside of the ship had turned into a blast furnace. W.H. "Swede'" Sandberg, the vice-president of the Texas City Terminal Railway, hurriedly called an engineer at a nearby chemical plant and asked whether burning ammonium nitrate was dangerous. Don't worry, came the reply, "It won't explode if it doesn't have a detonator." At 9:12 a.m., the Grandcamp disintegrated in a blast heard 150 miles away. Everyone standing near the ship was obliterated. Most of the spectators and all of the firefighters were killed, many instantly. Some survivors later recalled hearing two explosions. Investigators theorized that the ammonium nitrate in hold No. 4 exploded, propelling the 7,176-ton ship about 20 feet into the air. Heat from the first blast then detonated the rest of the fertilizer in hold No. 2. The blast pushed a 15-foot wave of water out of the harbor. It thrust Longhorn II, a 150-foot hydrochloric acid barge, about 500 feet inland and dumped it on railroad tracks. The shock wave from the blast destroyed several dockside warehouses. The large Monsanto plant, only several hundred feet from the Grandcamp, turned into a heap of twisted steel and burning rubble. Flying shrapnel fanned out over the city. Within a mile of the dock, almost every home either collapsed or was so badly damaged it was later declared a total loss. Chunks of debris, some weighing five tons, punctured steel tanks located in chemical tank farms. Streaking overhead like comets, flaming debris - including bales of sisal twine from the Grandcamp's cargo - torched hundreds of thousands of gallons of oil, gasoline, benzene and propane that flowed from ruptured reservoirs and pipelines. A 1½-ton anchor from the Grandcamp made a 10-foot hole in the ground at the Pan American Refinery Co. after a flight of 1.62 miles. This is an abbreviated description of the damage.

Changes resulting from the disaster were:

- An appreciation of the explosive potential of ammonium nitrate. Shipping it from Texas City is forbidden.

- The city developed an Industrial Mutual Aid System that has been copied all over the world. It incorporates resources from city government, police and fire departments and all of the industries at the port. Emergencies covered in the plan range from simple chemical spills and vapor leaks up to Category Five hurricanes and jetliner crashes.
- The refineries and the port base decisions on safety - not economics.
- The city's fire department now has 60 full-time employees, modern fire engines, foam-spraying trucks and other vehicles to combat hazardous materials spills and fires of all sizes.
- Texas City has a four-tiered warning system for emergencies. Level Four includes activation of the city's warning sirens, the signal for residents to tune into radio and TV broadcasts for additional emergency information. They have had 4 Level Four alerts in 5 years.
- Sheltering-in-place has become common throughout the country where there is industry like that at Texas City.

7.1.2.2 Chemical Explosion at Kyshtym - Southern Urals, USSR

September 29, 1957 an explosion occurred at Kyshtym, a nuclear weapons facility concerned with recovery of plutonium and its waste production. The facility began operation in 1949 using the sodium acetate process. Three large 80,000 gal. waste storage tanks were involved in the accident that held 2E7 Ci of waste in 100 gm/ℓ concentration with sodium acetate in 60-80 gm/ℓ concentration. The cooling of one tank failed and the instrumentation was faulty. Water in the solution evaporated, the temperature continued to rise, and at 380° C, a nitrate explosion occurred with about 10 tons of TNT yield to lift the 160 ton concrete shielding cover from the tank, releasing 2 million Curies (Chernobyl released 50 million Curies). The two other tanks were damaged. An evacuation was ordered for the 23,000 km^2 area about the facility involving 10,000 people. Before the accident, the operators fled from the heat emanating from the tank, which may be the reason that no fatalities were reported.

7.1.2.3 Twenty Eight Die at Flixborough[b]

On June 1, 1974, a vapor cloud explosion destroyed the 70,000 tons per year, Flixborough Nyprocyclohexane oxidation plant killing 28 people. Other plants on the site were seriously damaged or destroyed and the site was destroyed.

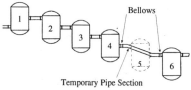

Fig. 7.1-6 Temporary Repair at Flixborough

The plant was a stage in the production of nylon. It manufactured a mixture of cyclohexanone and cyclohexanol (known as KA, for the ketone/alcohol mixture) by oxidizing cyclohexane, a hydrocarbon similar to gasoline, in a series of reactors. When cyclohexane at 155° C and 7.9 atm is depressurized to atmospheric pressure, it vaporizes. The liquid reaction mass, flows

---

[b] Adapted from the Flixborough web site: http://www.dyadem.com/papers/flixboro.htm and Crowl (1990).

by gravity, through the series of reactors. Reactor 5 developed a leak, was removed and replaced by a temporary jumper pipe (Figure 7.1-6). Each reactor normally contained about 20 tons of cyclohexane.

No professional mechanical engineer was on the site at the time. The workers designed and installed the pipe by judgment from experience and got the plant on line in a few days. However, they did not allow sufficiently for the stress on a 20 inch pipe at 150 psi pressure, or the weakening of the metal from operation at 150° C (302 °F). It was designed on a chalkboard in the workshop and installed with jerry-rigged scaffolding

Several months before the accident, reactor 5 in the series was found to be leaking. Inspection showed a vertical crack in its stainless steel. It was removed for repairs, but it was decided to continue operation by connecting reactor 4 directly to reactor 6 in the series. The loss of the reactor reduced the yield but production continued with unreacted cyclohexane being separated and recycled at a later stage.

The feed pipes connecting the reactors were 28 inches in diameter. Since only 20-inch pipe stock was available at the plant, the connections between reactors 4 and 6 used bellows sections. The bypass pipe section ruptured from inadequate support and over-flexing. When the bypass ruptured, an estimated 30 tons of cyclohexane volatilized into a large vapor cloud that ignited from an unknown source about 45 seconds after the release. The resulting explosion leveled the entire plant facility, including the administrative offices. Twenty-eight people died and 36 others were injured. Eighteen of these fatalities occurred in the main control room when the ceiling collapsed. Loss of life would have been greater if the accident had occurred during normal working hours. Damage extended to 1,821 nearby houses and 167 shops and factories. Fifty-three members of the public were reported injured. The fire in the plant burned for over ten days.

This accident could have been prevented or mitigated by safety procedures: 1) the jumper pipe was not designed for the operating conditions, 2) unsuitable material was used, 3) it was installed without a safety review or competent supervision, and 4) the plant site contained excessively large inventories of dangerous compounds. On site were: 330,000 gallons of cyclohexane, 66,000 gallons of naphtha, 11,000 gallons of toluene, 26,400 gallons of benzene, and 450 gallons of gasoline that contributed to the fires after the initial blast.

7.1.2.4 Dioxin Released at Seveso, Italy[c]

On July 10, 1976 a major chemical accident occurred outside of Seveso, near Milan, at a plant operated by Icmesa S.p.A., a company belonging to the Roche Group. A reaction vessel containing the intermediate trichlorophenol overheated. The resulting excessive pressure ruptured the disk of a safety valve. This released a mixture of chemicals, including highly toxic dioxin, which settled on large areas of the communes of Seveso, Meda, Cesano Maderno and Desio. In the days and weeks that followed, a number of people reported inflammation of the skin and chloracne.

---

[c] Adapted from the Seveso website: http://www.roche.com/roche/about/esevesg1.htm and Crowl (1990).

Chickens and rabbits died, and many people had to be evacuated from their homes.

Over the next weeks, months and years, Roche helped alleviate the consequences of the accident through: 1) compensation for the damage suffered by private citizens and public bodies, 2) medical examinations of the affected population, and 3) paying compensation for any damage to health. The injuries were few and transient, hence, the compensations were modest in comparison with those for property damage. The affected population is still undergoing long-term observation, including regular check-ups.

In cleaning up the site, dioxin-contaminated waste from the reaction vessel were packed in forty-one barrels that went astray in May 1983. They were eventually found in 1985 in the northern French town of Anguilcourt-le-Sart at an abattoir. Their contents were burned in a Ciba high-temperature incinerator.

The plant's product was hexachlorophene, a bactericide, with trichlorophenol produced as an intermediate. During normal operation, a very small amount of TCDD (2,3,7,8 tetrachlorodib-zoparadioxin) is produced in the reactor as an undesirable side product. TCDD is perhaps the most potent toxin known to man. Studies have shown TCDD to be fatal in doses as small as 1E-9 times the body weight. It insolubility in water makes decontamination very difficult. Nonlethal doses of TCDD result in chloracne, an acne-like disease that can persist for several years.

On July 10, 1976, the trichlorophenol reactor went out of control, resulting in higher than normal operating temperature and increased production of TCDD. An estimated 2 kg of TCDD was released through a relief system in a white cloud over Seveso. Subsequently, heavy rain deposited the TCDD contaminating about ten square miles of soil. Local authorities, due to poor communications, did not start civilian evacuation until several days later; by then, over 250 cases of chloracne were reported. Over 600 people were evacuated and an additional 2,000 were given blood tests. The most severely contaminated area immediately adjacent to the plant was fenced, the condition it remains in today.

A similar release occurred in Duphar, India in 1963, after which the plant was disassembled brick by brick, encased in concrete and dumped into the ocean. Less than 200 grams of TCDD were released at Duphar, and the contamination was confined to the plant. Fifty men did the clean up; four eventually died from the exposure.

These accidents would have been prevented by: 1) adequate process safety design, 2) proper procedures to prevent the initiation, 3) hazard evaluation procedures to identify and correct hazards before accident occurrence, and 4) containment systems to prevent release into the atmosphere.

7.1.2.5 Twenty-Five Hundred Die at Bhopal, India[d]

The Accident

At about 12:40 A.M. December 3, 1984, the accident occurred that took 2,500 lives at the Union Carbide (India) Ltd. (UCIL) operated pesticide plant. The first indication of process upset was when the control-room operator (CRO) noted that the temperature in a supposedly refrigerated tank storing methyl isocyanate gas (MIC), had risen to 77° F. The pressure, normally 2 to 25 psi was 55 psi.

---

[d] Adapted from Shrivastava (1987) and Crowl (1990).

While running to the storage tank area, the CRO heard a loud rumbling sound and saw a plume of gas rising from the stack. The supervisor, CRO, and several other operators, actuated safety devices including the storage tank refrigerator and the scrubber. When these failed, they ran. Nearby residents, feeling a burning sensation in eyes and throat, also ran.

The district collector and the superintendent of police, were awakened by telephones and rushed to the police control room to coordinate emergency relief efforts. The police and the army tried to evacuate affected neighborhoods, but were too slow, and gave improper directions, resulting in 200,000 residents fleeing into the night. Morning showed the devastation with thousands of injured victims streaming into the city's hospitals. Doctors and other medical personnel struggling with the chaos knew neither the cause of the disaster nor how to treat the victims. They treated the immediate symptoms by water washing eyes, applying eye drops, giving aspirin, inhalers, muscle relaxants, and stomach remedies. The official death toll was 2,700, more than 300,000 exposures to the poison, 2,000 dead animals and 7,000 injured animals. Aggravating the accident was industrial growth over the last thirty years, without the supporting infrastructure, resulting in dense slum housing near the plant. Narrow streets and the absence of transportation made evacuation ineffective.

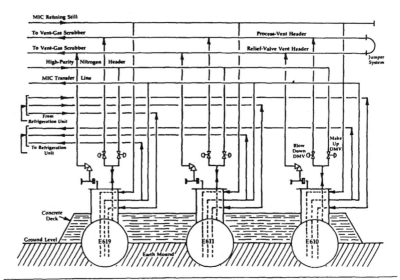

Fig. 7.1-7 Common Headers to the MIC Storage Tanks at Bhopal

MIC is a highly toxic substance used for making carbaryl, the active agent in the pesticide Sevin 12. It is highly unstable and needs to be kept at low temperatures. UCIL manufactured MIC in batches and stored it in three large underground tanks (Figure 7.1-7) until it was needed for processing. Two of the MIC tanks had met specifications, the third had not.

A schematic layout of the storage tanks and the various pipes and valves involved in the accident. A stainless-steel header branched to deliver MIC to each of the storage tanks as shown in Figure 7.1-8). MIC was transferred out of the storage tanks by pressurizing the tank with high-purity nitrogen. Once out of the tank, MIC passed through a safety valve to a common relief header that led to the production reactor unit. Another common line took rejected MIC back to the storage tanks for reprocessing or sent contaminated MIC to the vent-gas scrubber. Excess nitrogen was regulated by a blow-down valve. Though they served different purposes, the relief-valve pipe and the process pipe were connected by a jumper that had been installed about a year before the accident to simplify maintenance. Each storage tank had separate temperature and pressure gauges located locally and in the distant control room. Each tank had a high- temperature alarm, a level indicator, and high- and low-level alarms.

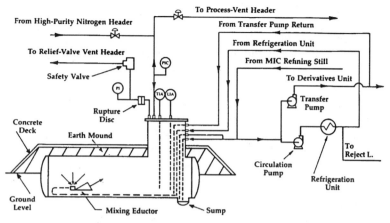

Fig. 7.1-8 MIC Storage Tank at Bhopal

The safety valve through which MIC escaped was redundant with an unmonitored graphite rupture disk. Checking it required manual inspection of a pressure indicator located between the disk and the safety valve.

A safety feature of the plant was the vent-gas scrubber that neutralized toxic exhausts from the MIC plant and storage system. Off-gas scrubbed with caustic soda was discharged at a height of 100 feet and routed to a flare. The tower flare burned normally vented gases from the MIC section and other units in the plant to detoxify the gases before venting them into the atmosphere. However, the flare was not designed to handle large quantities of MIC vapors. A few weeks before the accident the scrubber was placed in standby.

The vapor pressure of MIC was kept low by a thirty-ton freon refrigeration system that chilled the salt water coolant for the MIC storage tank. However, the refrigeration system was shut down June 1984, and its coolant was drained for use in another part of the plant, thus making it impossible to switch on the refrigeration system during an emergency. An additional safety feature was a water-spray to scrub escaping gases, or deluge over-heated equipment, or fires.

The last batch of MIC, before the accident, was produced in the interval October 7-22, 1984. At the end of the campaign the E610 storage tank contained about 42 tons of MIC, and the E611 storage tank contained about 20 tons. After which, MIC production was shut down, and parts of the plant were dismantled for maintenance. The flare tower was shut down to replace a piece of corroded pipe.

Forty-three days before the accident, the nitrogen pressure in E610 dropped from 17 (about normal) to 3.5 psig. Lacking pressure in E610, any MIC needed in the manufacturing process was drawn from E611. But four days before the accident, tank E611 also lost pressure because of a bad valve. An attempt to pressurize E610 failed, so the pressure system for E611 was repaired.

On the night before and early morning of the accident, a series of human and technical errors caused the water for flushing pipes to pass through several open valves and flow into the MIC tank. The water and MIC reacted to produce a hot and highly pressurized gas, foam, and liquid, that discharged through the plant's stack into the atmosphere.

Normally, water and MIC reacting in small quantities in the plant's pipes create a plastic called trimer which is periodically removed by water flushing. But this reaction of water and MIC was so volatile that a blind flange was used to block the discharge pipes. In the evening before the accident, the second-shift production superintendent ordered the MIC plant supervisor to flush out several pipes that led from the phosgene system through the MIC storage tanks to the scrubber. Although MIC unit operators were in charge of the flushing operation, insertion of a blind flange was the responsibility of the maintenance department. The position of second-shift maintenance supervisor had been eliminated several days earlier, hence, the blind flange was not inserted, nevertheless the flushing operation began at 9:30 P.M.

Downstream from the flushing, several clogged bleeder lines caused water to backup in the pipes. Many valves leaked, including the one for isolating the lines being flushed. This resulted in water passing through that valve into the relief pipe. Noting the absence of flow from the bleeder lines, the operator shut off the flow, but was ordered to turn it back on by the MIC plant supervisor.

Water flowed from the relief pipe (20 ft elevation) to tank E610. It first flowed through the jumper to the process pipe to the normally closed blow-down valve which was open (possibly inadvertently left open or failed to seat in the attempt to pressurize the tank). About 119 gallons of water flowed through this open blow-down valve, through another, normally open, isolation valve to enter tank E610, where it reacted with MIC.

At 11 P.M., after a shift change, the new control-room operator, noticed that the pressure in tank E610 was 10 psi, well within the operating range of 2-25 psi. One-half hour later, a field operator saw MIC near the scrubber; MIC and dirty water were discharging from a branch of the relief pipe downstream of the safety valve. They also found that a process-safety valve had been removed, and the open end of the pipe was not blocked for flushing. The control room was informed. By 12:15 A.M., the pressure in tank E610 had risen to 30 psi and rising. Within 15 minutes it pinned at 55 psi.

A hissing sound came from the popped safety valve; local temperature and pressure gauges showed values beyond their maximums of 77° F and 55 psi; loud rumblings and heat radiation came from the tank. An attempt was made to activate the scrubber from its standby mode, but caustic soda was not circulating. A cloud of gas discharged from the stack. The plant superintendent suspended operations and sounded the toxic-gas alarm to warn the community. A few minutes later the alarm was turned off with only the in-plant siren warning workers in the plant. Operators turned on the firewater sprayers to douse the stack, the tank mound, and the relief pipe to the scrubber. The water spray did not reach the gas, which were being emitted at a height of 30 meters. The supervisors tried to turn on the refrigeration system to cool the tanks, but failed because the coolant had been removed. The 400° F, 180 psi discharge continued for two hours. Because of proximity to the slums, thousands of people were exposed to this lethal mixture. Twenty-four hundred people died, thousands more were injured; 2,000 animals were killed; the environmental damage was considerable. The hospitals, dispensaries and morgues were overloaded and could not accommodate the flow of injured victims.

Causes of the Accident
- Insufficient staff quality and training
- Lack of safety culture
- Job insecurity
- Ongoing labor-management conflicts
- Carelessness in operations
- Lack compliance with rules
- Multiple responsibilities
- High turnover rate
- Elimination of the position of maintenance supervisor on the second and third shifts (this was a direct cause of the failure to install the blind flange)
- Lack of personnel for backup, inspections, control, and checking
- Lack of emergency planning and training
- Turning off the alarm that warned the public
- Failure to investigate causes of equipment inoperability and permitting the storage of hazardous material e.g. the failure of tank E610 pressurization
- Failure to make corrections that were found in a safety audit conducted two years before the accident
- Failure to maintain pressurization of MIC tanks resulted in contaminations that catalyzed the exothermic reaction.
- Poor design that stored MIC in large tanks
- The scrubber and offgas equipment were not designed to mitigate accidents
- Operation with the flare tower inoperable
- Lack of electronic emergency communications within the plant, and with neighbors
- Inability of the water spray to reach the discharge gases
- Inoperability of the refrigeration system
- Failure of a relief valve
- Lack of fixed water-spray protection in several areas of the plant
- Maximum allowable scrubber pressure of 15 psi while the rupture disk allowing gas to come into the scrubber was rated 40 psi.
- A spare tank was not empty to accept MIC transfer
- Tank E610 was filled to 75 to 80% of capacity; but 50% was recommended
- Failure to maintain a safe exclusion zone about the plant
- Flushing of pipes without investigating the blockage

7.1.2.6 Two Die in Ammonium Dichromate Explosion[e]

A January 17, 1986 explosion at a Diamond Shamrock Chemical plant near Ashtabula, Ohio, killed two employees, critically injured another, and caused lesser injuries to at least a dozen other workers. The blast involved 2,000 lb of ammonium dichromate in the unit for its production at the plant.

---

[e] Adapted from C&EN, January 27, 1986.

Ammonium dichromate is an intermediate in the manufacture of many important chromium chemicals, including pigments and magnetic recording materials. It is made by reacting chromium(VI) oxide in aqueous solution with ammonia. One of the final steps in the process involves drying the bright red-orange crystals in a rotary vacuum dryer. It was in that unit that the explosion took place. Ammonium dichromate decomposes on heating to yield nitrogen gas, water vapor, and chromium(III) oxide that was seen as a green cloud over the accident site. The wind carried the cloud over Lake Erie to dissipated. Diamond Shamrock notes that chromium(III) oxide can cause skin irritation but is otherwise nontoxic, however OSHA sets the threshold limiting value for chromium(III) oxide in air at 0.5 mg per cubic meter.

7.1.2.7 Three Thousand Evacuate Hydrofluoric Acid Leak at Texas City[f]

October 31, 1987 a hydrofluoric acid leak occurred at a Marathon Petroleum Co. refinery leading to the evacuation of 3,000 people over a 52-block area and hundreds of injuries. Hydrofluoric acid is used in many oil refineries as an alkylation catalyst. On the day of the accident, refinery workers were using a crane to lift a 40-ton convection section from a hydrofluoric acid heater. At about 5:20 PM, the crane failed and the piece fell severing pipes that were connected to a vessel containing the hot acid. Hydrogen fluoride gas spewed from the broken pipes to form a cloud that spread over the city. Cooling of the vessel began immediately; by about 6:30 PM, acid was being transferred from the vessel to transport trucks and railway tankcars. In three days, the vessel was empty and the facility was safe.

No plant employees were injured, many people were afflicted with inflamed eyes and lungs as a result of the fumes; 140 patients were admitted for observation and treatment for exposure to the gas. About 800 people were seen, treated, and released. Marathon set up a claims office; by the next day 600 people filed; from Texas City and nearby towns; filing went on for several days.

Hydrofluoric acid is highly corrosive to skin and mucous membranes. Even in fairly low concentrations, it causes painful skin burns and severe damage to eyes and the respiratory system. Exposure at higher levels results in destruction of tissues and death. No one in Texas City was exposed to more than trace concentrations of hydrofluoric acid. The acid vessel had a capacity of about 850 barrels of which a small fraction was released.

Other circumstances mitigating the accident were: the hydrogen fluoride gas was hot and rose rapidly; it was dissipated quickly by a brisk, dry wind; only several small pipes were broken, so the leak was fairly easily controlled.

---

[f] Adapted from C&EN, November 9, 1987.

7.1.2.8 Three Die in Explosion at Hoechst Celanese Plant[g]

November 14, 1987, three people were killed when two explosions and a fire caused extensive damage to the Hoechst Celanese acetic acid and acetic anhydride plant located near Pampa, TX. Also on the Pampa site was a specialty resins plant operated by Interez, a unit of RTZ Corp. which was interrupted because it depends on power supplied from the main Hoechst Celanese operation.

The cause of the explosion, which also left four people seriously injured and 31 others with minor injuries, is unknown. A Hoechst Celanese spokesman says that the first blast occurred near a gas-fired boiler and the second blast at a nearby reactor in which butane is reacted with steam to produce acetic acid. The 35-year-old plant employs 600 persons, 150 as contract maintenance and construction workers, but the blasts on Saturday only involved a weekend crew of 60.

A plant team working with the Pampa fire department brought the fire under control. The Chemical Manufacturers Association's Community Awareness & Emergency Response Program (CARE), developed after the Bhopal disaster was credited with effectiveness of their efforts in putting out the fire.

Hoechst Celanese officials said it was too early to decide whether to rebuild the butane reactor, install a newer methanol-to-acetic acid process, or start up a standby acetic acid facility at Bay City, TX, to replace the idled 350-million-lb per year acetic acid plant.

7.1.2.9 Two Workers Die and Solid Rocket Fuel Supply Destroyed[h]

May 4, 1988, explosions leveled a Pacific Engineering & Production Co. (PEPCO) plant, at Henderson, NV, one of only two U.S. plants producing 20 million lb/ year (maximum of 40 million lb/year - see Table 7.1-2) ammonium perchlorate for solid rocket fuel. It was the principal supplier for the space shuttle and sole supplier for the Titan rocket and several military missiles.

The explosion killed two workers, injured 350 local residents, destroyed property up to 20 miles away, and damaged 50% of Henderson's 17,000 residential and commercial buildings. The two largest blasts registered 3.0 and 3.5 on the Richter scale hundreds of miles away in California, cars were crumpled, doors blown off their hinges, and windows shattered for miles around. Uninsured losses were more than $20 million, plus $3 million more in insurance deductibles. Just as serious, were the psychological and emotional impact of the explosions, on schoolchildren and adults.

Table 7.1-2 Ammonium Perchlorate Use in Rockets (million lb/year)

User	1988	1989	1990
Army	9.3	14.0	11.7
Navy	3.7	6.5	5.4
Air Force	11.0	11.5	16.1
NASA	8.5	19.8	22.1
Commercial	3.5	3.7	5.2
European	1.8	2.6	5.3
Other	1.0	2.0	2.0
Total	38.8	60.1	67.8

---

[g] Adapted from C&EN, November 23, 1987.

[h] Adapted from C&NE, June 13 and August 8, 1988.

The fire department blamed the accident on: welders cutting in hazardous areas without a fire watch, highly combustible structural components (fiber-glass-resin), high-density storage of highly flammable and detonable material, spilled ammonium perchlorate about the plant, and high wind conditions.

The disaster brought an awareness for the potential hazards of the Kerr-McGee plant, just 0.7 mile from the center of town. Such was expressed at a hearing held by the Subcommittee on Investigations & Oversight of the House Committee on Science, Space & Technology.

The U.S. faced a crisis in its space and missile programs over the three years following the accident because of shortages of the solid fuel oxidizer ammonium perchlorate. A recovery strategy was formed by an interagency group, including members from DOD and NASA in liaison with the Departments of Commerce and Transportation. The first step is to restart Kerr-McGee and raise its capacity to 40 million/lb a year. Then, with federal aid, Kerr-McGee and PEPCO will build new plants with 20 million lb and 30 million lb annual capacities, respectively, but not in Henderson.

Before reopening, the plant adopted safety measures such as:

- Dispersal of the barrels of ammonium perchlorate stored outdoors about the site,
- An additional blender to expedite shipments and reduce stored inventory,
- Greater emphasis on safety, and
- Involvement with the public and their concerns.

7.1.2.9 One Dies in BASF Plant Explosion[i]

A reaction vessel explosion at BASF's resins plant in Cincinnati (July 19, 1990) killed one and injured 71. The BASF facility manufactures acrylic, alkyd, epoxy, and phenol-formaldehyde resins used as can and paper-cup liner coatings. The explosion occurred when a flammable solvent used to clean a reaction vessel vented into the plant and ignited. The cleaning solvent that was not properly vented to a condenser and separator, blew a pressure seal, and filled the 80-year-old building with a white vapor cloud.

7.1.2.11 Eight Die at Nitroparaffin Plant in Sterlington, LA[j]

The explosion and fire (May 1, 1990) caused eight deaths (including a vice-president and site manager), 120 injuries, extensive damage to buildings in the area, and evacuation of an entire small town. The cause of the explosion, at the Angus Chemicals Plant, was the blowout of a compressor line on the nitroparaffin unit.

---

[i] Adapted from C&EN, July 30, 1990.

[j] Adapted from C&EN, May 6 and May 13, 1991.

Across Louisiana State Highway 2, a hospital, local school, and nearby residences were damaged by the blast. On the site, the administration building was damaged but other processing units had little damage. IMC Fertilizer's two ammonia production units were not damaged. Emergency response and investigative teams moved in rapidly despite previous flooding and continued bad weather. Efforts initially concentrated on shutting down propane supply lines to the unit, stopping a leaking nitric acid tank perforated by the blast, and putting out fires that were limited to the plant site and were under control by early the next morning.

Within a few hours of the blast, all Sterlington was evacuated for fear of leaking toxic chemicals. Residents were allowed to return on May 3 after no leaks were found. Damage to the surrounding area was extensive. Many residences had windows blown out or were damaged by large pieces of piping and metal thrown by the explosion.

7.1.2.12 One Dies in Union Carbide Ethylene Oxide Plant Explosion[k]

An explosion and fire (March 13, 1991) occurred at an ethylene oxide unit at Union Carbide Chemicals & Plastics Co.'s Seadrift plant in Port Lavaca, TX, 125 miles southwest of Houston. The blast killed one, injured 19, and idled the facility, that also produces ethylene, ethylene glycol, glycol ether ethanolamines, and polyethylene. Twenty-five residents were evacuated for several hours as a safety precaution. The plant lost all electrical power, for a few days, because its cogeneration unit was damaged. The Seadrift plant, with 1,600 workers, is capable of making 820 million lb per year of ethylene oxide which is one-third of Carbide's worldwide production of antifreeze, polyester fibers, and surfactants; Seadrift produces two thirds of Carbide's worldwide production of polyethylene.

7.1.2.13 Six Die in Charleston Explosion[l]

An explosion and fire at an Albright & Wilson Americas phosphorus chemicals plant in Charleston, SC (June 17, 1991) killed six and injured 33. The damaged unit lost part of its walls and roof. Eight other units on the 200 employee site, sustained minor or no damage, but were shutdown for a few days. At the time of the accident, plant workers were mixing chemicals in the No. 2 reactor in the special products unit when an explosion and fire occurred. Five of the people killed were contract employees not directly involved with the reactor, but were installing insulation nearby.

Ironically, the reactor was used to produce Antiblaze 19, a flame retardant used in textiles and polyurethane foam. Antiblaze 19 is a cyclic phosphorate ester produced from a mixture of trimethyl phosphite, dimethyl methylphosphonate (DMMP), and trimethyl phosphate (TMOP). The final product is not considered flammable, but trimethyl phosphite is moisture sensitive and flammable, with a flash point of about 27° C.

---

[k] Adapted from C&EN, March 18, 1991.

[l] Adapted from C&EN, June 24, 1991.

7.1.2.13 Six Die in Benzoic Acid Explosion in Rotterdam[m]

December 13, 1991, a benzoic acid tank exploded causing six deaths and three injuries at the Dutch chemical firm: DSM's chemical complex in Rotterdam Harbor. Shipping traffic was halted for an hour in the world's busiest port while the fire was controlled.

A platform was being constructed from one storage tank to another, without the workers knowing that mixture of benzoic acid and air in the open storage tank could explode. DSM is the largest European producer of benzoic acid which is used in a range of applications from a plasticizer to food preservative. DSM's capacity here is 440 million lb per year.

7.1.2.15 Four Die at Fertilizer Plant Explosion[n]

An early morning explosion, Dec. 13, 1994, killed four employees and injured 18 at Terra Industries' Port Neal, Iowa, nitrogen fertilizer plant, 16 miles south of Sioux City. The blast leveled half the facility and forced evacuation of more than 2,500 people from nearby towns with an ammonia cloud released from a ruptured storage tank. Road and air traffic were diverted. Some were injured by falling debris and others by the impact of the explosion. Many suffered from ammonia inhalation.

The plant employed 119 people in three shifts, but only 30 were there when the explosion occurred. Terra's plant produced about 12% of the U.S. nitrogen-based liquid solution fertilizer. About one half of the plant was totally destroyed, the other half of the plant was damaged. The explosion was heard 50 to 60 miles away and shook Sioux City.

The explosion occurred in the ammonium nitrate production area. Shrapnel from the blast ripped open a 15,000 ton anhydrous ammonia storage tank, releasing an ammonia plume that could be smelled 30 miles away. A two-mph wind kept the ammonia from dissipating quickly. Metal fragments punctured a nitric acid tank, spilling up to 100 tons of 56% nitric acid. The ammonia tank, initially about one-third full, stopped leaking when the fluid level dropped below the puncture. Off-plant damage included blown windows and doors torn off their hinges. There was also structural damage to a nearby power plant. Emergency crews responded from Iowa, Nebraska, and South Dakota. A State Emergency Operations Center was activated. EPA, OSHA, the Coast Guard and the Department of Transportation were among the many federal, state, local, and private agencies participating the response.

---

[m] Adapted from C&EN, December 23, 1991.

[n] Adapted from C&EN December 19, 1994.

7.1.2.16 Four Die at New Jersey Chemical Plant Explosion

A slow response to a smoldering mix of chemicals at the Napp Technologies plant in Lodi, NJ is blamed for an April 21,1995 explosion and fire that killed four workers and injured dozens of others. The blast destroyed more than 70% of the plant, which made pharmaceutical and cosmetic intermediate products, and employed 110 workers. The explosion wrecked several stores housed in the Napp building, damaged nearby buildings, and forced evacuation of 400 residents for about 13 hours. Chemicals leaked into a nearby river, killing hundreds of fish.

The explosion occurred 12 hours after a gold precipitating process went awry. Problems began in the evening of Thursday, April 20, with clogging of a line that supplied benzaldehyde to a vacuum tumble dryer. A mix of 1,000 lb of aluminum powder and 8,000 lb of sodium hydrosulfite was already in the unit. Workers, trying to clear the blockage inadvertently introduced water into the vat. The water set off a smoldering reaction, but little was done to stop the reaction until the 6 AM shift change. The plant was evacuated at 6:30 AM; A chemist recommended nitrogen to smolder the reaction, but it was too late. At 8 AM, heat from the sodium hydrosulfite-aluminum water reaction reached the flashpoint of aluminum and it exploded and caused a fire.

7.1.2.17 More than Thirty Die at Chinese Dye Factory Explosion[o]

An explosion at a dye factory at Tianjin, near Beijing, killed more than 30 people and injured at least 60 workers and residents. The blast occurred June 26, 1996 at the China Petrochemical Factory (Dabua Huagongchang). The explosion was strong enough to shatter windows as far as 2 miles away and to create a crater about 10 feet deep and 90 feet in diameter. All buildings in an area of several thousand square yards were flattened; a steel beam from the factory's roof was thrown about 160 feet and a mixing furnace flew about 500 feet.

The incident started when sodium in a sulfurization process ignited. The blast occurred shortly after workers attempted to put out the fire with water.

7.2 Chemical Processes

7.2.1 The Topography of Chemical Processes

Chemical Reactors and Nuclear Reactors
Nuclear power reactors cause the transmutation of chemicals (uranium and plutonium) to fission products using neutrons as the catalyst to produce heat. Fossil furnaces use the chemical reaction of carbon and oxygen to produce CO_2 and other wastes to produce heat. There is only one reaction and one purpose for nuclear power reactors; there is one reaction but many purposes for fossil-burning furnaces; there are myriad chemical processes and purposes.

All chemical processes involve bringing atoms of selected types in close proximity for the

---

[o] Adapted from C&EN, July 8, 1996.

atomic forces to react and produce other chemicals. In addition to producing chemicals, heat is produced which may or may not be detrimental. Temperature and pressure are thermodynamic parameters that must be controlled to get the reaction started and to control the reaction. Thus, chemical processes consist of the ingredients listed in Table 7.2-1. Accidents result when the reaction rate becomes uncontrollable and the confinement fails.

Taxonomy of the Chemical Process Industry[p]

The chemical process industry is vast and varied. The value of chemicals and chemical products in 1993 was $0.5 trillion for the U.S,[q] involving 67,000 chemical engineers, 98,000 chemists. There were 5.5 non-fatal occupational injuries per 100 employees in 1995 involving chemical and allied products, and 4.8 per 100 workers in petroleum and coal products. There were 101 fatalities due to exposure to caustic, noxious or allergenic substances and 208 deaths from fires and explosions in 1995.

Table 7.2-1 Chemical Process Steps
- Get materials to be reacted
- Bring the chemicals together in confinement vessels called reactors
- Supply or remove heat
- Control temperature and pressure
- Use catalysts, light, and/or electricity
- Remove and process wastes.

The following is a list by major heading of processes in the Chemical Industry. While accidents can occur in any activity, the scope of the chemical process industry is so wide that processes are selected for description based on judged hazard. These are identified by the number of the section in which they are described. Process not identified by a three digit number are excluded on the basis of low perceived accident potential.

7.2.2	Inorganic chemicals	7.2.5	Explosives
7.2.3	Fertilizers		Rubber
7.2.4	Organic chemicals:	7.2.6	Plastics and resins
	Soaps and detergents		Man-made fibers
	Dyes	7.2.7	Paints and Varnishes
	Pharmaceuticals		Papermaking
		7.2.8	Petrochemical Processing

7.2.2 Inorganic Chemicals

Chlorine and Products

The Leblanc process for converting table salt (NaCl) to sodium carbonate ($NaCO_3$) for soap

---

[p] The following description of the chemical process industry is adapted from Britannica (1990).

[q] Information from the 1997 U.S. Statistical Abstracts website, http://www.census.gov. prod/www/abs/cc97stab.html.

and glass production was invented in 1793. In the first step, sodium chloride was treated with sulfuric acid to produce sodium sulfate and hydrogen chloride. The sodium sulfate was heated with limestone and coal to produce black ash which contained sodium carbonate mixed with calcium sulfide and unreacted coal. The sodium carbonate (soda ash) was removed by filtration, and crystallized. The hydrogen chloride gas damaged vegetation so it was converted to elemental chlorine, absorbed in lime to form bleaching powder. Thus, a liability was converted to an advantage.

Because calcium sulfide contained in the black ash had a highly unpleasant odor, methods were developed to remove the odor by recovering the sulfur, therein providing at least part of the raw material for the sulfuric acid required in the first part of the process. Thus, the Leblanc process demonstrated, at the very beginning, the typical ability of the chemical industry to develop new processes and new products, and often in so doing to turn a liability into an asset.

The Leblanc process was replaced by the ammonia soda (Solvay - 1860) process, in which sodium chloride brine is treated with ammonia and carbon dioxide to produce sodium bicarbonate and ammonium chloride. Sodium carbonate is obtained from the bicarbonate by heating. Ammonium chloride treated with lime gives calcium chloride and ammonia. The chlorine in the original salt becomes calcium chloride that is used for melting snow and ice. The ammonia is reused in the process (99.9% recovery).

The development of electrical power made possible the electrochemical industry. Electrolysis of sodium chloride produces chlorine and either sodium hydroxide (from NaCl in solution) or metallic sodium (from NaCl fused). Sodium hydroxide has applications similar to sodium carbonate. The advantage of the electrolytic process is the production of chlorine which has many uses such as production of polyvinyl chloride. PVC, for plumbing, is produced in the largest quantity of any plastic.

Sulfur and Products

Sulfuric acid is, by far, the most widely manufactured chemical. Burning sulfur, in air produces sulfur dioxide, which, when combined with water, gives sulfurous acid. Combining the dioxide with oxygen forms the trioxide which, when combined with water forms sulfuric acid. When this process take place in a chamber, it is called the "chamber" process.

The contact process which replaced the chamber process reacts the products using a platinum or a vanadium catalyst.

About half of the world's production of sulfuric acid goes to producing superphosphate and related fertilizers. Other uses are many, such as the manufacture of high-octane gasoline, of titanium dioxide (white pigment - a filler for some plastics and paper), explosives, rayon, uranium processing, and the pickling of steel. Sulfur was mined from volcanic deposits in Sicily, but is mined under by liquefaction. It is also obtained from the ore iron, pyrite, by burning to produce sulfur dioxide, and from some natural gases (sour gas), that contain hydrogen sulfide. It also is obtained from roasting zinc or copper sulfides to release sulfur dioxide. The sulfur present in low percentages in fossil fuels is a notorious source of air pollution in most industrial counties. Removal of sulfur from crude oil adds to the sulfur supply and reduces pollution. It is more difficult to remove sulfur directly from coal.

Ammonium sulfate fertilizer is made by reacting ammonia with sulfuric acid. In many parts of the world, calcium sulfate is in mineral form convertible to ammonium sulfate by combining it with ammonia and water - a virtually limitless source of sulfur.

Carbon disulfide is made by reacting carbon from natural gas, and sulfur from hydrogen sulfide, or sulfur dioxide. Carbon disulfide is used to make rayon and regenerated cellulose film.

7.2.3 Fertilizer Production

Two of the previously mentioned accidents and the Oklahoma City bombing showed the explosive potential of nitrogen fertilizer so this is a good place to start.

Fertilizers represent one of the largest markets for chemicals. Elements needed for agriculture are nitrogen, phosphorus, and potassium; calcium, magnesium, and sulfur are secondary nutrients; copper, zinc or molybdenum are needed in trace quantities.

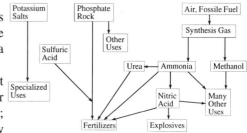

Fig. 7.2-1 Fertilizers and Uses

Nitrogen makes up 78 % of the atmosphere, hence, it is readily available. Ammonia is produced by fixing of atmospheric nitrogen with hydrogen. Mineral sources of phosphorus and potassium are converted to a suitable form for fertilizer. These three elements have other use than fertilizer; they are used and interact with other facets of the chemical industry, making a highly complex picture. A schematic of the interactions is presented in Figure 7.2-1.

Potassium

The seventh element in order of abundance in the Earth's crust is potassium - about the same as sodium with similar properties. While sodium is readily available from the ocean, potassium is found and extracted from many mineral formations. About 90 percent of the potassium that is extracted goes to the production of fertilizers. Other purposes for it are ceramics and fire extinguishers for which potassium bicarbonate is better than sodium bicarbonate.

Phosphorus

By far the largest source of phosphorus is phosphate rock, with some use of phosphatic iron ore, from which phosphorus is obtained as a by-product from the slag. Phosphate rock consists of the insoluble tricalcium phosphate and other materials. For use as a fertilizer, phosphate must be converted to the water soluble form, phosphoric acid (H_3PO_4) which has three hydrogen atoms, all of which are replaceable by a metal. Tricalcium phosphate, is converted to soluble monocalcium phosphate and to superphosphate. A fertilizer factory, typically, a large installation, characterized by large silos; produces year round, but peaks with the demands of the growing season. Phosphorus has many uses other than for fertilizer.

The weight of the superphosphate can be reduced by replacing sulfuric acid with phosphoric acid (obtained by sulfuric acid acting on phosphate rock followed by product separation, or by the electric furnace process). This process results in triple superphosphate, in which all the calcium originally in the phosphate rock appears as calcium monophosphate. The efficacy of the fertilizer is the high percentage of phosphoric oxide, which may be 45% in the triple variety, hence, twice as effective as fertilizer.

Instead of using sulfuric or phosphoric acid, nitric acid can be used to treat the phosphate rock to produce calcium nitrate fertilizer. Instead of neutralizing phosphoric acid with calcium which is useless, ammonia can be used to give ammonium phosphate, hence, two fertilizing elements.

A use other than fertilizer is phosphoric acid and the various phosphates derived from it for soft drinks. Disodium phosphate is used in: cheese processing, baking powder, flame proofing, treatment of boiler water in steam plants, and detergents. Elemental white phosphorus is used for rodent poisoning and for military smoke screens. Red phosphorus, is used in matches. Ferrophosphorus is an ingredient in high-strength low-alloy steel. Organic compounds of phosphorus are additives in gasoline, lubricating oil, plasticizers for plastics, insecticides, and related to nerve gas.

Nitrogen

In 1914, while abundant in the atmosphere, natural supplies of nitrogen in manure and in deposits of sodium nitrate were insufficient to meet the demand of war The solution was the Haber process for producing ammonia (NH_3) by heating hydrogen and nitrogen gas at high pressure. Hydrogen is obtained by decomposing water (H_2O) using heat or electricity. Burning coke and water produces steam, carbon monoxide, and hydrogen (water gas). Water gas reacts with steam and a catalyst to yield more hydrogen, and carbon dioxide which is removed by water dissolution. The mixture of carbon monoxide and hydrogen is the synthesis gas for methanol as shown in Figure 7.2-1 as the source of ammonia and methanol. The hydrogen for ammonia is obtained from the water-gas shift reaction. Originally the process used coke from coal or lignite (brown coal), but these has been replaced by petroleum products and natural gas.

The carbon dioxide removed in synthesis gas preparation can be reacted with ammonia, to form urea $CO(NH_2)_2$. This is an excellent fertilizer, highly concentrated in nitrogen (46.6%) and also useful as an additive in animal feed to provide the nitrogen for formation of meat protein. Urea is also an important source of resins and plastics by reacting it with formaldehyde from methanol.

Ammonia is used as a fertilizer by injecting it beneath the soil surface, or by mixing it with irrigation water. Granular ammonium nitrate is applied directly. Other uses for ammonia are making sodium carbonate using the ammonia soda process. It also is used for making rayon, as a refrigerant in commercial refrigerators and as a convenient portable source of hydrogen. Liquefied ammonia is used as a high density source of hydrogen for metallurgical processes in which ammonia is decomposed by heat to release the hydrogen and nitrogen.

Nitric Acid

The most important use of ammonia is in the production of nitric acid (HNO_3). Ammonia burns in oxygen, releasing hydrogen to form water and free nitrogen. With the catalysts platinum and rhodium, ammonia is oxidized and reacted with water to form nitric acid. Nitric acid treated with ammonia gives ammonium nitrate for use as a fertilizer. It is also a constituent of explosives. Three fundamental explosive materials are obtained by nitrating (treating with nitric acid, often in mixture with sulfuric acid): cellulose nitrate from wood, nitroglycerin from glycerol, and TNT from toluene. Another explosive ingredient is ammonium picrate (2,4,6-trinitrophenol) from picric acid. Detonating agents, or such priming caps are lead azide [$Pb(N_3)_2$], silver azide (AgN_3), and mercury fulminate [$Hg(ONC)_2$]. These are not nitrates or nitro compounds, but they contain nitrogen from the use of nitric acid in their manufacture.

Related to explosives are rocket propellants for missiles or spacecraft launch vehicles. Rocket propellants consist of an oxidant and a reductant. The oxidant is not necessarily a derivative of nitric acid, but may be liquid oxygen, ozone (O_3), liquid fluorine, or chlorine trifluoride. Other uses for nitric acid, not related to explosives or propellants, are in the production of cellulose nitrate for use in coatings. Without a pigment it forms a clear varnish used in furniture finishing. Pigmented, it forms brilliant shiny coatings called lacquers. Nitrating benzene yields nitrobenzene which can be reduced to aminobenzene, better known as aniline. Aniline also can be made by reacting ammonia with chlorobenzene. Similar treatment applied to naphthalene ($C_{10}H_8$) results in naphthylamine. Both aniline and naphthylamine are the parents of a large number of dyes, but today synthetic dyes are usually petrochemical in origin. Aniline, naphthylamine, and the other dye intermediates lead to pharmaceuticals, photographic chemicals, and chemicals for rubber processing.

Synthesis of ammonia must be done free of oxygen compounds that will poisons the catalyst, hence, all traces of carbon dioxide and carbon monoxide must be removed. A typical plant has large compressors and other equipment for the preparation and purification of the synthesis gas on the requisite scale. High pressure increases the yield at the expense of compression. Low temperature increases the yield, but as the temperature is lowered the reaction slows. The temperature that is chosen is about 500° C (930° F) to optimize results. Each time a fraction of the synthesis gas is converted to ammonia, therefore, after each pass the ammonia is removed and the remaining gas is recycled. Atmospheric nitrogen contains about 1% argon, a totally inert gas, which must be removed from time to time so that it does not build up in the system indefinitely. The nitric acid factory has equipment for producing ammonium nitrate in the grain size convenient for fertilizer. Growing plants do not receive all of their nitrogen from synthetic fertilizer. Some is supplied naturally by nitrogen-fixing plants, notably beans, in a symbiotic relationship with bacteria.

7.2.4 Halogens and Their Compounds

Chlorine

The first large-scale use of chlorine was for bleaching paper and cotton textiles; it also is widely used as a germicide for public water supplies. Presently it is used principally in production of the chemical compounds: sulfur chloride, thionyl chloride, phosgene, aluminum chloride, iron(III) chloride, titanium(IV) chloride, tin(IV) chloride, and potassium chlorate.

Organic chemicals made directly from chlorine include derivatives of methane: methyl chloride, methylene chloride, chloroform, carbon tetrachloride, chlorobenzene ortho- and para-dichlorobenzenes; ethyl chloride, and ethylene chloride.

The oldest process for chlorine production used hydrochloric acid reacting with manganese dioxide. This was superseded by the 1868 Deacon process that reacted atmospheric oxygen with the hydrochloric acid, from the Leblanc soda ash process. The chlor-alkali process uses simultaneous production of chlorine and sodium hydroxide by electrolytic decomposition of salt (sodium chloride) using brine is the electrolyte, and graphite rods as the anodes. An alternative process uses iron or mercury as the cathodes.

Passage of a direct current through brine causes chemical reactions where the electrodes contact the electrolyte. At the graphite anode chloride ions present in the dissolved salt are oxidized to elemental chlorine which goes to a vent. At the iron cathode, hydrogen gas is removed, while the hydroxide ions remain in the solution. The net result is that chloride ions and water are consumed and chlorine gas, hydrogen gas, and hydroxide ions are produced. Complete conversion of chloride to hydroxide is not practical, but as brine is continuously introduced at the top of the cell, a solution containing nearly equal amounts of salt and caustic soda is withdrawn at the bottom. Purification of this yields solid sodium hydroxide and a small amount of salt. Successful production of chorine and caustic soda in these cells requires that the two products be separated, because they would react with one another if mixed. Chlorine is separated from the caustic by a diaphragm between the electrodes (diaphragm cell).

The other main variant of the chlor-alkali process employs a mercury cell. The cathode is a shallow layer of mercury flowing across the bottom of the vessel. Graphite anodes extend down into the brine electrolyte. Direct current passes between the graphite rods and the mercury surface. At the anodes, chloride ions are converted to chlorine gas as in the diaphragm cell. The reaction at the mercury cathode is different from the reaction at an iron cathode. Positively charged sodium ions in the brine migrate to the mercury surface where the voltage is high enough to reduce them to sodium metal, but water is not reduced because of the mercury potential. The metallic sodium formed at the cathode dissolves in the mercury, and the amalgam flows out of the cell into another vessel, where it contacts water to forms sodium hydroxide and hydrogen. A mercury cell produces the same results as a diaphragm cell: sodium chloride and water are changed into sodium hydroxide, chlorine, and hydrogen. The mercury cell generates sodium hydroxide free of sodium chloride. This higher purity is needed for certain products such as the manufacture of rayon.

Fluorine

The fluorine industry is intimately related to aluminum production. Aluminum oxide, (Al_2O_3) is electrolyzed to metallic aluminum with a flux of sodium fuoroaluminate (Na_3AlF_6), called cryolite - a rare mineral found in commercial quantities only in Greenland with other uses: glass, enamels, and as a filler for resin-bonded grinding wheels.

Synthetic cryolite solved the supply problem, but synthetic cryolite requires fluorine which is actually more abundant in the Earth's crust than chlorine, but dispersed in small concentrations in rocks. Until the 1960s, fluorspar (CaF_2) a mineral long known and used as a flux in various metallurgical operations was the source. A source is phosphate rock that contains fluorine is 3% quantity.

Organic fluorine as a separate industry that began in the late 1920s with the discovery by Midgle of fluorocarbons for use as refrigerants. Ammonia was unsuitable because of the hazard and unpleasant smell from minute leaks. Sulfur dioxide had similar problems. The best refrigerant was

Freon 12 (CCl_2F_2) which is now banned by the ozone protection treaty. Also used is Dichlorodifluoromethane, Freon 22 ($CHClF_2$), and chloro-difluoromethane. Several analogous compounds containing carbon, fluorine, chlorine, and sometimes hydrogen are available.

Another application is uranium hexafluoride for enriching uranium in the fissile isotope uranium-235 by diffusion. Fluorine is known for being difficult to handle because of its intense chemical reactivity. The solution required the development and use of fluorine-resistant materials so that fluorine manufacture is now routine. Other uses developed: as a component in some rocket propellants, for the preparation of the extremely reactive interhalogen compounds such as chlorine trifluoride (ClF_3), used for cutting steel, and for the preparation of sulfur hexafluoride, an extremely stable gas that has been employed as an insulator in electrical applications. Other uses are teflon, a fluorocarbon resin such as poly-tetrafluoroethylene for coating frying pans to prevent food from sticking. There are several other fluorocarbon and fluorinated hydrocarbon resins; some have highly specialized applications in the aerospace industry. Fluorinated compounds are also used in textile treatments; some are soil-release agents that make fabric easy to wash. The salt: sodium fluoroacetate is an extremely powerful rodenticide. Sodium bifluoride is used to remove iron stains without weakening fabric. A minor but important use of fluorine in some countries is the fluoridation of drinking water for dental health.

Bromine

The properties of bromine are different from those of fluorine and chorine, and it is far less abundant. It was discovered in the early 19th century, as a salt (bromide) in the bitterns remaining after evaporating seawater and extracting sodium chloride for use as medicine. It was found in the production of potassium salts and from salt lakes. It gained industrial importance as an emulsion of silver bromide particles in gelatin for making photographic film.

Tetraethyllead was used to prevent knocking in car engines using ethylene dibromide, often in mixture with ethylene dichloride. The supply from brines from the Great Lakes and from the Dead Sea, at about 5 parts bromine per thousand, was insufficient for the demand. An alternative is the use of seawater at 70 parts per million in which a small amount of sulfuric acid is added for acidification. Then chlorine, which releases bromine, is absorbed by sodium carbonate and released by sulfuric acid. Besides the use of ethylene dibromide in gasoline as a lead scavenger, it is used as a fumigant, as a solvent for certain gums, and in further syntheses as is methyl bromide which also is used as a fumigant and fire extinguisher.

Iodine

On a smaller scale, the largest producer of iodine is Japan where it is extracted from seaweed containing more than 0.05 parts per million. The most important industrial iodine compound is silver iodide used with silver bromide in photography. Iodine is important in medicine for treating thyroid problems by adding it to table salt. It is used directly as a disinfectant, and a component of dyes. Crystalline silver iodide is used for cloud seeding.

7.2.5 Organic Chemicals

Originally the chemical industry was inorganic; in the 1960s organic chemicals (means they contain carbon) came into prominence with the compounds: benzene, phenol, ethylene, and vinyl chloride. The organic chemicals: benzene, phenol, toluene, and the xylenes compose the aromatic group.

7.2.5.1 Aromatic Hydrocarbons

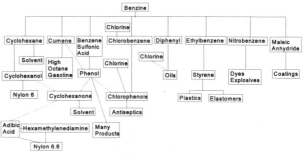

Fig. 7.2-2 Uses for Benzene

Benzene, the basis of the aromatics, has a closed, six-sided ring structure of carbon atoms with a hydrogen atom at each corner. Thus, a benzene atom is made up of six carbon (C) atoms and six hydrogen (H) atoms (C_6H_6). Originally benzene was obtained from heating coal to produce coke, combustible gas, and by-products, including benzene. Coke was used to manufacture iron and steel, but the supply was insufficient - each ton of coal produced only about two to three pounds of benzene. World War I shortage of toluene was for making TNT. The solution was toluene production from petroleum which today is the principal source of the aromatics.

Toluene

Toluene differs from benzene by the replacement of one hydrogen atom by a methyl group of carbon and hydrogen (CH_3), thus all aromatics are, to some extent, interchangeable. A use for toluene is the production of benzene by removing the methyl group (dealkylation). All of these hydrocarbons are useful as gasoline additives because of their anti-knock properties. Toluene also is used as a solvent for organic solutes such as coatings, adhesives, textiles, pharmaceuticals, inks, photographic film, for degreasing metal and in making polyurethane plastics and elastomers.

Xylene

Three isomeric (same number and kind of atoms but are arranged differently) xylenes occur together, and with the isomer, ethylbenzene, which has an ethyl group (C_2H_5) replacing a hydrogen atom in the benzene ring. These isomers may be separated with difficulty. The small letters: o-, m-, and p- (for ortho-, meta-, and para-) preceding the name xylene identify the three different isomers that vary in the ways that the two methyl groups displace the hydrogen atoms of benzene. Ortho-xylene is used to produce phthalic anhydride, an important intermediate for coatings and plastics. Meta-xylene is used to make coatings and plastics. Para-xylene leads to polyester fibers for clothing.

Benzene

Benzene achieves various uses through using other materials (Figure 7.2-2). For example, it is used with ethylene to produce styrene, with sulfuric acid to produce benzenesulfonic acid, and with chlorine, for several products. Many of the applications of benzene are not shown in this simplified diagram. For example, styrene is obtained from benzene by passing through ethylbenzene, a mixture with its isomers, the xylenes. Figure 7.2-2 shows coatings, plastics, and elastomers which are polymers (i.e., large molecules formed by bonding small molecules).

7.2.5.2 Aliphatic Hydrocarbons

The simplest organic chemicals are the saturated hydrocarbons methane (CH_4), ethane ($H_3C - CH_3$), propane ($H_3C - CH_2 - CH_3$), and others. These are burned but chemically they are rather unreactive. To use them as feed stock for other chemicals, they are cracked by heat to become unsaturated hydrocarbons. They contain less hydrogen than saturated hydrocarbons, and have one or more double or triple valence bonds connecting carbon atoms. Some of the most important unsaturated hydrocarbons are acetylene ($HC\equiv CH$), ethylene ($H_2C=CH_2$), propylene ($H_3C-CH=CH$), and butadiene ($H_2C=CH-CH=CH_2$). Figure 7.2-3 is a schematic of some of the products from petroleum cracking.

Ethylene

Ethylene is produced in quantity using acetylene or propylene as feedstock to make a large number of products (Figure 7.2-3) such as: acetaldehyde, acrylonitrile, acetic acid, and acetic anhydride. These are made generally from acetylene which is made from calcium carbide.

Acetylene

The raw materials for calcium carbide are: lime, coke, and electric power (Figure 7.2-3). Thus calcium carbide production is suitable for a country with hydroelectric power but lacks petroleum reserves. Calcium carbide generates acetylene when acted upon by water. The quantity produced may be small such as using the bright flame of acetylene for lighting, or oxyacetylene welding, or large such as chemical manufacturing.

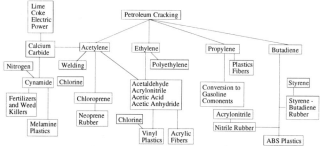

Fig. 7.2-3 Unsaturated Hydrocarbons from Petroleum

Acetylene and ethylene compete as a chemical raw material. Ethylene is generally more economical, resulting in declining use of acetylene as a raw material. Calcium carbide, a raw material for acetylene has other uses. Treated with nitrogen, it gives calcium cyanamide, valuable as a fertilizer and weed killer, and a raw material for the production of melamine, used in making some modern plastics.

Propylene

Propylene is produced in less volume than acetylene or ethylene. It is a raw material for detergents; derivatives used in antiknock gasoline additives. It can be polymerized to a product with uses similar to those of polyethylene. As a fibre, it is useful for carpets and rope.

Elastomers

Butadiene (Figure 7.2-3) is used for plastics and elastomers. Elastomers were originally synthetic substitutes for natural rubber. Research led to a copolymer of 75 parts butadiene and 25 parts of styrene (SBR) - styrene-butadiene rubber. It is often blended with other rubbers to get the best properties. Figure 7.2-3 shows that acrylonitrile can be copolymerized with butadiene (roughly 1/3 acryionitrile, 2/3 butadiene) to form nitrile rubber (NBR). This synthetic has different properties and is used for rubber hose, tank lining, conveyor belts, gaskets and wire insulation. Acrylonitrile and styrene, together with butadiene, form a terpolymer, called ABS, which is useful for high-impact-strength plastics.

Acrylonitrile contains nitrogen, and therefore is different chemically from natural rubber, which contains only carbon and hydrogen. Natural rubber's molecule is a repeating unit of five carbon atoms. By starting with the unsaturated hydrocarbon isoprene (C_5H_8), a synthetic natural rubber polymer can be made with the same spatial arrangement of atoms and properties as natural rubber. Another hydrocarbon elastomer starts with isobutylene (C_4H_8) and gives butyl, a rubber that is resistant to oxygen and impermeable to gases; it is used as cable insulation and as a coating for fabrics. Figure 7.2-3 shows acetylene as the raw material for chloroprene (C_4H_5Cl), which converts to neoprene. Rubber-like products that contain sulfur are known as thiokols. Related tough plastic materials containing carbon, sulfur, and oxygen are the sulfones.

Film materials

Most of the above can be made into plastic films - primarily used for wrapping. Film properties vary widely from permeable for food to impermeable to preserve dryness. Paper, treated or untreated, has been used for many years as a covering film, but has low strength when wet and is difficult to make transparent. In the 1920s, the transparency of cellophane revolutionized wrap. It is regenerated cellulose, like viscose rayon, except it is extruded in sheet instead of fiber, unfortunately it is sensitive to water and humidity.

Many other polymers compete with cellophane such as polyethylene which is extruded as a tough film or in greater thickness as a nonbreakable bottles. Vinyl products used in films are polystyrene, polyesters, and nylon. A chemical derivative from nature rubber, chlorinated rubber, gives films of extraordinary stretch ability.

Closely related to films are paints. Previously linseed oil, thinned with turpentine, formed the film into which were imbedded pigments. Turpentine, from pine trees, is still used, but competes with a petroleum distillate (less smelly). Water-based latex paints are a water emulsion of an organic paint that is thinned with water.

Carbon Black

Carbon black dominates as black pigment. It is a petrochemical made from natural gas or petroleum residues by incomplete combustion - cooking to split the hydrocarbon into hydrogen and carbon. Its primary use is in compounding rubber for making tires of which an average passenger car tire contains four pounds of carbon black. Carbon black is used in printing ink, but is difficult to remove for recycling. A carbon black is as an additive to phonograph records; a special form from acetylene, is used in electrochemical dry cells.

An important form of alcohols is methanol (Figure 7.2.1) synthesized from carbon monoxide and hydrogen. Methanol is called wood alcohol because it was obtained by distilling wood. About half of its production goes to making formaldehyde (CH_2O), a very reactive, widely used chemical. Methanol is used to produce various plastics and derivatives such as methyl chloride, a solvent for inks and dyes, and in the purification of steroidal and hormonal medicine. It is used to form important groups of plastics: the urea-formaldehyde resins; it is used as a fungicide and preservative, in paper and textile treatments, and in the synthesis of other products.

Ethanol and its Products

Methanol (CH_3OH) is the simplest form of alcohol; the next is ethanol (CH_3CH_2OH), with two carbon atoms. It is the intoxicating drink. For industrial use it is fermented from cheap material such as molasses, or made from ethylene by causing it to combine with water using a catalyst (sulfuric acid or phosphoric acid). Oxidation to acetaldehyde (CH_3CHO) is a major industrial use of ethanol. It could be in Figure 7.2-3 between ethylene and the block containing acetaldehyde and related chemicals. Ethanol is used to produce ethyl chloride (for tetraethyl lead), making various plastics, and in the further syntheses. Acetaldehyde may be made from ethanol in the same plant. Alcohol, for human consumption, is highly taxed, but this tax is avoided by denaturing with methanol to make it poisonous.

There are two possible structures (isomers) of three carbon atom alcohol. One is n-propyl alcohol (or l-propanol), the other is isopropyl alcohol (or 2-propanol). The former, not manufactured in large quantities is used in printing inks. The latter is manufactured in millions of tons to make propylene by a process similar to that used to convert ethylene to ethanol. The manufacture of 2-propanol by this process initiated the petrochemical industry in the 1920s.

A principal use of 2-propanol is to make acetone, a solvent, and as a starting material in making other organic compounds. Smaller amounts of 2-propanol are convened to other chemicals or used as a solvent, rubbing alcohol, or denaturing agent for ethyl alcohol.

Higher alcohols have more than three carbon atoms. Examples are the dihydric alcohol, ethylene glycol used for: antifreeze, brake fluids and as derivatives in: resins, paints, explosives, and polyester fibers. Reactions with propylene make propylene glycol, a moistening agent in foods and tobacco.

7.2.6 Explosives

Chemical explosives detonate, or deflagrate. Detonating explosives (e.g., TNT or dynamite) rapidly decompose to produce high pressure and a shock front (travels faster than the velocity of sound). Deflagrating explosives (e.g., black and smokeless powders) burn fast, produce lower pressures and no shock front. In large quantities and with confinement, deflagrating explosives may detonate.

Detonating explosives are primary or secondary. Primary explosives detonate by flame, spark, impact, or other heat sources of sufficient magnitude. Secondary explosives require a detonator and, in some cases, a supplementary booster. Some explosives are primary or secondary depending on conditions.

Black Powder

Black powder, a mixture of potassium nitrate, sulfur, and carbon, may have been invented in 10th century China for fireworks. There is written record they used it in bamboo tubes to propel stone projectiles. There is evidence that Arabs invented it around 1300 A. D. using a gun made of bamboo with iron bands to fire an arrow. Roger Bacon wrote its recipe in 1242, as a Latin anagram, but may have read it in Arabic.

The modern process for black powder begins with charcoal and sulfur in a hollow drum with heavy steel balls. This ball mill pulverizes the contents as the drum rotates through the tumbling of the steel balls. The potassium nitrate is crushed separately using heavy steel rollers. A mixture of several hundred pounds of potassium nitrate, charcoal, and sulfur goes onto a cooking pan where it is continuously turned over by plows. Next, it is ground and mixed by two rotating iron wheels, each weighing about 10 tons. The process takes several hours with water being added to keep the mixture moist. The product is put through wooden rolls to break up the larger lumps and formed into cakes under about 4,000 psi pressure. Coarse-toothed rolls crack the cakes into manageable pieces and the mixture is rolled to the desired size. Glazing (a graphite coating to make it flow better) tumbles the grains for several hours in large wooden cylinders to round the corners while forced air circulation adjusts the moisture content. After glazing it is graded by the size of the grains. Black powder is relatively insensitive to shock and must be ignited by flame or heat. In the early days, glowing tinder or heated rods ignited a powder train leading to the main charge to give the firer time to reach a safe place. Cannons used a touchhole in the breech filled with fine powder for ignition by slow-burning punk.

Black powder is still used in guns that cannot withstand the pressures of smokeless powder. The use of black powder in underground coal mines is no longer allowed in most counties. The low production makes black powder more expensive than dynamite. However, there is no substitute for black powder in certain military applications and nothing is equal to it for the manufacture of safety fuse. Black powder was used in constructing the Mont Cenis (eight-mile) railway tunnel through the Alps connecting France and Italy (1857-1871).

Nitroglycerin

Nitroglycerin was discovered by Ascanio Sobrero, in 1846. Its danger made it a laboratory curiosity until Alfred Nobel improved it along with other inventions such as the blasting cap using mercury fulminate.

The first large scale nitroglycerin factory in the U.S. was built by George Mobray to complete the Hoosac Tunnel (4 miles) at North Adams, Massachusetts in 1867. This plant, before closing, produced about 1-million pounds of nitroglycerin without accidents in either manufacture or shipment. An early use of nitroglycerin in the U.S. was blasting oil wells to increase their flow of oil. Another use was the construction of the Sutro silver mine in Nevada (1864-1874).

Frozen pure nitroglycerin (52°F) is relatively insensitive and easily kept frozen by packing it in ice. In cold climates, this presented a problem in thawing that was best overcome with different explosives.

Dynamite

In 1867, Nobel invented dynamite (Greek meaning "power") based on his discovery that kieselguhr earth absorbs nitroglycerin thereby making a safer explosive. The next step was replacement of the inert-earth absorber with combustible material (wood pulp) using a sodium nitrate oxidizer. Blasting action was controlled by adjusting the ratio of nitroglycerine to absorber.

Nobel invented gelatinous dynamite in 1875 by accident. He investigated the effect of nitroglycerine on the collodion (nitrocellulose in a mixture of ether and alcohol) that he used to treat a cut finger and found that it produced a tough plastic material with adjustable viscosity and high water resistance.

The high freezing point of nitroglycerin affects dynamite. Frozen dynamite is very insensitive, undependable and difficult to punch for blasting cap. Accidents occurred in thawing. In 1907, low-freezing dynamite was developed by adding 25% of the liquid isomer TNT (discussed later) to the nitroglycerin, or nitrated sugar in glycerin. In 1911, diglycerin (a glycerin polymer) was nitrated to produce tetranitrodiglycerin with a lower freezing point than nitroglycerine. The ultimate solution was nitrating ethylene glycol (a common antifreeze) to produce ethylene glycol dinitrate with properties practically identical to those of nitroglycerin, but low freezing. Dynamite made of it and nitroglycerin was stored in the open at Point Barrow, Alaska, for four years without freezing.

Ammonium Nitrate

An important advance in dynamite was the substitution of ammonium nitrate for part of the nitroglycerin to produce a safer and less expensive explosive. Nobel made this new dynamite successful by devising gelatins that contained from 20 to 60 percent ammonium nitrate. Ammonium nitrate was too hygroscopic, hence, work began to develop a nongelatinous form. , The solution, found in 1885, was coating ammonium nitrate with a little paraffin to produce a series of ammonia dynamites.

Major underground coal-mining countries regulate explosives by specifying those that are permitted. Practically all allowable explosives have ammonium nitrate as the major ingredient, because of its low explosion temperature. Nearly all contain a cooling agent such as table salt or ammonium chloride to prevent the heat of explosion in a mine from igniting methane or coal dust.

With inexpensive synthetic ammonia, and natural gas, explosives shifted from ammonium nitrate for nitroglycerin in the forms: 1) low-density ammonia dynamites and 2) semigelatins. Control of the blast and heat of explosion minimized nitroglycerin and a maximized ammonium nitrate, with dilution by low-density ingredients such as cane fibers to reduce the cost per stick. The semigelatins with a high nitroglycerin content are partially gelatinized with nitrocellulose. This gelatinization provides water resistance and plasticity that is desirable in loading holes prior to blasting. A moderate amount of water resistance in the ammonia dynamites, without gelatinization of nitroglycerin, can be achieved with water-repellent calcium stearate, or ingredients that form a water gel on the surface of the dynamite to slow the penetration of water.

Ammonium nitrate-fuel oil mixtures (ANFO) and ammonium nitrate-base water gels marked the biggest change in the explosives industry since dynamite. They now constitute more than 70% of the high explosives use in the United States because both ANFO and water gels are delivered in bulk by special trucks and loaded directly into boreholes.

In 1955 it was discovered that mixtures of ammonium nitrate and fine coal dust have satisfactory blasting capabilities in large (9-inch) holes used in open-pit coal mines to remove the rock and soil covering the coal. Polyethylene bags containing this material deform to fit the hole and provide moderate water resistance. ANFO is used in open-pit iron and copper mines and for construction such as road building. The mixture is air blown into 2-inch holes or less in many underground mines.

ANFO uses prilled rather than crystallized ammonium nitrate. Prills are free-flowing pellets developed for fertilizer as a coarse product with little setting tendency that can be spread easily and smoothly. A small amount of selguhr earth is added to improve the flow properties. Prills are made by tower dropping molten ammonium nitrate droplets. While falling, they dry as a slightly porous solid that absorbs oil to form a more sensitive product than in crystalline form. ANFO is a mixture of 94% prills and 6% No. 2 fuel oil that imparts water resistance which is enhanced by polyethylene bags.

Water gels, introduced in 1958, were mixtures of ammonium nitrate, TNT, water, and gelatinizing agents (guar gum and a cross-linking agent such as borax). Sometimes aluminum and other metals were used and better gelatinizers have been developed. In addition nonexplosive sensitizers were developed to dilute the TNT. Water gels have the advantages of: high yield, high degree of water resistance, plasticity, ease of handling and loading, and good safety characteristics.

Nitrocellulose

Schoenbein invented guncotton in 1845 by dipping cotton in a nitric and sulfuric acids in hope for a better propellant for military weapons, but it reacted too fast. A successful propellant was made in 1860, by nitrating wood in nitric acid, dissolving the acid, and impregnated the pieces with barium and potassium nitrates to supply oxygen for burning the incompletely nitrated wood. This powder worked in shotguns but was too fast for rifles. In 1884, smokeless powder was made by dissolved nitrocellulose in ether and alcohol until forming a gelatinous mass that was rolled into sheets and cut into flakes. The product, after the solvent evaporated, was satisfactory for all types of guns.

In 1887, Nobel invented "Ballistite" composed of 40% low nitrogen nitrocellulose and 60% nitroglycerin. Cut into flakes, it was a good propellant. The British ignored his patent and developed a similar product called "cordite."

DuPont in the U.S. developed about 1909, a smokeless powder from cotton of relatively low nitrogen that was quite soluble in ether alcohol. A small amount of diphenylamine was used as a stabilizer. After forming the grains and removing the liquid, a coating of graphite was added to make the smokeless powder that was used in the U.S. Other double-base types contain about 25% nitroglycerin. Cotton lint for nitration has been replaced by purified wood cellulose.

TNT Trinitrotoluene

TNT is the most useful military high explosive, although it was known for many years in the dye industry, it was not used as an explosive until 1904. The wide spread between its melting point and decomposition temperature gives it the ability to be safely melted and cast either alone or with other explosives. Its shortcomings are its extreme insensitive in the cast form, and the difficulty of casting without air holes. Its insensitivity is overcome by filling a 1-inch hole running the length of the explosive charge with tanitrophenylmethvinitramine (tetrvl); porosity is overcome by using a mixture of 40% trinitroxyiene (TNX) and 60% TNT. This mixture not only casts well but can be detonated with a smaller tetryl booster. TNX was replaced by PETN and RDX in World War II.

Picric Acid and Ammonium Picrate

Ammonium picrate (Explosive D) has exceptional value as a charge for armor-piercing projectiles. Loaded in a shell with a suitably insensitive primer, it can be fired through 12-inches armor plate and detonate on the far side. Early in World War I it was found that mixtures of molten TNT and ammonium nitrate were almost as effective for shell loadings as pure TNT. The mixtures most commonly used were 80-20 and 50-50 AN and TNT, known as amatol. Their advantage is stretching TNT use and reducing cost. Several explosives used in World War II were: RDX, PETN, and ethylenediaminedinitrate (EDNA). All were cast with about 50% TNT and used where the highest shattering power was needed. For example, cast 60-40 RDX-TNT, called cyclotol develops a detonation pressure of about 4-million psi. Mixtures of PETN and TNT have almost as much shattering effect. Compositions C-l to 4 are a series of plastic demolition explosives with great shattering power containing about 80% RDX combined with a mixture of various oils, waxes, and plasticizers. They differ in temperature ranges of usefulness.

7.2.7 Plastics and Resins

Plastics, in the modern meaning, are synthetic materials capable of being molded. The Greek word *"plastikos"* means to form. Natural products, while plastic, are usually excluded. Resins may be natural or synthetic. The distinction between plastics and resins is arbitrary - many synthetic materials are both. Historically, the term resin was applied to synthetic substitutes for the natural product; plastic was applied to compositions involving molding in their fabrication. Customarily, the fiber industry is considered to be distinct from the plastics industry, although it uses the same raw materials - polyamides (nylon), cellulose, and cellulose acetate.

Plastics, when heated, soften without melting, to attain a shape that is retained on cooling. They are composed of polymers, with giant molecules having immensely long chains of repeating units of chemically-bonded short molecules called monomers. The bonding forces are: a) the relatively weak London, Van der Waals, or dispersion forces, or b) the strong polar forces, hydrogen bonds or dipole-dipole interactions that result in higher softening temperature. Long polymers are usually tangled and considerable force is required for separation. When heated, the chains separate sufficiently to slide over one another to allow deformation.

Plastics are fabricated by molding, i.e., by melting plastic granules and pouring into a mold, extruding, rolling to form sheets, vacuum to form sheets to a mold or blowing from tubes into bottles at moderate temperatures (300 to 480° F). The results are tough, light, and transparent to take coloring. The polymers are either thermoplastic or thermosets which form a network of molecules that does not soften on reheating (e.g., phenol-formaldehyde).

The first plastic was celluloid, made of nitrocellulose softened by vegetable oils and camphor. It was used for car windshields and for movie film. The first completely synthetic plastic was bakelite (1910) produced from phenol and formaldehyde for use in the electrical industry.

Meanwhile, cellulose acetates were intensively studied to produce triacetate, a waterproof varnish to the fabric on World War I airplane wings. Triacetate also was made in powdered form for later melting and molding. With manufacturing development of methylene chloride, triacetate film came into use as a photographic base in 1935. Transparent acetate film was developed, and used for packaging.

Vinyl chloride (1835) formed by reacting acetylene with hydrochloric acid, was polymerized as polyvinyl chloride (PVC) in 1912. The theory of polymerization by Staudinger in the 1920s led to the advances that followed. The acrylate were polymerized as polymethylmethacrylate to come into production in 1927. Polystyrene was developed similarly and concurrently. Polyethylene came into production in 1939 for use in radar and now is ubiquitous.

Polymerization

The molecular chains of plastics are formed by condensation or addition polymerization. A condensation polymer forms by stepwise reacting molecules with each other and eliminating small molecules such as water. Addition polymer forms chains by the linking without eliminating small molecules.

The most important concept in condensation polymers is "functionality," i.e., the number of reactive groups in each molecule participating in the chain buildup. Each molecule must have at least two reactive groups, of which hydroxyl (-OH), acidic endings (-COOH), and amine endings (-NH) are the simplest. Hydroxyl is characteristic of alcohol endings that combine with an acid ending to give an ester which can polymerize to polyester. Examples are polyethylene terephthalate (ethylene glycol - hydroxyl ending and terephthalic acid - two acidic groups) and polycarbonate resins. Alcohols are a class of oxygen-containing chemical compounds with a structure analogous to ethyl alcohol (C_2H_5OH). Amines are compounds derived from ammonia by replacement of hydrogen by one or more hydrocarbon radicals (molecular groups that act as a unit). Esters are compounds formed by reacting an acid with alcohol or phenol and eliminating water.

Bulk addition polymerization of pure monomers is mainly confined to styrene and methyl methacrylate. The process is highly exothermic and heat must be removed to maintain chain length with intensive stirring of the viscous, partially polymerized mixture before passage down an increasing temperature tower. Alternatively, polymerization may be completed in containers that are small enough to avoid an excessive temperature. Methyl methacrylate is partially polymerized before being formed into clear acrylic sheet by molding between sheets of plate glass.

Suspension polymerization produces beads of plastic for styrene, methyl methacrylate, vinyl chloride, and vinyl acetate production. The monomer, in which the catalyst must be soluble, is maintained in droplet form suspended in water by agitation in the presence of a stabilizer such as gelatin; each droplet of monomer undergoes bulk polymerization. In emulsion polymerization, the monomer is dispersed in water by means of a surfactant to form tiny particles held in suspension (micelles). The monomer enters the hydrocarbon part of the micelles for polymerization by a water-soluble catalyst. X-rays or ultraviolet radiation may initiate polymerization for the manufacture of thermoplastics such as polyethylene and polyvinyl chloride.

Thermosetting Resins

Resins not softened by heating are the phenolics: furan resins, aminoplastics, alkyds, allyls, epoxy resins, polyurethanes, some polyesters, and silicones.

Phenolics or phenol-aldehydes include the important commercial phenolic resin: bakelite based on phenol and formaldehyde. A one-step process produces resol resin from more than one molecule of formaldehyde per phenol molecule. A two-step process uses an excess of phenol to produce novolacs - resins that have no reactive methylol groups and must be mixed with an aldehyde to undergo further reaction.

Resol resins thermoset on heating and are used for adhesives. Novolacs require a further source of formaldehyde in the form of hexamethylenetetramine to produce molding powders. Phenolic moldings are resistant to heat, chemicals, and moisture with good electrical and heat insulation qualities. Complex phenols from, e.g., cashew-nut shell liquid, are used in making brake linings.

Furan resins such as furfural is a five-member ring molecule consisting of four carbon and one oxygen atom. Carrying the aldehyde group, -CHO, it reacts like formaldehyde with phenols in the presence of an acid catalyst to give a rigid polymer with high chemical resistance for industrial coatings. It is used in semi-liquid form having low viscosity and high penetrating power for application to porous forms such as foundry sand cores.

Aminoplastics, such as urea resins, are made by condensing in aqueous solution of formaldehyde and urea in the presence of ammonia as an alkaline catalyst. The result is a colorless syrupy solution to which cellulose filler is added to make a molding powder. When heated, the powder melts in a mold and cools to a water-white solid that is strengthened by the filler to make cups and tumblers. Very large quantities of urea-formaldehyde resin are used in kitchens and bathrooms. Melamine behaves similarly to urea, but it is more moisture resistant, harder, and stronger for use in plates and food containers. Melamine moldings are glossy and harder than any other plastic. Solutions of the thermoplastic forms of urea-formaldehyde resins are widely used as bonding agents for plywood and wood-fiber products.

Alkyds

Alkyds are polyesters formed of phthalic acid (two acid groups) and glycerol (three hydroxyl groups). The solid resins mold fast under low pressure, cure quickly to produce material with good insulating properties, strength, and dimensional stability over a wide range of voltage, frequency, temperature, and humidity conditions for use as vacuum-tube bases, automotive ignition parts, switchgear and housings for portable tools.

Polyesters of Unsaturated Alcohols

The resins known as DAP and DAlP, are cross linked allyl esters of phthalic and isophthalic acid, respectively. They are notable for rigidity and excellent electrical properties at temperatures up to $450°$ F. Allylic resin-impregnated glass cloth is used in aircraft and missile parts. Other applications use the good storage life, and absence of gas evolution during polymerization. The optically clear resin: allyl diglycol carbonate is used for making cast objects that are more heat and abrasion resistant than other cast resins.

Epoxy Resins

Epoxy resins have outstanding mechanical and electrical properties, dimensional stability, resistance to heat and chemicals, and adhesion to other materials. They are used for casting, potting, encapsulation, protective coatings, and adhesives. Epoxy glues separate the resin from the curing agent to be mixed just prior to use.

Polyurethanes

Polyurethanes are formed by reacting diisocyanates and polyols (multihydroxy compounds) to form some of the most versatile of rigid to elastic plastics. A major use is for foams with good flexibility and high rigidity. Thermoplastic polyurethanes can be extruded as sheet of extreme toughness.

Polyesters of Unsaturated Acids

Certain esters can be polymerized to an infusible resin for use in glass-fiber-reinforced plastics. Unsaturated maleic acid anhydride is polymerized to a relatively short polymer chain by condensation with a dihydric alcohol such as propylene glycol. The chain length is determined by the relative quantities of the two ingredients. The resulting syrup of condensation polymer is diluted with a monomer, such as styrene, or an initiator for addition polymerization. This mixture is quite stable at room temperature. Frequently, a silicone compound is added to promote adhesion to glass fibers, and wax to protect the surface from oxygen inhibition of polymerization. Glass-fibers are impregnated with the syrup and polymerized by raising the temperature. Alternatively, the polymerization is done at room temperature by addition of a polymerization accelerator to the syrup immediately before impregnation. After a controlled induction period exothermic polymerization takes place to give a glass-fiber-reinforced cross-linked polymer.

Silicones

Silicon, unlike carbon, does not form double bonds or long chains. However, it forms long chains with oxygen such as in siloxanes in which hydrocarbon groups are attached to the silicon producing material for a wide range of oils, greases, and rubbers. By replacing certain atoms in the chain, silicon resins can be used for high- and low-pressure lamination with fibers to produce electrical material having high dielectric strength, and low dissipation over a wide temperature and humidity range. Silicones do not distort in temperatures up to 752° F.

Thermoplastic Resins

Resins that soften on heating include: cellulose derivatives, addition polymers, and condensation polymers. Cellulose is a naturally recurring chain carrying three hydrox groups, which can be esterified by acids. Cellulose nitrate (gun cotton) is formed by treating cellulose with a mixture of strong sulfuric and nitric acids. After washing, mixing with alcohol and camphor a plastic is produced for rolling into sheets, sliced, dried and polished by hot pressing between metal sheets. These sheets still contain alcohol and can be cut and repressed to give patterned material or it can be extruded as plain sheet. Nitrocellulose cannot be injection molded because of its flammability. It is tough and moisture resistant for use as eyeglass frames and table tennis balls.

If cellulose is pretreated with acetic acid, followed by more acid and its anhydride, acetate esters are formed giving cellulose triacetate. Complete acetylation would, produce an acetic acid content of 62.5%; in practice about 61% is usual. The material has a crystalline structure that is moisture-resistant, dimensionally stable, and a soluble in a limited range of solvents, especially methylene chloride with 10% methanol. Further treatment of the triacetate in solution in the presence of sulfuric acid splits off some acetic acid to give diacetate which is soluble in acetone, and compatible with a range of plasticizers that can be incorporated using a rugged mixer without solvent to yield molding powders especially suited to injection molding, e.g., tool handles.

If butyric or propionic acids are included in the esterifying mixture, acetate butyrates and acetate propionates are formed. These, requiring less plasticizer, are tougher and more moisture resistant material for making automobile steering wheels.

Polyethylene

Polyethylene has low density when polymerized at pressures 9,000 - 45,000 psi and high density when made with special catalysts at 250 - 500 psi. Low-density polyethylene softens 68° F lower than high-density polyethylene, which is more crystalline and stiffer. The rigidity characteristics and surface of high-density polyethylene are comparable with polystyrene. It feels like nylon, has a bursting strength three times that of low-density polyethylene, and withstands repeated exposure to 250° F, hence, it can be sterilized.

Low-density polyethylene is used as film to make bags with thicknesses ranging from 0.25 to 7 mils for agriculture, building-site protection, packaging, and electrical insulation. Because of its stiffness, high-density polyethylene is used for large drums, gasoline tanks, bottle crates, and domestic appliances. It can also be produced as a foam.

Polyethylene can be chlorinated in solution in carbon tetrachloride or in suspension in the presence of a catalyst. Below 55-60% chlorine, it is more stable and more compatible with many polymers, especially polyvinyl chloride, to which it gives increased impact strength. The low pressure process copolymerizes polyethylene with propylene and butylene to increase its resistance to stress cracking. Copolymerization with vinyl acetate at high pressure increases flexibility, resistance to stress cracking, and seal ability of value to the food industry.

Ionomers are ethylene polymers in which carboxyl groups are located along the ethylene chain. These acid groups are convened into salts of sodium, potassium, magnesium, or zinc to stiffen and toughen the plastic without destroying its thermoplastic character. As a result, unplasticized sheet has drawing characteristics superior to those of any other resin. Ionomers are highly transparent, resistant to organic solvents, and maintain toughness at low temperatures. They are used in composite films with polyethylene and nylon because of their puncture resistance. They have strong natural adhesion to aluminum and are used in food packaging and as injection moldings in the shoe industry.

Polypropylene is made by the polymerization of propylene. Stronger and more rigid than polyethylene, with a harder surface, it softens 50 to 70° F higher, and is the lightest common plastic. It is more prone to oxidation than polyethylene unless appropriate stabilizers and antioxidants are added. Injection molding of it makes luggage, automotive panels and trims. Its fatigue resistance makes it useful for integral hinges. Polypropylene is used for fibers and film. The latter is extruded and develops its greatest strength when oriented by stretching biaxially i.e. along two axes at right angles. Film clarity is obtained by minimizing the size of crystals and by various additives. It is made in thicknesses as thin as 0.2 mil. Copolymers of propylene with minor amounts of ethylene are also used for making film.

Methylpentene polymer, a light plastic, has a crystalline melting point of 464° F, form retention up to 392° F, transparency of 92%, and electrical properties similar to polytetrafluoroethylene. Its impact strength is greater than polystyrene and polymethyl methacrylate; it is resistant to alkalies, weak acids, and non-chlorinated solvents. It may be injection molded into implements for food packaging and preparation, medical care, and non-stick coatings.

Vinyl chloride is made from ethylene or acetylene and chlorine. The polymer, polyvinyl chloride is made plastic by calendering, extrusion, and molding. The most commonly used plasticizer is dioctylphthalate, but adipates may be used for special resistance to temperature change. Applications of flexible polyvinyl chloride are clothing, flooring, wire insulation, home furnishings, plumbing, toys, and steel sheet covering.

Copolymers of vinyl chloride, containing 5 to 40 percent vinyl acetate made by the inclusion of vinyl acetate in the polymerization process, have lower softening points and flow more easily than polyvinyl chloride. They are soluble in ketones, such as acetone, and certain esters for making film from solutions. They are used for phonograph records, rigid clear sheeting, and molding powders.

Acrylic

Acrylic is a generic name for derivatives of acrylic acid, of which methyl methacrylate is the most important. Polymerization is controlled to produce chain length of 800 to 3,000 monomer units. A small amount of plasticizer such as dibutyl phthalate may be added before bulk polymerization to assist in deep molding. The outstanding property of polymethyl methacrylate is 92% transparency; resistance to ultraviolet radiation from fluorescent lamps and ability to be fashioned by deep drawing or injection molding. It is used for airplane canopies, dentures, optics, and light covers of various colors.

A use is for internally illuminated signs in which any color can be introduced. Its dimensional stability recommends it for many optical uses. Acrylics are modified by copolymerization to improve impact strength at the loss their extreme transparency. An example is acrylic-modified polyvinyl chloride sheet, which is tougher than acrylonitrile-butadiene-styrene and polycarbonate and is suitable for corrosion-resistant pans, aircraft parts and materials handling equipment.

Polytetrafluoroethylene and fluorinated ethylene-propylene are the only resins composed wholly of fluorine and carbon. The polymer consists of fluorine atoms surrounding the carbon chain as a sheath, giving a chemically inert and relatively dense product from the strong carbon-fluorine bonds. Polytetrafluoroethylene must be molded at high pressure. Fluorinated ethylene-propylene can be injection molded and extruded as thin film. Both plastics have exceptional heat resistance and very low friction coefficients. They are used for seal rings, non-lubricated bearings, packings, and electrical applications.

Styrene

Styrene is polymerized to the clear resin: polystyrene by: mass-block, continuous mass process, or suspension polymerization in water. The mass-block method, polymerizes using cells alternately filled with monomer and a cooling medium. Continuous polymerization uses a tower with cooling zones and a heated base from which the polymer is extruded. Polystyrene is an injection material that is hard, brittle, flammable, transparent, good insulator, but softens in boiling water. High-impact polystyrene is made by dissolving styrenebutadiene rubber in styrene and polymerizing the styrene.

Acrylonitrile-butadiene-styrene polymers are similar in structure, but the acrylonitrile hardens the polymer. Minute rubber particles act as stress-relief centers, making it good for large objects: luggage or car body parts. It can be chrome plated, foamed, injection molded, blown, and alloyed with other plastics.

Formaldehyde Polymers

Formaldehyde polymerizes to paraformaldehyde, a stable polymer of high molecular weight. It is one of the strongest and stiffest thermoplastics. It can be injection molded and extruded. It is used in automotive safety steering columns because of fatigue resistance, and in snap-fit assemblies because of its impact strength and elastic recovery. It is moisture and abrasion resistant with good electrical properties over a wide temperature range. When glass-fiber-filled, it has exceptional resistance to creep.

Nylon and Condensation Polymers

Nylon-6.6 is made by heating amine and acid in an autoclave from which it is extruded onto a chilled drum and segmented. It has: high melting point, strength, stiffness, toughness, and chemical resistance. These properties make it a substitute for metal used for gears, oil seals, bearings, and structural components. Extruded film is used for boil-in-bag packaging.

Other nylons are made by varying the molecular length of the diamines and the dibasic acids: Nylon-6.10 uses sebacic acid (10 carbon atoms), nylon-11 uses an acid from castor oil, and nylon-12 uses butadiene. These variations decrease moisture absorption. Other variations use amines with a ring structure, e.g., the aromatic nylons to give polymers with softening points above 577° F.

Polyethyleneterephthalate

This is a polyester made from glycol and dimethyl terephthalate by ester interchange. The reaction produces methanol instead of water for displacement by ordinary esterification. The product is extruded as a transparent amorphous film. If it is stretched biaxially at 176° F, it is still amorphous and is used for shrink-wrapping food stuffs and medical accessories by momentarily reheating it above the stretch temperature. If the film is biaxially stretched to three or four times its original dimensions, at a higher temperature, it crystallizes and develops very high tensile strength (25 kpsi) and dimensional stability.

Polycarbonates

Are long-chain linear polyesters of carbonic acid and dihydric phenols, such as bisphenol A. This faintly amber, transparent plastic has toughness, tensile strength, ductility, and dimensional stability from 60 to 300° F. It is injection molded, blow molded, and extruded as a metal substitute with good electrical insulating properties. It is used for tool housings, safety helmets, computer parts, and vandal-proof windows.

Polyimides

Are made by a two-step reaction of dianhydride (tetrabasic acid with two molecules of water removed) with an aromatic diamine giving first a polyamic acid that is converted by heat or catalysts to high molecular weight linear polyimides. They are used as film, laminating resins, adhesives, and molding compounds. They are the best organic materials for high temperature insulators, capacitors, and bearings.

If one amino group in o-phenylenediamine is converted to an amide group by formic acid, the intermediate: benzimidazole is formed. This reaction, conducted with a wide range of reactants, produces resins (polybenzimidazoles) used as high-temperature adhesives for laminates in the aerospace industry. Heat insulation is made by including tiny bubbles of silica and alumina fibers for reentry vehicle fuel cells, and nuclear reactors. Nose cone ablative material is made by incorporating phenolic microballoons and carbon fibers in the plastic.

7.2.8 Paints and Varnishes

Paint is a pigment in a binder hiding what is underneath it. Inorganic pigments are insoluble, metallic compounds whose colors come from the energy levels of the pigment. Organic pigments are compounds containing molecules with atoms of the desired properties. Paint is manufactured in batches up to 2,500 gallons; the binders are composed of polymeric resins in a solvent. The pigments are ground to colloidal size in ball mills (rotating containers containing the material being ground and steel or ceramic balls), or sand grinders, which circulate a suspension of sand in paint through a rotor assembly at high speed.

Paint must be uniform in color, flow, and have other physical properties. Driers, fungicides, and tint bases are added during thinning. The paint is measured for hue, brightness, and chrome. White pigments and extenders provide hiding power by scattering light; hue is provided by pigments and toners that absorb light of certain wavelengths.

White Pigments

White lead ($2PbCO_3 \cdot Pb[OH]_2$) has been manufactured for over 300 years by subjecting lead sheet to acetic acid fumes and carbon dioxide for several months. Sulfate of lead ($PbSO_4 \cdot PbO$) is manufactured by vaporizing lead and sulfur. Zinc oxide (ZnO) results from the reaction of zinc vapor and oxygen. Zinc sulfide (ZnS) is formed by precipitation from an aqueous solution and heating to a high temperature. Lithopone ($ZnS + BaSO$) is formed by precipitating zinc sulfate ($ZnSO_4$) and barium sulfide (BaS) together. Antimony trioxide (Sb_2O_3) is produced by roasting antimony in air. Titanium dioxide (TiO_2) requires the removal of iron from ilmenite ore, $FeTiO_3$, after which it is heated with small amounts of aluminum, antimony, silicon or zinc oxide to produce white pigments with suitable properties.

Black Pigments

Most black pigments are made of carbon black formed by depositing carbon from a smoky flame of natural gas on a metal surface. Lampblack is made similarly by burning oil. Bone blacks are made from charred bones. Graphite occurs naturally or can be prepared from coal in electric furnaces. Mineral blacks come from shale, peat, and coal dust. Iron oxide blacks are found in nature or prepared. Blue lead sulfate is a pigment for priming. Of these, carbon black is superior.

Colored Pigments

Colored pigments are grouped according to their position in the visible spectrum. Mineral varieties dominate the red and blue ends with organic varieties in between. Common red pigments are red lead, iron oxide, cadmium reds, cuprous oxide, and various organics. Red lead (Pb_3O_4) is a protective coating for steel. It is made by heating metallic lead or lead oxide. Iron oxide (Fe_2O_3) occurs naturally by weathering of silicate rocks and was used prehistorically. Other ocres are the sennas composed of 30 to 75% iron oxide ranging from yellow-brown to reddish-orange depending on the variety and the degree of calcination. Timbers have high manganese dioxide content and tend toward brown coloration. Iron oxide ores include hematite ores, such as Spanish oxide and Persian Gulf oxide, magnetite (black), limonite (yellow), siderite (brown), and pyrites (dark brown). Synthetic iron oxide, made by calcinating ferrous sulfate ($FeSO_4$), ranges from light Turkey red to dark Indian red; calcination in lime (CaO) produces a mixture of ferric oxide and calcium sulfate ($CaSO_4$) known as Venetian red. Miscellaneous inorganic red pigments include cadmium reds, which rely on the mixture of cadmium sulfide (CDS) and selenide (CdSe) to produce colors ranging from light red to maroon, and cuprous oxide (Cu_2O), a fungicide. Organic red pigments include naturally occurring carmine and madder and synthetics. Synthetics are expensive but can have an infinite variety of hues by changing the molecular structure.

Yellow and orange pigments are obtained from metallic salts and from organic compounds. Chrome yellows and oranges are low-cost, bright, and stable pigments that display a considerable range of hues. Consisting of various proportions of lead chromate ($PbCrO_4$), lead sulfate ($PbSO_4$), and lead monoxide (PbO), all turn brown when exposed to sulfide gases because of the formation of lead sulfide (PbS). Molybdate oranges are made from lead chromate precipitated with lead molybdate ($PbMoO_4$) and lead sulfate ($PbSO_4$). Zinc yellow has a greenish cast because of a high content of chromic oxide. It is light fast, rust inhibitive, and sulfide resistant. Basic zinc chromate ($5ZnO\ CrO_3$) is an undercoat. Cadmium yellow and oranges are formed by diluting yellow cadmium sulfic (CDS) with white zinc sulfide (ZnS) and barium sulfate ($BaSO_4$), using small amounts of red cadmium selenide (CdSe) to impart the orange coloration.

Blue and green pigments are mostly synthetic soda produced by fusing china clay in a mixture of sulfur, sodium sulfate, and an organic reducing agent such as rosin. The glassy product is pulverized, leached free of sodium sulfate, and size-separated by sedimentation. The most widely used blue pigments are iron blue that contain various amounts of iron, ammonium and sodium ions, and hydration. Chinese blue has a greenish-blue tint; Milo blue tends toward the red. Phthalocyanine pigments, with a rather complicated organic structure, are reproducible, light-fast, brilliant but expensive to produce.

Binders

A surface coating protects the substrate against abrasion, moisture, light, and corrosion. The binder for the pigment and extenders is fluid before application and rigid soon after. Natural binders range from gum arabic to fish oil. The first varnishes were solutions of natural resins, having transparency, hardness, amorphous structure, and little permanence.

Synthetic resins form the heart of the paint industry. The two main types of synthetic resins are condensation polymers and addition polymers. Condensation polymers, formed by condensation of like or unlike molecules into a new, more complex compound, include polyesters, phenolics, amino resins, polyurethane, and epoxies. Addition polymers include polyvinyl acetate, polyvinyl chloride, and the acrylates.

Condensation Polymer Resins

Alkyds are the most widely used protective-coatings at 1-million pounds annual world consumption. Brittle and insoluble in common solvents, they are manufactured with a plasticizer. Manipulating the molecular structure produces alkyds with the desirable properties.

Phenol (C_6H_5OH) reacts readily with aldehydes to form oil soluble compounds called phenolic resins that dry rapidly, are hard, chemical and abrasion resistant, but they tend to yellow with age.

Condensation of urea (H_2NCONH_2) or melamine ($C_3H_6N_6$) with formaldehyde (HCHO) produces the amino resins which are non-yellowing, alkali resistant and have good adhesive properties for use in coatings. These amino resins contain an alcohol as a third component. Both polyurethanes and epoxy resins are produced using complex organic reactions. Polyurethane coatings dry hard and inert on a substrate; epoxy resins have qualities of toughness, adhesion, chemical resistance, rigidity, and thermal resistance. Silicones are produced by attaching organic groups to a silicon atom. They are hard, tough, clear, glossy, heat, weather and chemical resistant.

Addition Polymers

Ethylene, a simple, unsaturated organic molecule, is made into translucent, inexpensive, waxy polyethylenes at high pressures. When polyethylenes are applied to substrates, the resulting film is resilient and resistant to acids and oxidizers.

A synthetic addition polymers that has largely replaced natural resins is polyvinyl acetate. It is a clear, transparent resin that softens below the boiling point of water. It is sensitive to water, soluble in many organic solvents and has good adhesion properties. Vinyl resins are dissolved or suspended in an organic medium, like the natural resins, or polymerized and marketed as an aqueous latex emulsion. Polyvinyl chlorides are not desirable as coatings. Clear, non-yellowing, and heat stable acrylic compounds, are used to produce acrylic emulsion paints. Their manufacture consists of dispersing the pigment in water and adding the latex with only moderate agitation.

7.2.9 Petrochemical Processing

7.2.9.1 Overview

Many chemical products are produced from crude oil. Initially, little chemistry was involved therefore the petrochemicals were not considered part of the chemical process industry. Today, materials ranging from specialized fuels, plastics and synthetics makes it part of the chemical processing. The petroleum refinery is where the chemical processing of oil begins.

Topping and Hydroskimming Refineries

The simplest plant is a topping refinery that prepares feedstocks for petrochemical manufacture or for production of industrial fuels. It consists of tankage, a distillation unit, recovery facilities for gases and light hydrocarbons, and supporting utilities (steam, power, and water-treatment plants). The range of products is increased by the addition of hydrotreating and reforming units comprising a hydroskimming refinery, which can produce desulfurized distillate fuels and high-octane gasoline. About half of the production is fuel oil.

Conversion Refineries

More versatile is a conversion refinery consisting of the equipment found in topping and hydroskimming refineries with gas oil conversion operations such as catalytic cracking and hydrocracking units, olefin conversion plants such as alkylation or polymerization units, and, frequently, coking units for sharply reducing or eliminating the production of residual fuels. Two-thirds of the production of a modern conversion refinery may be unleaded gasoline, with the balance being high-quality jet fuel, LPG, low-sulfur diesel fuel, and a small quantity of petroleum coke. Many such refineries also incorporate solvent extraction processes for manufacturing lubricants and petrochemical units to recover high-purity propylene, benzene, toluene, and xylenes for further processing into polymers.

7.2.9.2 Basic Refinery Processes

A refinery separates the many types of hydrocarbons present in crude oils into related fractions of chemically converted, separated hydrocarbon products, and recovers the wastes that are produced.

Fractional Distillation

The primary process for separating the hydrocarbon components of crude oil is fractional distillation i.e. separation according to the boiling points of the components. These separated fractions are processed further by: catalytic reformers, cracking units, alkylation units, or cokers which have there own fractional distillation towers for its products.

Modern crude oil distillation units are larger than those in the chemical process industry, producing up to 200,000 barrels per day of product.

Figure 7.2-4 shows the operating principles of a modern crude oil distillation unit. Crude oil is pumped from tankage at ambient temperature through heat exchangers to raise its temperature to 248° F. A controlled amount of fresh water is introduced, and the mixture is pumped into the desalter, where it passes through an electrical field to separate the saltwater phase to prevent deposition on the tubes of the furnace and cause plugging. The desalted crude oil passes through additional heat exchangers and then through steel alloy tubes in a furnace. There it is heated to about 698° F - depending on the type of crude oil, and the end products desired. A

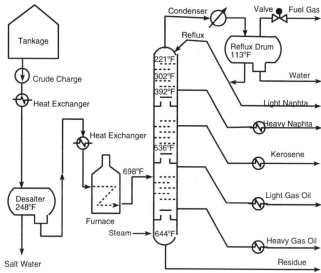

Fig. 7.2-4 Diagram of a Fractional Distillation Column

mixture of vapor and unvaporized oil passes from the furnace into the fractionating column, which is a vertical cylindrical tower up to 150 feet high containing 20 to 40 fractionating trays spaced at regular intervals. The most common fractionating trays are of the sieve or valve type. Sieve trays are simple perforated plates with small holes about 0.25 inch in diameter. Valve trays are similar, except the perforations are covered by small metal disks that restrict the flow through the perforations under certain process conditions. Oil vapors rise up through the column and are condensed by a water or air-cooled condenser at the top of the tower. A small amount of gas remains uncondensed and is piped into the refinery fuel-gas system. A pressure control valve on the fuel-gas line maintains fractionating column pressure (~ atmospheric). Part of the condensed liquid, called reflux, is pumped back into the top of the column and descends from tray to tray, contacting rising vapors as they pass through the slots in the trays. The liquid progressively absorbs heavier constituents from the vapor and gives up lighter constituents to the vapor phase. Condensation and reevaporation takes place on each tray. At equilibrium, there is a continual gradation of temperature and oil properties throughout the column, with the lightest constituents on the top tray and the heaviest on the bottom. The use of reflux and vapor-liquid contacting trays distinguishes fractional distillation from simple distillation columns.

As shown in the figure, intermediate products are withdrawn at several points from the column. In addition, modern crude distillation units employ intermediate reflux streams. Typical boiling ranges for various streams are: light straight-run naphtha (overhead), heavy naphtha (top sidestream), 195 to 330° F; crude kerosene (second sidestream), 300-475 °F; light gas oil (third sidestream), 420 to 600° F. Unvaporized oil entering the column flows downward over stripping trays to remove any light constituents remaining in the liquid. Steam is injected into the bottom of the column to reduce the partial pressure of the hydrocarbons and separation products. Typically a single sidestream is withdrawn from the stripping section to give a heavy gas oil, with a boiling range of 545 -700° F. The residue from the bottom of the column is suitable for blending into industrial fuels. Alternately, it may be further distilled under vacuum conditions to yield quantities of distilled oils for manufacture into lubricating oils or for use as a feedstock for a gas oil cracking process.

Vacuum distillation resembles fractional distillation except the column diameters are larger to give comparable vapor velocities at a pressure of 50 to 100 mm Hg (absolute) achieved by vacuum pump or steam ejector. Vacuum distillation reduces the boiling temperature for processing materials below that boiling temperature at atmospheric pressure, thus avoiding thermal cracking of the components. Firing conditions in the furnace are adjusted so that oil temperatures do not exceed 800° F. The residue (bitumen) remaining after vacuum distillation, is used for road asphalt, residual fuel oil, or a feedstock for thermal cracking or coking units.

Superfractionation is an extension of distillation using smaller diameter columns and 100 or more trays to achieve reflux ratios exceeding 5:1. This equipment separates a narrow range of components such as of high-purity solvents, e.g., isoparaffins or individual aromatic compounds for use as petrochemicals.

Absorption

Absorption recovers valuable light components such as propane/propylene and butane/butylene as vapors from fractionating columns. These vapors are bubbled through an absorption fluid, such as kerosene or heavy naphtha, in a fractionating-like column to dissolve in the oil while gases, such as hydrogen, methane, ethane, and ethylene, pass through. Absorption is effectively performed at 100 to 150 psi with absorber heated and distilled. The gas fraction is condensed as liquefied petroleum gas (LPG). The liquid fraction is reused in the absorption tower.

Solvent Extraction

Solvent extraction removes harmful constituents such as heavy aromatic compounds from lubricating oils to improve the viscosity-temperature relationship. The usual solvents for extracting lubricating oil are phenol and furfural.

Adsorption

Certain highly porous solid materials selectively adsorb certain molecules. Examples are: silica gel for separation of aromatics from other hydrocarbons, and activated charcoal for removing liquid components from gases. *Adsorption* is analogous to *absorption, but* the principles are different. Layers of adsorbed material, only a few molecules thick, are formed on the extensive interior area of the adsorbent - possibly as large as 50,000 sq. ft./lb of material.

Molecular sieves are an adsorbent that is produced by the dehydration of naturally occurring or synthetic zeolites (crystalline alkali-metal aluminosilicates). The dehydration leaves inter-crystalline cavities into which normal paraffin molecules are selectively retained and other molecules are excluded. This process is used to remove normal paraffins from gasoline fuels for improved combustion. Molecular sieves are used to manufacture high-purity solvents.

Crystallization

The crystallization of wax from lubricating oil fractions makes better oil. This is done by adding a solvent (often a mixture of benzene and methyl ethyl ketone) to the oil at a temperature of about -5° F. The benzene keeps the oil in solution and maintains fluidity at low temperature; the methyl ethyl ketone acts to precipitate the wax. Rotary filters deposit the wax crystals on a specially woven cloth stretched over a perforated cylindrical drum. A vacuum in the drum draws the oil through the perforations. The wax crystals are removed from the cloth by metal scrapers and solvent-washed to remove oil followed by solvent distillation to remove oil for reuse.

7.2.9.3 Conversion

The separation processes separate the constituents of crude-oil based on physical properties. Conversion of one molecule into another greatly extends the usefulness of petroleum by extending the range of hydrocarbon products.

Thermal Conversion

The initial process for molecular conversion was thermal reforming (late 1920s). Thermal reforming at 950 - 1050° F and 600 psi produced gasolines of 70 to 80 octane and heavy naphthas less than 40 octane. Products were olefins, diolefins, and aromatic compounds that were unstable in storage and tended to form heavy polymers and gums, which caused combustion problems.

According to the free-radical theory of molecular transformation, thermal cracking works, by heat-breaking the electron bond between carbon atoms in a hydrocarbon molecule to generate a hydrocarbon group with an unpaired electron. The negatively charged, free radical, reacts with other hydrocarbons, continually producing other free radicals via the transfer of negatively charged hydride ions (H^-). This chain reaction leads to a reduction in molecular size called "cracking," of components of the original feedstock.

Catalytic Conversion

By 1950 a reforming process was introduced using a catalyst to improve the yield of the desirable gasoline components while minimizing the formation of unwanted material. In catalytic as in thermal reforming, a naphtha-type material serves as the feedstock, but the reactions are carried out in the presence of hydrogen, to inhibit the formation of unstable unsaturated compounds that polymerize into higher-boiling materials. In most catalytic reforming processes, platinum on aluminum oxide is the catalyst. Small amounts of rhenium, chlorine, and fluorine are catalyst promoters. The high cost of platinum is overcome by its

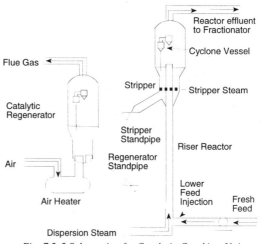

Fig. 7.2-5 Schematic of a Catalytic Cracking Unit

long life and effectiveness. The principal reactions involve breaking down the long-chain hydrocarbons into smaller saturated chains while forming isoparaffins. Formation of ring compounds by cyclization of paraffins into naphthenes, also takes place, and the naphthenes are then dehydrogenated into aromatic compounds. The desirable end products are isoparaffins and aromatics, both having high octane numbers.

In a typical reforming unit (Figure 7.2-5) the naphtha charge is first passed over a catalyst bed in the presence of hydrogen to remove any sulfur impurities. The desulfurized feed is mixed with hydrogen (about five molecules of hydrogen to one of hydrocarbon) and heated to 930 to 1000°F at 100 psi. Because heat is absorbed in reforming reactions, the mixture must be reheated in intermediate furnaces between the reactors. After leaving the final reactor, the product is condensed to liquid, separated from hydrogen and distilled to remove the light hydrocarbons. The product may be blended into gasoline without further treatment.

Process parameters are set to obtain the required octane level (~90). In the process, minute amounts of carbon are deposited on the catalyst which reduces the product yield, but can be removed by batch burning. Continuous regeneration avoids periodic shutdowns and maximizes the high-octane yield. This employs a moving bed of catalyst particles that is circulated through a regenerator vessel, for carbon removal, and returned to the reactor.

Polymerization and Alkylation

The light gaseous hydrocarbons produced by catalytic cracking are highly unsaturated and are usually converted into high-octane gasoline components by polymerization or alkylation. In polymerization, the light olefins, propylene and butylene, are induced to polymerize into molecules two or three times their original molecular weight using as the catalyst: phosphoric acid on pellets of kieselguhr. Pressures of 400 to 1,100 psi at 350 - 450°F produce polymer gasolines derived from propylene and butylene having octanes above 90.

The alkylation reaction achieves a longer chain molecule by combining two smaller molecules - an olefin and an isoparaffin (usually isobutane). During World War II, alkylation became the main process for the manufacture of isooctane, a primary component for blending aviation gasoline. Two industrial alkylation processes use different acid systems as catalysts. In sulfuric acid alkylation, concentrated sulfuric acid of 98% serves as the catalyst for a reaction that is carried out at 35 - 45°F. Refrigeration is necessary because heat is generated by the reaction. The octane of the alkylates is 85 - 95. The chemical reactions with hydrofluoric acid as the catalyst are similar to those with sulfuric acid as the catalyst but higher temperatures (115°F) can be used, thus avoiding refrigeration. Hydrofluoric acid is recovered by distillation. Stringent safety precautions must be exercised when using this highly corrosive and toxic substance.

Hydrocracking

The use of hydrogen from catalytic reforming was a far-reaching development. in the 1950s. By 1980, it had become so prominent that many refineries incorporate hydrogen manufacturing plants. Though using a feedstock similar to catalytic cracking, it offers even greater flexibility in product yields. The process can produce: gasoline, jet fuels, heavy gas oils, high-quality lubricating oils, or lighter oils by distilling the residues.

Hydrocracking is accomplished at lower temperatures (500 - 800°F) than catalytic cracking, but at higher pressures (1,000 - 4,000 psi). While hydrocracking catalysts vary widely, cracking reactions can be induced by silica-alumina materials. In units that process residual feedstocks, hydrogenation catalysts are: nickel, tungsten, platinum, or palladium. The catalysts do not require regeneration.

Isomerization

The demand for aviation gasoline during World War II was so great that isobutane from alkylation feedstock was insufficient. This deficiency was remedied by isomerization of abundant normal butane into isobutane using the isomerization catalyst: aluminum chloride on alumina promoted by hydrogen chloride gas.

Commercial processes isomerize low-octane normal pentane and normal hexane to the higher-octane isoparaffin, using a platinum catalyst in the presence of hydrogen. Hydrogen as a catalyst inhibits undesirable side reactions. This reaction is followed by molecular sieve extraction and distillation which excludes low-octane components from the gasoline blending pool, and does not produce a product of sufficiently high octane for unleaded gasoline.

Visbreaking, Thermal Cracking, and Coking

Since the war the demand for gasoline, jet, and diesel fuels has grown, while the demand for heavy industrial fuel oils has declined. Furthermore, many new oil finds have yielded heavier crudes, therefore the need to convert residue components into lighter oils for feedstock for catalytic cracking.

As early as 1920, large volumes of residue were processed in thermal cracking units. The furnace residence time was controlled to prevent clogging the furnace tubes. The heated feed was charged in a reaction chamber, at a pressure high enough to permit cracking of the large molecules but limit coke formation. Afterward, the process fluid was cooled to inhibit further cracking and charged to a distillation column for separation into components. These visbreaking units typically converted about 15% of the feedstock to naphtha and diesel oils and produced a lower-viscosity residual fuel.

Thermal cracking units provide more severe processing and often convert as much as 50 to 60% of the incoming feed to naphtha and light diesel oils.

Coking is severe thermal cracking. The residue feed is heated to about 890 - 970°F in a furnace with very low residence time and is discharged into the bottom of a large vessel called a coke drum for extensive and controlled cracking. The cracked lighter product rises to the top in the drum and is drawn off to the product fractionator for separation into naphtha, diesel oils, and heavy gas oils for further processing in the catalytic cracking unit. The heavier product settles in the drum and, because of the retained heat, cracks ultimately to coke, a solid carbonaceous substance akin to coal. Once the coke drum is filled with solid coke, it is removed from service and replaced by another coke drum.

Decoking is a routine daily occurrence accomplished by a high-pressure water jet. First the top and bottom heads of the coke drum are removed then a hole is drilled in the coke from top to bottom. A rotating stem is lowered through the hole, spraying a jet high-pressure water sideways to cut the coke into lumps that fall to the bottom of the drum for shipment to customers. Coking cycles consists of 24 hours to fill the drum and 24-hours cooling, decoking, and reheating.

Cokers produce no liquid residue but yield up to 30% coke. Much of the low-sulfur product is used for electrolytic electrodes for smelting of aluminum. Lower-quality coke is burned as fuel mixed with coal.

Purification

Petroleum products must be purified of sulfur compounds such as hydrogen sulfide (H_2S) or the mercaptans ("R"SH), where "R" is the hydrocarbon radical forming complex organic compounds. Besides the bad odor, sulfur compounds reduce the effectiveness of antiknock additives, interfere with the operation of exhaust-treatment systems, and corrode diesel engines. Other undesirable components are nitrogen compounds, that poison catalytic systems, and oxygenated compounds leading to color formation and product instability. The principal treatment processes are: 1) sweetening, 2) mercaptan extraction, 3) clay treatment, 4) hydrogen treatment, and 5) molecular sieves.

1) Sweetening processes originally used sodium plumbite as the catalyst to oxidize mercaptans into more innocuous disulfides, which remain in the product fuels. This inexpensive process increased the total sulfur content in the product, hence it was replaced by copper chloride slurry as the catalyst used for making kerosene and gasoline. The oil is heated and brought into contact with the slurry while being agitated in a stream of air that oxidizes the mercaptans to disulfides. The slurry is allowed to settle and is separated for reuse. The temperature is sufficient to keep the water, formed in the reaction, dissolved in the oil, so the catalyst remains properly hydrated. After sweetening, the oil is water-washed to remove traces of catalyst and dried by passing through a salt filter.

2) Mercaptan extraction is used to reduce the total sulfur content of the fuel. When potassium isobutyrate and sodium cresylate are added to caustic soda, the solubility of the higher mercaptans is increased and they can be extracted from the oil. To remove traces of hydrogen sulfide and alkyl phenols, the oil is pretreated with caustic soda in a packed column or other mixing device. The mixture is allowed to settle and the product water washed before storage.

3) Clay treatment uses natural clays, activated by roasting or treatment with steam or acids, to remove traces of impurities by an adsorption process to remove gum and gum-forming materials from thermally cracked gasolines in the vapor phase. A more economical procedure, however, is to add small quantities of synthetic antioxidants to the gasoline. These prevent or greatly retard gum formation. Clay treatment of lubricating oils is widely practiced to remove resins and other color bodies remaining after solvent extraction. The treatment may be by contacting the clay directly with the oil while heated and filtering the clay, or by percolating the heated oil through a large bed of active clay adsorbent. The spent clay is discarded, or regenerated by roasting.

4) Hydrogen treatment (hydrofinishing, hydrofining, or hydrodesulfurization), are processes for removing sulfur compounds. The oil and hydrogen are vaporized, and passed over a catalyst (e.g., tungsten, or nickel) at 500 - 800°F and 200 - 1,000 psi. Such parameters are chosen according to the sulfur removal requirements without degrading the oil. The sulfur in the oil converts to hydrogen sulfide, which is removed by absorption in diethanolamine which can be reused by heating

with the hydrogen sulfide and provide high purity sulfur.

5) Molecular sieves (dehydrated zeolite) purify petroleum products with their strong affinity for polar compounds such as water, carbon dioxide, hydrogen sulfide, and mercaptans. The petroleum product is passed through the sieve until the impurity is sufficiently removed after which the sieve may be regenerated by heating to 400 - 600° F.

7.3 Chemical Process Accident Analysis

Similarities between Nuclear and Chemical Safety Analysis

Nuclear power production involves bringing fissionable material together to react nuclearly, removing the heat, converting the heat to steam to drive a turbo-generator, and managing the wastes. Chemical process production involves bringing molecules together to react chemically, removing or providing heat, controlling and confining the reaction, and managing the wastes. Both nuclear and chemical reactions involve fluids, hence, both use very similar equipment, therefore similar safety analysis methods should apply to both.

For historical and regulatory reasons, a PSA of a nuclear power plant begins with the system analysis to determine the ways that an upset condition could occur, its probability of occurrence, and the consequences to the public if it occurred. A PSA of chemical processing tends to reverse this procedure[r] by using scoping calculations of the consequences to determine if the hazard to workers and public is significant. If so, then to work backward to determine the process upsets and their likelihood that could cause these consequences.

This approach has the advantage of reducing the number of process upsets that must be examined. In a nuclear plant, the hazard has one location - the core; a chemical plant, hazards have many locations. In a nuclear plant, the hazard is exposure to radiation and fission products; in a

Table 7.3-1 Hazard Criteria

Question	Comment
Is there a possibility of nitrating a compound?	Many of the explosions in chemical processing are the result of nitrating be it cellulose, tributylphosphate (to form red oil), or ammonia to form fertilizer.
Are toxic materials being processed?	This was the cause of the Bhopal accident, and the less disastrous Seveso accident.
Are combustible powders or vaporizing liquids processed?	If these can be released and dispersed as a cloud an explosion can result. For example, there have been severe explosions at bread flower mills.
Are high temperatures involved that may cause combustible liquids to vaporize?	If these are liquid at high pressure and temperature and the confinement fails they will flash to vapor and could potentially explode.
Are caustic materials processed?	If these are released and contact people either as a vapor or liquid severe health hazards result. For example, uranyl fluoride forms hydrofluoric acid in the lungs.

---

[r] Private telephone conversation between Lester Wittenberg, CCPS and Ralph Fullwood, 1997.

Analyzing Chemical Process Safety Systems

Table 7.3-2 Objectives for Chemical PSA
1. Determine Risk Potential. The objective is to determine the possibility of significant risk of injury to workers or the public, or risk to the company's good name. It is done by bounding, consequence analyses and approximate frequency estimates, primarily for company protection.
2. Protect Workers and the Public. The reasons may not be entirely altruistic because worker injury is detrimental to production. Furthermore, we are the workers and wish a safe working environment free of immediate or latent injury. This requires detailed analysis of the process systems to estimate frequency of failure and the consequences that could result. Generally if the workers are safe, the public is safe.
3. Detect Financial Risks. In addition to physical injuries, a company may be injured by the perception of injury that result in law suits, drop in stock price, and reduction in sales. A detailed analysis (2) will assess vulnerability but it does not directly address the perception factor. A case in point is TMI-2 which complied with 10 CFR100, had no injuries and no deaths, but the company was driven to near-bankruptcy. Similarly with Bhopal and Union Carbide, although there were deaths.
4. Prioritize Safety Improvements. This uses the detailed analysis (2) to identify items having high risk importance. Engineering analysis identifies and costs candidate improvements which are selected on the basis of risk reduction for a given cost.
5. Legal Compliance. Bounding (1) or detailed analyses (2) may be used to show compliance with national and regional measures, e.g., the PSM Rule.
6. Emergency Planning. A scoping (1) or detailed (2) PSA may be used to specify the emergency actions needed for emergency response zones about the plant. The use of on-site personnel, and their integration with local, state, and national groups and plans needs consideration.
7. Plant Availability. A detailed PSA (2) may be used to analyze the effects of test and maintenance on plant operability through improved scheduling, component reliability improvement, and improved procedures. The cost-benefit of stocking vs warehousing of feed materials and spare parts may be done to find an optimum.

chemical plant, the hazard is exposure to many different chemicals with different hazardous effects to the workers and public thus a much larger range of hazards and locations must be considered, hence, the benefit of screening the hazards.

Is a PSA Needed?

Table 7.3-1 lists questions, based on experience, which, answered in the affirmative, indicate hazards. The first question deals with the possibility of forming a nitrated compound. While not all explosives contain nitrogen, these have had a bad history. The basis for the second question is the Bhopal accident where an exothermic pesticide was being processed. Similarly, chlorine, while less toxic has been used with devastating results in warfare (Ypres). Communities are occasionally evacuated when chlorine is spilled. The third question is to point out that benign combustible materials, finely divided and dispersed, as a cloud, may conflagrate. The fourth question relates to the fact that liquids, under pressure and above their critical temperature, can flash to vapor if the pressure is released by vessel rupture or trip of pressure relief. Of course, an ignition source is needed to cause fire or explosion, but sources are usually present in one form or another. The fifth question deals with materials, while not poisonous, can injure flesh with which it comes in contact. This list, while not exhaustive, highlights hazards that may cause accidents.

Table 7.3-2 lists some objectives for performing PSA on chemical process systems. Objective 1 is to determine if a process or plant has sufficient risk to justify a detailed analysis. This scoping analysis may be performed with a HAZOP (Section 3.3.4) or an FMEA (3.3.5) with either analysis held to systems levels and bounding consequences. If it turns out that serious risk concerns exist, then the detailed analysis (2) is needed. Here the depth of analysis is controlled according to other objectives. If the purpose is legal (5) the depth is that required by the law; if the purposes are financial exposure, a detailed analysis may not be necessary and a scoping analysis may be sufficient. To use PSA for purposes 2, 4, 6, and/or 7, a detailed PSA is needed possibly supplemented in specific areas of interest.

7.3.1 Scoping Analysis

The purpose of a scoping analysis is to determine, under worst case assumptions, if there is a risk that can cause injury, death or financial impact to the public, workers, company, or environment. The PSA begins by identifying the hazards, their physical and chemical properties, the confinement, conditions and distance for transport to a target, estimating the effects on the target, and comparing these effects with accepted criteria.

7.3.1.1 A Nuclear Scoping Analysis: WASH-740

It was stated earlier that PSAs for nuclear and chemical processes differ by preceding a chemical PSA with a scoping analysis that is omitted for a nuclear PSA. This is not true. WASH-740 performed this purpose for a hypothetical plant.

Originally, nuclear power was a government monopoly in the U.S. With Eisenhower's "Atoms for Peace," electric utilities were encouraged to use nuclear energy, but the utilities were hesitant because of the unknown risk. The Price-Anderson Act of 1957 capped the utility liability. In an attempt to scope the hazard, WASH-740 (1957) was prepared by BNL which estimated, under worst conditions, for a medium sized plant in or near a large city, an accident could cause 43,000 injuries, 3,400 deaths and $7 billion property damage. A reanalysis was performed in 1965 that concluded that the containment building eliminated risk to the public. It was later argued that the molten core would melt through the containment and an atmospheric release would occur.[s] While WASH-740 indicated the accident potential, the question that it did not answer was the probability of an accident. The Reactor Safety Study (WASH-1400) was conducted for this purpose.

Nevertheless, WASH-740 indicated the worst-case hazard of nuclear power, and the need for detailed PSAs of nuclear power plants. Since plants proposed for construction are sufficiently like the hypothetical plant in terms of fission products subsequent scoping analyses are not needed before the detailed PSA.

---

[s] It may be noted that in the case of TMI-2, the core partially melted but did not get out of the pressure vessel. In the case of Chernobyl, there was no containment in the U.S. sense and what confinement there was, was disrupted by the nuclear excursion (like an explosion).

7.3.1.2 Performing a Process Scoping Analysis

Preparing a Spreadsheet for Toxic Hazards

Figure 7.3-1 outlines steps in the scoping process. However, if this information is to be gathered, it should be in a convenient form such as collecting the data as a computer file in a notebook personal computer. The file may be a word processor, a spreadsheet, or a database. If the input is into a word processor such as WordPerfect or Word, it may be in table format. Very similar but

Hazardous Material
1) Identification
2) Quantities and Locations
3) Environmental Conditions
4) Confinement
5) Type of Hazard: Toxicity, Explosion, and/or Fire
6) Comparison with Guidance
7) Need for Detailed PSA?

Fig. 7.3-1 Process Scoping Steps

with greater computational capabilities is a spreadsheet such as Excel or QuatroPro. A database program such as Access or dBase allows even greater flexibility in storage, formatting, and printing the report, but has less direct access to the data, and computational capabilities. A preferred method is inputting the data into a spreadsheet because of the ease in performing the calculations and the flexibility regarding the number of columns which may be off the screen but accessible by scrolling.

Table 7.3.1-1 A Spreadsheet for Scoping Analysis Data									
1	2	3	4	5	6	7	8	9	10
Hazard Name/ CAS	Quantity/ phase	Location	Confine-ment	Temper-ature	Pressure	Type of Hazard	Chemical Reaction	Location and Amount of Reactants	Hazard Criterion/ Scenario
Chlorine/ 7782-50-5	2,200 lb/ liquid	Plan 1234 Location BB12	0.25 inch thick steel vessel	Ambient	100 psi	Inhalation	None needed	NA	TQ: 1500 lb/ Vessel rupture - liquid flashes to vapor to injure workers by inhalation

Table 7.3.1-1 shows spreadsheet headings and a row of representative data.
Column 1 identifies the hazard and its CAS (Chemical Abstract Service number).[t] The

---

[t] The chemical companies maintain chemical properties sheets including hazardous quantities. A general listing of companies on the Internet can be reached at web site: http://www.neis.com/chemical_companies.html. The U.S. Department of Health and Human Services publishes the "NIOSH Pocket Guide to Chemical Hazards," available from NIOSH Publications Mail Stop C-13, 4676 Columbia Parkway, Cincinnati, OH 45226-1998, phone 1-800-35-NIOSH or on the Internet: http://www.cds.gov/niosh/homepage.html.

hazard identification involves judgment as to what can be a hazard. In this example, chlorine was identified as a hazard. It is a well known poison because it was used at Ypres in World War I with devastating effect. It is listed in Table 1.9-1 that also gives its CAS.

Column 2 shows the quantity of material at a particular location in the plant and the phase it is in (solid, liquid, or gas).

Column 3 gives the floor plan coordinates of the location of the material.

Column 4 indicates how the hazard is confined. If it is a gas it will be in a closed container of sufficient strength to withstand the pressure. This may be expressed in terms of the wall thickness and material. If it is condensed to liquid phase, it will be under pressure; if it is a liquid at atmospheric pressure, it may be in a tank that is open at the top. Solids may be confined depending on whether or not they are powdered, granular or block.

Column 5 lists the normal temperature that the chemical will experience as well as the upper and lower ranges. Temperature is important because chemical reactions go exponentially with temperature. The range of variation may indicate possible changes of state

Column 6 provides the operating pressure and peak pressures that the process may go to.

Column 7 lists the type of hazard that is being considered. If the chemical is toxic then the release and dispersion to the workers and public is the hazard. Another simple hazard is a pressure vessel rupture due to over pressure. A steam vessel rupture may scald workers and injure them with shrapnel. Many people died in the last century from boiler explosions. The ves-

Table 7.3-2 Some Toxicology Sources

Source	Comments or Location
Table 1.9-1 in this book	Thresholds for triggering the PSM rule
"Limits for Human Exposure to Air Contaminants"	CRC Handbook, 1979
"NIOSH Pocket Guide to Chemical Hazards"	See footnote t, page 298 in this book
Material Safety Data Sheets for the chemical's manufacturer	These are available on the Internet. Usually you type http://www.xxx.com, where xxx is the manufacturer's name. If this doesn't work go to a search engine or try http://siri.org/msds/ for a general directory to chemical manufacturers.
Oral reference doses by chemical name for 535 substances	EPA's IRIS (Integrated Risk Information System) http://www.epa.gov/ngispgm3/iris/
DOE Chemical Safety Program	http://dewey.tis.eh.doe.gov/web/chem_safety/ This web site provides principles, chemical incidents in the DOE and hypertext connection to toxicity data.
Permissible Exposure Limits (PELS)	http://www.osha-slc.gov/SLTC/PEL/index.html
ACGIH	American Conference of Governmental Industrial Hygienists, http://www.acgih.org/ This web site connects with a number of publications some of which are listed below.
Hazardous Chemicals Desk Reference	Sax, 1987
Casarett and Doull's Toxicology : the Basic Science of Poisons	Casarett, 1991
Dangerous Properties of Industrial Materials	Sax, 1989

sel may have pressure valves, but if the contents are poisonous, as was the case at Bhopal, the discharge through relief channels may be hazardous. A complex example may occur if the contents react chemically with something else to release energy or form a poisonous substance. An example is the release of sodium into water.

Column 8 gives the chemical reaction that may occur, whether it is exothermic and poisonous.

Column 9 indicates the location and names of other reactant materials

Column 10 is used to list hazard criteria from sources such as given in Table 7.3.1-2. Inhalation limits often are given in either units of parts-per-million (ppm) in air or milligrams (mg) of the chemical per cubic meter. Equation 7.3-1 gives the connection between these units.

$$ppm = \frac{mg/m^3 * 22.41}{molecular\ weight\ of\ chemical} \quad (7.3\text{-}1)$$

Including Fire and Explosive Hazards in the Spreadsheet

Some chemicals can react with themselves or with other chemicals to burn or explode to release their chemical energy. The distinction between burning or exploding depends on the rate of reaction according to the following definitions:

- Fire, burning, or combustion is a chemical reaction that takes place so slowly that no pressure wave is produced. A flame may be visible, temperatures reaching 1,000s of degrees may be produced. The damage mechanisms are: burns, suffocation by depleting oxygen, poisoning from combustion product production (e.g., carbon monoxide), or mechanical damage from falling structure or steam formation in closed vessels.

- Deflagration is burning that takes place subsonically, hence, no shock front is produced. The damage mechanisms are: physical destruction from the pressure wave, collateral damage from falling structures, and burns of humans. Suffocation and poisoning are not usually associated with a deflagration because of its short duration. A substance that normally burns in air can produce a deflagration if it is finely suspended in air. Examples are: suspended coal dust, flour, gasoline, kerosene oil, and metal powder. In these examples the common reactant chemical is oxygen in the air. A mixture of chemicals such as compose black powder can produce a deflagration without air. Deflagrations may be caused by mixing one chemical with a reactant chemical. In such cases, safety is provided by identifying the reactant and isolating it from the reactee. Percent limits of inflammability of gases and vapors in air are given in CRC (1979, and later editions).

- Explosion is burning that takes place sonically, hence, a shock front is produced. The damage mechanisms are: physical destruction from the shock wave, collateral damage from falling structures, and human burns. Suffocation and poisoning are not usually associated with an explosion because of its short duration. A chemical explosives such as nitroglycerine, trinitrotoluene, dynamite, and others were discussed in Section 7.2.6.

Chemical Process Accident Analysis

Evaluation of the Scoping Analysis Spreadsheet

The MSDS from the chemical manufacturer identifies hazards for entry in the spreadsheet in columns 8 and 10. This is performed for all chemicals that are associated with the process, if the analysis is limited to a process, or for a plant. The spreadsheet may be filled out variously according to convenience and effectiveness. It is practically impossible to get all needed information from documentation alone. A plant walk-through is advised for viewing operating conditions as they exist, for interviewing operators about the risk concerns that they have, and about the operability of safety and mitigation systems. These results are entered into the spreadsheet.

The spreadsheet is reviewed to see if the quantity of hazardous material triggers a PSM Rule site evaluation, if the release of materials would result in exposures that exceed exposure limits. If either or both of these are true for any chemical, a detailed PSA is in order. The next step is to prepare the spreadsheet information as a report that details the investigation, provides the data and the bases for recommending a detailed PSA. This report is sent to the responsible management for their action.

7.3.2 Performing a Detailed Probabilistic Safety Analysis

Table 7.3.2-3 lists investigations to be conducted and documented for a detailed chemical plant PSA. The steps are similar to those required for a nuclear plant PSA except the hazards are more varied, and dispersed regarding concentration and location. Many of the steps previously described in Section 6.3.2 are applicable for the chemical PSA as well.

Figure 7.3-2, from CCPS (1989) details the work and

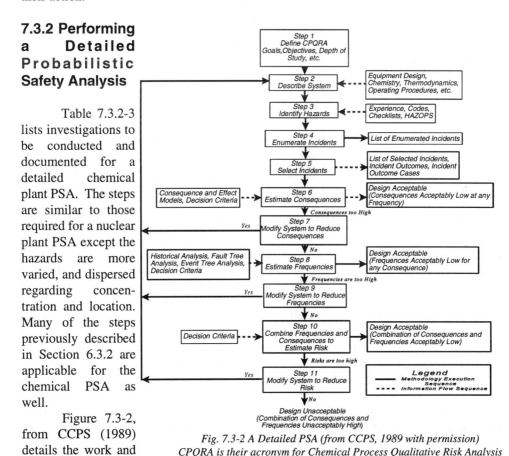

Fig. 7.3-2 A Detailed PSA (from CCPS, 1989 with permission) CPQRA is their acronym for Chemical Process Qualitative Risk Analysis

299

information needs for a quantitative chemical process PSA. Upon completion of the scoping analysis, management decides whether or not to embark on a detailed quantitative analysis. If the decision is to perform a PSA, management provides the general objectives of the study, funding, schedule, and personnel assignments.

Step 1 Goals etc.

This step defines the goals, objectives, and depth of study based on the strategic instructions from management and the resources provided. The organization and responsibilities of the team are defined along with project milestones, quality assurance, reviews and reporting. This information is reported to management for approval. Upon resolution, and with management go ahead, the project begins.

Even, limited PSAs use and contain much information. This information may come as memos and process reports and flow sheets, equipment layout, system descriptions, toxic inventory, hazardous chemical reactions, test, maintenance and operating descriptions. From this, data and analyses are prepared regarding release quantities, doses, equipment reliability, probability of exposure, and the risk to workers, public, and environment. An executive summary is prepared, the analysis is detailed, and recommendations made for risk reduction. Thus the information will be text, calculations of envelope fracture stresses, temperatures, fire propagation, air dispersion, doses, and failure probabilities - primarily in tabular form.

Recording the PSA Study

Some of the information may be in the scoping hazard analysis that indicated a need for a PSA. This information is conveniently prepared using a spreadsheet. Systems analysis to determine the probability of failure may be prepared using Section 3.6 computer codes for fault tree analysis. Better, code suites that were described in Section 3.6.2 may be used. Engineering calculations using graphs and formulas will be used to calculate criteria for and the consequences of equipment failure. Computer aided drawing (e.g., AutoCAD) may be used to dimension the layout of equipment, room configurations and determine distances for toxic transport, also to show recommended risk reduction measures.

Clearly the software for managing the detailed PSA must be able to import spreadsheets, engineering text including graphics and formulas, tables of data, performs simple mathematical operations and to present the results in graphical, tabular and text printed form. A team working on this project must be able to communicate with each other and transfer information. A way to do this is a network that couples the personal computers of all the team members. An analyst will use software on their computer for performing their specialized calculations; the results of which are shared with all of the other team members. The computer program with the most versatility in accepting a wide variety of input formats is the modern word processor that can perform calculations, rank order data, prepare graphical illustrations, and to link the disparate files through hypertext. Hypertext links words or phases to bookmarks in the master document or to ancillary documents. Hypertext is useful for preparing the report by giving quick access to the supporting documents as well as the ability to jump to relevant parts of the master PSA report. A hardcopy final report would lack these hypertext linkages but they could be available in an electronic form of the

Chemical Process Accident Analysis

Table 7.3.2-1 Investigations in a Detailed PSA (from CCPS, 1989)

Initiating Events/ Process Upsets	Propagation	Mitigation	Consequences

Process Deviations
  Pressure, Temperature, Flow rate, Concentration, Phase/state change, Impurities, Reaction rate/heat of reaction
Spontaneous Reactions
  Polymerization, Runaway reaction, Internal explosion, Decomposition
Envelope Failures
  Pipes, tanks, vessels, gaskets, seals
Equipment Failures
  Pumps, valves, instruments, sensors, interlock failures
Loss of Utilities
  Electrical nitrogen, water, refrigeration, air, heat transfer fluids, steam, ventilation
Management Failures
Human Error
  Design, Construction, Operation, Maintenance, Testing and inspection
External Events
  Extreme weather, Earthquakes, Near Accidents, Vandalism/Sabotage
	Equipment Failures
	  Safety system
	Ignition Sources
	  Furnaces, Flares, Incinerators, Vehicles, Electrical switches, Static electricity, Hot surfaces, Cigarettes
	Human Failures
	  Omission, Commission, Fault diagnosis, Decisions
	Domino Effects
	  Other containment failures, Other material release
	External Conditions
	  Meteorology, Visibility
		Risk Reduction Factors
		  Control/operator responses, Alarms, Control system response, Manual and automatic ESD, Fire/gas detection system
		Safety System Responses
		  Relief valves, Depressurization system, Isolation systems, High reliability trips, Back-up systems
		Mitigation System Responses
		  Dikes and drainage, Flares, Fire protection systems (active and passive), Explosion vents, Toxic gas absorption
		Emergency Plan Responses
		  Sirens/warnings, Emergency procedures, Personnel safety equipment, Sheltering, Escape and evacuation
		External Events
		  Early detection, Early warning, Specially designed structures
		Training
		Other Management Systems
			Analysis
			  Discharge Flash and evaporation, Dispersion by neutral or positive buoyancy, Dense gas
			Fires
			  Pool fires, Jet fires, BLEVES, Flash fires
			Explosions.
			  Confined explosions, Unconfined vapor cloud explosions (UVCE), Physical explosions (PV), Dust explosions, Detonations, Condensed phase, detonations
			Missiles
			Consequences
			  Effect analysis, Toxicity, Thermal effect, Overpressure effects,
			Damage Assessments
			  Community, Workers, Environment, Financial

final report.

Step 2: Describe the System

Returning to Figure 7.3-2, step 2, define the system. These are text and graphic files describing critical systems that were identified in the scoping analysis and in the PSA goals. Making this information available in electronic form makes it accessible to all of the team members and provides expert auditing of the understanding of the systems. Access is facilitated through hypertext and searchable catalogs. The electronic form facilitates annotation and correction of the text as well as the ability to annotate the drawings with indicators and text to indicate risk significant artifacts.

Step 3: Identify the Hazards

The scoping analysis identified the hazards for use in the detailed PSA.

Step 4: Enumerate Incidents

Instances of chemical accidents, like to those of Section 7.1, relative to the hazards listed in Step 3 are listed. Information for these must come from various sources since there is not a single database for chemical incidents. If the plant has operating experience, this is valuable information, although the data may be sparse.

General searching may be done from internet website: http:// dewey.tis.eh. doe.gov/web/ oeaf/oe_weekly/ for weekly descriptions of chemical accidents in the DOE. Most of the chemical incidents in Section 7.1.2 were taken from Chemical and Engineering News, published by the American Chemical Society (ACS). It has a website at: http://pubs.acs.org/cen/index.html. A subject search may be conducted from the ACS website at: http://www.acs.org/acsearch.htm Accident searches may be made on the New York Times at: http://search.nytimes.com/search/daily/, and on the Washington Post at: http://www.washingtonpost.com/wp-adv/archives/front.htm. Searches may be requested from professional organizations, chemical manufacturers and from chemical manufacturers' websites. Companies that insure chemical manufacturers and users are another source of information (such as M&M). Information of a general nature may be obtained from the OSHA website at: http://www.osha-slc.gov/. OSHA website: http:// www.osha.gov/ oshstats/ has a search for reports on: 1) Establishments, SIC (particlular industries) and 2) inspection number. Unfortunately these are closed to the public. General and specific international information is available from the Organization for Economic Cooperation and Development (OECD) at website: http://www.oecd.org/. Some of these data are reserved.

Step 5: Select Incidents

The incidents from Step 4 are sorted for relevance to the facility being analyzed on the basis of similarity of product, process, equipment, siting, organization and environment (physical and political). From this, incidents are selected according to their relationship to the PSA objectives.

The PSA should be as complete as possible, consistent with the PSA objectives. To this end, the incidents should be taken as suggestions of what might happen in your plant - not the only things that may happen. Further completeness may be sought using the techniques for completeness discussed in Section 3.3.

Step 6: Estimate Consequences

Chapter 9 details the process of estimating the consequences. In general it consists of:

1) Estimating the break size in whatever was containing the hazardous material.
2) Estimating the amount of material ejected through the break in the time that it being released with consideration for the flow parameters and the pressure driving the discharge.
3) Categorizing the material hazard according to poison, fire, or explosion.
    i) If a poisonous liquid, estimate the consequences if it contacts workers, public, and the environment. If it is a poisonous gas, consider it buoyancy to determine whether it will rise or blanket, building confinement, and its atmospheric dispersion. Rain has a large effect on the dispersion. The effects on humans depends on the poison concentration in the air, duration of injection, rate of breathing, and breath filtration.
    ii) If it is a flammable liquid, estimate the size of the fire and the evacuation of people from exposure to the fire, and the possibility of death from fighting the fire.
    iii) If it is explosive, estimate the explosive effect of amount released. This includes the effects of the pressure and shock wave on personnel and on equipment that may cause secondary hazards.

Step 7: Modify the System to Reduce Consequences

If the hazard is liquid, it should be confined by geometry. If drains are present, they should be designed to close automatically, manually from a central, safe location blocked, or quickly if individual and the toxicity is low. The enclosure may be designed with drains the go to a safe container. If the hazard is a gas, it is necessary beforehand to provide a containing structure. The mitigation are alarms, evacuation, and distribution of prophylactics if there is protection. If the hazard is a vaporizing liquid, keep it confined by closed drains (unless it drains into a container that can be closed). Keep it cool to reduce the rate of evaporation. Cover with a lighter inert material or absorptive material to isolate it as much as possible.

Fire hazards are minimized by minimum inventories of chemicals. Control or minimize ignition sources, provide a confining area, reduce the temperature of the material as much as possible, blanket the material to eliminate air contact. Have available fire protection equipment breathing apparatus, and protective clothing for the fire fighters. Use several hour fire walls to prevent the spread of fire to other process areas.

Explosive hazards are reduced by using similar techniques and by structures for explosions through the use of blow-out doors to direct the pressure wave in harmless directions.

Step 8: Estimate Frequencies

The frequency of an initiating event is usually based on industrial experience. If the process is new or rare, it may be estimated by a system model of the process steps (e.g., a fault tree) and using data from similar experience to give the probability of failure of the steps. Either of these estimates should consider the possibility of mitigating actions to prevent the hazard from having detrimental effects.

Step 9: Modify System to Reduce Frequencies

The model used to estimate initiator frequency provides suggestions for reducing the frequency. Further reduction may be achieved by considerations not in the model. Such could be modifying the process to avoid the use of hazardous materials or conditions. The use of automatic blanketing materials such as foams; automatic inerting atmospheres such as CO_2.

Step 10: Combine Frequencies and Consequences to Estimate Risk

This step takes the information from Steps 6 and 8. The frequency of an accident multiplied by the consequences is the risk. The consequences need to be in common units to get a measure of the risk. Of course, multiple consequence measures may be used and give multiple risk measures: frequency of fatalities, frequency of injuries, frequency of fishkill, frequency of monetary loss. Judgment must be used to rank there relative significance.

Step 11: Modify System to Reduce Risk

If Step 7 minimizes consequences, and Step 9 minimizes accident frequency, it would seem perforce the risk would be minimized and such is generally the case. However, there is a synergism when frequency and consequences are combined into risk. While the risk of a low-frequency high-consequence accident may be the same as a the risk of a high-frequency low-consequence accident, action may be needed to limit the liability in high consequence accidents. Frequent low-consequence accidents may have detrimental morale and public relations effects.

7.4 Example: Analyzing a Chemical Tank Rupture

This chapter shows that chemical process systems may fail and have serious consequences to the workers, public and the environment. Comparing with Chapter 6, chemical processes are similar to the processes in a nuclear power plant, hence, they may be analyzed similarly because both consist of tanks, pipes heat exchangers, and sources of heat. As an example of analysis, we analyze a storage tank rupture.

7.4.1 Defining the Problem with Fault Tree Analysis

The fault tree[u] (Figure 7.4-1) has "Pressure Tank Rupture" as the top event (gate G1). This may result from random failure of the tank under load (BE1), OR the gate G2, "Tank ruptures due to overpressure" which is made of BE6 "Relief valve fails to open" AND G3, "Pump motor operates too long." This is made of BE2, "Timer contacts fail to open," AND G4, "Negative feedback loop inactive" which is composed of BE3, "Pressure gauge stuck," OR BE4, "Operator fails to open switch," OR "BE5, "Switch fails to open."

Notice that one event has units of per-demand and the others have a per-unit-time dimension. From elementary considerations, the top event can only have dimensions of per-demand (pure probability) or per-unit-time dimensions. Which dimensions they have depends on the application. If the fault tree provides a nodal probability in an event tree, it must have per-demand dimensions. if the fault tree stands alone, to give a system reliability, it must have per-unit-time dimensions. Per-unit-time dimensions can be converted to probability using the exponential model (Section 2.5.2.6). This is done by multiplying the failure rate and the "mission time" to give the argument of the exponential which if small may be

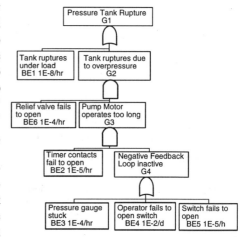

Fig. 7.4-1 Fault Tree for Tank Rupture

Table 7.4-1 Parameters for Figure 7.4-1

	Basic Event		Mission Time	Probability
1	tank	1E-8/hr	4E4 hr	4E-4
2	timer	1E-5/hr	8E3 hr	0.08
3	gauge	1E-5/hr	4E3 hr	0.04
4	operator	1E-2/d	na	0.01
5	switch	1E-5/hr	8E3 hr	0.08
6	rvalve	1E-4/hr	4E3 hr	0.4

---

[u] This example is taken from Dunglinson and Lambert (1983).

Example: Analyzing a Chemical Tank Rupture

```
** FTAP INPUT
** Chemical Process Vessel Rupture
g1   +   tank    g2
g2   *   rvalve  g3
g3   *   timer   g4
g4   +   gauge   operator  switch
ENDTREE
PUNCH
*XEQ
ENDJOB
```

Fig. 7.4-2 FTAP Input file: pv.fi

```
NAMES      6
1 - gauge    2 - operator   3 - rvalve   4 - switch
5 - tank     6 - timer
IMPBE    1
EVENT g1          4       14
 1  5
 3  1  3  6
 3  3  4  6
 3  2  3  6
```

Fig. 7.4-3 FTAP Punch Output: pv.pch

expanded in a Taylor's series giving 1 - the product as the probability of working when needed. Table 7.4-1 lists the data in the fault tree, mission times and the failure probabilities needed to quantify the tree. Using the logic and the values in Table 7.4-1, the tree is calculated as: 4E-4+0.4*(0.08*(.04+.01+.08)) = 4.56E-3. This problem is simple enough for hand calculation. Such is not always true so it is also calculated using the FTAPSUIT.

7.4.2 Applying FTAPSUIT to Chemical Process Tank Rupture

To do this go to FTAPSUIT (refer to Section 6.4.2.[v] Go to the FTAPSUIT main menu and select 1 to go to FTAPlus to input the data. The first selection is the title, the second is the file name (use "pv" for this example), and the third selection is the input. Type:g1,+,tank,g2 (the commas are needed) and type enter to get the results shown in Figure 7.4-2. Do the same for the second through fourth lines (enter) and type "#" to tell FTAPlus you are through inputting. FTAPlus puts the input in the proper columns and adds the additional lines. These lines provide working program input. FTAP has much

```
** Chemical Process Tank Rupture
 5
 1
BE FV
CS FV
ENDIM
DETAILCS     100   1.0E-2
NOPTION
gauge      4.0E-2    3.0  pressure gauge stuck
operator   1.0E-2    3.0  operator fails to open switch
rvalve     4.0E-1    2.0  relief valve fails to open
switch     8.0E-2    5.0  switch fails to open
tank       4.0E-4    5.0  tank ruptures under load
timer      8.0E-2    3.0  timer contacts fail to open
NDATA
NAMES      6
1 - gauge    2 - operator   3 - rvalve   4 - switch
5 - tank     6 - timer
IMPBE    1
EVENT g1          4       14
 1  5
 3  1  3  6
 3  3  4  6
 3  2  3  6
```

Fig. 7.4-4 Appearance of File pvn.ii

---

[v] I am writing this book using WordPerfect 6.1 in Windows 3.1. I found it convenient to make an "FTAP.bat" file to exit using the MSDOS icon and then type "FTAP" to see the main menu for FTAPSUIT.

more power which can be done using an editing program. Nevertheless FTAPlus is handy to put things in the correct column.

FTAP outputs two files: pv.fo and pv.pch. Figure 7.4-3 is pv.pch which is used for BUILD. This files says there are 6 names and assigns numbers to identify each. It found 4 cutsets: 1 first order and 2 third order.

Since dependency analysis is not needed, we can go on to the BUILD program. Go to FTAPSUIT and select 5 "Run Build." It asks you for the input file name including extender. Type "pv.pch." It asks you for name and extender of the input file for IMPORTANCE. Type, for examle, "pv.ii". It next asks for the input option. Type "5" for basic event failure probabilities. This means that any failure rates must be multiplied by their mission times as shown in Table 7.4-1. (FTAPlus was written only for option 5 which uses probabilities and error factors. Other options will require hand editing of the pvn.ii file. The switch 1 is for failure rate and repair time, switch 2 is failure rate, 0 repair time, switch 3 is proportional hazard rate and 0 repair time, and switch 4 is mean time to failure and repair time.)

After running BUILD, its output file (pv.ii) is ready for editing to insert the failure

```
THE IMPORTANCE COMPUTER CODE DEVELOPED BY FTA ASSOCIATES
INPUT FILE NAME: pvn.ii TIME AND DATE OF RUN -- 10:33:45 APR 17, 1998
                  ** chemical process tank rupture
OPTION 5 -- BASIC EVENT PROBABILITIES AND ERROR FACTORS
BIRNBAUM (NO), CRITICALITY (NO), UPGRADING FUNCTION (NO)
FUSSELL-VESELY (YES), INITIATOR (BARLOW PROSCHAN) (NO),
ENABLER (CONTRIBUTORY) (NO), STRUCTURAL (NO)
MIN CUTSET OPTIONS USED: INITIATOR (NO), FUSSELL-VESELY (YES)
INFORMATION ON DETAILED CUTSET OUTPUT -- NM =100 AND FACTOR
=1.000E-02
NAME   PROBABILITY  FACTOR  DISTRIBUTION    DESCRIPTION
gauge     4.000E-02   MEAN    20.0  LOGNORMAL  pressure gauge stuck
operator  1.000E-02   MEAN    30.0  LOGNORMAL  operator fails to open switch
rvalve    4.000E-01   MEAN    40.0  LOGNORMAL  relief valve fails to open
switch    8.000E-02   MEAN    50.0  LOGNORMAL  switch fails to open
tank      4.000E-04   MEAN    60.0  LOGNORMAL  tank ruptures under pressure
timer     8.000E-02   MEAN    70.0  LOGNORMAL  timer contacts fail to open
TOP EVENT PROBABILITY = 4.55383E-03
RANK    NAME              IMPORTANCE
1       rvalve            9.125E-01
1       timer             9.125E-01
2       switch            5.622E-01
3       gauge             2.811E-01
4       tank              8.784E-02
5       operator          7.027E-02
FUSSELL VESELY MEASURE OF SYSTEM UNAVAILABILITY
GROUP RANK    CUMULATIVE IMPORTANCE    RESIDUAL IMPORTANCE
1              .562164                  4.378E-01
2              .842526                  1.575E-01
3              .930027                  6.997E-02
4             1.000000                  0.000E+00
RANK IMPORTANCE   1 5.622E-01 CUTSET   3
   MIN CUT SET PROBABILITY = 2.560E-03          5.622E-01/ 4.378E-01
   NAME   MEAN PROB      E. F.  DISTRIBUTION DESCRIPTION
   rvalve    4.000E-01    4.00   LOGNORML  relief valve fails to open
   switch    8.000E-02    5.00   LOGNORML  switch fails to open
   timer     8.000E-02    7.00   LOGNORML  timer contacts fail to open
2 2.811E-01 CUTSET   2
   MIN CUT SET PROBABILITY = 1.280E-03          8.425E-01/ 1.575E-01
   gauge     4.000E-02    2.00   LOGNORML  pressure gauge stuck
   rvalve    4.000E-01    4.00   LOGNORML  relief valve fails to open
   timer     8.000E-02    7.00   LOGNORML  timer contacts fail to open
3 8.784E-02 CUTSET   1
   MIN CUTSET PROBABILITY = 4.000E-04           9.300E-01/ 6.997E-02
   tank      4.000E-04    6.00   LOGNORML  tank ruptures under pressure
4 7.027E-02 CUT SET   4
   MIN CUTSET PROBABILITY = 3.200E-04           1.000E+00/ 0.000E+00
   operator  1.000E-02    3.00   LOGNORML  operator fails to open switch
   rvalve    4.000E-01    4.00   LOGNORML  relief valve fails to open
   timer     8.000E-02    7.00   LOGNORML  timer contacts fail to open
CUTSET  NO.  ORDER  BASIC EVENTS
         1     1     tank
         2     3     gauge   rvalve   timer
         3     3     rvalve  switch   timer
         4     3     operator rvalve  timer
```

Fig. 7.4-5 Edited Output of IMPORTANCE

```
                ** chemical process tank rupture
MONTE CARLO SIMULATION RESULTS ( 1000 TRIALS )
INDEX OF SORTED VALUE          10    50    100    500    900    950    990
CONFIDENCE LEVEL IN PERCENT         (98)  (90)   (80)          (80)   (90)   (98)
PERCENTILE                      1    5     10     50     90     95     99
      MEAN FROM IMPORTANCE/ COMPUTED MEAN
PROB OF TOP EVENT   .455E-02/ .350E-02  .126E-03  .265E-03  .391E-03  .160E-02  .756E-02  .119E-01  .318E-01
  .642E-03  .423E-02
RANK BASIC EVENT
  1 rvalve      .913E+00/ .738E+00  .363E-01  .195E+00  .324E+00  .830E+00  .979E+00  .988E+00  .998E+00
  1 timer       .913E+00/ .738E+00  .363E-01  .195E+00  .324E+00  .830E+00  .979E+00  .988E+00  .998E+00
  2 switch      .562E+00/ .389E+00  .102E-01  .503E-01  .945E-01  .374E+00  .721E+00  .805E+00  .892E+00
  3 gauge       .281E+00/ .278E+00  .136E-01  .479E-01  .765E-01  .253E+00  .515E+00  .583E+00  .729E+00
  4 tank        .878E-01/ .262E+00  .231E-02  .121E-01  .211E-01  .170E+00  .676E+00  .801E+00  .960E+00
  5 operator    .703E-01/ .713E-01  .258E-02  .888E-02  .127E-01  .527E-01  .153E+00  .195E+00  .338E+00
```

Figure 7.4-6 Monte Carlo Calculation of the Chemical Process Tank Rupture

probabilities. Go to FTAPSUIT and type 6 "Prepare Input for IMPORTANCE." This takes you to FTAPlus where 5 is selected. It asks for a title, filename, presents the name of a component used in preparing FTAP, and asks for the failure probability in E-format. You enter this and it asks for the uncertainty in F-format. You enter this and it asks for an event name. After this is entered, it presents the name of another basic event and so on until it presents a filename (pvn.ii - the program adds "n" to the filename) and asks if you want to change it. Select 6 to exit and you are back in FTAPSUIT. Figure 7.4-4 shows file pvn.ii.

IMPORTANCE must be run to find the probability of the fault tree, even if the importances are not of interest. Go to FTAPSUIT and type "7." It asks for the input file name; type "pvn.ii." The output is pvn.io which after editing (for format) is presented as Figure 7.4-5. It has calculated the Fussel-Vesely importances because of the "FV" in Figure 7.4-3. The probability of the top event is 4.5538E-3 whereas 4.56E-3 was calculated by the previous hand calculation. It says that the cutset consisting of the relief valve-switch-timer is the most important.

A Monte Carlo calculation of the tree results from running MONTE by selecting "8" from the FTAPSUIT main menu. It asks for a file name (and extender); type "pvn.mi." It takes the most time to run of all of the programs; its output is: "pvn.mo" as shown in Figure 7.4-6.

7.5 Summary

Chemical accidents were identified to consist of: initiation, propagation, and termination. Eliminating the initiation of an accident is the most effective step. In spite of the best efforts, some accidents will be initiated then their propagation must be suppressed. If that cannot be done, the accident must be terminated in as benign manner as possible. To appreciate the magnitude of chemical safety, statistics from the hydrocarbon-chemical industry placing the losses at about $2 billion/year. To make this point many deadly accidents were examined. The causes ranged from not knowing the explosive potential (Texas City) to cost-saving and diffuse control (Bhopal). Chemical safety, of course, involves the chemical industry which is much larger economically, broader in range of process and more complex technologically than the nuclear industry. This examination was followed by describing the steps for a chemical process PSA. The chapter ends by returning to FTAPSUIT to calculate the failure probability of rupturing a chemical process tank.

7.6 Problems

1. Prepare an event tree diagram of the Bhopal accident.
2. Use FTAPSUIT to calculate the fault tree shown in Figure 3.4.4-7 using 1.0E-2/d for any valve failing to open, 1.0E-4/d for failure of the SIS signal, 2.0E-2/d for a pump failing to start, and 5E-3/d for a pump failing to run after it starts.

Chapter 8

Nuclear Accident Consequence Analysis

Chapter 6 was concerned, with determining the probability of various failures leading to insufficient core cooling of a nuclear reactor. This chapter describes how the accident effects are calculated as the accident progresses from radionuclide release, radionuclide migration within the plant, escape from retaining structures, atmospheric radionuclide transport and the public health effects.

8.1 Meltdown Process

Any release of radioactive material affecting the public requires temperature above the melting point of the materials to deform the reactor core and confining structures This section lists the barriers preventing release, presents scoping calculations that illustrate the conditions and time scale of concern. Conjectures are presented as to how core melt might happen. The section concludes with information about the partial core melt that occurred at TMI-2.

8.1.1 Defense in Depth - Barriers Providing Public Protection

The analyses of system failures which could challenge the containment or lead to the release of radioactivity form the licensing process. The design basis analyses are deterministic, and degraded core accidents are not considered. PSA determines the probabilities of the numerous sequences that could lead to core degradation and how the core behaves.

A reactor core's fission product inventory is the primary source of radioactivity from which the public is protected by the following independent barriers:

- Fuel matrix - Fission products, bound in a ceramic matrix, may escape only by slow diffusion or melting of the matrix.
- Fuel clad - If the fission products escape the matrix, they are sealed in a zircalloy or stainless steel (the cladding).
- Cooling water - The cooling water, that provides cooling and moderation, also retains fission products - especially the chemically active semivolatiles, as demonstrated at the TMI-2 accident.
- Reactor Cooling System - The coolant and fission products are confined within the piping and vessels of the reactor cooling system. This boundary may fail by high pressure relief,

Nuclear Accident Consequence Analysis

PORV operation, or rupture. A reactor is designed to accomodate these releases and still keep the radioactivity confined. Even in a damaged state, a power plant affords a tortuous path for release and a large surface for plate-out.

- Containment: Commercial reactors have a low-leak-rate concrete building to contain radioactivity which is the last barrier to an atmospheric release.
- Siting: Any release, should it occur, is greatly attenuated by the distance it must travel before reaching the public.
- Emergency Action: If it appears that the public will be affected by a radioactive release, people may evacuate from the expected area, remain indoors, or use iodine-blocking prophylaxis.

In calculating the system and barrier failures, consideration should be given to radiation embrittlement, chemical reactions, thermal shock, and metal fatigue.

8.1.2 Qualitative Description of Core Melt

While great public protection is provided by these barriers, accidents can happen. Regardless of the cause of an accident, the core cannot overheat while in contact with liquid water. Furthermore it cannot be critical in the absence of water (because of the low enrichment of the fuel); thus, any accident involves a subcritical core that is heated by decaying radionuclides with inadequate cooling. Figure 8.1-1 shows the rate of heat evolution as a function of time after shutting down a 3,000-MW reactor (Cohen, 1982). Even after an hour, the heat production is about 40 MW.

Temperature stabilization occurs when the rate of heating balances the rate of heat removal. Otherwise, the temperature rises at a rate dependent upon the integrated net heat evolution (Figure 8.1-2). Cohen (1982) considers decay heat removal by an RHR capable of 84 MW (8E4 BTU/s), which is the rate of evolution 200 seconds after shutdown (Figure 8.1-1). Before this time, Figure 8.1-2 shows 2.1E7 Btu has evolved, but only 1.6E7 is removed by the RHR (84E6*200/950). 87.5 tons of material of

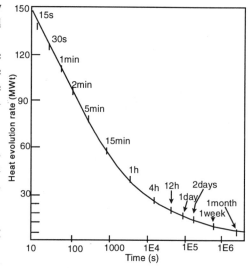

Fig. 8.1-1 Decay Heat from a 3,000 MWe Core

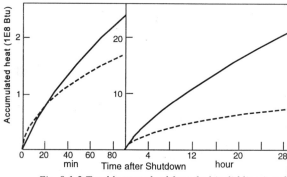

Fig. 8.1-2 Total heat evolved from fuel (solid lines) and absorbed by concrete (broken lines).

nominal specific heat of 0.07 Btu/lb° F changes temperature by 400° F to 1000° F, if it had been operating at 600° F before shutdown. If the RHR failed and there were no other cooling mechanism, the temperature would rise by 1700° F to about 2300° F in 3.3 minutes. Of course, this could not happen if the core is contact with liquid water.

To put the temperature in perspective, Table 8.1-1 presents the melting and boiling points of the various materials involved in a core melt. Many of the materials are alloys and the melting point of the alloy is not the same as the base materials. For example, zircalloy, the usual cladding material, melts at 3308° F (stainless steel at 2600° F). Returning to the numerical example and assuming linearity, the heat needed to raise the core from 600°F to zircalloy melting is 3.3E7 Btu. The time required for this amount of heat to be evolved, according to Figure 8.1-2, is about 7 minutes after shutdown, assuming no heat removal. Having established the physical possibility and minimum time scales, the following core melt scenario is discussed for comparison with the TMI-2 accident presented in Chapter 6.

As an accident progresses through event tree sequences, the blowdown phase is reached with sudden pressure release that causes some of the water to flash to steam. Depending on the size of the hole and the performance of emergency cooling systems, a considerable portion of the core may uncover (be exposed to steam and air). This uncovered period may be brief, if the injection systems function as designed, but if the core never recovers with liquid water, degradation will occur on the time scale just estimated. The water lost from the reactor vessel is first replenished from storage tanks such as the RWST, but in less than one hour, the emergency core cooling systems must switch to the recirculation mode as the sources of injection water are depleted.

Water may leave the vessel faster than it is resupplied

Table 8.1-1 Melting and Boiling Points of Materials

Material	Melting point (°F)	Boiling point (°F)
Volatile fission product		
I_2	237	365
CsI	1159	2336
CsOH	500	1814
Te	842	1810
Refractory fission products		
BaO	3493	5086
Ru	4082	7502
SrO	4406	5880
La_2O_3	4199	7230
Control rods		
Ag	1762	3925
In	315	3763
Cd	610	1413
B_4C	4478	6330
Hf	4032	8042
Zircalloy		
Zr	3366	7968
Sn	450	4717
Stainless steel		
Fe	2795	5184
Cr	3434	4842
Ni	2647	5277
Mn	2271	3744
Fuel		
UO_2	5144	5959

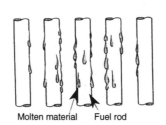

Fig. 8.1-3a Molten droplets and rivulets begin to flow downward

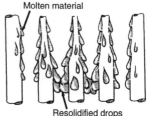

8.1-3b Formation of a local blockage in colder rod regions

Nuclear Accident Consequence Analysis

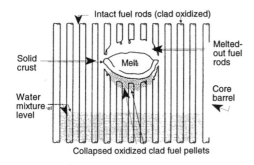

Fig. 8.1-3c Formation of the molten pool

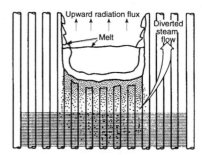

Fig. 8.1-3d Radial and axial growth of the molten pool

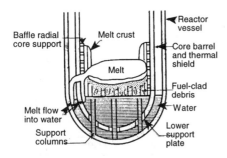

Fig. 8.1-3e Melt migrates to the side and fails the core barrel

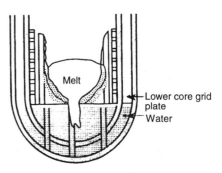

Fig. 8.1-3f Melt migrates downward into remaining water

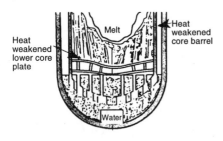

Fig. 8.1-3g Downward progress of coherent molten mass as the below core structure weakens

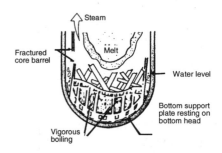

Lower plenum
- Failure of below-core structures
- Melt/water interactions in lower plenum (quenching steam explosion debris bed formation)
- Vaporization of remaining water

Fig. 8.1-3h Failure of support

or it may escape as steam if the core cooling is insufficient. In either case, the water level falls and

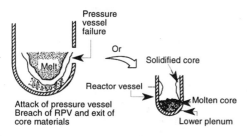

Fig. 8.1-3i Two possible modes of accident progression

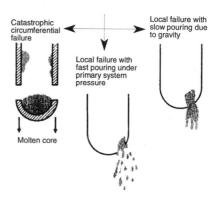

Fig. 8.1-3j Three possible failure modes

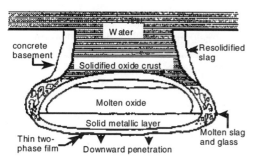

Fig. 8.1-3k Concept of solidified core in concrete after meltdown

the core uncovers. When enough of the active part of the fuel rods is uncovered, steam cooling may be inadequate to prevent core degradation and to prevent a reaction of zirconium and steam which forms zirconium oxide and releases hydrogen

If the core temperature continues to increase, the cladding is hypothesized to form rivulets (Figure 8.1-3a). Some of this flow may resolidify as it reaches lower power regions of the core and cause channel blockages (Figure 8.1-3b). These blockages stop the steam cooling and increase the rate of temperature rise, thus causing a regenerative effect on the accident until a molten pool is formed, supported by a solid crust and blocked channels (Figure 8.1-3c). The molten region migrates downward with steam explosions occurring as the debris falls into the remaining water (Figure 8.1-3d). The melt is unstable and may flow sideways into the region between the heat shield and the reactor vessel or it may migrate downward (Figures 8.1-3e and f). Figures 8.1-3g and h depict more details about the weakening and damaging of the core support structure and its collapse into the remaining water with quenching, steam explosions, and debris formation. The high temperature may weaken the pressure vessel with ensuing failure (Figure 8.1-3i, left), or the vessel may remain intact while the core slumps in the bottom (Figure 8.1-3i, right), only to result later in circumferential failure of the vessel or hot-spot formation with fast or slow flow from the vessel (Figure 8.1-3j). This flow falls into the sump (well) below the vessel with further steam explosions, hydrogen formation, and concrete ablation. Figure 8.1-3k shows the molten core penetration of the basemat.

These descriptions are conjecture and assume that the molten core is not dispersed by the steam explosions or geometry. The more it disperses, the better the heat transfer and the sooner the accident stabilizes.

While there is active core melt accident research, the TMI-2 accident demonstrated what happens in an accident that ends in partial core melt. Such a partial melt is in contradiction to the RSS assumption that if a melt starts, it will propagate to a full melt which were the only accidents considered in the RSS.

8.2 Source Terms for In-Plant Radionuclide Transport[a]

The amounts of material released from a damaged plant are usually expressed in fractions of the isotopic quantities in the core. These source terms (meaning source for the ex-plant transport) depend on accident physics, amount of core damage, time at elevated temperatures, retention mechanisms, and plate-out deposition of material as it transports from the damaged core to release from containment. This section gives an outline of early source term assessments, computer codes used in calculations, and some comparisons of results.

8.2.1 Background on the Source Term

A first attempt to estimate the potential consequences from severe LWRs accidents was the BNL report WASH-740 (1957). The authors of WASH-740, to overcome the lack of information and methods, estimated "Hazard States" based on the core state, radioactive inventory, fuel cladding, reactor coolant system, and containment conditions.

- Hazard State 1 (clad and fuel damage, no release outside of the RCS) was estimated to have a frequency ranging from l0-2 to 10-4 per plant-year;
- Hazard State 2 (core melt and release outside the RCS) was estimated at 1E-3 to lE-4 per plant-year; and
- Hazard State 3 (core melt, release outside RCS and containment) at lE-5 to lE-9 per plant-year.

In 1962 the report, TID-14844 was published presenting analysis and assumptions concerning the behavior of containment (essentially Hazard State 2). The TID report postulated the release of all of the noble gas, 50% of the iodine, and 1% of the radioactive solids to the containment. In addition, TID-14844 provided assumptions for containment leakage (the TMI-2 containment is intact) and for atmospheric transport of the fission products. These results form the basis for Regulatory

---

[a] From NUREG-0956.

Guides l.3 and l.4 which were used by the RSS. The RSS presented results of accident progression and fission products in release categories associated with various plant damage states.

Table 8.2-1 lists the RSS source terms consisting of nine release categories for the PWR, and five for the BWR. Each category is represented by several parameters that describe the release in terms of frequency, timing, and amount of radioactive material. The first column in Table 8.2-1 is the release category designation; the second column gives the cumulative probability (frequency/year) from all reactor accident sequences in the particular release category, and the third column gives the time from accident beginning to the time of release from containment. The fourth column is the duration of the release. The next three columns are, respectively, warning time, elevation of release (used to calculate the atmospheric diffusion) and energetics of the release as it affects the containment. The last eight columns are the source term release fractions used in the risk calculations. Because of the importance of these results and the fact that these source terms were contradicted at TMI-2, the source terms were reexamined.

Table 8.2-1 Reactor Safety Study Accident Release Categories

						Fraction of inventory released[a]							
Category	Frequency (r/y)	Time after accident (hr)	Duration of release (hr)	Warning time for evacuation (hr)	Elevation of release	Containment energy release (Mbtu/hr)	Xe-Kr	Org. I	I	Cs-Rb	TC-Sb	Ba-Sr	Ru[b]	La[c]
Pressurized Water Reactors (PWR)														
1	9E-7	2.5	0.5	1.0	25	520[d]	0.9	6E-3	0.7	0.4	0.4	0.05	0.4	3E-3
2	8E-6	2.5	0.5	1.0	0	170	0.9	7E-3	0.7	0.5	0.3	0.06	0.02	4E-3
3	4E-6	5.0	1.5	2.0	0	6	0.8	6E-3	0.2	0.2	0.3	0.02	0.03	3E-3
4	5E-7	2.0	3.0	2.0	0	1	0.6	2E-3	0.09	0.04	0.03	5E-3	3E-3	4E-4
5	7E-7	2.0	4.0	1.0	0	0.3	0.3	2E-3	0.03	9E-3	5E-3	1E-3	6E-4	7E-5
6	6E-6	12.0	10.0	1.0	0	NA	0.3	2E-3	8E-4	8E-4	1E-3	9E-5	7E-5	1E-5
7	4E-5	10.0	10.0	1.0	0	NA	6E-3	2E-5	2E-5	1E-5	2E-5	1E-6	1E-6	2E-7
8	4E-5	0.5	0.5	NA	0	NA	2E-3	5E-6	1E-4	5E-4	1E-6	1E-8	0	0
9	4E-4	0.5	0.5	NA	0	NA	3E-6	7E-9	1E-7	6E-7	1E-9	1E-11	0	0
Boiling Water Reactor (BWR)														
1	1E-6	2.0	2.0	1.5	25	130	1.0	7E-3	0.4	0.4	0.7	0.05	0.5	5E-3
2	6E-6	30.0	3.0	2.0	0	30	1.0	7E-3	0.9	0.5	0.3	0.1	0.03	4E-3
3	2E-5	30.0	3.0	2.0	25	20	1.0	7E-3	0.1	0.1	0.3	0.01	0.02	3E-3
4	2E-6	5.0	2.0	2.0	25	NA	0.6	7E-4	8E-4	5E-3	4E-3	6E-4	6E-4	1E-4
5	1E-4	3.5	5.0	NA	150	NA	5E-4	2E-9	6E-11	4E-9	8E-12	8E-14	0	0

a From WASH-1400 Appendix VI; background on isotope groups and release mechanisms in ibid. Appendix VII.
b Includes HO, Rh, To, Co.
c Includes Y; Ce, Pr, La, Nb, Am, Cm, Pu, Np, Zr.
d A lower energy release rate applies to part of the release period; the effect of lower energy release rates on the consequences is found in ibid. Appendix VI.

The NRC Chairman Ahearne, received (August 1980) a letter from three members of LANL and ORNL indicating that the RSS may have overestimated the risks from iodine by orders of magnitude. In September of the same year, Chauncey Starr of EPRI sent a letter to Commissioner

Hendrie with attached analyses indicating a similar conclusion. In response to these questions concerning the correctness of WASH-1400 in estimating fission product behavior, and applicability to the TMI-2 accident, the NRC began a review of procedures for predicting fission product release and transport. NUREG-0772 conclusions that depart significantly from the RSS are:

1. The data suggest that iodine will be released, predominantly, as cesium iodide under most postulated light water reactor accident conditions. However, formation of more volatile iodine species (e.g., elemental iodine and organic iodines) is not impossible under certain accident conditions.
2. The assumed form of iodine is not substantially retained in early containment failure, but may be retained in the reactor coolant system, where cesium iodide is more strongly retained than the elemental iodine assumed by the RSS.
3. The retention of radionuclides within the containment was little accounted for by the RSS, but ranges from little to very substantial because of agglomeration and deposition. This leads to a large over prediction of the iodine risk, but substantial agreement with RSS for some other isotopes.
4. Certain engineered safety features (containment sprays, suppression pools, ice condensers) will effectively remove fission products regardless of form. Other ESF such as filters are less effective.

The Sandia Siting Study, working with the results of NUREG-0771 and 0772, release fractions from RSS, rebaselining, and information on dominant accident sequences from PSAs, constructed source terms for a spectrum of accidents that better reflected the understanding of fission product behavior during reactor accidents (NUREG-0773). Five release categories were designated as representative of the full spectrum of potential accident conditions, each representing a different degree of core degradation and of failure of containment safety features.

The siting source terms described in NUREG-0773 were used to calculate accident consequences at 91 U.S. reactor sites using site-specific population data and a combination of site-specific and regional meteorological data, for an assumed 1,120-MWe reactor. These calculations, reported in NUREG/CR-2239, treated siting factors such as weather conditions and emergency response probabilistically but postulated the siting source term release.

8.2.2 Computer Codes for Fission Product Release and In-Plant Transport

Section 8.1 provided a description of a core melt. This section backs up to describe thermal-hydraulic calculations of the phenomena before, during, and after the accident, and other calculations to estimate the radioactive release from containment. In this accident physics cannot be analyzed separately from in-plant transport.

An accident sequence source term requires calculating temperatures, pressures, and fluid flow rates in the reactor coolant system and the containment: to determine the chemical environment to which fission products are exposed; to determine the rates of fission product release and deposition; and to assess the performance of the containment. All of these features are addressed in the computer codes that together form the BMI-2104 suite of codes used to calculate source terms for selected plants and accident sequences. This suite of codes represents an evolution of the codes used in the Reactor Safety Study.

MARCH calculates the thermal hydraulic behavior leading up to and following the core overheating. It couples to the release codes with ORIGEN providing the fission product inventory, CORSOR providing the release from fuel, and TRAPMELT providing the transport in the RCS. This path and an alternative path from the core debris and VANESA, couple with NAUA, SPARC, or ICEDF to provide the source terms to an ex-plant transport code (e.g., CRAC 2). While Battelle Columbus Laboratories (BCL) led calculating the source terms for the RSS and in the development and application of the BMI-2104 suite, BCL drew support from a number of institutions; primarily: Oak Ridge National Laboratory (ORNL), Sandia National Laboratories (SNL), Brookhaven National Laboratory (BNL), Pacific Northwest Laboratories (PNL), and Kernforschungszentrum Karlsruhe (KfK).

a) MARCH (NUREG/ CR-1711, 2285 and 3988) calculates the behavior of the reactor during a severe accident with 100 subroutines and is the largest code in the BMI-2104 suite (Figure 8.2-1) It evaluates the following:

1. Heatup of the reactor coolant inventory and pressure rise to the relief or safety valve settings with subsequent boiloff.
2. Initial blowdown of the coolant from the reactor coolant system.
3. Generation and transport of heat within the core, including boiloff of water from the reactor vessel.
4. Heatup of the fuel following core uncovery, including the effects of metal-water reactions.
5. Melting and slumping of the fuel onto the lower core support structures and into the vessel bottom head.
6. Interaction of the core debris with residual water in the reactor vessel.
7. Interaction of the core debris with the reactor vessel bottom head and subsequent failure of the head.
8. Interaction of the core debris with the water in the reactor cavity.
9. Attack of the concrete basemat by the core and structural debris.

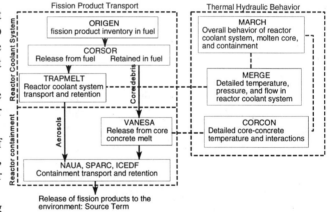

Fig. 8.2-1 BMI-2104 suite of codes used in the source term assessment

10. Relocation of the decay heat source as fission products are released from the fuel and transported to the containment.
11. Mass and energy additions to the containment associated with all the foregoing phenomena and their effect on containment temperature pressure, and steam condensation.
12. Effects of the burning of hydrogen and carbon monoxide on the containment pressure and temperature.
13. Leakage of gases to the environment.

Of these phenomena, the first three in particular, involve thermal hydraulics beginning with the pre-accident conditions. Items 4 through 7 address the meltdown of the core and its influence on (1) hydrogen production, which affects containment loads, (2) fuel temperatures, which affect in-vessel fission product releases, (3) thermal-hydraulic conditions, which influence fission product transport and retention, and (4) the types and quantities of materials in the core debris, which affect ex-vessel fission product releases. Figure 8.2-2 outlines the meltdown model in MARCH. The reactor core is divided into 240 annular rings (nodes) arranged in 10 radial and 24 axial segments. It is assumed that fuel begins to melt when the temperature of a node reaches the liquefaction temperature (4131° F). Molten fuel begins to slump out of the fuel region, the support grids fail when their temperature nears the melting temperature of stainless steel (2552° F), and the entire core falls into the vessel's bottom head when 75% of the core is molten. Item 8 addresses the boiling of water in the reactor cavity beneath the vessel and the further oxidation of zirconium metal in the core debris. Items 9 and 10, respectively, address core-concrete interactions and fission product release from the fuel.

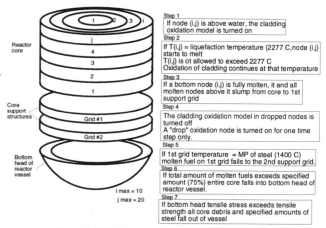

Fig. 8.2-2 Meltdown model in the MARCH code

The final three phenomena, items 11 through 13, are addressed in the containment performance models of MARCH, accounting for mass and energy additions to the containment, the burning of combustible gases, the effects of core sprays, ice condensers, and suppression pools. MARCH calculates only the containment loads; it does not model the containment failure.

The principal unresolved uncertainties in MARCH involve core melting and containment behavior. Important details that are not well known are: (1) the core liquefaction temperature (the core acts like a two-phase slurry), (2) the maximum temperature, (3) changes in the zircalloy oxidation rate because of geometry changes and reaction with UO_2, and (4) the amount of corium and steel that is expelled from the vessel and the rate of expulsion.

b) MERGE (NUREG/CR-4172), uses a more detailed control volume scheme than MARCH, but uses the MARCH-calculated values of system pressure, gas flow rate, gas temperature, and hydrogen-to-steam mass ratio at the core exit as boundary conditions. MERGE calculates gas flow rates, gas temperatures, gas conditions, and structure temperatures in each of its control volumes. to produce the parameters needed by TRAPMELT for calculating of fission product transport and deposition in the primary system.

c) CORCON (NUREG/CR-2142) calculates the rate of erosion of the concrete cavity, the temperature and composition of the molten layers, and the temperature, flow rate, and composition of the gases (CO_2, CO, H2, and steam) being evolved from the melt if corium falls from the reactor vessel into the concrete cavity. These temperatures, flows, and compositions are input for the VANESA code to calculate the release of fission products from the molten core debris.

CORCON initially assumes that the molten core debris is stratified as a dense oxidic layer on the bottom and a less dense metallic layer on the top. Later, when molten concrete slag dilutes the heavy oxide layer, the lighter oxide layer than the metal layer rises to the top. Each layer is assumed to be isothermal and heat is exchanged between: (1) the melt and the concrete, (2) layers of the melt, and (3) the top surface of the melt and the atmosphere above it. When the concrete heats up to about 2500° F, CORCON predicts the release of steam and CO_2 from concrete decomposition. The heat of reaction of the gases reacting with the materials of the melt are calculated.

d) ORIGEN (Croff, 1983) is a reactor physics code that calculates the change in composition of reactor fuel as a function of: (1) power history and (2) time after-reactor shutdown. ORIGEN performs a large number of straightforward physics calculations using an extensive database of three principal types of data: radioactive decay (half lives and branching fractions), photon energies per decay, and cross sections for neutron absorption (including fission product yields for the fissioning species).

e) CORSOR (NUREG/CR-4173) calculates the release of fission products from the fuel and the generation of inert aerosols from structural and control material in the core. CORSOR provides release rates for eleven fission products (Cs, I, Xe, Kr, Te, Ag, Sb, Ba, Ru, Mo, Sr), two cladding components (Zr, Sn), one structural component (Fe), and the UO_2 fuel, and it provides release fractions for the Ag-In-Cd control rod materials, used in many PWRs.

f) TRAPMELT (NUREG/CR-4205) calculates the rate of deposition of vapors and aerosols moving through the reactor coolant system (e.g, the upper plenum or the pressurizer). It uses the MERGE-provided temperature of these structures, flow rate, composition. TRAPMELT analyzes CsI, CsOH, Te, "other aerosols," or these four in addition to noble gas, I_2, and the less volatile Sr, Ru, and La fission products. Three species, CsI, CsOH, and Te, account for all the volatile fission products of interest in a TRAPMELT calculation. These three forms are treated as vapors leaving the core that can condense, evaporate, deposit, or be absorbed on a surface.

The rest of the less volatile fission products along with constituents of zircalloy, stainless steel, and the control rods are assumed to be in condensed form as inert aerosols that are treated together in TRAPMELT as "other aerosols." The aerosols are modeled as agglomerating and depositing on surfaces by several mechanisms (e.g., gravitational settling).

g) VANESA (Powers et al., 1985) calculates fission product release and aerosol production from the core-concrete melt using data such as: melt temperature, gas generation rates (CO_2 and steam), and other input from CORCON. It contains a library of thermodynamic properties (free energies from which vapor pressures are calculated) for chemical species (mostly elements, oxides, and hydroxides) that might be formed by fission products and other melt constituents.

h) NAUA -Mod 4 (Bunz et al., 1983) calculates the behavior of aerosols in containment with condensing steam atmospheres. Given an aerosol source rate and a containment leakage rate, NAUA determines: (1) the suspended mass concentration (particles and water), as a function of time, (2) the time-dependent size distribution of airborne material (mass concentration of water and particles in each size class), (3) the cumulative settled-out quantity, (4) the cumulative plated-out quantity, and (5) the cumulative leaked mass. The phenomena treated are: (1) agglomeration (random movement, gravity, turbulence), (2) removal (random movement, gravity, movement in a condensing steam flow [Stefan flow], movement in a temperature gradient [thermophoresis], and sprays), (3) steam condensation onto aerosols, and (4) homogeneous nucleation of water droplets.

i) SPARC (NUREG/CR-3317) evaluates the effects of suppression pools on aerosols. The code describes the deposition of aerosol particles on bubble walls as the gases transporting the aerosols bubble up through a pool of water. The calculation of aerosol retention includes the following deposition mechanisms: (1) random movement, (2) gravity, (3) inertia, and (4) movement in a condensing steam flow. SPARC also includes the effects of soluble particle growth by water vapor uptake.

j) ICEDF (NUREG/CR-32481) models thermal hydraulics and aerosol particle deposition in a PWR ice compartment. The particle deposition model includes the following mechanisms: (1) random movement, (2) gravity, (3) turbulence, (4) inertia, (5) movement in a condensing steam flow, and (6) movement in a temperature gradient (thermophoresis). The ICEDF code was not used in its standard form for the BMI-2104 calculations, but the ICEDF particle deposition mechanisms were incorporated into the NAUA code for that purpose.

8.2.3 Comparison with WASH-1400

Refinements have been made on the source term calculational methods since WASH-1400. The general trend has been toward source term reduction (e.g., the iodine release for the large LOCA) as compared with the RSS results. In cases such as Station Blackout, the RSS underestimated the tellurium and barium compared with results from the BMI-2104 suite. Comparisons with WASH-1400 are particularly relevant, because other PSAs have closely followed the RSS. Figure 8.2-3 presents CCDF comparisons for the RSS reference PWR (Surry), using WASH-1400 information and BMI-2104 procedures.

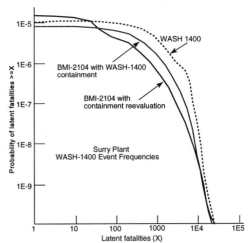

Fig. 8.2-3 Complementary cumulative distribution for latent cancer fatalities

8.3 Ex-Plant Transport of Radionuclides

Radioactivity reaches the public by ex-plant transport. If the completeness arguments presented in Section 3.2 are accepted, the only way the transport can happen is by fluid or gaseous transport. Published PSAs treat atmospheric transport as the only significant mechanism.

8.3.1 Atmospheric Transport Models[b]

The transport and dilution of radioactive aerosols, vapors, and gases released into the atmosphere from a nuclear power station are a function of the the atmosphere along the plume path, the topography of the region, and the characteristics of the effluents themselves. The concentration of radioactive material transported to the surrounding region depends on the rate, duration, and amount released; the height of the release; the momentum and buoyancy of the emitted plume; the windspeed, atmospheric stability, and airflow patterns at the site; and various effluent removal mechanisms. Geographic features such as hills, valleys, and large bodies of water strongly influence turbulent mixing and airflow patterns, as does surface roughness, including vegetation cover.

Two basic approaches have been used in atmospheric diffusion modeling: gradient-transport theory and statistical theory. Gradient-transport theory holds that diffusion at a fixed point in the atmosphere is proportional to the local concentration gradient; this theory attempts to determine momentum or material fluxes at fixed points. The statistical (i.e., Gaussian) approach is a diffusion model of atmospheric transport. Input data for either approach include: windspeed, atmospheric stability, and patterns in the region of interest. Several basic models have been developed using these approaches and vary according to their treatment of the spatial changes of input data and the consideration of either, a variable trajectory model, or a constant mean wind direction model. All PSA models have used the statistical approach.

At ground-level beyond a few miles from the plant, the concentrations of effluents are essentially independent of the release mode. However, for ground-level concentrations within a few miles, the release mode is very important. Normally, gaseous effluents are released from tall stacks and produce peak ground-level air concentrations beyond the site boundary; near-ground releases will produce concentrations that monotonically decrease from the release point to all locations downwind. Under certain conditions, the effluent plume may become entrained in the aerodynamic wake of the building and mix rapidly down to ground level. Methods have been developed to estimate the effective release height for calculation of effluent concentrations at all downwind locations. The important parameters in these methods include the initial release height, the location of the release point in relation to obstructions, the size and shape of the release point, the initial vertical velocity of the effluent, the heat content of the effluent, the ambient windspeed and temperature, and the atmospheric stability. As the effluent travels from its release point, several mechanisms can work to reduce its concentration beyond that achieved by diffusion alone. Such removal mechanisms include radioactive decay and dry and wet deposition.

---

[b] Adapted from Regulatory Guide 1.145.

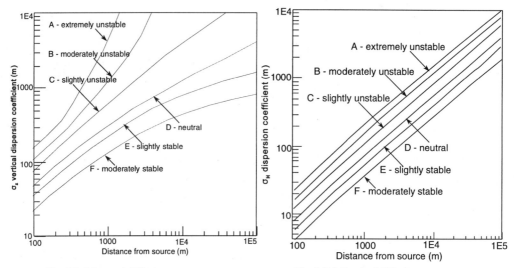

Fig. 8.3-1 Lateral diffusion parameter vs distance from the source. This does not consider meander and building effects

Fig. 8.3-2 Vertical diffusion parameter vs distance from the source. This does not consider meander and building effects

Radioactivity decays exponentially according to the half-life and time. All effluents can undergo dry deposition by sorption onto the ground surface. However, the dry deposition rate for noble gases, tritium, carbon-14, and nonelemental radioiodine is so slow that this depletion mechanism is negligible within 50 miles of the release point. Elemental radioiodine and other particulates are readily deposited. This transfer can be quantified as a transfer velocity (where concentration * transfer velocity = deposition rate). The transfer velocity is proportional to windspeed and, as a consequence, the rate of deposition is independent of windspeed since concentration in air is inversely proportional to windspeed.

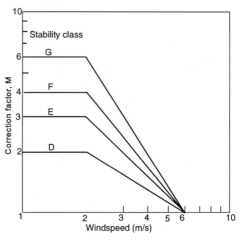

Fig. 8.3-3 Correction factors for σ_y by atmospheric stability class (USNRC Regulatory Guide 1.145)

Dry deposition, although not as efficient as the wet removal process, is continuous while wet deposition occurs only during periods of precipitation. At most sites, precipitation occurs during a small fraction of the hours in a year and, hence, is less likely to be as significant as dry deposition.

The diffusion of individual plume elements, according to Gifford (1959) can be determined from the general Gaussian diffusion model. The model is usually in two basic forms: a "puff" release - appropriate for an accident, and a continuous release for "routine" release or an accident of long duration.

Equation 8.3-1 is a "puff" release model (Start and Wendel, 1974), where $r^2 = (x - \bar{u}*t)^2 + y^2$ and $\sigma_h = \sigma_x = \sigma_y$. The other symbols are: h_e = effective release height (m), Q = amount effluent emitted during the release time, t = the travel time (s), $\bar{u}$ = the mean windspeed at the height of the effective release point, x is the distance from center of puff along the direction of flow (m), y is the distance from center of puff in the crossflow direction (m), σ_x (sigma x) is the plume spread along the direction of flow; σ_y (sigma y) is the lateral plume spread, σ_z (sigma z) is the vertical plume spread; and: χ (chi) is the atmospheric concentration of effluent in a puff at ground level and at distance x from the puff center.

The reason for calling equation 8.3-1 a "Gaussian diffusion model" is because it has the form of the normal/Gaussian distribution (equation 2.5-2). Concentration averages for long time intervals may be calculated by averaging the concentrations at grid elements over which the plume passes.

. Regulatory Guide 1.145 provides corrections to the sigmas to correct for the effects of wind meander at low windspeed - much the same effect as achieved in the CRACIT code. Figure 8.3-1, taken from this guide, shows how the horizontal dispersion coefficient varies with distance from the source. To calculate χ/Q (the fractional attenuation) use formula (8.3-1) for a particular distance from the plant, say 1 km. For a given stability condition (e.g., D), the abscissa intersects the "D" curve at 74 m. Figure 8.3-2 is a similar curve for the vertical dispersion coefficient, and Figure 8.4-3 provides the factor, M, by which the horizontal dispersion is multiplied to correct for the effects of wind meander at low windspeed.

$$\chi/Q = \frac{2*\exp[-\frac{1}{2}*(r^2/\sigma_h^2 + h_e^2/\sigma_z^2)]}{[(2*\pi)^{3/2}*\sigma_h^2*\sigma_z]} \quad (8.3\text{-}1)$$

These sigma values are for unrestricted flow over relatively flat, uniform terrain. They may require modification for application to rough terrain or restricted flow conditions (e.g., within the confines of a narrow valley, coastal or desert areas - see IAEA, 1980).

For purposes of estimating dispersion under extremely stable (G) atmospheric stability conditions, without plume meander or other lateral enhancement, equations 8.3-2 and 8.3-3 may be used.

$$\sigma_y(G) = 2/3 * \sigma_y(F) \quad (8.3\text{-}2)$$
$$\sigma_z(G) = 3/5 * \sigma_z(F) \quad (8.3\text{-}3)$$

8.3.2 Health Effects

Once the radionuclide concentration in the air, and the ground deposition at a certain location have been computed, the next step is to consider how the radiation reaches the people at that location. The four major pathways are:

- Immersion (external exposure to air),
- Inhalation (internal lung exposure from breathing),
- Ingestion (eating and drinking), and
- Radiation from the ground (external).

The relative importance of these pathways varies with the distance from the accident, weather conditions, and the exposure duration. The ingestion dose is usually ignored in computing acute (early) effects. It is impossible to generalize about the relative importance of the other three pathways because their importance varies greatly depending on the radionuclide composition of the release and the organ for which the dose is computed. The doses from immersion, inhalation, and the ground are computed to find the dominant pathway.

For example, given a BWR-1 release (Table 8.2-1), typical weather conditions, no rain and less than one mile from the accident immersion, inhalation, and ground produce about the same magnitude doses for exposure times of several hours or less. The exposure from a cloud ceases once the cloud has passed, but exposure may continue from the ground until the area is decontaminated or evacuated. If the exposure time is long, the ground dose can eventually dominate. Tables 9-6 and 9-7 in NUREG/CR-2300 show examples of the relative importance of these pathways.

The radiation dose from being in or near a "cloud" of airborne radioactivity can be calculated if the radionuclide concentration in the cloud is known. While radioactive noble gases may be inhaled, they are not retained in the body, hence, most of their dose contribution is by cloud radiation.

Ground radiation is from deposited radioactive particles. The deposition rate from a radioactive cloud without rain (dry deposition) is so low that the ground radiation dose is about the same as the inhalation dose. A heavy rain, however, may wash out enough particles from the plume to make ground radiation the dominant contributor to the total dose in a limited area. Rain will also attenuate radiation by leaching the radioactivity to be shielded by the soil and by moving it to streams for further removal.

Inhalation brings the radioactive material into the lung where it may be retained. The International Commission on Radiation Protection (ICRP) lung model is generally used to model absorption into the body. The final location of the radionuclides depends on the element. Iodine is concentrated in the thyroid, calcium in the bones, and so on for other nuclides. The largest part of the inhalation dose comes from particles contained in the plume as it passes over a populated area. However, particles deposited on the ground during passage of the plume may later become resuspended and then inhaled. The importance of inhaled resuspended particles increases with the length of the exposure period.

The ingestion pathway is much more complicated than the others since the radionuclides, except those in drinking water, have to be taken up by the plants and then consumed by humans in either vegetable or animal form. Of the various pathways, the milk pathway is particularly important because a dairy cow consumes a large amount of vegetation and concentrates radionuclides (e.g., ^{137}Cs, ^{131}I, and ^{90}Sr) in her milk which is usually consumed fairly near the cow within a few days after production. Other foodstuffs may be stored for months, allowing the short-lived radionuclides to decay away. Moreover, milk is a major food for children who are more susceptible to radiation because of their small size and rapid growth. Milk is the primary pathway for ^{131}I which is concentrated in the thyroid. Its half life is 8.5 days, so except for fresh milk, most of the radioactivity decays before ingestion.

The ingestion dose contributes very little to the dose from a severe reactor accident and is usually not computed. However, the food pathway is a major determinant of how the exposed area must be treated in the months and years following the accident. If the ground concentration is high, the land may be interdicted from agricultural use or grazing.

8.3.3 Radiation Shielding and Dose

8.3.3.1 Geometric Attenuation of Radiation

If an accident occurs, people are protected from radiation by materials, distance, and limited exposure times.

Dose from a Point Source

Dose is related to the amount of radiation energy absorbed by people or equipment. If the radiation comes from a small volume compared with the exposure distance, it is idealized as a point source (Figure 8.3-4). Radiation source, S, emits particles at a constant rate equally in all directions (isotropic). The number of particles that impact the area is: $S*t*Tr_g$ where Tr_g is a geometric effect that corrects for the spreading of the radiation according to ratio of the area exposed to the area of a sphere at this distance i.e. the "*solid angle - Ω* subtended by the receptor (equation 8.3-4).

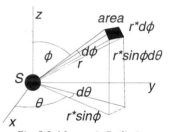

Fig. 8.3-4 Isotropic Radiation

Dose, D, is the amount of energy deposited in a mass of material which is given by the energy absorption coefficient μ_E (cm^2/gm) for the type and energy of radiation times the average energy of the radiation $<E>$ times the source strength, S, times the time of exposure, t, times the geometric attenuation, Tr_g (equation 8.3-5). This shows that the dose from a point source is inversely proportional with the square of the distance between source and receptor. Thus, twice the distance reduces the dose by $1/(2)^2 = 1/4$.

Numerical Example: Calculating the Interception of Radiation

Suppose you are 1.8 m tall and 0.4 m wide, your area is 1.8*0.4 = 0.072 m². If you are standing 10 m from a source emitting 1E10 n/sec, the flux density on you is: 1E10*0.072/(4*π*100) = 5.7E5 n/sec.

$Tr_g = Area/(4*\pi*R^2)$ (8.3-4)

$Tr = \mu_E*<E>*S*t*A_g$ (8.3-5)

Dose from a Line Source

The dose from a line of radiation can be easily calculated from the dose from a point source by adding the dose from many

$Tr_g = *1/(2*z)$ (8.3-6)

of point sources along a line. This is called a Green's function solution. Let S_L be the source strength per unit length of the line source. The rate of radiation into the receptor is $\int S_L * dx / [4 * \pi * (x^2 + z^2)]$ where the integral is for x from $-\infty$ to ∞ which gives $1/(2*z)$ (equation 8.3-6), where z is the closest distance to the line. The geometric attenuation from the line source is given by equation 8.3-6. Using this in equation 8.3-5 gives the dose from a line source to show that the dose from a line source is inversely related to the distance from the source. Doubling the distance from a line source reduces the exposure by ½.

Dose from a Plane Source

The attenuation factor for a receptor at the center of a finite plane uniform source of radiation is given by equation 8.3-7, where b is the radius and z is the height of the receptor above the plane. If there is no attenuation in the air Tr_g becomes ∞ as the area increases. Substituting this into equation 8.3-5 gives the dose at the center of a plane source.

$$Tr_g = 1/2 * \int_0^b r*dr/(z^2+r^2) = 1/2 * \ln[1+(b/z)^2] \quad (8.3\text{-}7)$$

8.3.3.2 Material Shielding

This discussion of geometric effects ignored the attenuation of radiation by material through which the radiation must travel to reach the receptor. The number of particles, *dN*, penetrating material, equals the number of particles incident N_o times a small penetration distance, *dx*, divided by the mean free path length $<x>$ of the type of particle in the type of material (equation 8.3-8). Integrating gives the transmission coefficient for the radiation (equation 8.3-9).

$$dN = -N*dx/<x> \quad (8.3\text{-}7)$$
$$Tr_m = N/N_o = \exp(-x/<x>) \quad (8.3\text{-}8)$$

Combining $Tr_g + Tr_m$ for a point source gives equation 8.3-10. Radiation is attenuated by the distance from the source, the shielding material for the type of radiation and the thickness of shield that it must penetrate.

$$Tr_t = \exp(-x/<x>)/(4*\pi*r^2) \quad (8.3\text{-}9)$$

Buildup Factor

Equation 8.3-9 assumes that a radiation particle impacts a nucleus and disappears. The collision processes depend very much on the type of particles involved in the collision. Heavily ionizing particles such as, alpha particles or protons, are very easily stopped by a small amount of material by their dense trail of ions. On the other hand, electrons scatter off other electrons losing energy and producing a gamma. Subsequently, the gamma may react with another electron to produce an electron and gamma. This process is called a *gamma cascade*. Neutrons lose energy by collisions with nuclei until they are captured or become thermalized.

These complex processes are treated as an attenuation process with a *build-up factor, B(E,r)*, in equation 8.3-9 to give equation 8.3-11.

$$Tr_m = B(E,r) * \exp(-x/<x>) \quad (8.3\text{-}10)$$

Large computers calculated theoretical models of the secondary processes to produce tables of build-up factors. These tables are for neutrons and gammas of various energies in many geometries and material combinations. Tabulated buildup factors depend on the type of primary radiation, the energy, E, of the primary radiation, the charge, Z, atomic number, A, and thickness of the shielding material.

Table 8.3-1 Units of Radioactivity

Name	Description
Curie	3.7E10 disintegrations per second
Becquerel	1 disintegration per second

A Numerical Example of a Build-Up Factor

For example, a 1 MeV point isotropic source of gamma-radiation has a buildup factor of 2.1 when penetrating a mean-free thickness of water. If the build-up factor is ignored, equation 8.3-11 is exp(-1) = 0.36. Hence, 36% of the radiation passes through the shield. But when the buildup factor is included, 2.1*0.36 = 76% of the radiation penetrates the shield.

Table 8.3-2 Units of Energy Deposition

Name	Description
Roentgen	1 esu of charge in one cc of air at STP
Rad	100 ergs/gm (0.01 Gr)
Gray (Gr)	1 J/kg of material

Table 8.3-3 Units of Health Effects

Name	Description
REM	Q*Rad
Sievert (Sv)	Q*gray

8.3.3.3 Dose, Dose Rate and Health Protection

Units of Radioactivity

The activity of a sample, source or contaminated material is the rate at which radioactive disintegrations are taking place. The initial term, named by Madam Curie for her husband was, the Curie (Table 8.3-1). The modern unit is the Becquerel named after the discoverer of radioactivity.

Dose

The number of disintegrations per second gives no information about the radiation. The early measure of this was the Roentgen that measures the ionization density in ion chambers (Table 8.3-2).

More closely related to health effects and material damage is the energy deposited in a mass of material. The Rad was an early unit; the SI unit is the *Gray*.

Table 8.3-4 Q Factors for Various Radiation

Type	Q
Gamma ray	1
X-ray	1
Beta/electron	1
Neutron	10
Proton	10
Alpha	20
Heavy ions	20

Health Effects: Dose Equivalent - Rem/Sievert

Around the beginning of this century, cancer and illness was associated with excessive use of X-rays. Watch dial painters got mouth cancer from radium in the paint. It soon was realized that radiation has health effects. The measures of energy deposition concepts introduced as the rad (gray) do not adequately describe the impact of energy on tissue because absorption of a given amount of energy in a given mass does not describe the dose effect. Tissue damage increases with the linear energy transfer (LET), i.e., the density of the ionization along the track. The health effect is estimated by correcting the gray by the relative biological effectiveness (RBE or "Q" factor) for radiation relative to 200 keV X-rays (equation 8.3-12). With the Q correction, Rad becomes REM (Roentgen equivalent mammal) and gray becomes Sievert (Table 8.3-3). Table 8.3-4 gives the Q-factors for several types of radiation; reflecting that Q increases with LET. Table 8.3-5 summarizes some guideline limits for radiological protection.[c]

$$H = D * Q \quad (8.3\text{-}11)$$

Everyone receives small radiation doses every day; Figure 8.3-5 illustrates some of the doses received from background and other types of radiation. Note that the scale is logarithmic, and that background and cosmic-ray doses vary over an order of magnitude just with location and elevation. In addition to these natural sources, most people receive some medical and dental doses each year.

Health effects are divided into early, late, and genetic. Early effects may appear in the first days after exposure and include injuries and fatalities. The dose to the bone marrow usually determines the number of early fatalities, although the dose to the lung and gastrointestinal tract may also be important. Early injuries (morbidities) are usually due to large doses to the thyroid, lungs, or gastrointestinal tract. Late effects are primarily cancer. Genetic effects are more difficult to detect and evaluate and are often excluded from consideration.

Radiation deposits energy in tissue to ionize molecules in the cells. The ionization forms free radicals which break chemical bonds, to alter other molecules in the cells. If the radiation dose is sufficiently large, enough molecules are ionized and enough free radicals are formed that the cell dies. At lower energies only cell damage may result. Some of the molecules altered by the free radicals will be DNA, and a change in the DNA molecule may cause any new cells created from the damaged cell to be different from the original cell, possibly lacking in normal growth restraints, hence, cancerous. The body possesses some repair mechanisms which reduce the effect of radiation at low doses and dose rates.

At high doses (hundreds of rem) and high dose rates, the effects are fairly well known (Glasstone and Dolan, 1977). The problem is at low doses where the correlation between exposure and cancers is poor because of experimental difficulties from competing effects. The procedure presented in BIER II linearly related data from large exposures to cancer. This relationship is about 1.5E-4 cancers/person-rem for adults and is extrapolated to low exposures. (Since the population of the U.S. is about 2.65E8 and the background radiation is 0.1 rem/yr, then 2.65E8 * 0.1 * 1.5E-4

---

[c] From the National Council on Radiation Protection and Measurements, "Recommendations on Limits for Exposure to Ionizing Radiation," NCRP report No. 91, 7910 Woodmont Ave., Bethesda, MD, June 1987.

Ex-Plant Transport of Radionuclides

Table 8.3-5 Some Maximum Doses from NCRP-91

Type of Exposure	Dose (mSv)	Dose (rem)
A. Annual occupational exposure		
1. Equivalent stochastic effects	50	5
2. Dose equivalents for tissues and organ (non-stochastic)		
a. Lens of eye	150	15
b. All others (e.g., bone marrow, breast, lung, gonads, skin etc.)	500	50
3. Guidance: cumulative exposure	10*age	1*age
B. Public (annual)		
1. Effective dose: continuous or frequent	1	0.1
2. Effective dose: infrequent	5	0.5
3. Remedial action recommended when:		
a. Effective dose equivalent	>5	>0.5
b. Exposure to radon and its daughters	>0.007 J/m^{-3}	>2 WLM
4. Dose equivalent limits for lens of eye, skin and extremities	50	5
C. Education and training (annual total)		
1. Effective dose	1	0.1
2. Dose limit for lens of eye, skin, and extremities	50	5
D. Embryo-fetus exposures		
1. Total dose equivalent limit	5	0.5
2. Dose equivalent limit in a month	0.5	0.005
E. Negligible individual risk level (annual) per source or practice (excluding background)	0.01	0.001

= 3975 cancers per year in the U.S. from background radiation.) The complete truth of the linear hypothesis is doubtful because it says that the effects are independent of dose rate, that is, a dose received abruptly is just as cancer-forming as the dose received over a long time. This ignores repair mechanisms for example, chemical hazards are generally believed to be related to dose rate. An aspirin a day does not have the same effect as 365 aspirin once a year, but there is little doubt that the linear hypothesis is conservative. WASH-1400 used a linear-quadratic hypothesis that assumes the relationship is linear in the range where it is known to be linear but is quadratic at low doses so that it goes to zero with a zero slope. Other hypotheses are possible, but data do not support any.

8.3.4 Computer Codes for Consequence Calculation

The calculation of atmospheric transport with its many parameters and frequency weighting of the wind rose is practical only with a computer code. Table 8.3-6

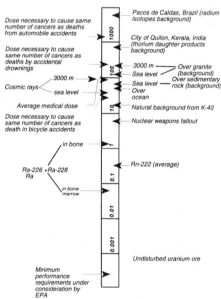

Fig. 8.3-5 Common Doses from Various Sources (Hinga, 1982). Reprinted with permission from Environmental Science & Technology, Vol. 16., copyright American Chemical Society

lists some codes developed for PSA purposes. The first code, used by the RSS, appeared in WASH-1400 draft, but it was severely criticized; a new code called CRAC was prepared, which was used in WASH-1400 Final. It is a dispersion model with a straight Gaussian plume. Evacuation is assumed to take place radially from the reactor, beginning immediately upon notification using very slow average speeds for conservatism. When the people in a sector are overtaken by a cloud while evacuating, they are assumed to remain at the location where they were overtaken until the cloud completely passes over them (a further conservatism).

CRAC determined the consequences of a reactor accident using meteorological data for six reactor sites. Each site was taken to represent an entire region, and all of the 68 sites of the first 100 reactors were assigned to one of these six regions. Wind speed and stability for each of the six sites was assumed to be representative of the entire region out to a distance of 500 miles. Wind direction because with so many sites, the wind direction was assumed random, but the population distribution for all 68 sites was used.

CRACIT considers preferential wind directions. The major changes from CRAC were:

Table 8.3-6 Codes for Nuclear Accident Consequences

Name	Description
CRAC	Original RSS code
CRAC2	Derivative of CRAC
CRACIT	Modified version of CRAC
NUCRAC	Modified version of CRAC
UFOMOD	German consequence code
TIRION	United Kingdom code
ALICE	French code
ARANO	Finnish code

- The straight-line plume model is not used; instead the model computes the trajectory of each segment of the plume as the wind changes with time. The actual terrain is modeled for the region within about nine miles of the plant.
- Specific evacuation routes were identified for each sector within the evacuation radius.
- About ten sources of meteorological data were used, each assumed to be representative of a region in the vicinity of the plant.
-

NUCRAC improves on the health effects model by a reexamination of Hiroshima and Nagasaki data. The dry deposition model was much improved by the inclusion of a particle-size distribution, a detailed settling model, and a detailed chronic exposure model via the food pathway. However, it does not include a rainout model.

It is important to consider the uncertainties in the consequence evaluation. Many of the events and occurrences modeled in the consequence evaluation are random or highly variable processes which can only be described statistically, hence, the results have statistical uncertainties (Section 2.7). Secondly, the models may contain modeling inaccuracies. Finally, to make the evaluation manageable, it is necessary to group by categories, which may be as misleading as it was in the RSS.

8.3.5 Aquatic Transport

Regulatory Guide 1.113 provides procedures for estimating the aquatic transport of accidental nd routine releases from nuclear reactors. Because these methods are complex and have not been included in PSAs, they are not discussed.

8.4 Summary

The analysis of the consequences of nuclear accidents began with physical concepts of core melt, discussed the mathematical and code models of radionuclide release and transport within the plant to its release into the environment, models for atmospheric transport and the calculation of health effects in humans. After the probabilities and consequences of the accidents have been determined, they must be assembled and the results studied and presented to convey the meanings.

8.5 Problems

1. Suppose an interstate highway passes 1 km perpendicular distance from a nuclear power plant control room air intake on which 10 trucks/day pass carrying 10 tons of chlorine each. Assume the probability of truck accident is constant at 1.0E-8/mi, but if an accident occurs, the full cargo is released and the chlorine flashes to a gas. Assume that the winds are isotropically distributed with mean values of 5 mph and Pasquill "F" stability class. What is the probability of exceeding Regulatory Guide 1-78 criteria for chlorine of 45 mg/m^3 (15 ppm).

2a. How long can you remain in a very large flat field that is contaminated with radioactivity at a density of 1 curie m^2 emitting gamma rays of 0.6 MeV mean energy before you exceed 10CFR100 limits for whole body? Assume the receptor is at 1 m.

2b. What whole body dose will you get from a 100 megacurie puff release passing at a perpendicular distance of 1 km traveling in a straight line at 3 mph?

3. Suppose that the RHR system in the example in Section 8.1.2 lost half of its heat removal capability after 90 s. When would it reach 2,300° F (a temperature criterion once suggested by Steve Hanauer regarding ATWS)?

Chapter 9
Chemical Process Accident Consequence Analysis

The previous chapter described the consequences of a nuclear reactor accident. Chemical process accidents are more varied and do not usually have the energy to melt thick pressure vessels and concrete basemats. The consequences of a chemical process accident that releases a toxic plume, like Bhopal did, are calculated similarly to calculating the dose from inhalation from a radioactive plume but usually calculating chemical process accidents differ from nuclear accidents for which explosions do not occur.

9.1 Hazardous Release

Chemical process materials, regardless of toxicity are benign as long as they are confined. They may release either by rupturing the confinement through designed exits such as release valves.

9.1.1 Pipe and Vessel Rupture

Hoop Stress

A fluid (gas or liquid) exerts uniform pressure on the walls of its confinement vessel. Should the pressure become excessive either by heating the fluid or by additional flow into the vessel, the pressure in the material used to construct will cause the metal to first elastically stretch. Further increase in pressure will exceed the elastic limit of the wall material and it will yield, i.e., take a permanent deformation. Further pressure will cause it to fail and release the process fluid. The yield stress is taken as a design criterion beyond which failure will occur. Table 3.1-1 provides some representative values.

We wish to calculate the relationship between the fluid pressure and the yield stress. Such a relationship can be looked up in handbooks but for simple geometries such as a

Table 9.1-1 Yield Strength of Some Metals

Material	Temp °F	Y.S.(ksi)
4340 (500° F temper) steel	70	217-238
4340 (800° F temper) steel	70	197-211
D6AC (1,000° F) temper	70	217
D6AC (1,000° F) temper	-65	238
A538 steel		250
2014-T6 aluminum	75	64
2024-T351	80	54-56
7075-T651 aluminum	70	75-81
7075-T7351 aluminum	70	56-88
Ti-6A1-4V titanium	74	119

pipe or spherical tank, the calculation is simple. Figure 9.1-1a shows the first quadrant of a section of pipe of length ℓ. Imagine that the pipe is cut and a strain gage is inserted in a longitudinal slit at the top. It measures a force, f, acting in a tangential direction. Dividing this by the area of the metal in the band gives the "hoop" stress so called because of the resemblance to the iron hoops used to encircle wooden barrels.

Since f is a tangential force, only the vector projection of the area is effective. Integrating these forces from 0 to $\pi/2$, f is given by equation 9.1-1 which is more easily done by a change of variables giving $f = \ell*p*r$, and the stress is $s = f/(t*\ell)$. Equation 9.1-2 gives the simple form of the hoop stress, where p is the pressure, r is the inside radius and t is the wall thickness.

Longitudinal Stress

Longitudinal stress is the force pulling on the ends of the pipe divided by the cross sectional area of the pipe wall. The pull is: $p*\pi*r^2$. The area of the wall is $\pi*(r+t)^2 - \pi*r^2 \approx 2*\pi*r*t$, and the longitudinal stress is given by equation 9.1-3. If $r>>t$, then $s_{long} = p*r/(2*t)$, and the longitudinal stress is half the hoop stress (equation 9.1-4). This is the reason why hoop stress dominates over the longitudinal stress.

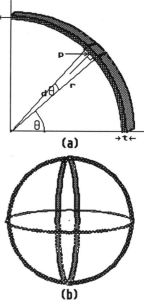

Fig. 9.1-1 Diagram for Hoop Stress

$$f = \int_0^{\pi/2} \ell*p*\sin\theta*r*d\theta = \int_0^1 \ell*p*r*d\cos\theta \quad (9.1\text{-}1)$$

$$S_{hoop} = p*r/t \quad (9.1\text{-}2)$$

$$S_{long} = pull/area = p*r^2/[t*(2*r+t)] \quad (9.1\text{-}3)$$

$$S_{long} = \tfrac{1}{2}*S_{hoop} \quad (9.1\text{-}4)$$

Numerical Example: Pipe Stress

In the Haber process hydrogen and nitrogen are compressed to high pressure say, 50 MPa which is 7,250 psig (pounds per sq. inch gage). If a pipe is 6 in. i.d. and must not be stressed above 200 ksi (thousand pounds/ sq. in.), how thick must the wall be? Answer: From equation 9.1-2, $t = p*r/s = 7250*3/200E3 = 0.10$ in.

Spherical Tank Rupture

A spherical shell is the best geometry for the storage of chemical fluids because it has the lowest surface to volume ratio. The larger the surface area, the more steel required and the greater surface area for corrosion. The wall thickness of a spherical vessel is calculated the same way that the hoops tress is calculated. Figure 9.1-1(b) shows a sphere from which a band of small width ℓ is sliced but still connected by a flexible joint to retain the fluid. A strain gage is inserted to measure the force being exerted. This is the same problem as before so the stress is $s = p*r/t$. If another slice were taken at right angles, the answer would be the same, because of symmetry, and hoop stress and longitudinal stress are equal in a spherical vessel.

Hazardous Release

Numerical Example: Spherical Tank Rupture
   Consider a 50 m radius tank of steel with yield strength 200 ksi. What must be the wall thickness to contain a fluid at 1,000 psi pressure? Answer: $t = p*r/s = 1000*50*3.28*12/2E5 = 9.8$ in.

9.1.2 Discharge through a Pipe

Newtonian Fluid
   Newton performed experiments with a viscous fluid sandwiched between plates (Figure 4.9-1). The force, f to slide one plate over the other is proportional to the contact areas, A, the spacing between the plates, t, the sliding velocity, v, and the viscosity η of the fluid are related as: $f = \eta*A*v/t$ or in terms of the shear pressure: $\tau = f/A$ as equation 9.1-5. A fluid that obeys equation 9.1-5 is called a "Newtonian fluid." The velocity the fluid is zero at the stationery plate, and equal the velocity v at the top plate This is shown as the straight line velocity profile in Figure 9.1-2.

$$\tau = f/A = \eta*v/t = \eta*dv/dy \qquad (9.1-5)$$

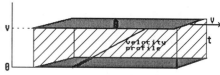

Fig. 9.1-2 Viscous Sliding

Flow in a Pipe
Figure 9.1-3 shows a cutaway of a tube of radius a, length ℓ, in which a fluid of viscosity η is flowing. The velocity goes to zero at the wall of the tube and reaches a maximum in the center in a parabolic shape. The flow is laminar (straight line, parallel to the axis), hence, an imaginary cylinder of radius r may be inserted as well as another at $r+dr$. The shear force between these cylinders may be calculated from Equation 9.1-5. For the inner cylinder, the viscous force is, $f = \eta*A*dv/dr$, where $A = 2*\pi*r*\ell$, giving equation 9.1-6, where the negative sign comes from the force being downward. Similarly, the shear force on the cylinder of radius $r+dr$ is equation 9.1-7, where the bracket contains the first two terms of a Taylor's expansion for the change in dv/dr resulting from going from r to $r+dr$. If the inlet pressure is p_1 and the outlet pressure is p_0, the net force on the cylinder is pressure times area (equation 9.1-8) which equals the sum of

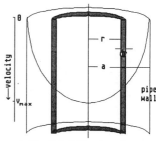

Fig. 9.1-3 Diagram for Viscous Flow in a Pipe

$$f(r) = -\eta*2*\pi*r*\ell*(dv/dr) \qquad (9.1-6)$$
$$f(r+dr) = \eta*2*\pi*(r+dr)*\ell*[(dv/dr)+(d^2v/dr^2)*dr] \qquad (9.1-7)$$
$$fnet = -(p_1-p_0)*2*\pi*r*dr = f(r+dr)+f(r) = $$
$$\eta*2*\pi*(r+dr)*\ell*[dv/dr+(d^2v/dr^2)*dr]-r*dv/dr$$
$$f_{net} = 2*\pi*\eta*\ell*[dv/dr+r*(d^2v/dr^2)*dr] \qquad (9.1-9)$$
$$r*(p_1-p_0) = \eta*\ell*(dv/dr+r*d^2v/dr^2) = \eta*\ell*d/dr(r*dv/dr) \qquad (9.1-10)$$
$$dv/dr = -(p_1-p_0)*r/(2*\eta*\ell)+C1/r. \qquad (9.1-11)$$
$$v = -(p_1-p_0)*r^2/(4*\eta*\ell)+C1*\log(r)+C2 \qquad (9.1-12)$$

335

forces (equation 9.1-6 plus 9.1-7). Simplifying, results in equation 9.1-9, where dr^2 is ignored since it is second order. Equating this to force being equal to the pressure differential times the area and canceling, results in equation 9.1-10. Integrating once gives equation 9.1-11; integrating again gives equation 9.1-12. Since $v=0$ at $r=0$, $C1=0$, similarly $v=0$ at $r=a$, then $C2 = (p_1-p_0)*a^2/(4*\eta*\ell)$ giving 9.1-13 which is the flow velocity.

The total flow rate through the pipe is found by integrating equation 9.1-13, over the cross sectional area to give Poiseuille's equation (9.1-14).

$$v = (p_1-p_0)*(a^2-r^2)/(4*\eta*\ell) \quad (9.1\text{-}13)$$

The total flow, Q, is inversely proportional to the length of the pipe and to the viscosity. It is proportional to the pressure differential and to the

$$Q = \int (p_1-p_0)*(a^2-r^2)/(4*\eta*\ell)*2*\pi*r*dr$$
$$= \pi*(p_1-p_0)*a^4/(8*\eta*\ell) \quad (9.1\text{-}14)$$

fourth power of the radius of the tube. Fluid flow, Q, is like electric current; pressure differential is analogous to voltage, hence, equation 9.1-14 is like Ohm's law for electricity.

Ohm's Law for Fluid Flow

Ohm's law, $V=I*R$ (voltage equals current times resistance), electricity has the same form as equation 9.1-14 which may be written as equation 9.1-15, where ΔP is the pressure differential, Q is the flow rate and resistance is given by equation 9.1-16, where η is the viscosity of the fluid. Table 9.1-2 shows that the viscosity of liquids is highly temperature-dependent. Gases are much less temperature dependent because of the greater separation between molecules. If there are multiple discharge paths the equivalent resistance is the same as electrical resistors in parallel (equation 9.1-17). If the discharge is through pipes connected together (series) of different sizes, equation 9.1-18 is used.

$$\Delta P = Q*R \quad (9.1\text{-}15)$$
$$R = 8*\eta*L/(\pi*a^4) \quad (9.1\text{-}16)$$

$$1/R_{parallel\ eq.} = \sum 1/R_i \quad (9.1\text{-}17)$$
$$R_{series\ eq.} = R_i \quad (9.1\text{-}18)$$

Table 9.1-2 Temperature Dependent Viscosity of Some Gases and Liquids ($N*s/m^2$)

Material\Temperature °C	0	20	40	60	80	100
water	1.79E-3	1.00E-3	.653E-3	.467E-3	.355E-3	.282E-3
ethyl alcohol	1.77E-3	1.20E-3	.834E-3	.592E-3		
carbon tetrachloride	1.33E-3	.969E-3	.739E-3	.585E-3	.468E-3	.384E-3
mercury	1.69E-3	1.55E-3	1.45E-3	1.37E-3	1.30E-3	1.24E-3
air	1.71E-9	1.83E-9	1.90E-9	2.00E-9	2.13E-9	2.24E-9
hydrogen	.84E-9	.875E-9	1.11E-9	1.14E-9	1.18E-9	1.21E-9

Numerical Example Discharge through a Pipe

At what rate will a process fluid with approximately the viscosity of water discharge through a 2 in. i.d. pipe 100 ft. long under a differential pressure of 100 psi?

First the pipe resistance must be calculated. To convert to British units, multiply the viscosity in N*sec/m² by 47.86 to get units of lb*sec/ft². Using dimensional analysis, it is found that R(lb*sec/ft^5) = 0.05321*η (N*s/m²)*L(ft)/a(ft)^4. From Table 4.9-1 the viscosity of water at 20°C is 1.0E-3 N*s/m². The radius of the pipe is 1 in = 0.0833 ft and it is 100 ft long. Thus, R = 0.05321*1.0E-3*100/(0.0833)^4 = 110. Since Q = P/R = 100*144/110 = 131 lb/sec = 943 gpm.

This shows that hazardous material can discharge at a high rate from a long pipe provided the pipe diameter is fairly large. The rate of discharge is proportional to the pressure differential, the fourth power of the pipe diameter, and the length of the pipe. Factors that affect the rate of discharge from a pipe are the end geometry if the pipe is broken or the orifice if it is through a relief valve.

$$r_{crit} = \left[\frac{p_{hi}}{p_{lo}}\right] = \left[\frac{\gamma+1}{2}\right]^{\gamma/(\gamma-1)} \quad (9.1\text{-}19)$$

9.1.3 Gas Discharge from a Hole in a Tank

Flow Regime

Gas may flow through an opening of arbitrary shape in one of two flow regimes: *sonic* (choked) flow for high pressure drop and *subsonic* for low pressure drops. The transition between them occurs at the critical pressure ratio, r_{crit}, which is related to the ratio of specific heat for the gas, γ (equation 9.1-19), where p_{hi} is the absolute high pressure on one side of the opening, p_{lo} is the pressure on the other side. All units are SI, e.g., pressure is in N/m² (1 bar = 1.013E5 N/m²). γ is given in Table 9.1-3 for some gases, it ranges from 1.1 to 1.67, hence, r_{crit} ranges from 1.71 to 2.05. Thus, most situations lead to sonic flow.

Table 9.1-3 Gamma for Several Gases

Gas	γ
Carbon dioxide (CO_2)	1.34
Carbon monoxide (CO)	1.43
Hydrogen (H_2)	1.43
Noble gases (He, A,...)	1.667
Nitrous oxide (N_2O)	1.22
Oxygen (O_2)	1.51
Water (H_2O)	1.22

Gas Flow Rate through an Orifice

The rate of gas flow through an opening is given by equation 9.1-20, where Q is the flow rate (kg/s), A is the area, c is the velocity of sound in the gas before expansion (equation 9.1-21) where R is the gas constant (8,210 J/kg*mol*°K), and M is molecular weight (kg/mol). ψ is the dimensionless flow factor which is given by equation 9.1-21a for subsonic flow and equation 9.1-21b for sonic flow.

C_d, the dimensionless discharge coefficient, is a complicated function of the

$$Q = C_d * A * p_{hi} * \psi/c \quad (9.1\text{-}20)$$
$$c = (\gamma * R * T/M)^{1/2} \quad (9.1\text{-}21)$$
$$\psi = \{2 * \gamma^2/(\gamma-1) * (p_{hi}/p_{lo})^{2/\gamma} * [1 - (p_{hi}/p_{lo})^{\gamma-1/\gamma}]\}^{1/2} \quad (9.1\text{-}22a)$$
$$\psi = \gamma * (1/r_{crit})^{(\gamma+1)/2*\gamma} \quad (9.1\text{-}22b)$$

Reynolds number for the fluid being released. For sharp edged orifices with Reynolds numbers approaching 30,000, it approaches 0.61. For these conditions, the exit velocity of the fluid is independent of the size of the hole. For a rounded opening, C_d approaches 1. For a short section of pipe (length/diameter > 3), $C_d \approx 0.81$. For cases where C_d is unknown or uncertain use $C_d = 1$ to conservatively over-estimate the flow.

Numerical Example: Hydrogen Release

A container of hydrogen at 2 bar at 27° F is damaged causing a 1 cm² sharp-edged opening. What is the rate of hydrogen discharge into the atmosphere?

γ for hydrogen is 1.43; for which equation 9.1-19 gives $r_{crit} = 1.9$. The velocity of sound is 1,888m/s, for sonic flow $\psi = 0.828$, and $C_d = 1$. Using these numbers in equation 9.1-20, gives Q = 4.4E-3 kg/sec.

9.1.4 Liquid Discharge from a Hole in a Tank

Liquid discharge from the work of Bernoulli and Torricelli is given by equation 9.1-23, where ρ is the liquid density (kg/m³), g is the acceleration of gravity (9.81 m/s²), and h is the liquid height above the hole (CCSP, 1989). The discharge coefficient for fully turbulent flow from sharp orifices is 0.61 -0.64.

$$Q = C_d * A * \rho * \sqrt{[2*(p_{hi}-p_{lo})/\rho + 2*g*h]} \quad (9.1\text{-}23)$$

Numerical Example: Break in a Pressurized Tank holding Benzene

A tank of benzene (specific gravity 0.88) pressurized to 2 bar (1 bar above atmospheric) is damaged causing a 10 cm² sharp opening located 2 m below the liquid level. What is the discharge rate?

Q = 0.64*(1/1000)*.88*1000*√(2* 1.013E5/880 + 2*9.81*2) = 9.24 kg/s. It may be noted that the gravity head only contributed about 10% of the effect of the internal pressure.

9.1.5 Unconfined Vapor Cloud Explosions (UVCE)[a]

Explosions

Chemical explosions are uniform or propagating explosions. An explosion in a vessel tends to be a uniform explosion, while an explosion in a long pipe is a propagating explosion. Explosions are deflagrations or detonations. In a deflagration, the burn is relatively slow, for hydrocarbon air mixtures the deflagration velocity is of the order of 1 m/s. In contrast, a detonation flame shock front is followed closely by a combustion wave that releases energy to sustain the shock wave. A

---

[a] The material in this section is adapted from Lees (1986) and CCPS (1989).

steady state detonation shock front travels at the velocity of sound in the hot products of combustion. For hydrocarbon-air mixtures the detonation velocity is typically of the order of 2,000 to 3,000 m/s[b].

A detonation generates greater pressures and is more destructive than a deflagration. While the peak pressure caused by the deflagration of a hydrocarbon-air mixture in a closed vessel at atmospheric pressure is of the order of 8 bar, a detonation may have a peak pressure of the order of 20 bar. A deflagration may turn into a detonation if confined or obstructions to its passage.

A basic distinction is made between confined unconfined explosions. Confined explosions occur within some sort of containment such as a vessel pipework, or a building. Explosions in the open air are unconfined explosions.

If a large amount of a volatile flammable material is rapidly dispersed to the atmosphere, a vapor cloud forms. If this cloud is ignited before the cloud is diluted below its lower flammability limit, a UVCE occurs which can damage by overpressure or by thermal radiation.[c,d] Rarely are UVCEs detonations; it is believed that obstacles, turbulence, and possibly a critical cloud size are needed to transition from deflagration to detonation.

Even if a flammable vapor cloud is formed, ignition is necessary for a UVCE. However, there are numerous ignition sources such as fired heaters, pilot flames smoking, vehicles, electrical equipment, etc. A site associated with many ignition sources tends to prevent a cloud from reaching its full hazard extent. Conversely, at such a site, there is less likelihood for a cloud to disperse without ignition. A PSA should take account of the ignition probability. Early ignition, would result in a flash fire or an explosion of smaller size; late ignition could result in the maximum possible effect being experienced.

$$W = \frac{\eta * m_c * E_c}{E_{TNT}} \quad (9.1\text{-}24)$$

Model of UVCE

UVCEs are modeled by equivalence of the flammable material to TNT by correlations with observed UVCEs (TNO model), or by computer modeling. Only the simple TNT model is discussed here.

Equation 9.1-24 presents this model, where W is the equivalent weight of TNT, m_c is the weight of material in the cloud, η is the empirical explosion yield (from 0.01 to 0.1) E_c is the heat of combustion of the cloud and E_{TNT} is the heat of combustion of TNT (Table 9.1-4 for other units, 1 kJ/kg = 2.321 BTU/lb = 4.186 calories/gm).

Table 9.1-4 Some Heats of Combustion (from Baumeister, 1967)

Name	kJ/kg	Name	kJ/kg
acetylene	5.0E4	hydrogen	1.4E4
butane	5.0E4	methane	5.5E4
ethane	5.2E4	methanol	2.3E4
ethanol	3.0E4	octane	4.7E4
ethylene	5.0E4	propane	5.0E4
hexane	4.8E4	TNT	4.7E3

---

[b] The velocity of sound in air at 0° C is 330 m/s.

[c] UVCEs were used in Vietnam, rather unsuccessfully, to send an overpressure into the underground tunnels.

[d] The Flixborough accident (Section 7.1.2.3) is an often-cited example of a UVCE.

Explosion Efficiency

If the explosion occurs in an unconfined vapor cloud, the energy in the blast wave is only a small fraction of the energy calculated as the product of the cloud mass and the heat of combustion of the cloud material. On this basis, explosion efficiencies are typically in the range of 1-10%.

In some cases, however, only the part of the cloud which is within the flammable range is considered to burn. This may be a factor of 10 less than the total cloud. For further discussion of explosion efficiency see CCPS (1989) or Leeds (1986).

Physical Parameters from the Explosion

Explosion effects have been studied extensively. Figure 9.1-4 from Army (1969) uses 1/3 power scaling of the TNT equivalent yield obtained from equation 9.1-24. This graph is for a hemispherical blast such as would occur if the cloud were lying on the ground. To find the blast effects at a certain distance, calculate the scaled ground distance from the blast using: $Z_g = R_g/W^{1/3}$, where R_g is the actual ground distance, Z_g is the distance for scaling, and W is the equivalent TNT weight in pounds. Using a vertical line from this location on the abscissa, from the point of interception on a curve of interest go horizontally left or right according to the type of information and read the value.

Calculating the Peak Positive Pressure from a Propane Explosion

If 10 tons of propane exploded with an explosive efficiency 0.05, 1,000 ft from you, what would be the peak positive overpressure? Referring to equation 9.1-24, W = 0.05*2E4*5E4/4.7E3 = 1E4 lb, where the heats of combustion are from Table 9.1-4, and 10 tons is 2E4 lb. The scaled range is $Z_g = 1000/1E4^{1/3}$ = 45.6 from which a positive overpressure of 2

Table 9.1-5 Damage Produced by Blast Overpressure

psi	Damage
0.02	Annoying noise (137 dB if of low frequency 10-15 Hz)
0.03	Occasional breaking of large glass windows already under strain
0.04	Loud noise (143 dB), sonic boom glass failure
0.1	Breakage of small windows under strain
0.15	Typical pressure for glass breakage
0.3	"Safe distance" (probability 0.95 no serious damage beyond this value); projectile limit; some damage to house ceilings; 10% window glass broken
0.4	Limited minor structural damage
0.5-1	Large and small windows usually shattered; occasional damage to window frames
0.7	Minor damage to house structures
1	Partial demolition of houses, made uninhabitable
1-2	Corrugated asbestos shattered; corrugated steel or aluminum panels, fastenings fail, followed by buckling; wood panels (standard housing) fastenings fail, panels blown in
1.3	Steel frame of clad building slightly distorted
2	Partial collapse of walls and roofs of houses
2-3	Concrete or cinder block walls, not reinforced, shattered
2.3	Lower limit of serious structural damage
2.5	50% destruction of brickwork of houses
3	Heavy machines (3,000 lb) in industrial building suffered little damage; steel frame building distorted and pulled away from foundations
3-4	Frameless, self-framing steel panel building demolished; rupture of oil storage tanks
4	Cladding of light industrial buildings ruptured
5	Wooden utility poles snapped; tall hydraulic press (40,000 lb) in building slightly damaged
5-7	Nearly complete destruction of houses
7	Loaded train wagons overturned
7-8	Brick panels, 12 in. thick, not reinforced, fail by shearing or flexure
9	Loaded train boxcars completely demolished
10	Probable total destruction of buildings; heavy machines tools (7000 lb) moved and badly damaged, very heavy machine tools (12,000 lb) survived

Hazardous Release

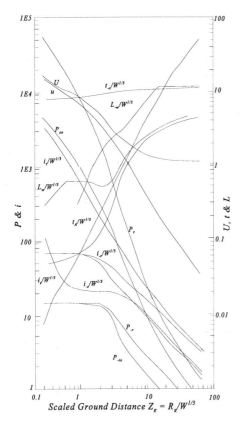

Fig. 9.1-4 Shock Wave Parameters for a Hemispherical TNT Surface Explosion at Sea level. Left ordinate: P_{so} - peak positive incident pressure, P_{-so} - peak negative incident pressure, P_r - peak positive normal reflected pressure, P_{-r} - peak negative normal reflected pressure, $i_s/W^{1/3}$ positive incident impulse, $i_{-s}/W^{1/3}$ - negative incident impulse, $i_r/W^{1/3}$ - positive normal reflected impulse, $i_{-r}/W^{1/3}$ - negative normal reflected impulse, $i_A/W^{1/3}$ - time of arrival, $i_d/W^{1/3}$ - positive duration of positive phase, $i_{-d}/W^{1/3}$ - negative duration of positive phase, $L_w/W^{1/3}$ - wavelength of positive phase, $L_{-w}/W^{1/3}$ - wavelength of negative phase. Pressure is in psi, scaled impulse is in psi-ms/lb$^{1/3}$, arrival time in ms/lb$^{1/3}$, wavelength is in ft/lb$^{1/3}$, R_g is the scaled ground distance from the charge. W is the charge weight in lb.

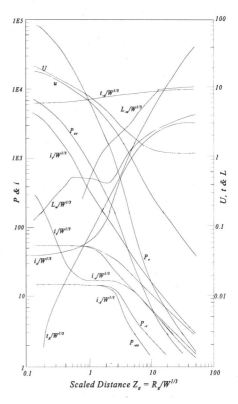

Fig. 9.1-5 Shock Wave Parameters for a Spherical TNT Surface Explosion at Sea level. Left ordinate: P_{so} - peak positive incident pressure, P_{-so} - peak negative incident pressure, P_r - peak positive normal reflected pressure, P_{-r} - peak negative normal reflected pressure, $i_s/W^{1/3}$ positive incident impulse, $i_{-s}/W^{1/3}$ - negative incident impulse, $i_r/W^{1/3}$ - positive normal reflected impulse, $i_{-r}/W^{1/3}$ - negative normal reflected impulse, $i_A/W^{1/3}$ - time of arrival, $i_d/W^{1/3}$ - positive duration of positive phase, $i_{-d}/W^{1/3}$ - negative duration of positive phase, $L_w/W^{1/3}$ - wavelength of positive phase, $L_{-w}/W^{1/3}$ - wavelength of negative phase. Pressure is in psi, scaled impulse is in psi-ms/lb$^{1/3}$, arrival time in ms/lb$^{1/3}$, wavelength is in ft/lb$^{1/3}$, R_g is the scaled distance from the charge. W is the charge weight in lb.

psi. is read. Table 9.1-5 indicates that this will cause partial collapse of walls and roofs of houses, concrete or cinder block walls, not reinforced, shattered, lower limit of serious structural damage.

9.1.6 Vessel Rupture (Physical Explosion)[e]

If a vessel containing a pressurized gas or vapor ruptures, the release of the stored energy of compression produces a shock wave and flying vessel fragments (CCSP, 1989). Previously we modeled explosive reaction of the gas with air if such energy also is involved, the effects will be more severe. Such vessel ruptures may occur from: 1) over-pressurization of the vessel which involves failure of the pressure regulators, relief valves and operator error, and 2) weakness of the pressure vessel from corrosion, stress corrosion cracking, erosion, chemical attack, overheating material defects, cyclic fatigue or mechanical damage. The energy that is released is in the form of: 1) kinetic energy of fragments, 2) energy in shock wave, 3) heat (e.g., the surrounding air), and 4) the energy of deformation. Estimates of how the energy is distributed are 40% goes into shock wave energy with the remainder in fragment kinetic energy if the vessel brittle fractures. In general the explosion is not isotropic but depends on the tank geometry and crack propagation. If the tank is filled with gas, the energy comes from the compression of the gas into the full volume; if the tank contains a liquid with vapor space, it is the energy of compression in the vapor space; if the tank contains a superheated liquid, the energy comes from flash evaporation.

Empirical Model

It is convenient to calculate a TNT equivalent of a physical explosion to use the military results of Figures 9.1-4 and 5. Baker et al. (1983) give a recipe for the rupture of a gas filled container assuming expansion occurs isothermally and the perfect gas laws apply (equation 9.1-25), where W is

$$W = 1.4E-6 * V * (P_1/P_0) * (T_0/T_1) * R * T_1 * \ln(P_1/P_2) \quad (9.1\text{-}25)$$

the energy in TNT lb equivalence, V is the volume of the compressed gas (ft^3), P is the initial pressure of the compressed gas (psia), P_2 is the pressure of expanded gas (psia), P_0 is sea level pressure (14.7 psia), T_1 is the temperature of compressed gas (°R), T_0 is standard temperature (492° R), R is the gas constant (1.987 Btu /lb-mol-°R, and 1.4E-6 is a conversion factor assuming 2,000 Btu per l lb of TNT. This calculates TNT equivalence to estimate shock wave effects, but it is not valid in the near field.

The blast pressure, P_s, at the surface of an exploding pressure vessel is estimated by Prugh (1988) with equation 9.1-26 where P_s is the pressure at surface of vessel (bara), Pb is the burst

$$P_b = P_s * \{1 - [3.5*(\gamma-1)]*(P_s-1)/[(\gamma*T/M)*(1+5.9*P_s)]^{0.5}\}^{-2\gamma/(\gamma-1)} \quad (9.1\text{-}26)$$

---

[e] Adapted from CCSP, 1989.

pressure of vessel (bara), γ is the ratio of specific heats, (C_p/C_v), T is the absolute temperature, (°K), and M is the molecular weight of gas, (lb/lb mole). This assumes that expansion occurs into air at STP. Notice equation 9.1-25 does not directly give what we want, the pressure at the surface of the bursting vessel. Instead of an analytic inversion, Program 9-1 iterates to find P_s given P_b. After finding P_s, and the scaled distance, Z, the explosion parameters are found from Figure 9.1-4.

Program 9-1 Calculating the Surface Pressure

```
1  CLS : INPUT "Burst pressure (bara)", pb
2  gamma = 1.4: temp = 300
3  INPUT "Molecular weight of gas (lb/lb-mole)
4  ", mole
5  DO
6  ps = ps +0.1
7  pbc = ps * (1 - (3.5 * (gamma - 1) * (ps - 1))
8  / ((gamma * temp / mole) * (1 + 5.9 * ps)) ^
9  .5) ^ (-2 * gamma / (gamma - 1))
10 LOOP UNTIL pbc > pb
   PRINT "Surface pressure (bara) "; ps
```

Line 1 clears the screen and requests the input of the burst pressure of the vessel. Line 2 sets gamma to 1.4 and the absolute temperature to 300. If your pressure or temperature is different, edit the program Line 3 requests the molecular weight of the gas in the vessel. Lines 5-10 loop to perform the iteration. Line 6 iterates the surface pressure in steps of 0.1 bara. Lines 7-9 are equation 9.1-25. Line 10 stops the iteration when pbc>pb, and line 11 prints ps.

Example: Pressures from a Bursting Tank

A 6 ft³ compressed air spherical tank at 15° C bursts at 546 bara. What is the side-on over pressure at 25 and 50 ft? From equation W = 1.4E-6*V*(P_1/P_0)*(T_0/T_1)*R*T_1*ln(P_1/P_2), = 1.4E-6*6*(7917/14.7)*(492/518)*1.987*518*ln(7917/14.7) = 27.8 lb TNT. By running program 9-1, using Pb = 546 bara and M = 28.9 we find 10.3 bara = 149.3 psia. From Figure 9.1-4, we read Z_g = 3.1 for this pressure. Using the TNT equivalence, we find R_g 9.4 ft. For V = 6 = 4/3*π*r³, r = 1.13 ft. R_g corrected to the center of the sphere is 9.4-1.1 = 8.3 ft. The blast pressure at 25+8.3 = 33.3 ft. has a Z_g = 33.3/27.8$^{1/3}$ = 11.0 at which Figure 9.4-1 gives P_r = 8 psig. At 50 ft., Z_g = 58.3/27.8$^{1/3}$ = 19.2 and the figure shows P_r = 3.1 psig.

9.1.7 BLEVE, Fireball and Explosion[f]

A boiling liquid expanding vapor explosion (BLEVE) may result from a fire that impinges on the vapor space of a tank of flammable liquid to weaken the tank material and build vapor pressure to such an extent that the tank ruptures and the superheated liquid flashes to vapor and explodes generating a pressure wave and fragments (CCSP, 1989). BLEVEs while rare, are best known for involving liquefied petroleum gas (LPG). The vessel usually fails within 10 to 20 minutes from the time of impingement. Incidents occurred at: San Carlos, Spain (July 11, 1978), Crescent City, Illinois (June 21, 1970), and Mexico City, Mexico (November 19, 1984). The fireball modeling is separate from the projectile modeling.

---

[f] Adapted from CCSP, 1989.

Chemical Process Accident Consequence Analysis

BLEVE Fireball Physical Parameters

Useful formulas for BLEVE fireballs (CCSP, 1989) are given by equations 9.1-27 thru 9.1-30, where M = initial mass of flammable liquid (kg). The initial diameter describes the short duration initial ground level hemispherical flaming-volume before buoyancy lifts it to an equilibrium height.

$$Diameter(m) \; D_{max} = 6.48 M^{0.325} \quad (9.1\text{-}27)$$
$$Duration(s) \; t = 0.825 M^{0.26} \quad (9.1\text{-}28)$$
$$Height(m) \; H = 0.75 D_{max} \quad (9.1\text{-}29)$$
$$Hemi.dia.(m) \; D_{initial} = 1.3 * D_{max} \quad (9.1\text{-}30)$$

Thermal Radiation from the Fireball

The thermal radiation received from the fireball on a target is given by equation 9.1-31, where Q is the radiation received by a black body target (kW/m²) τ is the atmospheric transmissivity (dimensionless), E = surface emitted flux in kW/m², and F is a dimensionless view factor.

$$Q = \tau * E * F \quad (9.1\text{-}31)$$
$$\tau = 2.02 * (p_w * x)^{-0.09} \quad (9.1\text{-}32)$$

The atmospheric transmissivity, τ, greatly affects the radiation transmission by absorption and scattering by the separating atmosphere. Absorption may be as high as 20-40%. Pietersen and Huerta (1985) give a correlation that accounts for humidity (equation 9.1-31), where τ = atmospheric transmissivity, p_w = water partial pressure (Pascals), x = distance from flame surface to target (m).

$$E = \frac{f_{rad} * M * h_c}{\pi * D_{max}^2 * t} \quad (9.1\text{-}33)$$

$$F = \frac{D^2}{4 * x^2} \quad (9.1\text{-}34)$$

The heat flux, E, from BLEVEs is in the range 200 to 350 kW/m² is much higher than in pool fires because the flame is not smoky. Roberts (1981) and Hymes (1983) estimate the surface heat flux as the radiative fraction of the total heat of combustion according to equation 9.1-32, where E is the surface emitted flux (kW/m²), M is the mass of LPG in the BLEVE (kg) h_c is the heat of combustion (kJ/kg), D_{max} is the maximum fireball diameter (m) f_{rad} is the radiation fraction, (typically 0.25-0.4). t is the fireball duration (s). The view factor is approximated by equation 9.1-34, where D is the fireball diameter (m), and x is the distance from the sphere center to the target (m). At this point the radiation flux may be calculated (equation 9.1-30).

Problem: Size Duration and Flux from a BLEVE (CCSP, 1989)

Calculate the size duration, and flux at 200 m from 100,000 kg (200 m³) tank of propane at 20° C, 8.2 bara (68° F, 120 psia). The atmospheric humidity has a water partial pressure of 2,710 N/m² (0.4 psi). BLEVE parameters are calculated from equations 9.1-26 to 9.1-29: D_{max} = 6.48*1E5$^{0.325}$ = 273 m. H = 0.75*273 = 204 m. $D_{initial}$ = 1.3*273 = 354 m.

Using a radiation fraction f_{rad} = 0.25, the view factor from equation 9.1-33 is: F = 273²/(4*200)² = 0.47. The path length for the transmissivity is the hypotenuse of the elevation and the ground distance minus the BLEVE radius: $\sqrt{(204^2 + 200^2)}$ - 273/2 = 150 m. The transmissivity from equation 9.1-31 is: τ = 2.02*(2820*150)$^{-0.09}$ = 0.63. The surface emitted flux from equation 9.1-32 is: 0.25*1E5*46,350/(π*273²*16.5) = 300 kW/m², and the received flux from equation 9.1-30 is 0.63*300*0.47 = 89 kW/m².

Hazardous Release

9.1.8 Missiles[g]

Number of Missiles

The explosion from an explosion generates heat, pressure and missiles. Studies have been performed by: Baker (1983), Holden and Reeves (1985) and the Association of American Railroads, AAR (1972) AAR (1973). AAR reports out of 113 major failures of tank cars in fires, about 80% projected fragments. The projections are not isotropic but tend to concentrate along the axis of the tank. 80% of the fragments fall within a 300-m range (Figure 9.1-6). Interestingly, BLEVEs from smaller LPG vessels have greater range.

The total number of fragments is a function of vessel size and perhaps other parameters. Holden (1985) gives a correlation based on 7 incidents as equation 9.1-35, where F is the number of fragments and V is the vessel volume in m^3 for the range 700 to 2,500 m^3.

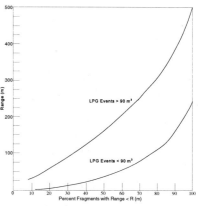

Fig. 9.1-6 Range of LPG Missiles (Holden, 1985)

$$F = -3.77 + 0.0096 * V \quad (9.1\text{-}35)$$

Example: Explosion of a Propane Tank (CCSP, 1989)

Fire engulfs a 50 ft-dia. propane tank filled to 60% capacity. The tank fails at 350 psig. Estimate the energy release, number of fragments, and maximum range of the missiles. The volume of the tank is: $V = \pi*D^3/6 = 6.5E4\ ft^3$. The volume of the liquid is $6.5E4 * 0.6 = 3.9E4\ ft^3$ and vapor occupies $6.5E4 * 0.4 = 2.6E4\ ft^3$ which is considered to cause the explosion. Equation 9.1-25 estimates W = 2.8E3 lb TNT (using T1 = 614). The number of fragments may be estimated from equation 9.1-35 using the vapor space (2.6E4/35.28 = 737 m^3) to be F = 4 (rounding up).

The surface area of the spherical shell is $A = \pi*D^2 = 7854\ ft^2$, and the volume is 490.8 ft^3 (for a thickness t = 0.75 in.). The weight of the shell is $490.8*489\ lb/ft^3 = 2.4E5$ lb so the average fragment weight is 6E4 lb. The area of a fragment is 7854/4 = 1,964 ft^2 which, if circular, the diameter is 50 ft. = 600 in.

The velocity of a fragment is estimated from Moore (1967) equation 9.1-36, where u is the initial velocity in ft/s, P is the rupture pressure psig, D is the fragment diameter in inches, and W is the weight of fragments in lb.

$$u = 2.05 * (P*D^3/W)^{1/2} \quad (9.1\text{-}36)$$

Substituting into equation 9.1-36 gives the fragment initial speed as 2,300 ft/s.

---

[g] Adapted from CCSP, 1989.

Chemical Process Accident Consequence Analysis

The next step is to estimate the lift/drag ratio for a fragment: C_L*A_L/C_D*A_D. For a "chunky" fragment $C_L*A_L/C_D*A_D = 0$. The drag coefficient, C_D, ranges from 0.47 to 2.05 for a long rectangular fragment perpendicular to the air flow. Take $C_D = 0.5$ for this case and use Figure 9.1-7 to estimate the range. The axes are scaled by $B = \rho_o*C_d*A_d/M$ where ρ_o is the density of air (0.0804 lbm/ft^3), C_d is the drag coefficient, A_d is the area of a fragment, ft^2, M is mass in lbm. The ordinate is $B*R$, where R is maximum fragment range, ft. The abscissa is $B*u^2/g$ where u is fragment velocity ft/s, and g is the acceleration of gravity, 32.2 ft/sec^2.

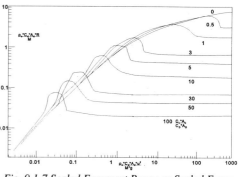

Fig. 9.1-7 Scaled Fragment Range vs Scaled Force (from Baker et al., 1983)

For this case, B = 0.0804*0.5*1964/6E4 = 1.32E-3. The abscissa is B*5.29E6/32.2 = 217. The corresponding ordinate for the "0" curve is 5.5 = B*R giving R = 4166 ft.

9.2 Chemical Accident Consequence Codes

The chemical and physical phenomena involved in chemical process accidents is very complex. The preceding provides the elements of some of the simpler analytic methods, but a PSA analyst should only have to know general principles and use the work of experts contained in computer codes. There are four types of phenomenology of concern: 1) release of dispersible toxic material, 2) dispersion of the material, 3) fires, and 4) explosions. A general reference to such codes is not in the open literature, although some codes are mentioned in CCPS (1989) they are not generally available to the public.

However, around 1995, U.S. DOE through its Defense Programs (DP), Office of Engineering and Operations-Support, established the Accident Phenomenology and Consequence (APAC) Methodology Evaluation Program to identify and evaluate methodologies and computer codes to support accident phenomenological and consequence calculations for both radiological and nonradiological materials at DOE facilities and to identify development needs. APAC is charged to identify and assess the adequacy of models or computer codes to support calculations for accident phenomenology and consequences, both inside and outside the facility, associated with chemical and/or radiological spills, explosions, and fires; specify dose or toxic reference levels for determining off-site and on-site radiological/toxicological exposures and health effects.

Six working groups were formed of these, the reports of four are used: 1) the Spills Working Group, Brereton 1997; 2) the Chemical Dispersion and Consequence Assessment Working Group, Lazaro, 1997; the Fire Working Group, Restrepo, 1996; and 4) the Explosions and Energetic Events Working Group, Melhem, 1997. In addition to this, material that I found on the Internet is presented as well as material that was requested via the Internet.

9.2.1 Source Term and Dispersion Codes

Some codes just calculate source terms (release quantities) and others just calculate the dispersion after release, and some codes do both. Each heading in the following identifies the code's capability.

9.2.1.1 Adam 2.1 (Release and Dispersion)

The Air Force Dispersion Assessment Model (ADAM -1980s) calculates the source term and dispersion of accidental releases of eight specific chemicals: chlorine, fluorine, nitrogen tetroxide, hydrogen sulfide, hydrogen fluoride, sulfur dioxide, phosgene, and ammonia. It treats a wide spectrum of source emission conditions (e.g., pressurized tank ruptures, liquid spills, liquid/vapor /two-phase) and a variety of dispersion conditions (e.g., ground-based jets or ground-level area sources, relative cloud densities, chemical reactions and phase changes, and dense gas slumping) including reactions of the chemical with water vapor. It has been validated using a number of field experiment data.

The pool evaporation model is Ille and Springer (1978) for non-cryogenic spill and Raj (1981) for cryogenic spills. Liquid release from a tank uses Bernoulli's equation but does not account for pressure variation or level drop. Gas release from a tank uses 1-d compressible flow equations for the release rate and temperature, pressure variation or level drop is not treated. Adiabatic flashing and rainout are not included. The turbulent jet model is Raj and Morris (1987). ADAM would require source code modification to include any but its 8 chemicals. . It assumes that the jet is ground-based and horizontally-directed; the vertical component of the jet trajectory in the transport and dispersion calculations is not included; it does not treat rainout and assumes all droplets remain airborne until evaporated.

Output: mass rate and vapor temperature of release, mass rate of air entrained, density of mixture, aerosol mass fraction in cloud. *Limitations:* single chemical source terms, limited chemical database, only steady-state release, based on spill data of rocket propellants. *Sponsor/Developer*: AL/EQS, Tyndall AFB, FL. *Developer*: TMS, Burlington, MA. *Custodian:* Captain Michael Jones, AL/EQS, 139 Barnes Drive, Tyndall AFB, FL 32403-5319, Phone: (904) 283-6002 *Developer*: Dr. Phani Raj, TMS Inc., 99 South Bedford St., Suite 211, Burlington, MA 01803-5128 Phone: (617) 272-3033. *Cost*: None. The source and executable versions of the code may be downloaded from EPA's web site: http://www.epa.gov/scram001/t22.htm. Look in "Non-EPA Models (277kB ZIP). *Computer*: IBM compatible, math co-processor recommended, 512 kB RAM, runs in DOS. menu driven.

9.2.1.2 AFTOX (Version 3.1 Dispersion)

AFTOX is a Gaussian dispersion model that is used by the Air Force to calculate toxic corridors in case of accidental releases. Limited to non-dense gases, it calculates the evaporation rate from liquid spills. It treats instantaneous or continuous releases from any elevation, calculates buoyant plume and is consistent with ADAM for passive (neutrally-buoyant) gas releases.

Dispersion calculations use standard Gaussian formulas. Stability class is treated as a continuous variable with averaging time.

Output: Maps of toxic corridors, estimates of concentration at specific positions, and estimates of the magnitude and location of the maximum concentration occurring a certain time after release. *Sponsor/Developing Organization:* Phillips Laboratory Directorate of Geophysics, Air Force Systems Command, Hanscom AFB, MA 01731-5000. *Custodian*: Steven Sambol, 30 WS/WES, Vandenberg AFB, CA 93437-5000, Phone: (805) 734-8232. *Cost*: None. May be downloaded from EPA's website: http://www.epa.gov/scram001/t22.htm. Look in "Non-EPA Models" for files AFTOX Toxic Chemical Dispersion Model (173 kB ZIP), and AFTOXDOC - AFTOX Model User's Guide, ASCII (26 kB ZIP)

9.2.1.3 Areal Locations of Hazardous Atmospheres (ALOHA) 5.2 - Release and Dispersion)

ALOHA is an emergency response model for rapid deployment by responders, and for emergency preplanning. It computes time-dependent source strength for evaporating liquid pools (boiling or non-boiling), pressurized or nonpressurized gas, liquid or two-phase release from a storage vessel, and pressurized gas from a pipeline. It calculates puff and continuous releases for dense and neutral density source terms. Model output is provided in both text and graphic form and includes a "footprint" plot of the area downwind of a release, where concentrations may exceed a user-set threshold level. It accepts weather data transmitted from portable monitoring stations and plots footprints on electronic maps displayed in a companion mapping application, MARPLOT. The modified time-dependent Gaussian equation is based on *Handbook on Atmospheric Diffusion* and the heavy gas dispersion is based on *Development of an Atmospheric Dispersion Model for Heavier-Than Air Gas Mixtures*. The dense gas submodel of ALOHA is a simplified version of the DEnse GAs DISpersion (DEGADIS) model.

It calculates pool evaporation using conservation of mass and energy; an analytical solution of the steady-state advection-diffusion equation calculates the evaporation mass transfer from the pool (Brighton, 1985). The liquid release from a tank is calculated with the Bernoulli equation. Leaked material is treated with the pool evaporation model, puddle growth is modeled (Briscoe and Shaw, 1980). It treats heat conduction from the tank and evaporative heat transfer into the tank vapor. Gas release rate and temperature from a tank is calculated using one-dimensional compressible flow equations. A virial equation of state handles high pressure non-ideal gas behavior. (Belore and Buist, 1986; Shapiro, 1953; Blevins; 1985, and Hanna and Strimaitis, 1989). Two-phase tank release assumes quick evaporation, thus, no liquid pool is formed. Releases through a hole are modeled with Bernoulli's equation modified for the head and pressure of the liquid gas mixture. Release though a short pipe is modeled with a homogenous non-equilibrium model while release through a long pipe uses a homogenous equilibrium model (Fauske and Epstein, 1988; Henry and Fauske, 1971; Leung, 1986; Huff, 1985; and Fisher, 1989).

ALOHA has a comprehensive chemical source term library (>700 pure chemicals). The code can address many types of pipe and tank releases, including two-phased flows from pressurized/ cryogenic chemicals. The user may enter a constant or variable vapor source rate and duration of release. Vertical wind shear is included in the pool evaporation model as well as the Gaussian (passive) and heavy gas dispersion models. ALOHA only models the source-term and dispersion of single-component, nonreactive chemicals. It does not account for terrain steering or changes in wind speed and horizontal direction. It does not model aerosol dispersion, nor does it account for initial positive buoyancy of a gas escaping from a heated source. Its straight-line Gaussian model prevents addressing scenarios in complex terrain.

Sponsors: National Oceanic and Atmospheric Administration, Hazardous Materials Response and Assessment Division, 7600 Sand Point Way N.E. Seattle, WA 98115; United States Environmental Protection Agency Chemical Emergency Preparedness and Prevention Office, Washington, D.C. 20460, National Safety Council. Custodians: Dr. Jerry Galt, NOAA/HAZMAT, 7600 Sand Point Way N.E., Seattle, WA 98115, Phone: (206) 526-6323, E-mail: jerry_galt@hazmat.noaa.gov. Mark Miller Phone: (206) 526-6945, E-mail: mark_miller@hazmat.noaa.gov. *Cost*: ~$375. *Computer:* IBM compatible, 386 with math coprocessor desired, Windows 3.0 or higher, 1 MB RAM, 2.5 MB hard drive, or Macintosh, 1 MB RAM and 2 MB hard drive runs under Finder or Multifinder.

9.2.1.4 CALPUFF (Dispersion)

CALPUFF is a multilayer, multispecies non-steady-state puff dispersion model that simulates the effects of time- and space-varying meteorological conditions on chemical transport, transformation, and removal. It uses 3-dimensional meteorological fields computed by the CALMET preprocessor or simple, single-station winds in a format consistent with the meteorological files used to drive the ISC3 or AUSPLUME steady-state Gaussian models. CALPUFF contains algorithms for near-source effects such as building downwash, transitional plume rise, partial plume penetration, sub-grid-scale complex terrain interactions and longer range effects such as pollutant removal (wet scavenging and dry deposition) chemical transformation, vertical wind shear, and overwater transport. It has mesoscale capabilities that are not needed for chemical spill analysis. Most of the algorithms contain options to treat the physical processes at different levels of detail, depending on the model application. Time-dependent source and emission data from point, line volume, and area sources can be used. Multiple option are provided for defining dispersion coefficients. Time-varying heat flux and emission from fire can be calculated with the interface to the Emissions Production Model. A graphical users interface includes point and click model setup and data entry, enhanced error checking of model inputs and online help files.

Although CALPUFF is comprehensive, it was not designed for simulating dense gas dispersion effects or for use in evaluating instantaneous or short-duration releases of hazardous materials. The model is best applied to continuous releases rather than short-duration releases from accidents. Averaging times are long (minimum of one hour) for input data (release rate and meteorology) and output. Terrain effects on vapor dispersion may be modeled in the code. However, dispersion of heavier-than-air releases is not considered. It does not have a built in library of chemicals or a spills front end. Version 6 has provisions for time steps of less than one hour. The easy interface, documentation, detailed final results and intermediate submodel output comprise a useful code.

Chemical Process Accident Consequence Analysis

Sponsor: U.S. Environmental Protection Agency, U.S. Department of Agriculture, Forest Service, California Air Resources Board, and several industry and government agencies in Australia. *Developer*: Earth Tech (formerly Sigma Research). *Custodian:* John Irvin, U.S. Environmental Protection Agency. *Cost:* None. May be downloaded from EPA's Website: http://www.epa.gov/scram001/t26.htm. Look under CALPUFF for files READ1ST 8/01/96 CALPUFF MUST READ FIRST (12kB TXT), CALPUFFA 7/11/96 CALPUFF Modeling System-File A (1,146K,ZIP), CALPUFFB 7/11/96 CALPUFF Modeling System-File B (456K,ZIP), CALPUFFC 7/11/96 CALPUFF Modeling System-File C (1,040kB ZIP), CALPUFFD 7/11/96 CALPUFF Modeling System-File D (1,082 kB ZIP).

9.2.1.5 CASRAM/Beta 0.8 (Release and Dispersion)

CASRAM calculates the source emission rate for accidental releases of chemicals, as well as the transport and dispersion of these chemicals in the atmosphere. The evaporation of liquid pools of spilled chemicals is treated in great detail; however, no algorithms are included for the emission rate from pressurized jet releases. The specifications of atmospheric boundary layer parameters and the vertical dispersion of plumes are modeled with state-of-the-art methods. CASRAM does not treat dense gas releases, and it is assumed that the height of the release is near the ground. The Monte Carlo version allows for variability in key parameters pertaining to accident location, time and spill characteristics. Output parameters include the maximum concentration and plume width as a function of downwind distance. Two versions of the model exist: 1) the main CASRAM calculates mean, variance and other statistics of model predictions, thus, accounting for known variabilities; and 2) the deterministic version (CASRAM-CD) calculates only the mean. CASRAM's technical documentation and code are in draft (Brown et al.,1977a, 1977b, 1994a, 1994b).

The model contains a surface energy method for parameterizing winds and turbulence near the ground. Its chemical database library has physical properties (seven types, three temperature dependent) for 190 chemical compounds obtained from the DIPPR[h] database. Physical property data for any of the over 900 chemicals in DIPPR can be incorporated into the model, as needed. The model computes hazard zones and related health consequences. An option is provided to account for the accident frequency and chemical release probability from transportation of hazardous material containers. When coupled with preprocessed historical meteorology and population densities, it provides quantitative risk estimates. The model is not capable of simulating dense-gas behavior.

---

[h] The DIPPR file contains textual information and pure component numeric physical property data for chemicals. The data are compiled and evaluated by a project of the Design Institute for Physical Property Data (DIPPR) of the American Institute of Chemical Engineers (AIChE). The numeric data in the DIPPR File consist of both single value property constants and temperature-dependent properties. Regression equations and coefficients for temperature-dependent properties are also given for calculating additional property values. All of the data are searchable, including regression coefficients, percent error, and minimum/maximum temperature values.

It can simulate a wide variety of release scenarios but is particularly well suited to assessing health consequence impacts and risk.

CASRAM predicts discharge fractions, flash-entrainment quantities, and liquid pool evaporation rates used as input to the model's dispersion algorithm to estimate chemical hazard population exposure zones. The output of CASRAM is a deterministic estimate of the hazard zone (to estimate an associated population health risk value) or the probability distributions of hazard-zones (which is used to estimate an associated distribution population health risk).

The release of a liquid from an unpressurized container or the liquid fraction from a pressurized release is assumed to be instantaneously spilled on the ground (e.g., Bernoulli's equation is not employed) thus, overestimating the initial liquid pool evaporation rate. Currently, CASRAM-SC models near-surface material concentrations from near-surface releases. Spills of multiple chemicals are treated in CASRAM-SC. In these instances, no interaction between chemicals is considered, but the combined toxicological effects are modeled as outlined in the American Congress of Government and Industrial Hygienists Threshold Limit Value (ACGIH-TLV) documentation. For stochastic simulations, accident times (year, day of year and hour) and spill fractions are determined by sampling distributions obtained from the HMIRS database. Relative spill coverage areas are sampled from a heuristic distribution formulated using average chemical property and pavement surface data.

Sponsor: Argonne National Lab, Environmental Assessment Division, Atmospheric Science Section, 9700 S. Cass Ave., Argonne, IL 60439. ***Developer:*** University of Illinois, Dept. of Mechanical Engineering, 1206 W. Green, Urbana, IL 61801. ***Custodians:*** D.F. Brown University of Illinois, Dept. of Mechanical Engineering, 1206 W. Green, Urbana, IL 61801. ***Computer:*** It runs on a 486, Pentium PC, or any workstation. The deterministic version runs (slowly) on a 386. The stochastic version of the code was originally written for a Sun Workstation. The PC version requires at least a 486, 33 MHz machine with a minimum of 10 MB of free hard disk space. ***Cost***: None?.

9.2.1.6 DEGADIS (Dispersion)

DEGADIS (Dense GAs DISpersion) models atmospheric dispersion from ground-level, area sources heavier-than air or neutrally buoyant vapor releases with negligible momentum or as a jet from pressure relief into the atmosphere with flat unobstructed terrain. It simulates continuous, instantaneous or finite-duration, time-invariant gas and aerosol releases, dispersion with gravity-driven flow contaminant entrainment into the atmosphere and downwind travel. It accounts for reflection of the plume at the ground, and has options for isothermal or adiabatic heat transfer between the vapor cloud and the air. Variable concentration averaging times may be used and time-varying source-term release rate and temperature profiles may be used. It has been validated with large-scale field data. It models induced ground turbulence from surface roughness.

The jet-plume model only simulates vertical jets. Terrain is assumed to be flat and unobstructed. Application is limited to surface roughness mush less than the dispersing layer. User expertise is required to ensure that the selected runtime options are self-consistent and actually reflect the physical release conditions. Documentation needs improvement; there is little guidance

Chemical Process Accident Consequence Analysis

for effectively varying numerical convergence criteria and other program configuration parameters to obtain the final results.

Sponsors: U.S. Coast Guard and the Gas Research Institute. **Reference:** Spicer, T. and J. Havens, 1989. **Cost**: Free. The source and executable versions of the code may be downloaded from EPA's website: http://www.epa.gov/scram001/t22.htm. Look in "other models."

DEGATEC 1.1

DEGATEC is a user-friendly program that calculates the hazard zone surrounding releases of heavier than-air flammable and toxic gases. It incorporates the FORTRAN program DEGADIS 2.1, (Havens, 1985). Its Windows implementation inputs and outputs easier than the original DEGADIS2.1. DEGATEC was developed for the Gas Research Institute (GRI) by Tecsa Ricerca and Innovazione, s.r.1. (TRI) and Risk and Industrial Safety Consultants, Inc. (RISC).

Custodian and Developer: Tecsa, Via Oratorio, 7, 20016 Pero (Ml), Italy, and Risk and Industrial Safety Consultants Inc., 292 Howard Street, Des Plaines, IL 60018, ***Hardware:*** 386 or better CPU with a math coprocessor. Calculations require about 2 MB hard drive and 4 MB of RAM. The resident file requires 3 MB of storage in the hard drive. ***Software:***Windows 3.0 or higher. **Cost**: $900 for User's Manual, the executable program, protection key, quarterly bulletin, and telephone and fax support. GRAPHMAP, software that allows superimposition of DEGATEC concentration contours over a map, is available for an additional $200.

9.2.1.7 EMGRESP (Release and Dispersion)

EMGRESP is a source-term and dispersion emergency response screening tool for calculating downwind contours with a minimum of user input and computational expense in the event of a release of a hazardous chemical. The program provides hazardous contaminant information, calculates toxic concentrations at various distances downwind of a release, and displays the information on the screen compared to threshold exposure levels such as the Threshold Limit Values (TLV), Short Term Exposure Limits (STEL), and Immediately Dangerous to Life and Health (IDLH). For a combustible release, the code gives an estimate of the mass of vapor within the flammable limits. Model Description References: Model description and references are provided in the user's manual (Diamond et al., 1988). References for EMGRESP are: Chong (1980); Eidsvik (1980); Fryer and Kaiser (1979); and Opschoor (1978).

The source-term models consist of: gas release from a reservoir, a liquid release from a vessel, and liquid pool evaporation. The dispersion models are: neutrally buoyant Gaussian plume, instantaneous puff model, and an instantaneous dense gas "box" model. It has a large database of physical/chemical properties; the mass of vapor calculations within flammable concentration limits was confirmed with large-scale field data.

EMGRESP is overly conservative for passive gas dispersion applications. No time-varying releases may be modeled. Dense gas dispersion may be computed for only "instantaneous" releases conditions.

Sponsor/Developing Organization: Ontario Ministry of the Environment Air Resources Branch, 880 Bay St., 4th Floor, Toronto, Ontario M5S 1Z8, Phone: (416) 3254000. Website is:

http://www. cciw. ca/glimr/agency-search/other/omee.html. ***Custodians:*** Dr. Robert Bloxam, Science & Technology Branch, 2 St. Clair Ave. W., Floor 12A, Toronto, Ontario M4V 1L5, Phone: (416) 323-5073, Sunny Wong Science & Technology Branch 2 St. Clair Ave. W. Floor 12A Toronto, Ontario M4V 1L5 Phone: (416)323-5073. ***Computer***: IBM PC compatible computer (8086 processor or greater), CGA or greater graphics support, 640 kB of RAM, DOS 3.0 or higher with GWBASIC. ***Cost***: None.

9.2.1.8 FEM3C (Dispersion)

FEM3C is a three-dimensional finite element model for simulation of atmospheric dispersion of denser-than-air vapor and liquid releases. It is useful for modeling vapor dispersion in the presence of complex terrain and obstacles to flow from releases of instantaneous or constant-rate vapors or vapor/liquid mixtures. The algorithm solves the 3-dimensional, time-dependent conservation equations of mass, momentum, energy, and chemical species. Two-dimensional calculations may be performed, where appropriate, to reduce computational time. It models isothermal and non-isothermal dense gas releases and neutrally buoyant vapors, multiple simultaneous sources of instantaneous, continuous, and finite-duration releases. It contains a thermodynamic equilibrium submodel linked with a temperature-dependent chemical library for analyzing dispersion scenarios involving phase change between liquid and vapor state. It models the effects of complex terrain and obstructions to flow on the vapor concentration field using a number of alternate turbulence models. The user can model turbulence parameterized via the K-theory or solving the k-ϵ transport equations. The code cannot accept typical vapor/aerosol source terms (e.g., pressurized jets, time-varying vapor emissions). Although the code treats complex terrain, modeling diverse vegetation in the same computation is difficult. The aerosol model does not model all relevant physical behavior (e.g., droplet evaporation and rainout). FEM3C is configured specifically for the Cray-2 (a parallel-processing computer), and, therefore, porting the code to another computing architecture would be a big job. Significant postprocessing of the output by the user is required to obtain concentration values at specific locations and averaging times.

Sponsor: U.S. Army Chemical Research, Development and Engineering Center (CRDEC). ***Developer:*** Stevens T. Chan Lawrence Livermore National Laboratory, P.O. Box 808, Livermore, CA 94551-9900, Phone: (501) 422-1822 ***Custodian:*** Diana L. West, L-795 Technology Transfer Initiative Program, Lawrence Livermore National Laboratory Livermore, CA 94551, Phone: (510) 423-8030.

9.2.1.9 FIRAC (Release)

FIRAC is a computer code designed to estimate radioactive and chemical source-terms associated with a fire and predict fire-induced flows and thermal and material transport within facilities, especially transport through a ventilation system. It includes a fire compartment module based on the FIRIN computer code, which calculates fuel mass loss rates and energy generation rates within the fire compartment. A second fire module, FIRAC2, based on the CFAST computer code, is in the code to model fire growth and smoke transport in multicompartment structures.

It is user friendly and possesses a graphical user interface for developing the flow paths, ventilation system, and initial conditions. The FIRIN and CFAST modules can be bypassed and temperature, pressure, gas, release energy, mass functions of time specified. FIRAC is applicable to any facility (i.e., buildings, tanks, multiple rooms, etc.) with and without ventilation systems. It is applicable to multi species gas mixing or transport problems, as well as aerosol transport problems. FIRAC includes source term models for fires and limitless flow paths, except the FIRIN fire compartment limit of to no more than three

The PC version runs comparatively slow on large problems. FIRAC can perform lumped parameter/control volume-type analysis but is limited in detailed multidimensional modeling of a room or gas dome space. Diffusion and turbulence within a control volume is not modeled. Multi-gas species are not included in the equations of state.

9.2.1.10 GASFLOW 2.0 (Dispersion)

GASFLOW models geometrically complex containments, buildings, and ventilation systems with multiple compartments and internal structures. It calculates gas and aerosol behavior of low-speed buoyancy driven flows, diffusion-dominated flows, and turbulent flows during deflagrations. It models condensation in the bulk fluid regions; heat transfer to wall and internal structures by convection, radiation, and condensation; chemical kinetics of combustion of hydrogen or hydrocarbons; fluid turbulence; and the transport, deposition, and entrainment of discrete particles.

The level of detail GASFLOW is changed by the number of nodes, number of aerosol particle size classes, and the models selected. It is applicable to any facility regardless of ventilation systems. It models selected rooms in detail while treating other rooms in less detail.

It calculates one-dimensional heat conduction through walls and structure; no solid or liquid combustion models are available. The energy and mass for burning solids or liquids must be input. It has no agglomeration model nor ability to represent log-normal particle-size distribution.

9.2.1.11 HGsystem 3.0 (Release and Dispersion)

HGSystem is a general purpose PC-based, chemical source term and atmospheric dispersion model. It is comprised of individual modules used to simulate source-terms, near-field, and far-field dispersion. The chemical and thermodynamic models available in HGSystem include reactive hydrogen fluoride and nonreactive, two-phase, multicompound releases. It accounts for HF/H_2O/air thermodynamics and cloud aerosol effects on cloud density, pressurized and non-pressurized releases, predicts concentrations over a wide range of surface roughness conditions and at specific locations for user-specified averaging periods, steady-state, time-varying, and finite duration releases, and crosswind and vertical concentration profiles.

HGSystem offers the most rigorous treatments of HF source-term and dispersion analysis available for a public domain code. It provides modeling capabilities to other chemical species with complex thermodynamic behavior. It treats: aerosols and multi-component mixtures, spillage of a liquid non-reactive compound from a pressurized vessel, efficient simulations of time-dependent dispersion by automating the selection of output times and output steps in the HEGADAS-T., improved dosage calculations and time averaging (HEGADAS-T/HTPOST), ability to model dispersion on terrain with varying surface roughness, effects of dike containments (EVAP), upgrade of evaporating pool model, user friendliness of HGSystem, upgrading of post-processors HSPOST and HTPOST, shared input parameter file. It is one of the few codes with a rigorous treatment of thermodynamic effects and near field transport in vapor dispersion. The underlying computational models and the user input and processing instructions are well documented. It has been validated with large-scale field data.

HGSystem cannot be used for analyzing releases where there is pipework present between the reservoir and the outlet orifice. There are a limited number of pool substrate materials which can be modeled by HGSystem, specifically sand, concrete, plastic, steel, and water. It is difficult to extend the physical/chemical database utility DATAPROP to include additional chemical species. Such modifications would require an extremely good knowledge of the HGSystem program structure and some consultation with the code developers. User expertise is needed for the range of computational options, extensive input data and large number of models. Source term improvement is needed for modeling highly stable meteorological conditions (e.g., pool evaporation for a 1 m/s, F stability wind field). Low momentum plumes, in general, are difficult to model. Releases from pipes cannot be calculated by HGSYSTEM. Liquid aerosol rainout is not modeled; aerosols are assumed to remain airborne until complete evaporation.

The user's manual is Post (1994a); model references are: Clough (1987), Colenbrander (1980) DuPont (1988) Post (1994b), Schotte (1987) Schotte (1988) Vanderzee (1970), and Witlox (1993).

Sponsor: Shell Component models Industry Cooperative HF Mitigation/Assessment Program, Ambient Impact Technical Subcommittee (HGSYSTEM Version 1.0), American Petroleum Institute (HGSYSTEM Version 3.0). Developer: Shell Research Limited Shell Research and Technology Centre Thomton, P.O. Box 1, Chester, CHI 3SH. ***Custodian***: HGSYSTEM Custodian Shell Research and Technology Centre Thornton, P.O. Box 1, Chester, CH1 3SH, UK, Phone: 44 151 373 5851 Fax: 44 151 373 5845, E-mail: hgsystem@msmail.trctho. simis.com. The program and technical reference manuals are available through the American Petroleum Institute (API), National Technical Information Service (NTIS), and Shell Research Ltd. ***Computer***: The code is intended for compilation under Microsoft FORTRAN Powerstation with PC execution, but may be adapted to any FORTRAN 77 compiler. It runs on a PC with a 386 CPU, 4 MB of RAM, 2 MB hard disk, math coprocessor recommended. ***Cost:*** None.

9.2.1.12 HOTMAC/RAPTAD (Dispersion)

HOTMAC/RAPTAD contains individual codes: HOTMAC (Higher Order Turbulence Model for Atmospheric Circulation), RAPTAD (Random Particle Transport and Diffusion), and computer modules HOTPLT, RAPLOT, and CONPLT for displaying the results of the calculations. HOTMAC uses 3-dimensional, time-dependent conservation equations to describe wind, temperature, moisture, turbulence length, and turbulent kinetic energy.

Its equations account for advection, Coriolis effects, turbulent heat, momentum, moisture transport, and viscosity. The system treats diurnally varying winds such as the land-sea breeze and slope winds. The code models solar and terrestrial radiation, drag and radiation effects of a forest canopy, and solves thermal diffusion in soil. RAPTAD is a Monte Carlo random particle statistical diffusion code. Pseudoparticles are transported with the mean wind field and the turbulence velocities.

HOTMAC/RAPTAD requires very extensive meteorological and terrain data input. The program user's guide and diagnostics are inadequate. HOTMAC does not model multiple scale eddy turbulence and does not provide for dispersion of gases that are denser-than air. It must be tailored to reflect the climatic characteristics of specific sites.

Sponsor: U.S. Army Nuclear and Chemical Agency. ***Developer:*** Los Alamos National Laboratory (LANL). ***Custodian:*** Michael D. Williams, Los Alamos National Laboratory Technology & Assessment Division TSA-4, Energy and Environmental Analysis, Los Alamos, New Mexico 87545, Phone: (505) 667-2112, Fax: (505) 665-5125, E-mail address: mdw@lanl.gov

9.2.1.13 KBERT, Knowledge+Based System for Estimating Hazards of Radioactive Material Release Transients (Release)

The current prototype code, is included because of being knowledge-based and its potential relevance for chemical releases. Presently, it calculates doses and consequences to facility workers from accidental releases of radioactive material. This calculation includes: specifying material at risk and worker evacuation schemes, and calculating airborne release, flows between rooms, filtration, deposition, concentrations of released materials at various locations and worker exposures.

Facilities are modeled as a collection of rooms connected by flowpaths (including HVAC ducts). An accidental release inside the facility is simulated using the incorporated Mishima database as a relational database along with databases of material properties and dosimetry and mechanistic models of air transport inside the facility that use user input flow rates. Aerosol removal is currently gravitational settling. Filters may be modeled in any flowpath. Evacuation plans for individual workers or groups of workers can be specified. Radiological dose models include cloudshine, inhalation, groundshine, and skin contamination pathways with allowance for protective equipment. Output includes airborne and surface concentrations of released material throughout the facility, whole-body and organ-specific worker doses. A simulation begins by specifying the hazardous materials and scenario in an event dialogue. An intuitive dialogue box interfaces with the Mishima database to provide either a one time airborne release fraction or a continuous airborne release rate and respirable fractions for the materials involved in accidents such as fires, explosions, and spills. The code tests and limits input to "only realistic combinations" of release data, materials at risk, and accident stresses.

This program is an integrating tool. Other than the dynamics of the air flow and settling of material, the program makes heavy use of databases for information concerning characteristics of releases, health effects and health models, therefore it could be adapted to chemical releases. The program is menu driven with easy change of input and default parameters. Modeling of a facility is relatively easy and the facility model can be stored. Once a facility has been modeled, part of it can be used for any particular accident scenario. As it stands, this program calculates the expected worker dose from an accidental release of radioactive material within the confines of a facility. The code assumes that released material is mixed instantaneously in a room, so concentration gradients are not available. However a room can be broken into smaller volumes to get around this. Aerosols settle only by gravitational settling without agglomeration. The manual indicates, that other deposition mechanisms can be "readily" implemented. The model is described in KBERT (1995) the release fraction reference is Mishima (1993).

Sponsor: DOE/EM, *Developer*: Sandia National Laboratories, *Custodians* Dr. S.K. Sen, U.S. Department of Energy, Office of Environment, Safety, and Health, EH-34, 19901 Germantown Road Germantown, MD 2087-1290 Phone: (301) 903-6571 Fax: (301) 903 4672, E-mail: subir.sen@hq.doe.gov. Dr. K.E. Washington, Sandia National Laboratories, P.O. Box 5800, Albuquerque, NM 87185, Phone: (505) 844-0231, Fax: (505) 844-3296, E-mail: kewashi@sandia.gov. *Computer*: an IBM compatible PC with 386 CPU, 486 preferred, Windows 3.1, VGA graphics display (SVGA preferred) mouse and a hard disk with at least 10 MB free space. It is written in Microsoft Visual C++. *Cost*: unknown.

9.2.1.14 MISM, Modified Ille-Springer Model (Release)

The MISM model is an evaporation and dispersion model specifically for hydrazine (N_2H_4), monomethylhydrazine (MMH), and unsymmetrical dimethyl hydrazine (UDMH) ground spills. However, it is useful for modeling the release of other toxic, flammable, or explosive materials. It predicts the rate of evaporation from a pool of spilled liquid by estimating a mass transfer rate which is determined from the vapor pressure of the pure liquid, the diffusive properties of the vapor at the air-liquid interface, and the wind generated turbulence over the pool. The model assumes: the spill occurs without atomization; the spill is adiabatic and the liquid is chemically stable; the spill occurs onto a flat non-porous surface; evaporation occurs at a steady state pool temperature; the spill is a pure liquid; and ideal gas behavior of the vapor film. The steady-state pool temperature is determined by iterating energy balance that equates heat inputs (convective heating from the ground and atmosphere, solar insolation and radiative heat transfer from the atmosphere) with heat losses (radiative emission from the pool and evaporative cooling) to establish the equilibrium vapor pressure of the liquid. The separation of this equilibrium vapor pressure from the steady-state vapor concentration that develops at the pool interface establishes the concentration driving force for evaporation. A mass transfer coefficient equates the driving force to the mass transfer rate. The mass transfer coefficient is a complex function of the wind speed, the degree of atmospheric turbulence, and the pool diameter. The properties of the evaporating vapor enter through a Schmidt number. A derivation of the model and examples of input development are given in Appendix A to the user manual. A simple, point source Gaussian model without any buoyancy or plume rise effects is used to calculate dispersion and downwind concentration of the releases. The dispersion coefficients, σ_y and σ_z, are based on Turner, 1970.

The user's manual is Ille (1978). Other references are Mackay (1973), Kunkel (1983), Schmidt (1959), and Bird, (1966).

Chemical Process Accident Consequence Analysis

Sponsor/Developing Organization: Department of Defense, Civil and Environmental Engineering Development Office, Tyndall Air Force Base, Panama City, Florida, 32403. ***Computer***: The code is 200 lines of FORTRAN, hence, it should compile and run on most PC. ***Cost***: Unknown.

9.2.1.15 Army ORG40/TP10 (Part EVAP4.FOR of D2PC)

This is a subroutine that calculates an evaporation rate from a pool of spilled liquid in presence of wind (ORG-40), or in still air (TP-10). It was developed by the U.S. Army for downwind hazard prediction following release from smoke munitions and chemical agents. The code calculates the evaporation rate of a liquid pool, given the physical state variables, wind speed, and diameter of pool. ORG-40 and TP-10 models are coded as a Fortran 77 subroutine, EVAP4.FOR, in D2PC. The user's manual is Whitacre (1987).

Sponsor/Developing Organization: U.S. Army Chemical Research and Development Engineering Center, Aberdeen, MD ***Custodian***: Commander, U.S. Army CRDEC, Aberdeen Proving Ground, MD 21010-5423.

9.2.1.16 PHASTProfessional (Commercial - Release and Dispersion)

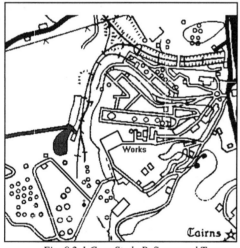

Fig. 9.2-1 Case Study Refinery and Town

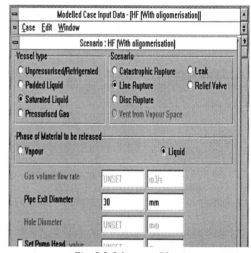

Fig. 9.2-2 Input to Phast

PHASTProfessional uses a user-friendly Windows interface to link source-term, dispersion, fire, and explosion models. The output is text and color contour. It models pool boiling evaporation, two-phase releases (flashing, aerosol formation, and rainout), free jet plumes and fires, BLEVEs, and vapor cloud explosions. It facilitates data entry with complete model references and user guide documentation. Some model validation has been conducted against large-scale field and wind tunnel data. Graphical results (e.g., cloud footprints) can overlay on a map background, for assessing affected zones in a consequence assessment.

Case Study: A refinery is five kilometers from a small town on a flat plain at 10 meters elevation surrounded by arable farmland (Figure 9.2-1). Summer temperatures are about 20° C with 70% relative humidity. The refinery contains a large alkylation unit with hydrogen fluoride, LPG and alkylates in large quantities. The concern is the control room and safety in the town. Previous studies of the hydrogen fluoride risk have not accounted for its oligomeric behavior. The LPG consists of 40% butane and 60% propane. The HF is stored in a 150 tonne tank with an alkylation reactor over the tank. Both the LPG and the HF are stored at ambient temperature without dikes or building enclosure. While all possible scenarios, identified by HAZOP, must be considered, a leak of HF from a thermowell in the acid circulation leg of the reactor (3 cm) and a cold release of LPG from a feed pipe to the reactor (6 cm) are of particular interest. The LPG is maintained at atmospheric pressure and the thermowell is situated toward the bottom of the reactor at a liquid head pressure of about 4 meters. Various measures can be used to gauge the threat posed by toxic (HF) and flammable (LPG) consequences of these releases. Emergency Response Planning Guideline (ERPG) concentrations levels are designed to assist the preparation of off-site emergency plans. A number of toxic correlations can be used to determine the incidence of fatality resulting from exposure to the cloud. Heat radiation effect levels can be set to different levels to model different levels of damage. It contains these toxic and flammable data in its database.

Figure 9.2-2 shows a data input screen in which general characteristics are input by radio buttons and numerical data is typed. The program calculates distances to specified toxic concentrations and other requested consequence levels automatically. Results are available in a variety of formats including cloud footprints, sideview, cross section, pool evaporation rate, concentration vs distance and heat flux contours. Figure 9.2-3 shows the calculated results as a toxic plume superimposed on the map with and without oligomerization.

Although the code is based on well-recognized models referenced in the literature, some of the underlying models are based on "older" theory which has since been improved. The code does not treat complex terrain or chemical reactivity other than ammonia and water. The chemical database in the code is a subset of the AIChE's DIPPR database. The user may not modify or supplement the database and a fee is charged for each chemical added to the standard database distributed with the code. The code costs ~ $20,000 and requires a vendor supplied security key in the parallel port before use.

Sponsor/Developer: Det Norsk Veritas, http://www.dnv.com/technica. *Custodian:* DNV Technica, Palace House, 3 Cathedral St., London SE1 9DE, UK, Phone +44(0)171 716 6545, or DNV Technica Inc., 40925 Country Center Dr. Suite 200, Temecula, CA 92591, (909) 694-5790., *Computer:*
Windows 3.x, '95, or NT, 40 MB hard drive, 16 MB RAM. *Cost:* Depends on type perpetual license, fixed period license, non-commercial academic are available. For example, one-year lease including upgrades, training, and technical support is $11,500.

9.2.1.17 Pspill (Release)

PSpill determines the mass airborne from a spill of solid powder. It models the shearing effect of the air on the powder as it falls. The shearing effect is enhanced by the lengthening of the

Chemical Process Accident Consequence Analysis

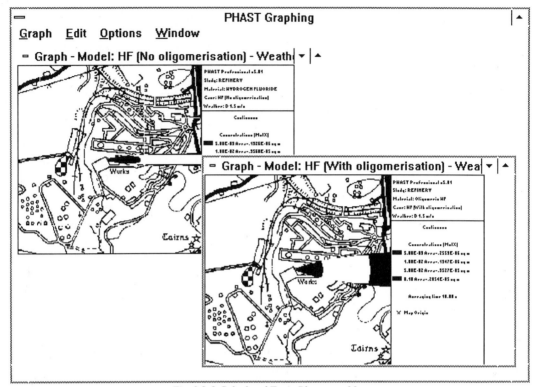

Fig. 9.2-3 Calculated Toxic Plumes on Map

spill column caused by the acceleration of gravity. The mass of material that becomes airborne from a spill of powder is assumed proportional to the drag force on the powder as it falls. The growth rate of the powder front is constant and can be characterized by an angle of dispersion. The diameter of the powder front at the start of the spill is equal to the diameter of the container from which it was spilled. The model assumes that any airborne material created as a result of impact with a surface is negligible, and ignores airborne particulate generated as a result of impact. Its model applies only to powders for bulk spills, not continuous pours using experimental data from NUREG/CR-2139 NUREG /CR-4658. The model is described in its users manual: NUREG/CR-4997.

Sponsor /Developing Organization: Pacific Northwest Laboratory, P.O. Box 999, Richland, WA 99352, *Custodian*: Marcel Ballinger, Pacific Northwest Laboratory, P.O. Box 999, Richland, WA 99352. Phone: (509) 373-6715 *Computer:* The original code was written in FORTRAN, to run on a VAX minicomputer, but can be rewritten to run on most any platform. *Cost*: None.

9.2.1.18 SLAB (Dispersion)

SLAB is one of the most widely-used dense gas models in the public domain. The model accepts evaporating pool sources, jet releases at any elevation, and instantaneous volume sources.

SLAB calculates chemical concentrations at positions downwind and heights above the ground. The plume may be denser-than-air, neutrally-buoyant, or less dense than air. Thermodynamics effects are accounted for, including latent heat exchanges due to the condensation or evaporation of liquids. Time averaged results may be calculated. SLAB is the easiest of the publicly-available dense gas models to set up and run. It has been extensively validated against large-scale field data.

It does not calculate source emission rates. While it handles jets, it does so simply and does not calculate the details of the jet motions and thermodynamics. It should not be used for strongly buoyant plumes. The error diagnostics are limited to checking the consistency of input parameters. Run time error diagnostics are missing but are rarely needed due to its robustness.

Sponsor/Developing Organization: Department of Energy (DOE)/Lawrence Livermore National Laboratory (LLNL). It is distributed and supported by the EPA and by Bowman Environmental Engineering. ***Custodians:*** EPA 11N Bulletin Board, Phone (919) 541-5742. An enhanced and user supported commercial version of SLAB is available from Bowman Environmental Engineering, P.O. Box 29072, Dallas, TX 75229, (214)233-5463. ***Cost:*** None. May be downloaded from EPA's web site: http://www.epa.gov/scram001/t22.htm. Look under "non-EPA files.for SCREEN2C - SCREEN2C - contains Schulman-Scire cavity algorithm (150kB ZIP)

9.2.1.19 TSCREEN (Release and Dispersion)

TSCREEN (A Model of Screening Toxic Air Pollutant Concentrations) analyzes toxic emissions and subsequent dispersion from many different types of releases. It is a simple, conservative screening tool for obtaining maximum downwind vapor concentrations at select receptor locations for releases resulting from a) particulate releases, b) gaseous releases, c) liquid releases, and d) superfund releases. A particular release scenario is selected via input parameters, and TScreen automatically selects and executes the appropriate dispersion model to simulate that scenario. The model to be used and the worst case meteorological conditions are automatically selected. It has four imbedded air toxin dispersion screening models, SCREEN2, RVD, PUFF, and the Britter McQuaid model. The release scenario is defined by the input parameters; it automatically selects and executes the dispersion model for the scenario. SCREEN2 is a Gaussian dispersion model applicable to continuous releases of particulate matter and nonreactive, nondense gases that are emitted from a point, area, volume, and flared sources. The RVD model provides short-term ambient concentration estimates for screening pollutant sources that release vertically-directed, denser-than-air gases and aerosols. The PUFF model is used where the release duration is smaller than the travel time. The Britter-McQuaid model is used for continuous and instantaneous denser-than-air scenarios. The program is a screening tool, some of the models are overly conservative in the source-term and downwind dispersion. The most conservative models are superheated flashing releases and unconfined liquid pool evaporation. Furthermore, it only computes steady-state release rates, assuming ideal state conditions. Releases must fit into one of several scenarios that are defined in the workbook. If input parameters are not consistent with the selected category, a new category may be suggested and the model starts over with the input requests. The user cannot define meteorological conditions and cannot select which of the models to use. The

Sponsor/Developing Organization: EPA, Research Triangle Park, NC ***Custodian:*** Lawson, Jr., U.S. Environmental Protection Agency, Research Triangle Park, NC 27711, Phone: (919) 541-1199, or Jawad Touma, Project Officer US EPA, (3AQPS, TSD (MD-14) Source Receptor Analysis Branch, Research Triangle Park, NC 27711, Phone: (919) 541-5381, ***Computer:*** PC with 286 or higher CPU, 570 kB RAM and 2 MB on the hard drive. ***Cost:*** None. May be downloaded from EPA's website: http://www.epa.gov/scram001/t22.htm. Look under "non-EPA files" for SCREEN2C-SCREEN2C, contains Schulman-Scire cavity alg.(150kB ZIP)

9.2.1.20 VDI 3783 Parts 1 & 2 (Dispersion)

VDI Part 1 models the dispersion of vapor plumes with output consisting of vapor concentration as a function of time and downwind distance and denser-than-air vapor releases. VDI Part 2 determines the downwind distance to the lower flammable limit of a combustible vapor. Part 2 may also be used in conjunction with Part 1 to model a toxic gas emission.

Part 1 is a computerized implementation of the Gaussian plume equations for continuous ground-level or elevated release. The release rate may be time varying within specific prescribed constraints on variability. Reflection of the plume off the mixing layer lower boundary is also modeled. Point, line, area, and volume source geometries may be modeled using the code. Part 2 predicts the dispersion of denser-than-air gases based on empirical data obtained from wind-tunnel studies for puff and continuous releases into boundary layer shear flow. The model is applicable for level, unobstructed dispersion as well as more complex flow and turbulence structure due to the presence of downwind obstacles. Part 1 assumes complete reflection of the plume at ground level and in the mixing layer. Volume sources are initially approximated as right parallelpipeds. The modeling of line, area, and volumetric vapor sources is not an accurate solution of the diffusion equation and may give inaccurate results in downwind concentration estimates when the dimensions of the source are not much less than the downwind distance of interest. Part 2 assumes dense gas releases are ground-level point sources released into an ambient temperature of 20 ° C. Heat transfer effects typically present in dense gas dispersion are not modeled in the code.

Sponsor/Developing Organization: Verein Deutscher Ingenieure (VDI) Kommission Reinhaltung der Luft, D-4000 Dusseldorf, Deutschland. ***Custodian:*** Dr. Michael Schatzmann Meteorologisches Institut Universitat Hamburg, Bundesstrase 55 D-20146 Hamburg, Germany. Telephone: 49-40-41235040, Telefax: 49-40-41173350.

9.2.2 Explosions and Energetic Events

9.2.2.1 EXPAC

EXPAC analyzes an interconnected network of building rooms and ventilation systems. A lumped-parameter formulation is used that includes the effects of inertial and choking flow in rapid gas transients. The latest version is specifically suited to calculation of the detailed effects of explosions in the far field using a parametric representation of the explosive event. A material transport capability models the effects of convection, depletion, entrainment, and filtration of explosions in the far field using a parametric representation of the explosive event. A material transport capability models the effects of convection, depletion, entrainment, and filtration of material. The effect of clogged filter systems on the gas dynamics is accounted for in the model. The air-transported contaminant material is restricted to a single phase for a single species. No phase transitions or chemical reactions are modeled in the code. If the contaminant is an aerosol (solid or liquid), then it is treated as a monodisperse and homogenous spherical particle. Material transport initiation may be user-specified or calculated via aerodynamic entrainment. Setting up the problem is tedious. Flow resistance coefficients must be estimated for the ventilation components. The user's manual and model reference is Gregory and Nichols (1991).

Sponsor/Developing Organization: LANL. *Developer:* W.S. Gregory and B.D. Nichols, *Custodian*: LANL/W.S. Gregory Telephone (505) 667-1120. *Hardware:* CRAY mainframe computer, 32K words of storage for the executable module, run time: several minutes (real time). *Language*: FORTRAN. *Portability:* Designed for portability to other computer platforms if the word length is less than 64 bits, double precision in the solution modules of the code may be needed to retain comparable accuracy. Comment statements regarding the input and output formats and a complete glossary of variables are included in the source code to facilitate later modifications. *Cost:* None.

9.2.2.2 GASFLOW

GASFLOW models in-facility transport and vapor space combustion by solving the 3-D compressible Navier Stokes equations and the transport equations for internal energy and chemical species. The geometry is made discrete by a mesh in Cartesian or cylindrical geometry. The primary hydrodynamic variables in the final analysis are the cell-face-centered nominal velocity and cell-volume-centered density, internal energy, and pressure. Three different turbulence models are available for modeling turbulent diffusivity: algebraic, subgrid-scale, and k-ϵ. The model that accounts for heat transfer between the convective flow streams and the facility walls or structures can treat enhanced transfer rates associated with high mass fluxes toward the walls. No solid or liquid combustion model is available. The energy and mass source for burning solids and liquids must be input. Assumptions are: each cell is well mixed, all gases are assumed to be ideal, and choke flow in the ventilation system components is not considered. It treats multidimensional flow multispecies diffusion, including chemistry models with capability for modeling diffusion flame and flame propagation through vapor spaces having complex geometry. The user's manual is: Travis et. al. (1994, Volume 2), the model description is in Volume 1 and comparisons with field experiments is in Volume 3.

Sponsor/Developing Organization: DOE-DP, DOE-EM, and NRC. *Custodian*: Kin Lam M/S K575 LANL, Fax: (505) 665-0879, E-mail: klam@lanl.gov. *Hardware:* CRAY (UNICOS) mainframe computer, SGI (UNIX) workstations, and SUN (UN) workstations. Computer memory requirements depend on the problem size and can be large for complex problems. Runtimes on the CRAY vary from a few seconds to several hours. Typical runtimes for Hanford waste tank burp and burn problems. *Software*: FORTRAN NAMELIST feature specifies all model option selections and numerical data. Input is tested for valid range and self-consistent user options. The post-processor graphics library assessed by GASFLOW is only available at LANL. Therefore, porting the code to other platforms would require, in part, disabling any program calls to the graphics routines and providing alternate graphics capabilities. *Cost*: Not presently available.

9.2.2.3 HOTSPOT/Version 8.0

HOTSPOT provides a fast, field tool for evaluating accidents involving radioactive materials. It consists of 4 main computational modules: radioactive source-terms generated by explosions, radioactive source-terms generated by fires, resuspension of radioactive material, and a Gaussian dispersion model to estimate the transport of radioactive material in a continuous plume (puff) release. The blast overpressure models are based on TNT-equivalency methods or empirical correlations derived from experimental tests involving the detonation of nuclear materials. Blast overpressure calculations are performed for conventional and nuclear weapons. Radiation exposure due to an energetic release of plutonium, uranium, and tritium is also calculated. The later is based on a user-supplied release fraction where the energy of the blast is used to determine the loft height of the material and the Gaussian dispersion equations are used to determine the airborne exposure. It is easy to use for estimating radiation, thermal, and blast overpressure exposure as a function of distance from the source event, but simplicity does not handle energetic events in complicated scenarios, e.g., partial containment by a structure, blast wave enhancement or damping due to the presence of nearby structures. The user's manual and reference is Homann (1994). Other documentation is Nadeau (1995a-c).

Sponsor/Developing Organization: LLNL. *Developer*: Steven G. Homann, LLNL L-380, P.O. Box 808, Livermore, CA 94551 Phone (501) 423-4962, Fax: (501) 423-3090, E-mail: shomann@llnl.gov. *Custodian*: Same. *Hardware:* IBM PC compatible with a 80286 processor or greater, 512 kB of RAM, Microsoft DOS version 3.0 or later, Math coprocessor chip is desirable optional, but highly recommended, 0.5 MB hard disk space. *Software*: Pascal, typical run time several seconds. *Cost:* No charge from LLNL.

9.2.2.4 CHEETAH I.40

CHEETAH is a thermo-chemical code that implements the Chapman/Jouget (C-J) detonation theory in a PC based calculational model. It consists of a standard model for evaluating explosives and propellants. In C-J theory the detonation point is in thermodynamic and chemical equilibrium; Cheetah solves thermodynamic equations between product species to find chemical equilibrium. It can calculate thermodynamic states where the pressure and temperature are not explicitly indicated. From these properties and elementary detonation theory, detonation velocity and detonation energy are predicted. It includes a gun and a rocket model for evaluation of propellant formulations, and predicts concentrations and properties of reaction products. It is recommended for evaluating simple fuels to high explosives for predicting energy release and reaction by-products concentrations. CHEETAH does not consider finite size effects in explosives predictions; condensed products are poorly parameterized for far from ambient conditions which may overestimate the detonation velocity for materials producing large concentrations of condensed products. References are Fried a and b.

Sponsor/Developing Organization: LLNL. ***Developer***: Laurence E. Fried LLNL, P.O. Box 808 Livermore, CA 94551, E-mail: cheetah@llnl.gov. ***Hardware***: IBM-PC or clone, Windows 3.1, Windows 95, Mac OS 7.x or later, SUN and SGI workstations, 4.3 MB of hard disk. ***Software:*** ANSI C. Run execution time for typical problem (CPU or real time). Standard run: About 30 seconds on a Power Macintosh 6100/80. ***Cost:*** None from LLNL. Source code is available, with the stipulations that all modifications be preapproved and forwarded to the sponsor for tracking

9.2.3 Fire Codes

9.2.3.1 FIRAC/FIRIN

FIRAC/FIRIN predicts the distribution of temperatures, pressures, flows, and toxic materials within a facility. It models the movement of hazardous materials between gloveboxes, rooms, and cells connected by ventilation systems to assess the release within such compartments. It uses a source submodule called FIRIN for modeling of user specified parameters such as energy, temperature, pressure, or mass addition. As such, it can be used to model the release concentration of hazardous material within a burning room and in connecting compartments to determine the building source term. The FIRIN submodule calculates the mass generation rate and the size distribution of airborne material released by fire accident scenarios. FIRIN calculates the mass and energy input to FIRAC. The FIRAC/FIRIN combination evaluates the source terms within and from a facility.

It has a graphical user interface for drawing a representation of a particular fire scenario and the ventilation system involved. As picture elements are added, input windows open automatically to allow the input of information. When the graphical model is complete, including the required data, the preprocessor automatically generates the required input file. FIRAC/FIRIN capabilities and models are divided into the following major categories: Gas dynamic models, material transport models, heat transfer models, fire models, and source term models. The user's manuals are: Nichols and Gregory (1986) Chan et al. (1989). Benchmarking references are: Allison (1992), Gregory (1989), Alvares, (1984), Nichols (1986), and Gregory (1991). Validation against experiment: Claybrook (1992)

Sponsor: U.S. Nuclear Regulatory Commission (NRC). ***Developing Organizations***: the NRC with Los Alamos National Laboratory, Westinghouse-Hanford Company, and New Mexico State University. The fire compartment model within FIRAC/FIRIN (called FIRIN) was originally developed by Battelle Pacific Northwest Laboratories and extensively modified before incorporation. ***Original Developer***: B. D. Nichols and W.S. Gregory, LANL. ***Current Custodian***: W.S. Gregory, LANL, phone (505) 667-1120. ***Hardware Requirements***: IBM-PC or comparable VGA graphics, math coprocessor with an 80386 or better CPU, 5.25 or 3.5 inch high density disk drive to install the distribution disk, a mouse recommended. Software DOS 3.X or higher, SVS FORTRAN to change the source code. ***Cost***: None through LANL.

9.2.3.2 CFAST

CFAST, within zonal model limitations, solves the mass and energy conservation equations to estimate compartment temperatures and species concentrations in a burning facility. The front-end module, Cedit, allows the user to develop the input deck via as series of menus. CPlot, a post-processing module, generates on-screen and hard copy plots of the data. It lacks a fire growth model, however, specific fire-related physics dealing with hot gas layers, fire plumes, door jets, radiative heat transfer, and ignition of secondary fires are modeled. In addition, a building ventilation system (ducts, vents, and fans) can be included in a facility model. CFAST can generate tabular (ASCII) and HPGL graphics output. The variables that are output are: layer interface height, upper and lower layer temperature, surface (walls, floors, ceilings) temperatures; pressure; fire locations; optical density; some flow rates, and species concentrations. Limitations are: it requires a priori specification of the temporal behavior of the fire; does not calculate radiological source terms; assumptions in the zone model formulation of the conservation equations; does not output all important variables; does not model mitigative systems, and deficient documentation. The user's manual and methodology is Portier (1992); the methodology, source code verification and benchmarking reference is Peacock (1993a). Validation against experimental data is reported in: Peacock (1993), Jones and Peacock (1994) and Jones (1994).

Sponsor: NIST Building and Fire Research Laboratory. ***Developing Organization:*** Same. ***Original Developer***: Same. ***Current Developer and Custodian***: Walter W. Jones ,NIST Building and Fire Research Laboratory, Gaithersburg, MD 20899. ***Hardware***: IBM-PC compatible 386 or better, with at least 2.5 MB of free extended memory, VGA graphics; or Silicon Graphics machine. ***Software:*** MS-DOS 5.0, or higher, or Windows 3.xx or Silcon Graphics IRIX 4.0.5 or 5.xx. ***Cost***: None, it is supplied on two 3.5 inch disks.

9.2.3.3 FPEtool

FPEtool consists of the modules:

SYSTEM SETUP is a utility for changing file destination, source directories, operating units, and screen colors.

FIREFORM is a collection of quick procedures that solve single-parameter questions, e.g. ceiling jet temperature, mass flow through a vent, upper layer temperature, etc. It treats temperature variation and smoke layer descent, atrium smoke temperature, buoyant gas head, ceiling jet temperature, ceiling plume temperature, egress time, fire/wind/stack forces on a door, lateral flame speed, effective time of thermal resistance, mass flow through a vent, plume filling rate, radiation ignition of a near fuel, smoke flow through an opening, sprinkler/detector response, energy requirements for flashover, upper layer temperature, and ventilation limits.

MAKEFIRE models the growth, steady state, and decay phases of the each fuel element in the compartment. It consists of routines that create and edit fire files that specify the fire heat release rate and fuel pyrolysis rate as a function of time.

FIRE SIMULATOR predicts the effects of fire growth in a 1-room, 2-vent compartment with sprinkler and detector. It predicts: temperature and smoke properties ($O_2/CO/CO_2$ concentrations and optical densities), heat transfer through room walls and ceilings, sprinkler/heat and smoke detector activation time, heating history of sprinkler/heat detector links, smoke detector response, sprinkler activation, ceiling jet temperature and velocity history (at specified radius from the fire), sprinkler suppression rate of fire, time to flashover, post-flashover burning rates and duration, doors and windows which open and close, forced ventilation, post-flashover ventilation-limited combustion, lower flammability limit, smoke emissivity, and generation rates of CO/CO_2 pre- and post-flashover.

CORRIDOR calculates the characteristics of a moving smoke wave.

3rd ROOM predicts smoke conditions developing in a room and the time to human untenability from heat and inhalation exposure.

FPEtool is described in Deal (1995). No formal benchmarking has been performed, nor has any source code validation or validation against experimental data but the code is based on a number of experiments: Notarianni (1993), Walton (1993), Nelson (1992), Peacock (1993b), Heskestad (1986) and Hinkley (1975).

Sponsor: GSA, Public Building Service, Office of Real Property Management. ***Developing Organization***: NIST Building and Fire Research Laboratory, Gaithersburg, MD 20899. ***Original Developer***: Same. ***Current Custodian for Development***: Walter W. Jones, Group Leader Fire Modeling and Applications Group, Building and Fire Research Laboratory, NIST, Gaithersburg, MD 20899, Phone: (301) 975-6887. ***Distribution***: Richard Bukowski, Fire Research Information Services, Building and Fire Research Laboratory, NIST Gaithersburg, MD 20899-001 Phone: (301) 975-6853. ***Hardware***: IBM PC compatible computer with an 8086 or higher CPU, a minimum of 640 kB RAM, and 3 MB of hard-disk space. CGA graphics or better. A math coprocessor is recommended. ***Software*** TSCREEN needs DOS version 3.1 or higher. It runs under Windows. To modify and compile FPEtool, Microsoft QuickBasic v7.0, or equivalent is needed as well as the BASIC source code.

9.2.3.4 COMPBRN III

COMPBRN III is a single-room zone fire model for probabilities risk assessment calculations. It models fires in an open or closed compartment using a 2-layer zone model approach. Thermal radiation and flame propagation are included. It requires a large amount of input: room geometry ventilation, doorway information, fuel bed geometry, orientation, thermal and combustion properties, ignition fuel location and properties, ignition temperatures of fuel, burning rate as a function of incident heat flux (surface-controlled burning) or ventilation (ventilation-controlled burning), and a definition of which fuel elements exchange thermal radiation with each other and with ceiling and walls. It outputs: total mass burning rate, total heat release rate, hot gas layer temperature and depth, indication of fuel cell damage and burning, radiative and total heat fluxes to targets, fuel cellmass, flame height over each fuel cell, flame temperature over each fuel cell, fuel cell temperature, and heat transfer coefficient (convective) for each cell. It only is applicable to single-room, pre-flashover fire compartments. Quasi-steady state assumptions are made; it is highly

Chemical Process Accident Consequence Analysis

dependent on experimentally determined parameters. It has no spatial resolution of temperature gradients nor fire suppression models. Flame spread occurs along discrete elements, and requires user input of burning rate parameters. It has a flame spread model, runs quickly, and calculates thermal radiation to targets. It applies to small fires for which correlations are generally applicable, simple compartment geometries, and not to fast-transient fires. The user's manual is Ito (1985). Related references are: Chung (1985), Siu (1980), Siu (1982), Sui and Apostolakis (1982), Siu (1983), Lambright (1989), and Nicolette (1989).

Sponsor: USNRC. *Developing Organization*: Mechanical Engineering Department, University of California at Los Angeles. *Original Developer*: G. Apostolakis, N. Siu, V. Ito, G. Chung *Current Custodians*: EPRI has latest utility version, COMPBRN 3-e; Prof G. Apostolakis, MIT, has COMPBRN III (presumably public domain). *Hardware*: IBM PC or clone with 386 CPU or better or a Workstation. *Software*: FORTRAN Compiler, Text Editor if code modifications are needed. *Cost*: Nominal for disks and manuals, may be more restricted from EPRI.

9.2.3.5 VULCAN

VULCAN is applicable to internal compartment fires, external pool and jet fires. It uses finite difference methods to solve governing equations. Included are combustion, soot, turbulence, and thermal radiation models. Its "first-principles" approach solves the basic conservation equations. The model reference is: Holen (1990); user's manual reference is: Kameleon and Kameleon. The source code verification is available for review from V. Nicolette at SNL. References for validation against experimental data: Holen (1990); Gritzo (1995a); Gritzo (1995b). Current plans are to extend the model to multiple fuels, including solid material, and to include suppression models (both gaseous inerting and water sprays). The model has been applied extensively to fires in compartments, on offshore oil platforms, and pool fires. This model is at the forefront of the science of fire modeling.

Sponsor: Norwegian Oil and Gas Industries and U.S. Defense Nuclear Agency. *Developing Organization*: SINTEF (Norway) and SNL. *Original Developer*: SINTEF (Norway) and Norwegian Institute of Technology (NTH). *Current Custodian*: SINTEF and SNL. *Hardware*: Workstation preferably silicon graphics to take advantage of integrated graphics and calculations. The larger (i.e., faster) the computer the better. *Software*: FORTRAN Compiler, C Compiler, Text Editor. See Holen, et al, previous reference. *Cost*: Not for sale, commercially, arrangements must be made with SNL and SINTEF.

9.3 EPA's Exposure Model Library and Integrated Model Evaluation System

9.3.1 IMES

The U.S. Environmental Protection Agency concentrates on modeling the dispersion of toxic material in the environment and the health effects thereof rather than accident prevention. The reason for this is that most of the environmental problems are not due to accidents but to intentional or careless releases in a time when there was less concern for the environment. However in EPA's area of emphasis they are very comprehensive and most if not all of their models are in the public domain although contractor variations may not be.

EPA, 1996,[i] is the third edition of EML/IMES on CD-ROM for distributing exposure models, documentation, and the IMES about many computer models used for exposure assessment and other fate and transport studies as developed by the EPA's Office of Research and Development (ORD).

The disk contains over 120 models in files that may contain source and executable code, sample input files, other data files, sample output files, and in many cases, model documentation in WordPerfect, ASCII text or other formats. The disk contains IMES with information on selecting an appropriate model, literature citations on validation of models in actual applications, and a demonstration of a model uncertainty protocol.

The IMES software is an MS-DOS application capable of running on a network. Model codes and documentation can be downloaded from the CD-ROM to a hard drive. An MS-DOS interface is included to provide easy access to IMES and to the model directories, although such is not required to access the files. This third edition provides an HTML Interface for viewing the model directories and Internet sources of some the models.

Note: The CD-ROM is provided as a means of model distribution. With a few exceptions, the models are not designed to run on the CD-ROM (although the IMES is). Many of the model files were originally designed for distribution from floppy diskettes or bulletin board systems; keep this in mind as you read the installation instructions which are provided in the model directories. These instructions have not been rewritten for installation from CD-ROM. In most cases, you can simply copy the files directly to a hard drive for execution of the model. Many files have been stored in a compressed format using PKZIP® from PKWARE, Inc. The directory PKZIP contains the pkunzip utility for extracting these files. For models distributed and maintained by the EPA Center for Exposure Assessment Modeling (CEAM), an installation directory may be provided; use the supplied installation program to install these models, because these programs check your hardware and software for possible conflicts.

Getting Started

The Exposure Models Library may be accessed either with the DOS EML interface program or with an HTML browser program such as Netscape or Microsoft Internet Explorer. To use the DOS EML interface and to access the IMES, enter the drive letter of your CD ROM drive, set the default directory to EML and then enter EML (e.g., D: CD\EML EML). From the menu, run or download the available models of IMES to your hard disk. Also you may access EML directories with your HTML browser by opening the EMLINTRO.HTM flle in the root directory. The HTML

---

[i] For copies of the CD or for technical assistance, call Richard Walentowicz at the EPA Office of Research and Development at 202-260-8922 [waientowicz.rich@epamail.epa.gov]. Information on individual models should be obtained from the model developers.

files contain links to sites for many of the models that were on the Internet when the CD was produced. With Internet access, these links may be checked for model updates.

Table 9.3-1 Computer Models on the EPA Disk

Air models		Groundwater models		Surface Water Models		Mulitmedia models
BLP	MPTER	AT123D	PHREEQE	CEQUALICM	PLUMES	MULTIMED
CAL3HQC	OCD	BALANCE	PHREEQER	CEQUALR1	QUAL2E	PATRIOT
CALINE3	PLUVUE2	BIOPLUMII	PRZM	CEQUALRIV1	RECOVERY	PCGEMS
CALPUFF	PTPLU	BTEX	RANDWALK	CEQUALW2	RESTMF	*Other models*
CDM2	RAM	CHEMFLO	RETC	CORMIX	RIVMOD	FGETS,
CMB7	RPM	DBAPE	RITZ	EXAMS	SEDDEP	NRDAM
COMPLEX1	RTDM	EPA-WHPA	RUSTIC	FATE	SMPTOX	PIRANHA
CRSTER	RVD2	FEMWATER	STF	HEC5Q	THERMS	THERdbASE
CTDMPLUS	SCREEN3	GEOPACK	STFBASE	HEC6	WASP5	
CTSCREEN	SDM	GLEAMS	SUTRA	MICHRIV	WQAM	
DEGADIS	SHORTZ	HSSM	SWANFLOW	MINTEQA2	WQRRS	
EKMA	TOXLT	HST3D	SWIFT 11	PCPROUTE	WQSTAT	
FDM	TOXST	MOC	TETRANS	*Non-point source models*		
ISC3	TSCREEN	MOCDENSE	TRIPLOT	AGNPS	GWLF	SWMM
LONGZ	UAM	MODFLOW	TWQM	ANSWERS	HELP	SWRRBWQ
MESOPUFF	VALLEY	MOFAT	VLEACH	CREAMS	HSPF	USGSREG
MPRM	VISCREEN	MT3D	WATEQF	EPIC	P8-UCM	WEPP
		OASIS	WHEAM	EXPRES	PREWET	WMM
		PESTAN		FHWA	SWAT	

System Requirements

Needed are an Intel, or equivalent, personal computer with an ISO 9660 compliant CD ROM, hard disk with 8 MB available (to run IMES on the hard disk), at least 540 kB of free RAM; VGA monitor or better. Operating System: MS DOS 3.1 or higher (for DOS EML). Windows 3.x or higher for Windows-based nonpoint source model selection module. The Exposure Models Library/IMES consists of a selection of fate transport and ecological models which can be used for exposure assessments in various environmental media. The models were developed primarily by various EPA offices and other federal agencies and are in the public domain. Models included are listed in Table 9.3-1. Model codes and documentation are contained within subdirectories for each category.

IMES was developed to assist in the selection and evaluation of exposure assessment models and to provide model validation and uncertainty information on various models and their applications. IMES is composed of 3 elements: 1) Selection - a query system for selecting models in various environmental media, 2) Validation - a database containing validation and other information on applications of models, and 3) Uncertainty - a database demonstrating application of a model uncertainty protocol.

9.3.2 PIRANHA

This directory contains the USEPA Pesticide and Industrial Chemical Risk Analysis and Hazard Assessment system. Documentation for PIRANHA is contained in a MANUALS subdirectory; enter PIRANHA C: where C: is a hard disk to receive the output files to run the system. For efficient operation of PIRANHA, transfer the files from the CD-ROM to your hard disk (it requires 28 MB). Data files are accessed from the CD-ROM when running PIRANHA.

9.3.3 REACHSCN

ReachScan is a PC program package that aids an exposure assessor by bringing to PC's the Hydrologically Linked Data Files (HLDF) maintained by the Assessment and Watershed Protection Division (AWPD) of the Office of Wetlands, Oceans, and Watersheds (OWOW) of USEPA. HLDF in addition to others contains the files: REACH, GAGE, the Industrial Facility Discharge File (IFD), and the Water Supply Database (WSDB). These files contain information on rivers in the United States (their names, lengths, and flows), industrial dischargers, and drinking water supplies. Although capable of several functions, ReachScan's primary purpose is to:

- Report public drinking water utility intakes located down river from a discharge point (an industrial facility).
- Report the distance between a discharge point and intake.
- Estimate the concentration of a discharged chemical at the intake point by: 1) simple dilution, or by a simplified degradation equation with dilution.
- Report the name of the down river water utility and the number of people served by it.

The program can also search upriver and downriver for industrial facilities starting from an industrial facility, water utility, or river segment. This directory also contains PDM3, a PC-based program to perform screening level exposure assessments. PDM3 predicts how many days per year a chemical at a concern level in an ambient water body will be exceeded after being discharged from an industrial facility.

9.3.4 SIDS

The Select Industrial Discharge System (SIDS) assists exposure assessors and other analysts by consolidating current data contained in various EPA water-related environmental information systems in one electronic database. SIDS contains data that have been selectively downloaded from nine existing information systems and is designed to provide a comprehensive database of facility, discharge, and receiving stream information for select industrial dischargers to surface waters in the United States.

9.3.5 THERDCD

THERdbASE (Total Human Exposure Risk database and Advanced Simulation Environment) is a PC-based computer software that manages data and models. It has exposure, dose, and risk related data files organized in a Database Engine. Exposure, dose, and risk related models, that draw information from the Database Engine, are organized in a Modeling Engine. THERdbASE comes from an ongoing project with EPA's Office of Research and Development (National Exposure Research Laboratory, Las Vegas, NV) to develop a tool for making exposure, dose, and risk assessments.

The Database Engine in THERdbASE has the following data groups: 1) Population Distributions, 2) Location/Activity Patterns, 3) Food Consumption Patterns, 4) Agent Properties, 5) Agent Sources (including use patterns), 6) Environment Characterizations, 7) Environmental Agent Concentrations, 8) Food Contamination, 9) Physiological Parameters, 10) Risk Parameters, and 11) Miscellaneous Data Files.

The Modeling Engine in THERdbASE has the following model groups: 1) Population Distributions, 2) Location/Activity Patterns, 3) Food Consumption Patterns, 4) Agent Releases Characteristics, 5) Microenvironment Agent Concentrations, 6) Macroenvironment Agent Concentrations, 7) Exposure Patterns and Scenarios, 8) Dose Patterns, and 9) Risk Assessment.

A list of the data related functions in THERdbASE is:

- Data View for viewing contents of data files in tabular form; coded/decoded information can be viewed with a button click -columns of data can be set in "show" or "hide" mode; multiple data files can be viewed at the same time; data files can also be edited when in view mode.
- Data Query using simple query structures (filters) can be built on field values or on coded and decoded information.
- Statistics on the data fields: summary statistics (mean, std dev, min, max), percentile values at desired intervals, and linear regression on two numerical data fields.
- Graphics- model generated results can be viewed as: single/multiple bar, pie, line, and XY.
- Print- output to any printer: full contents of a data file, query results, graphs.
- Data Save/Export/Import- subsets of data files can be saved data can be exported to or imported from "dbf" format.
- Data Create- data files can be edited in advanced mode- new files can be created and information imported.

The model related functions in THERdbASE are:

- Data Input models are achieved through a standardized procedure as single values, custom distributions (normal, lognormal, etc.), distributions based on data files present in THERdbASE, or specific percentile values. Models have relational access to input data files.
- Model Execution uses efficient algorithms to access input data, to perform numerical simulations, to generate appropriate output data using a batch process.
- Data Output- output from models is done in the following two ways: as THERdbASE data files or as pre-set graphs.

Data files on the THERdbASE CD are: 1990 Bureau of Census Population Information, California Adult Activity Pattern Study (1987-88), AT&T-sponsored National Activity Pattern Study (1985), Chemical Agents from Sources, Chemical Agent Properties, Air Exchange Rates, Information from EPA's TEAM (Total Exposure Assessment Methodology) Studies, Information from EPA's NOPES (NonOccupational Pesticides Exposure Study) Studies, Information from EPA's AIRS (Aerometric Information Retrieval System), and Human Physiological Parameters.

The models in the THERdbASE CD are: Chemical Source Release, Instantaneous Emission, Chemical Source Release, Timed Application, Indoor Air (2-Zone), Indoor Air (N-Zone), Exposure Patterns for Chemical Agents, Benzene Exposure Assessment Model (BEAM), Source Based Exposure Scenario (Inhalation + Dermal), and Film Thickness Based Dermal Dose.

Minimum computer requirements to run THERdbASE are a 486 CPU, IBM or clone, at least 8 MB of RAM, at least 40 MB of disk space, color VGA monitor, a mouse, Microsoft Windows 3.1. To install THERdbASE, execute the File/Run option from within Windows and specify SETUP.EXE found in the THERDCD directory. This program leads you through the installation. Updates for THERdbASE can be obtained over the World Wide Web from the Harry Reid Center for Environmental Studies at the University of Nevada, Las Vegas (http://www.eeyore.lv-hrc.nevada.edu).

9.4 Summary

This chapter concerned calculating the release and dispersion of hazardous materials. Section 9.1 discussed "hands-on" calculations to give the reader a better "feel" of the physics. In this, the hoop stress for the failure of tanks and pipes was presented, gas and liquid flow through a hole in a tank was calculated as were: UVCE (unconfined vapor cloud explosions), vessel rupture, BLEVE (boiling liquid expanding vapor explosion), and missiles. Atmospheric dispersion of chemical releases can be hand calculated using the Gaussian diffusion model but it was not discussed since it had been discussed previously (Section 8.3). While these basic methods are interesting, the modern approach to consequence calculations use computers. Section 9.2 summarized codes for release and dispersion, explosion and fire that were reviewed by the DOE APAC (Accident Phenomenology and Accident Assessment) program. Most of these codes are free and some can be downloaded over the Internet. Section 9.3 overviewed the EPA CD ROM entitled "Exposure Models Library and Integrated Model Evaluation Systems" which includes models for atmospheric, aquatic, point source, multiple source and multimedia models. It is likely that some of the information on this disk will be valuable to the reader.

Chapter 10

Assembling and Interpreting the PSA

10.1 Putting It Together

At this point, following the chapters, the objectives have been defined, the effect of government regulations and standards are known, accidents have been identified and analyzed by various methods to determine the probability of an accident, and the accident consequences have been calculated. These parts must be assembled to present the risk and the analysis of the risk according to its various contributors.

10.1.1 Integrated and Special PSAs

The scope and complexity of a PSA depends on its purpose. Large, expensive PSAs of a whole plant or group of plants (e.g., Canvey Island, Indian Point 2) are designated as *integrated*, meaning that the objective is the assessment of the overall accident[a] risk of a complex process. Integrated PSAs may be done to determine the financial exposure of a process as it is presently being conducted, or it may be in response to regulatory pressure. A *special* PSA is an accident risk analysis that is limited to a component, system, group of systems or a type of problem (e.g. DB-50 contactors, chlorine rail-loading facility, stress corrosion cracking, thermal shock, etc). We will address the problem of assembling an integrated PSA with the understanding that a special PSA is a subset of this and steps that are not needed for a smaller analysis may be excluded.

10.1.2 Assembling the PSA

The assembly process (Figure 10-1) brings together all of the assessment tasks to provide the risk, its significance, how it was found, its sensitivity to uncertainties, confidence limits, and how it may be reduced by system improvements. Not all PSAs use fault trees and event trees. This is especially true of chemical PSAs that may rely on HAZOP or FMEA/FMECAs. Nevertheless the objectives are the same: accident identification, analysis and evaluation. Figure 10-1 assumes fault tree and event tree techniques which should be replaced by the equivalent methods that are used.

---

[a]There is little point in calculating the routine risk of an operating plant because it can be measured, however, there is good reason to calculated the routine risk of a plant before it is constructed or before restarting after major modifications to determine if the risk is tolerable.

Assembling and Interpreting the PSA

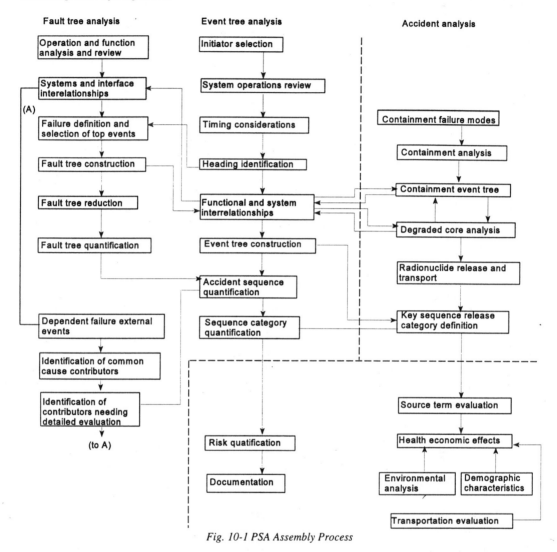

Fig. 10-1 PSA Assembly Process

- **Event Tree Development** delineates the various accident sequences. This activity includes: an identification of initiating events and the systems that respond to each initiating event. Systems that serve to mitigate, but do not contribute to the prevention of a core melt accident, may be excluded, depending upon the purposes of the PSA. Separate event trees are generally constructed for each initiating event or class of initiating events having a unique event tree structure.
- **System Analysis** consists of five subtasks: 1) Link system event tree sequences to a containment event tree, 2) Eliminate unnecessary combinations, 3) Group remaining sequences by

376

release category, 4) Calculate sequence probabilities, and 5) Calculate release category frequencies. In the RSS method of PSA construction, a major task links system event trees with the for containment response. As illustrated in Figure 10-1 different end-points in the event tree may require different containment trees. It is also clear that the number of discrete sequences when the containment failure modes are considered can become quite large. This is treated in several ways. First, the end points in which the containment does not fail or fails by containment by-pass through the basemat are not considered further, because the offsite consequences are small compared with the consequences of containment. Also, end points of low relative probability and low consequences may be grouped with similar low-risk end points. Finally, the sequences must be grouped into release categories to reduce the number of consequence calculations to practicality.

- *Analysis of External Events* uses the models developed in the plant system analysis with considerations for seismic, fire, flood, high winds and missiles on the plant. Additional event trees or their equivalent may be needed for the external events.

- *Source Terms and In-Plant Transport* the fraction of the inventory that makes it to the environment must be estimated. Computer models are to track the hazardous materials that are released from their process confinement through transport and deposition inside the plant to their release into the environment as a source term for atmospheric and aquatic dispersion.

- *Consequence Analysis* the effects of the in the plant on the workers and the dispersed hazardous materials on the public and environment is assessed using computer models.

- *Uncertainty Analysis* determines the effects on the overall results from uncertainties in the database, assumptions in modeling, and the completeness of the analysis. Sensitivity analyses determine the robustness of the results; importance calculations are useful for identifying and prioritizing plant improvements.

Assembling this material is a large data management task to assure the availability and traceability of all needed information with the output of one task fitting the input of another. This requires that quality assurance has been adhered to throughout the PSA project.

10.2 Insights and Criticisms

Whether constructing a new PSA, modifying or using an old PSA it is worthwhile to ask what has been learned from previous works and some of the criticisms that have been levied against them. Fortunately the insights were largely supplied by NUREG-1050 from which the bulk of the following information is taken. The criticisms cover some of the same ground presented in Chapter 1 except with the bulk of PSA completed, the criticisms are more relevant.

10.2.1 Insights from Past PSAs

1. The estimated probability of accidents leading to a hazardous release is higher than thought before PSAs were applied.
2. PSAs estimate that the frequency of reactor damage cover about two orders of magnitude: from about 1E-5/y to 1E-3/y. This variation is attributable to: plant design, construction, and operation, to site characteristics, scope of the PSAs, and methods and analytical assumptions. Such comprehensive studies of comparable chemical process plants do not exist.
3. For accidents to affect the public, containment and off-gas processing must fail, such as occurred at Bhopal and Chernobyl and did not occur at TMI-2.
4. Plants meeting all applicable NRC regulatory requirements have been found to vary significantly in calculated risk and in terms of the key accident sequences.
5. PSAs calculate that accidents more severe than those postulated in the design basis are the principal risk contributors. This indicates that safety designed with a specific goal is successful.
6. Latent cancer is calculated to be the primary risk from a nuclear accident (this may be due to the conservatism in the low-dose models). At Chernobyl, most of the deaths were from fire and impact. Chemical process risk depends on the chemicals being processed. Experience shows that processing poisons poses the highest risk to public and workers.
7. Most of the financial risk of a process accident is suffered by the plant, not by offsite property, although unemployment may be a major contributor to the financial loss.
8. Demography has a large effect. At Bhopal, the surrounding population was dense, considering the hazardousness of the operation, and impeded emergency action.

10.2.2 Criticism of Past PSAs

This section reflects on the limitations of the PSA process and draws extensively from NUREG-1050. These subjects are discussed as plant modeling and evaluation, data, human errors, accident processes, containment, fission product transport, consequence analysis, external events, and a perspective on the meaning of risk.

10.2.2.1 Completeness

Nuclear PSAs tend to accept a set of initiators such as the WASH-1400 set or the EPRI set and proceed with little introspection as to completeness of the set for a specific plant. A well-known omission is sabotage. In general, earthquake and fire is treated poorly; tornado is usually not addressed nor are other aspects of adverse weather such as freezing and ice storms which may have common cause potential. The record has not been too good. After the Browns Ferry fire, it was claimed that such was addressed in WASH-1400. If so, attention was only brought after the fact. The TMI-2 sequence was not addressed in the PSAs to an extent that action was taken to prevent it.

10.2.2.2 Representativeness

Representativeness can be examined from two aspects: statistical and deterministic. Any statistical test of representativeness is lacking because many histories are needed for statistical significance. In the absence of this, PSAs use statistical methods to synthesize data to represent the equipment, operation, and maintenance. How well this represents the plant being modeled is not known. Deterministic representativeness can be answered by full-scale tests on like equipment. Such is the responsibility of the NSSS vendor, but for economic reasons, recourse to simplified and scaled models is often necessary. System success criteria for a PSA may be taken from the FSAR which may have a conservative bias for licensing. Realism is more expensive than conservatism.

Certainly errors should be conservative, but cumulative conservatism reduces accuracy and usefulness while raising costs.

10.2.2.3 Data Adequacy

The nuclear equipment failure rate database has not changed markedly since the RSS and chemical process data contains information for non-chemical process equipment in a more benign environment. Uncertainty in the database results from the statistical sample, heterogeneity, incompleteness, and unrepresentative environment, operation, and maintenance. Some PSAs use extensive studies of plant-specific data to augment the generic database by Bayesian methods and others do not. No standard guidance is available for when to use which and the improvement in accuracy that is achieved thereby. Improvements in the database and in the treatment of data requires substantial industrial support but it is expensive.

10.2.2.4 Human Errors

Human actions can initiate accident sequences or cause failures, or conversely rectify or mitigate an accident sequence once initiated. The current methodology lacks nuclear-plant-based data, an experience base for human factors probability density functions, and a knowledge of how this distribution changes under stress.

The analysis of human actions is complicated because a human is a responsive system like a servo. Such analysis does not lend itself to simple models as do inanimate components. Classifying human actions into the success or failure states used in logic models for plant equipment does not account for the wide range of possible human actions. A generally applicable model of the parameters that affect human performance is not yet available.

The uncertainties in human error rates may be within the stated uncertainty bounds, but such is not demonstrated from sparse experiments. Both the qualitative description of the human interaction logic and the quantitative assessment of those actions rely on the virtually untested judgment of experts.

10.2.2.5 Accident Analysis

Nuclear PSAs contain considerable uncertainty associated with the physical and chemical processes involved in core degradation, movement of the molten core in the reactor vessel, on the containment floor, and the response of the containment to the stresses placed upon it. The current models of these processes need refinement and validation. Because the geometry is greatly changed by small perturbations after degradation has commenced, it is not clear that the phenomena can be treated.

Full-scale core melt-water and core melt-concrete experiments are not feasible, but considerable progress has been made since the RSS in theoretical work, model development, and small-scale experiments using nonradioactive molten metal. There is increasing evidence that steam explosions in the reactor vessel and in the containment are incapable of failing the containment. There is still much to be done in characterizing the dispersal of the debris bed that may follow reactor vessel failure, the size of the corium particles formed, and the coolability of the resulting debris bed.

10.2.2.6 Containment Analysis

Chemical process plants do not have a large domed containment but rely on vessel integrity and offgas processing. Most nuclear plants have such containment for which the stresses may be calculated if a scenario is defined. However such calculations are uncertain regarding the time of failure, the hole size, and location.

Computer sensitivity studies show that hole size strongly affects the fraction of fission products released from the containment. The failure location determines mitigation due to release into another building in which condensation and particulate removal occur. The quantity released depends on the time of containment fails relative to reactor vessel failure. If containment integrity is maintained for several hours after core melt, then natural and engineered mechanisms (e.g., deposition, condensation, and filtration) can significantly reduce the quantity and radioactivity of the aerosols released to the atmosphere.

It has also been shown that heat radiation may heat and weaken the containment. The failure of penetration seals due to high temperature has been investigated, but full scale containment failure test have not been done.

10.2.2.7 Hazardous Material Transport Analysis

Chapters 8 and 9 presented computer codes that are available for computer hazardous material release and transport. Many of these codes have been tested using controlled experiments with varying agreement depending upon the code's applicability to the phenomena. In the author's opinion, the accuracy of the consequence calculation is not much better than the calculation of accident probabilities.

10.2.2.8 External Events

The methodology for assessment of external events is qualitatively satisfactory but not quantitatively. Little confidence should be placed in any estimates of the risk from external initiators compared to those from internal initiators. This is exacerbated by the fact that the external risk is the larger of the two in many cases.

10.2.2.9 Risk

The validity of the risk of nuclear that was estimated by the Reactor Safety Study was questioned when the TMI-2 accident occurred. This showed a misunderstanding of the probabilistic nature of risk. A single event neither proves nor disproves a statistical result; it may, however, question the usefulness of risk as a safety measure.

Chapter 11

Applications of PSA

11.1 U.S. Commercial Nuclear PSAs

11.1.1 Commercial Nuclear PSAs before IPE

The primary motivation of PSAs is to assess the risk of the plant to the public. The immediate purpose of the RSS was to support the Price-Anderson hearings on liability insurance (i.e., assess the financial exposure of a nuclear power reactor operator) a purpose which, even today, is beyond PSA technology. However, PSA is sufficiently precise to provide relative risk comparisons of reactor designs and sites. These uses of PSA were presented at the Indian Point hearings, and in defense of Shoreham. The PSAs for the high-population-zone plants (Limerick, Zion, and Indian Point) were prepared to show that specific features of these plants compensate for the higher population density relative to plants studied in the RSS.

The RSS was used by pro- and anti-nuclear power advocates in the Proposition 15 (nuclear power ban) voting in California; the Barsebek and Ringhals PSAs were prepared as part of the nuclear controversy in Sweden; the German Risk Study, 1981, had similar applications as did the Sizewell-B studies in Great Britain.

After publication of WASH-1400, the Reactor Safety Study Methodology Applications Program (RSSMAP) was developed to extend PSA to other plants with other types of containment. RSSMAP addressed four plants: Oconee, a B&W PWR in a large dry containment; Calvert Cliffs, a CE PWR also in a large dry containment; Sequoyah, a W (Westinghouse, see Glossary) with an ice condenser containment; and Grand Gulf, a GE BWR in a Mark III containment.

Following the TMI-2 accident and the endorsement of extended uses of PSA by the Kemeny Commission and the Rogovin Inquiry Group (Bernero, 1984), the Interim Reliability Evaluation Program (IREP) began with a study of Crystal River. This was followed by similar studies of four more plants (Table 11.1-1). These studies were limited to internal events leading to core melt with modest uncertainty and sensitivity analyses. The Accident Sequence Evaluation Program (ASEP) modeled 18 BWRs and 40 PWRs to represent the U.S. Nuclear Power Industry using a number of modules for different plant systems. The idea was to assemble fault trees of the modules that distinguish a plant and that this assembly will be the plant PSA. This work is largely unpublished, but NUREG/CR-3301 reports results at the system level. The NRC chose to test the methodology

Table 11.1-1 PSA Studies of U.S. Plants Completed before IPE

Complete PSAs	Surry I (W)	Peach Bottom 2 (GE)	Big Rock Point (GE)	Zion 1 & 2 (W)
Indian Point 2 & 3 (W)	Yankee Rowe (W)	Limerick 1 & 2 (GE)	Shoreham (GE)	Susquehanna (GE)
Oconee 3 (B&W)	Clinch River Breeder Reactor (W)	Millstone 3 (W)	Millstone 1 (GE)	Oyster Creek (GE)
Sequoyah 1 (W)	Browns Ferry 1 (GE)	Midland (B&W)	Seabrook (W)	McGuire 1 (W)
Bellefonte (B & W)	TMI 1 (B&W)	Diablo Canyon (W)	Crystal River 3 (B&W)	
IREPs or RSSMAPs	Oconee 3 (B & W)	Sequoyah 1 (W)	Grand Gulf 1 (GE)	Calvert Cliffs (CE)
Crystal River 3 (B & W)	Browns Ferry 1 (GE)	Arkansas 1 (B & W)	Millstone 1 (GE)	Calvert Cliffs 2 (CE)
NUREG-1050	Peach Bottom 2 (GE)	Grand Gulf (GE)	Surry (W)	Sequoyah (W)

with in-depth PSAs, reported in NUREG-1150, with the technical bases in NUREG/CR-4550.[a] Independently, the utilities owning Limerick, Zion, and Indian Point sponsored studies in response to the concern of the NRC regarding plants in high population zones. The Limerick PSA was performed by GE/SAIC in the style of RSS with alternative realistic analyses where justified. The Zion and Indian Point analyses (ZIP) were performed by one prime contractor (PL&G), using basically the same methods for all four plants. Other utilities have also contracted for and participated in the preparation of their own PSAs to estimate their public and financial risks, to obtain PSA-guided operational and safety improvements, and to provide technical rebuttals to NRC-ordered changes. In some cases, utilities have sponsored PSAs in response to local opposition to their license, e.g., Shoreham and Seabrook. Table 11.1-1 summarizes these PSAs.

11.1.2 ATWS

A long evolving use of PSA was for Anticipated Transients without Scram (ATWS) which extended over 15 years to culminate in NUREG-0460 which was upset by the Salem failure-to-scram incident and the subsequent SECY Letter 83-28. Other special studies have been: (a) value-impact analysis (VIA) studies of alternative containment concepts (e.g., vented containment, NUREG/CR-0165), (b) auxiliary feedwater studies, (c) analysis of DC power requirements, (d) station blackout (NUREG/CR-3220), and (e) precursors to potential core-damage accidents (NUREG/CR-2497), to name a few of the NRC sponsored studies.

---

[a] Summaries of these results were reported in a special session of the American Nuclear Society ANS trans. 53, pp. 315-325, 1986.

Industry response is increasingly couched in PSA terms. In fact, the previously mentioned Big Rock Point PSA was not primarily to determine the public risk but to respond to post-TMI requirements. These were generic so that the low power of Big Rock Point in relation to the containment and unique design of this plant evoked a PSA-based utility response to show that the requirements applied to this plant could not be supported on the basis of a value-impact analysis (VIA). The PSA showed this conclusion and indicated areas of very effective improvements. Some other utility PSA responses have related to diesel surveillance testing, pipe rupture, sabotage, and hazardous gas releases near a plant. Many requests for technical specification (Tech Spec) changes are based on PSA.

11.1.3 Risk-Based Categorization of NRC Technical and Generic Issues

Perhaps the first well-known use of PSA insights in the regulatory process occurred in 1978 when the Probabilistic Analysis Staff performed a study (SECY/78-7516) to categorize the existing technical and generic issues facing the NRC. The primary objective was to assist in identifying, for action plans, those issues that have the greatest relative risk significance. One hundred thirty-three task action items were reviewed and assigned to four broad categories ranging from those having high-risk significance to those not directly relevant to risk. Of these, only 16 fell into the high-risk categories. The ranking aided the selection of the generic issues that would be designated "unresolved safety issues." This effort was redone by the NRC's Office of Nuclear Reactor Regulation (NRR) to include all TMI action plan issues and post-TMI issues.

A most important product of a PSA is the framework of engineering logic generated in constructing the models. The patterns, ranges, and relative behaviors can be gained only from the integrated consistent approach PSA to develop insights into the design and operation of a plant.

In reaching decisions, the regulator must compare the credibility of the information gained from PSA with the credibility of other sources of information. Thus, consider the magnitude of the estimated risk, the degree of uncertainty associated with it, the results of sensitivity analyses, and the net estimated effect of proposed alterations on the overall safety of the plant. Considered in this way, the PSA insights provide the decision maker with a comprehensive, contextual definition of the problem and a better quantitative understanding of the public risk with its uncertainties. The PSA objectively, instead of subjectively, judges aspects of risk.

11.1.4 Use of PSA in Decision Making

The decision maker must weigh the qualitative and quantitative PSA insights and other information. There is no cookbook answer to this question, because it will depend heavily on the nature of the issue, the results of the PSA, the nature of other information, and other factors affecting overall judgment. The PSA should provide for the following needs:

- Scope and depth of the PSA should match the needs of the decision maker.
- Peer reviews should attest the credibility of the PSA results.
- Qualitative insights from the PSA should be plausible.

- The effect of alternative regulatory actions on the estimated risk, including costs and impact.

11.1.5 Value-Impact Analysis (VIA)

PSA techniques were used in the 1978 study of underground siting of nuclear power plants. NUREG/CR-0165, using insights from the RSS, considered nine alternative designs found to be the "logical alternatives." Filtered atmospheric venting was judged to be the most promising design alternative by value impact. This study contributed to the subsequent focusing of containment research on filtered vents and a diminution of interest in the underground siting of nuclear power plants. Other VIAs are the NRC-sponsored study of steam generators (Morehouse et al., 1983); three EPRI-sponsored VIA studies; a study of VIA applied to eight NUREG-0737 items specific to BWRs (NUREG-0626), Fullwood et al. (1984); a methodological development and application to shutdown heat removal, Fiksel et al. (1982); and a study of three auxiliary feedwater alternatives for PWRs, Stamatelatos et al. (1982).

11.1.6 Response to the TMI-2 Accident

The NRC response to the TMI-2 accident (NUREG-0737) impacted the four U.S. LWR manufacturers. Consumers Power Company (a midwest utility) estimated that cost of retrofitting their Big Rock Point plant to be $125 million hence, not economical, given its age and size (240 MWt), nor were the requirements appropriate for unique characteristics of the plant and its siting. The utility initiated a PSA in which their plant personnel fully participated with the PSA contractors to identify and correct some areas of weakness. The NRC accepted the results without the major modifications that would otherwise have been needed. Furthermore, the understanding of the plant and its PSA by plant personnel made this a model for using PSA in plant improvement.

11.1.7 Special Studies

11.1.7.1 Analysis of DC Power Supply Requirements

This study was undertaken as part of the NRC's generic safety task A-30, "Adequacy of Safety Related DC Power Supplies" (NUREG-0737). The issue stemmed from the dependence of decay-heat-removal systems on DC power supply systems. It was found that dc power-related accident sequences could represent a significant contribution to the total core-damage frequency. It was also found that this contribution could be substantially reduced by the implementation of design and procedural requirements, including the prohibition of certain design features and operational practices, augmentation of test and maintenance activities, and staggering test and maintenance activities to reduce human errors.

11.1.7.2 Station Blackout

Two studies resolved the Unresolved Safety Issue A-44, "Station Blackout." The first study, "The Reliability of Emergency AC Power Systems in Nuclear Power Plants," when combined with the relevant loss-of-offsite-power frequency, provides estimates of station-blackout frequencies for 18 nuclear power plants and 10 generic designs. The study also identified the design and operational features most important to the reliability of AC power systems. The second study, "Station Blackout Accident Analysis" (NUREG/CR-3226), focused on the relative importance to risk of station-blackout events and the plant design and operational features that would reduce this risk.

11.1.7.3 Precursors to Potential Severe Core-Damage Accidents

This study (NUREG/CR-2497) applies PSA techniques using operating experience to identify the high-risk features of plant design and operation. The operating experience base is derived from the licensee event reports (LERs) to find multiple events that, when coupled with postulated events, lead to plant conditions that could eventually result in severe core damage.

In the first 2.5 years of the precursor study, 169 significant precursors were identified for the 432 reactor-years of operating experience represented by LERs submitted from 1969 to 1979; preliminary findings show 56 precursors for 126 reactor-years of operating experience for 1980-1981. The results were used to analyze accident sequences to estimate core melt frequencies for operating plants and to compare these results with the estimates in existing PSAs.

11.1.7.4 Pressurized Thermal Shock

In addressing pressurized thermal shock, probabilistic assessments were used to derive screening criteria to identify operating plants needing modification. The owners' groups associated with the different PWR designs submitted estimates of frequencies of severe overcooling events. Analytical efforts using PSA techniques are exemplified in NUREG/CR-4183 and NUREG/CR-4022.

11.1.7.5 BWR Water Level--Inadequate Core Cooling

PSA techniques were used in the analysis of TMI Action Item II.F.2, BWR water level inadequate core cooling. The results indicate that there is no need for additional instrumentation for detecting inadequate core cooling in the BWRs. The study showed that improvements in existing systems for water-level measurement and in operator performance (Shoreham and Limerick) could make the predicted core-damage frequency due to failure in water-level measurements in the plants analyzed much smaller than the total core-damage frequency predicted in recent PSAs for BWRs.

11.1.7.6 Other TMI Action Plan Items

PSA provided tools for the analysis of TMI action items II.K.3.2 and II.K.3.17 (i.e., the frequencies of LOCAs caused by stuck-open pressurizer PORVs and outages of ECCS. respectively). The results of II.K.3.2 indicated that the frequency of small LOCAs from stuck-open PORVs was in the range of the small-LOCA frequency in the RSS and that no additional measures to reduce the PORV-LOCA frequency are required. The purpose of the data collection under item II.K.3.17 was to determine whether cumulative outage requirements are needed in the technical specifications and which plants had a significantly greater than average cumulative ECCS outage time.

11.1.7.7 LWR Shutdown Risk

Usually, nuclear power plant PSAs are performed for what is thought to be the most dangerous operating condition: full power operation. The following, from Chu et al. (1995), analyzes the risk when a reactor is not only shutdown but the primary is opened and partially drained called mid-loop operation (MLO). Chu et al. found that the mean core damage frequency during MLO from internal events is about an order of magnitude less than during full power operation, for internal fire it is about twice that of power operation, and for internal floods it is about an order of magnitude less.

In 1989, the NRC questioned the premise that most risk is in full power operation and requested an investigation of non-full power risk by BNL using Surry (two 788 MWe PWRs) as a model and SNL using Grand Gulf for a BWR model. NUREG-1150 results for the same plants in power operation was used for comparison. The investigation was Level 3 for internal events and Level 1 for internal fire, seismic, and flood induced core damage.

Phase 1 identified vulnerable plant CDFs as high, medium, or low and provided core damage accident scenarios and risk information for a more detailed Phase 2 analysis in which MLO[b] was selected because it is the most hazardous shutdown condition. Phase 1, plant outages are: refueling, drained maintenance, non-drained maintenance with the residual heat removal (RHR) system, and non-drained maintenance without the RHR system. Plant operational states (POSs) were defined and characterized for each outage type. Each POS is characterized by a set of operating conditions (e.g., temperature, pressure, and configuration). For example, 15 POSs are used to represent a refueling outage. Each POS used Surry-specific data from reviewing operating and abnormal shutdown procedures, shift supervisor logs, monthly operating reports, and by performing thermal-hydraulic calculations (details are in Volumes 3 and 4 of NUREG/CR-6144).

Internal Events

The high risk POSs: R6, R10, and D6 have different decay heats and plant configurations:

---

[b] In MLO, the reactor is shutdown, and the coolant is drained until it is level with the center of the hot leg of the primary cooling system

- R6, for eddy current testing of the steam generator tubes, is performed in a refueling outage by quickly draining the RCS loops, hence, the decay heat rate is high.
- R10, also in a refueling outage is for additional test and maintenance after some decay of the residual heat.
- D6 is MLO with the highest rate of decay: heat among the three mid-loop POSs.

Four time segments were defined for decay heat shutdown with unique sets of success criteria corresponding to the decay heat rate. R6 and D6 require all four time segments, but R10 only uses time segments 3 and 4. A statistical analysis determined the probability that an accident occurs in a particular time segment and an accident occurred. An event tree was developed for each initiating event, POS and time segment; 160 event trees were developed for 16 initiating events.

Internal Fires

The internal fire analysis is conventional. Surry-specific fire frequencies were assessed for important plant areas involving cable fires, transient fires, and equipment fires as a local phenomenon. The analysis locates the fire source and vulnerable equipment and/or cables. Cables of important systems were traced for vulnerability to a fire in various locations. Fire growth calculations (COMPBRN-IIIe) were performed for the most vulnerable locations. A transition-diagram type of suppression model was used, which, in conjunction with the fire growth model, gives the probability of damage given a fire in that location. The damage fraction with event tree and fault tree models gives the core damage frequency (CDF) for a fire scenario. The scenario-dependent human error probabilities (HEPs) were estimated using the same human reliability analysis (HRA) method that was used in the internal event analysis.

Fire Assumptions

- Multiple initiators were not considered.
- Fire barrier failure was not considered because of its small probability.
- All cables were not traced if the risk they posed was small.
- It was assumed fire frequency data from plants to Surry may be used.
- Relay fires were screened probabilistically.
- Both power and control cable were traced and considered.

Internal Floods

The internal flood analysis was conventional and consistent with the NUREG 1150 and IPE studies. The steps were:

1. Identification of Risk Susceptible Flood Areas was performed in the Phase I study using information from a plant visit and from NUREG-1150 and the IPE.
2. Flood Event Frequency Estimates were developed from flooding events in nuclear power plants with adjustments for plant-specific features and data. The data were from the IPE Surry flood analysis, industry sources, and licensing event reports (LERs). Some plant specific models were developed for the circulating water (CW) and service water (SW) lines

in the Turbine Building whose flood dominates the risk. The dominant failure mechanism is large pipe rupture due to water hammer. Generic pipe failure data and Thomas (1981) were used when plant-specific data was unavailable. Flood barriers and other mitigators were considered in the damage states.

The equipment damage due to internal flood is a function of flood level which depends on the flooding rate, confluence, drainage and mitigations. Flood damage states were developed to take these factors into account by incorporating the time dependency of the flood event and mitigations. The damage states represent a flood-exceedance frequency at a location with equipment damage criteria. The flood initiating frequencies were developed to incorporate simple recovery actions as expected from experience.

Location and Scenario Identification

Potential accident scenarios and flood locations were identified from plant drawings and the RHR system fault tree that identifies the equipment and support needed for RHR system operation. The equipment location was correlated with flood areas with consideration for plant features which may impede or divert the flow. The flood scenarios identify the effect on systems required to prevent core damage. Quantification accounts for the rate of rise of the flood relative to the critical level in each specific plant area. The time available for any recovery action is calculated from the volume and the flow rate.

Results

Table 11.1-2 presents the CDFs of different events in MLO, summarizes results, and compares them with NUREG-1150 and IPE results for the same plants. Also, presented are the level-1 uncertainty analysis, results. The MLO mean core damage frequency from internal events is about an order of magnitude lower than that of full power operation. The mean core damage frequency due to internal fire is comparable to that of power operation. Like internal events, the mean core damage frequency due to internal floods is comparable to that of power operation as estimated by IPE.

Table 11.1-2 Shutdown CDF (Mid-Loop Operation) and Full Power CDF

Study			Mean	5th%	50th%	95th%	Error
Internal Events	Full Power NUREG-1150/y		4.0E-5	6.8E-6	2.3E-5	1.3E-4	4.4
	Full Power IPE/y		7.4E-5				
	MLO/y (while in MLO)		4.9E-6	4.8E-7	2.1E-6	1.5E-5	5.7
Internal Fires	Full Power NUREG-1150/y		1.1E-5				
	MLO/y (while in MLO)		2.2E-5	1.4E-6	9.1E-6	7.6E-5	7.2
Internal Flood	Full Power IPE/y		5.0E-5				
	MLO/y (while in MLO)		4.8E-6	2.2E-7	1.7E-6	!.8E-5	9.0
Seismic Events	Full Power NUREG-1150/y	LLNL	1.2E-4				33
		EPRI	4.0E-5				4.4
	MLO/y (while in MLO)	LLNL	3.5E-7				32
		EPRI	8.6E-7				37

Insights

Internal Event and Operator Response in MLO - dominated CDF because of operator failure to mitigate the accident with a very large uncertainty.

Procedures for Shutdown Accidents - there are few procedures for accidents in MLO. Procedure for loss of decay heat removal, AP 27.00, is the only one written specifically for the shutdown scenarios analyzed in this study. The procedure is conservative regarding equipment needed to establish reflux cooling and feed-and-bleed. This study treated fewer than the number of steam generators specified in the procedure for reflux cooling as a recovery action with more realistic success criteria for feed-and-bleed with high decay heat. Some procedures are wrong for MLO conditions, e.g., procedure for loss of offsite power, AP 10.00, states that when the EDG is the only source of power, the Component Cooling Pump should not be in service. During shutdown, CCW flow to the RHR heat exchanger is necessary for decay heat removal.

Instrumentation: level instrumentation for MLO (standpipe level and ultrasonic) has limited use in a shutdown accident. Standpipe level is correct in the absence pressure in the system. Ultra-sonic is correct only when the level is within the reactor coolant loops.

Thermal-Hydraulic Analysis must be extended to at least 24 hours into the accident. This study used calculations for feed-and-bleed operation with a charging pump, and with gravity feed from the refueling water storage tank (RWST).

Maintenance Unavailability: shift supervisor's logs and minimum equipment lists so unavailability of equipment needed to mitigate a shut-down accident.

Isolation of Reactor Coolant Loops - contributes to the core damage frequency by being isolated in a refueling outage to disable the affected steam generators for decay heat removal upon loss of RHR. In a cold shutdown, the steam generators are maintained in a wet lay-up condition with the secondary side filled with water. During mid-loop operation, reflux cooling with the steam generators is a possible mitigation for loss of RHR especially in a station blackout.

Valve Arrangement of Auxiliary Feedwater System and Main Steam System during Shutdown - the auxiliary feedwater system has six MOVs in the flow path to the steam generators and the main steam nonreturn valves are normally closed during shutdown. These valves depend on offsite power and are difficult to manually open.

Potential for Plugging the Containment Sump When Recirculation Is Needed: transient material and equipment are brought into containment during shutdown. This could increase the potential for plugging the containment sump if an accident requiring recirculation from the containment sump occurs.

Internal Fire CDF is less likely during shutdown than under power but welding or cable fires may be more likely. A fire at shutdown may be detected and extinguished more readily because of floor traffic. However cable routing to support RHR or to mitigate an accident during mid-loop operation can result in fire critical intersections. Risk significant scenarios involved in the emergency switchgear room (ESGR), the cable vault and tunnel (CVT), and the containment. In the ESGR, the most risk significant ESGR scenarios occur in locations where many cables for the H and the J emergency divisions come together in close proximity. The tunnel part of the CVT is a constrained space, where damage would quickly propagate to both divisions. In containment, the risk is from the relatively high fire frequency and non-separation of the two RHR divisions.

Internal Flood is dominated by Turbine Building flood events which may be initiated by valve or expansion joint failures in the main inlet lines of the circulating water system. At Surry the circulating water system is gravity fed from a large capacity intake canal and isolation is not quick and cannot be accomplished in a timely manner. At other plants, stopping dedicated cooling water pumps would stop the flood. The potential draining of the intake canal into the Turbine Building is a dominant risk because of the spatial interdependence of this plant. For both units, the Emergency Switchgear Rooms are located in the Service Building on the same elevation as the Turbine Building basement, but separated by a fire door and 2 foot flood dikes. A large flood could overflow the dikes, and the two-unit ESGR to lose emergency power for both units. Normal off-site power to the plant would not be affected since the normal switch gear room is at a higher elevation.

11.1.8 Individual Plant Evaluation PSAs

11.1.8.1 Overview

On August 8, 1985, the U.S. Nuclear Regulatory Commission (NRC)[c] requested the operators of nuclear power plants in the U.S. to perform Individual Plant Examinations (IPE) on their plants. IPEs are probabilistic analyses that estimate the core damage frequency (CDF) and containment performance for accidents initiated by internal events (including internal flooding, but excluding internal fire). Generic Letter (GL) 88-20 was issued to implement the IPE request to "identify any plant-specific vulnerabilities to severe accidents and report the results to the Commission."

75 IPEs for 108 plants (Table 11.1-3) were prepared by licensees and/or consultants in varying depth and methodology. They were reviewed by the NRC Office of Nuclear Regulatory Research (NUREG-1560) regarding four major areas:

1. Risk of U.S. nuclear power,
2. Plant improvements indicated by their IPE,
3. Plant-specific features and modeling assumptions affecting risk, and
4. Use of IPEs for risk-based regulation.

Core damage and containment performance was assessed for: accident sequences, component failure, human error, and containment failure modes relative to the design and operational characteristics of the various reactor and containment types. The IPEs were compared to standards for quality probabilistic risk assessment. Methods, data, boundary conditions, and assumptions are considered to understand the differences and similarities observed.

---

[c] "Policy Statement on Severe Accidents Regarding Future Designs and Existing Plants" (Federal Register, 50FR32138).

Table 11.1-3 IPE Submittals
(BWR is GE, B&W is Babcock & Wilcox, CE is Combustion Engineering, and W is Westinghouse)

Group	Plants / Description
BWR 1/2/3	Big Rock Point; Dresden 2,3; Millstone 1; Oyster Creek; Nine Mile Point 1
	These plants generally have separate shutdown cooling and containment spray systems and a multi-loop core spray system. With the exception of Big Rock Point, which is housed in a large dry containment, these plants use an isolation condenser
BWR 3/4	Browns Ferry 2; Brunswick 1,2; Fermi 2; Fitzpatrick; Limerick 1 ,2; Monticello; Quad Cities 1, 2;· Susquehanna 1, 2; Cooper; Duane Amold; Hatch 1, 2; Hope Creek; Peach Bottom 2, 3; Pilgrim 1;· Vermont Yankee
	These plants are designed with two independent high-pressure injection systems, namely reactor core isolation cooling and high-pressure coolant injection (HPCI). The associated pumps are each powered by a steam-driven turbine. These plants also have a multi-loop core spray system and a multi-mode residual heat removal (RHR) system that can be aligned for low-pressure coolant injection, shutdown cooling, suppression pool cooling, and containment spray functions.
BWR 5/6	Clinton; Grand Gulf 1; Perry 1; River Bend; LaSalle 1, 2; WNP 2; Nine Mile Point 2
	These plants use a high-pressure core spray (HPCS) system that replaced the HPCI system. The HPCS system consists of a single motor-driven pump train powered by its own electrical division complete with a designated diesel generator. These plants also have a single train low-pressure core spray system, as well as a multi-mode RHR system similar to the system design in the BWR 3/4 group.
B&W	ANO 1; Crystal River 3; Davis Besse; Oconee 1, 2 & 3; TMI 1
	The B&W plants use once-through steam generators. Primary system feed-and-bleed cooling can be established through the pressurizer power relief valves using the high-pressure injection (HPI) system. The HPI pump shutoff head is greater than the pressurizer safety relief valve setpoint. Emergency core cooling recirculation (ECCR) requires manual alignment to the containment sumps. The reactor coolant pumps (RCPs) are generally a Byron Jackson design.
CE	ANO 2; Calvert Cliffs 1,2; Fort Calhoun 1; Maine Yankee; Millstone 2; Palisades; Palo Verde 1, 2, 3;· San Onofre 2, 3; St. Lucie 1, 2; Waterford 3
	The CE plants use U-tube steam generators with mixed capability to establish feed-and-bleed cooling. I Several CE plants are designed without pressurizer power-operated valves. The RCPs are a Byron I Jackson design. West nghouse
W 2-loop	Ginna ; Kewaunee;· Point Beach 1, 2; Prairie Island 1, 2
	These plants use U-tube steam generators and are designed with air-operated pressurizer relief valves. Two independent sources of high-pressure cooling are available to the RCP seals. Decay heat can be removed from the primary system using feed-and-bleed cooling. ECCR requires manual switchover l to the containment sumps. The RCPs are a Westinghouse design.
W 3-loop	Beaver Valley 1, 2; Farley 1, 2; North Anna 1, 2; Robinson 2; Shearon Harris 1; Summer; Surry 1, 2; Turkey Point 3, 4
	This group is similar in design to the Westinghouse 2-loop group. The RCPs are a Westinghouse design.
W 4-loop	Braidwood 1, 2; Byron 1, 2; Callaway; Catawba 1, 2; Comanche Peak 1, 2; DC Cook 1, 2; Diablo Canyon 1, 2; Haddam Neck; Indian Point 2, 3; McGuire 1, 2; Millstone 3; Salem 1, 2; Seabrook; Sequoyah 1, 2; South Texas 1, 2; Vogtle 1, 2; Watts Bar 1; Wolf Creek; Zion 1, 2
	The Westinghouse 4 loop group includes nine plants housed within ice condenser containments. Many of these plants have large refueling water storage tanks such that switchover to ECCR either is not needed during the assumed mission time or is significantly delayed. The RCPs are a Westinghouse design.

11.1.8.2 Impact of the IPE Program on Reactor Safety

The primary licensee goal was to: "identify plant-specific vulnerabilities to severe accidents" that could be reduced with low-cost improvements. However, GL 88-20 did not define what constitutes a vulnerability; hence, the IPEs are diverse in the vulnerability criteria, hence, a problem considered to be a vulnerability at one plant may not be identified as a vulnerability at another plant. Less than half of the licensees identified "vulnerabilities" in their IPE submittal; but nearly all of the licensees identified areas warranting investigation for potential improvements. Thus, the IPE program provided impetus for improving the overall safety of nuclear power plants.

Four boiling water reactor (BWR), and 15 pressurized water reactor (PWR) licensees acknowledged plant vulnerabilities. Some BWR vulnerabilities are failure of:

- Water supply to isolation condensers
- High-pressure coolant injection and reactor core isolation cooling when residual heat removal has failed
- Control of low-pressure injection during an anticipated transient without scram (ATWS)
- Drywell steel shell to prevent melt-through in a Mark I containment

Some PWR vulnerabilities are:

- Reactor coolant pump (RCP) seal failures that lead to a loss of coolant accident (LOCA)
- Turbine-driven auxiliary feedwater pump reliability·from design and maintenance problems
- Internal flooding caused by component failures
- Operator failure to switch over from the coolant injection phase to the recirculation phase
- Critical switchgear ventilation equipment failure leading to loss of emergency buses
- Operator failure to depressurize during steam generator tube rupture
- Valve surveillance that leads to interfacing system LOCAs
- Loss of certain electrical buses
- Compressed air system failures
- Inability to crosstie buses during loss of power

Most than 500 plant design or operation improvements were identified by licensees: 45% of the improvements were procedural/operational changes; 40% were design/hardware changes. Few of the improvements change maintenance. Procedural or design changes require training revision. Approximately 45% of the plant improvements had been implemented; 25% had been implemented and were credited in the IPEs. Other improvements were planned or under evaluation. Utility activities and other requirements (primarily the station blackout rule) are reasons for change. While these changes may not be due to the IPE, the IPE prioritizes the implementation. Specific improvements vary with plants, but common improvements were: changes to AC and DC power, coolant injection systems, decay heat removal systems, heating ventilation and air conditioning systems, and PWR reactor coolant pump seals.

11.1.8.3 IPE Core Damage Frequency Assessment

IPE results genrally are consistent with the results of previous NRC and industry risk studies in indicating that the CDF is often determined by many accident sequence combinations, rather than a failure. The largest contributors to CDF vary among the plants (e.g., LOCAs dominate some IPE while station blackout [SBO] dominates others). Support systems, whose design varies considerably with plant, are important to all plants because they can cause multiple front-line system failures. This may account for much of the variability in the IPE results.

Consistent with previous risk studies, the average CDFs reported for BWRs is less than for PWR as shown in Figure 11.1-1. BWR and PWR results are strongly affected by the support system considerations discussed above, but some differences between the two types of plants explain why the average is less for BWRs:

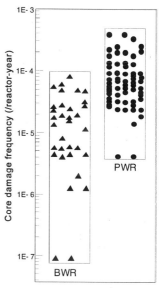

Fig. 11.1.-1 Summary of BWR and PWR CDFs from IPEs

- BWRs have more injection systems than PWRs and can depressurize more easily using low-pressure injection (LPI) systems which reduces the BWR LOCA vulnerability. Figure 11.1-1 shows some BWR CDFs are higher than some PWR CDFs The variation in the CDFs is driven primarily by a combination of the following factors:
- Plant design differences (primarily in support systems such as cooling water, electrical power, ventilation, and air systems)
- Variability in modeling (e.g., accounting for alternative accident mitigating systems)
- Differences in data used in quantifying the models

11.1.8.4 Containment Performance

IPE accident progression analyses generally are consistent with typical analyses of containment performance. Failure mechanisms identified in the past as important are important in the IPEs. In general, the IPEs confirmed that the larger volume PWR containments are more robust than the smaller BWR pressure suppression containments in meeting the challenges of severe accidents (Table 11.1-4). Containment performance is important to reducing the risk from early releases. Analyses in IPE submittals emphasize phenomena, mechanisms, and accident scenarios associated with early

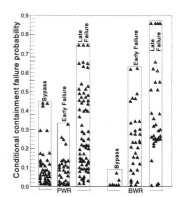

Fig. 11.1.-2 Conditional Containment Failure Probabilities from the IPEs

	Table 11.1-4 IPE Submittals' Containment Types (MK I, MK II, and MK III are GE Mark I, II, and III resp.; LDSA are large dry sub-atmospheric and ice uses ice for heat absorption)
MK I	Browns Ferry 2; Brunswick 1, 2; Cooper; Dresden 2, 3; Duane Arnold Fermi 2; Fitzpatrick; Hatch 1, 2; Hope Creek; Millstone 1; Monticello; Nine Mile Point 1; Oyster Creek; Peach Bottom 2, 3; Pilgrim 1; Quad Cities 1, 2; Vermont Yankee
	The Mark I containment consists of two separate structures (volumes) connected by a series of large pipes. One volume, the drywell, houses the reactor vessel and primary system components. The other volume is a torus, called the wetwell, containing a large amount of water used for pressure suppression and as a heat sink. The Brunswick units use a reinforced concrete structure with a steel liner. All other Mark I containments are free-standing steel structures. The Mark I containments are inerted during plant operation to prevent hydrogen combustion.
MK II	LaSalle 1, 2; Limerick 1, 2; Nine Mile Point 2; Susquehanna 1, 2; WNP 2
	The Mark II containment consists of a single structure divided into two volumes by a concrete floor. The drywell volume is situated directly above the wetwell volume and is connected to it with vertical pipes. Most Mark II containments are reinforced or post-tensioned concrete structures with a steel liner, but WNP 2 uses a free-standing steel structure. These containments are also inerted during plant operation to prevent hydrogen combustion.
MK III	Clinton; Grand Gulf I; Perry I; River Bend
	The Mark III containment is significantly larger then Mark I and Mark II containments, but has a lower design pressure. It consists of the drywell volume surrounded by the wetwell volume, with both enclosed by the primary containment shell. The drywell is a reinforced concrete structure in all Mark III containments, but the primary containment is a free-standing steel structure at Perry and River Bend, and a reinforced concrete structure with steel liner at Clinton and Grand Gulf. These containments are not inerted, but rely on igniters to burn off hydrogen and prevent its accumulation during a severe accident.
LDSA	ANO1, 2; Beaver Valley 1, 2; Big Rock Point; Braidwood 1, 2; Byron 1, 2; Callaway; Calvert Cliffs 1, 2; Comanche Peak 1, 2; Crystal River 3; Davis Besse; Diablo Canyon 1, 2; Farley 1, 2; Fort Calhoun 1; Ginna; Haddam Neck; Indian Point 2, 3; Kewaunee; Maine Yankee; Millstone 2, 3; North Anna 1, 2; Oconee 1, 2, 3; Palisades; Palo Verde 1, 2, 3; Point Beach 1, 2; Prairie Island 1, 2; Robinson 2; Salem 1, 2; San Onotre 2,3; Seabrook; Shearon Harris 1; South Texas 1, 2; St. Lucie 1, 2; Summer; Surry 1, 2; TMI-1; Turkey Point 3, 4; Vogtle 1, 2; Waterford 3; Wolf Creek; Zion 1, 2
	The large dry and subatmospheric containment group includes of 65 units, of which 7 have containments kept at subatmospheric pressures. These containments rely on structural strength and large internal volume to maintain integrity during an accident. Most of these containments use a reinforced or post-tensioned concrete design with a steel liner. A few units are of steel construction.
Ice	Catawba 1, 2; DC Cook 1, 2; McGuire 1, 2; Sequoyah; Watts Bar 1
	The ice condenser containment is a pressure suppression containment that relies on the capability of l the ice condenser system to absorb energy released during an accident. The volumes and strength of l these containments are less than those of the large dry containments. Ice condenser containments l also rely on igniters to control the accumulation of hydrogen during an accident. Seven of the ice l condenser units have a cylindrical steel containment surrounded by a concrete secondary containment. The remaining two units have a concrete containment with a steel liner and lack secondary l containments.

Table 11.1-5 Findings from the IPEs
Transients (other than station blackouts and ATWS) Important contributor for most plants because of reliance on support systems; failure of such systems can defeat redundancy in front-line systems. Both plant-specific design differences and IPE modeling assumptions contribute to variability in results: • Use of alternative systems for injection at BWRs • Variability in the probability that an operator will fail to depressurize the vessel for LPI in BWRs • Availability of an isolation condenser in older BWRs for sequences with loss of decay heat removal (DHR) • Susceptibility to harsh environment affecting the availability of coolant injection capability following loss of DHR • Capability to use feed-and-bleed cooling for PWRs • Susceptibility to RCP seal LOCAs for PWRs • Ability to depressurize the reactor coolant system in PWRs affecting the • Ability to use LPI ability to cross-tie systems to provide additional redundancy
SBOs Significant contributor for most plants, with variability driven by: • Number of redundant and diverse emergency AC power sources • Availability of alternative offsite power sources • Length of battery life • Availability of firewater as a diverse injection system for BWRs • Susceptibility to RCP seal LOCAs for PWRs
ATWS Normally a low contributor to plant CDF because of reliable scram function and successful operator responses BWR variability mostly driven by modeling of human errors and availability of alternative boron injection system PWR variability mostly driven by plant operating characteristics, IPE modeling assumptions, and assessment of the fraction of time he plant has an unfavorable moderator temperature coefficient
Internal floods Small contributor for most plants because of the separation of systems and compartmentalization in the reactor building, but significant for some because of plantspecific designs. Largest contributors involve service water break.
LOCAs (other than interfacing system LOCAs [ISLOCAs] and steam generator tube ruptures [SGTRs]) - Significant contributors for many PWRs with manual switchover to emergency core cooling system recirculation mode BWRs generally have lower LOCA CDFs than PWRs for the following reasons: • BWRs have more injection systems • BWRs can more readily depressurize to use low-pressure systems
ISLOCAs Small contributor to plant CDF for BWRs and PWRs because of the low frequency of initiator Higher relative contribution to early release frequency for PWRs than BWRs because of low early failure frequency from other causes for PWRs
SGTR Normally a small contributor to CDF for PWRs because of opportunities for the operator to isolate a break and terminate an accident, but important contributor to early release frequency

release such as early structural failure of the containment, containment bypass, containment isolation failures and, for some BWR plants, deliberate venting of the containment.

As a group, the large dry PWR containments have significantly smaller conditional probabilities of early structural failure (given core melt) than the BWR pressure suppression containments. Nonetheless, containment bypass and isolation failures generally are more significant for the PWR containments. As seen in Figure 11.1-2, these general trends are not necessarily true for individual IPEs. Conditional probabilities for both early and late containment failure for a number of large dry PWR containments were higher than those reported for some of the BWR pressure suppression containments.

BWR results, grouped by containment type, follow expected trends and indicate that, in general, Mark I containments are more likely to fail during a severe accident than the later Mark II and Mark III designs. The range of failure probabilities is large for all BWR containment designs, and there is a large variability from different failure modes within BWR containment groups. Plants in all three BWR containment groups showed a significant probability for early or late structural failure, given core damage.

PWRs indicate that most of the containments have relatively low conditional probabilities of early failure, although a large variability exists in the contributions of the different failure modes for both large dry and ice condenser containments.

11.1.8.5 Findings from the IPE

These are presented in Table 11.1-5.

11.1.8.6 Plant Improvements from the IPEs[d]

The Severe Accident Policy Statement formulates systematic safety examinations for detection of accident vulnerabilities and implementation of cost-effective changes. The NRC issued Generic Letter 88-20 to implement this plan through IPEs. While the primary goal was the identification of plant vulnerabilities, no definition of vulnerability was provided. Only 4 operators of BWRs identified vulnerabilities and only 16 operators of PWRs did so. Over 500 plant improvements were identified, but few vulnerabilities were.

Many of the plant improvements are generic. The most frequently cited BWR improvements addressed station blackout; the PWR improvements addressed loss of power and loss of reactor coolant pump seal cooling. Changes to improve core cooling or injection reliability, under a loss of AC power are identified for both PWRs and BWRs. Improvements to prevent internal flooding and interfacing system LOCAs are identified more often in PWR than in BWR IPEs. A summary of the commonly identified plant improvements is provided in Table 11.1-6. Most of the plant improvements are procedural/operational changes (~45%), design/hardware changes (~40%), or both. Some of the improvements are associated with other requirements (primarily the station blackout rule). Improvements may be prioritized using their IPE.

The BWR improvements related to station blackout improve AC and/or DC power reliability, extend battery operable life and address other system weaknesses. AC power system modifications consisted of: adding or replacing diesel generators, and/or redundant offsite power capabilities to improve the AC power recovery. BWR 5s and 6s were improved by a high pressure core spray diesel. Some licensees improved diesel generator cooling. DC power system improvements included: alternate battery charging, battery upgrades, load shedding and improved cross-tie capabilities. These changes were made to assure at least one train of DC during loss of AC power. Other changes are to reduce the vulnerability to station blackout and reduce core damage from other types of accidents.

---

[d] From Drouin (1996).

Table 11.1-6 Plants Improvements from IPE from Drouin, 1996

Area	Improvements (b- BWR, p - PWR)	Area	Improvements (b- BWR, p - PWR)
AC power	-Added or replaced diesels (b,p) -Added or replaced gas turbine (b,p) -Redundant offsite power capabilities (b,p) -Improved bus/unit cross-tie (b,p)	Support Systems	-Procedures and portable fans for alternate room cooling on loss of HVAC (b,p) -Temperature alarms in rooms to detect loss of HVAC (b,p) - Revise procedures and training for loss of support systems (b,p)
DC power	-New batteries, chargers, invertors (p) -Alternate battery charging capabilities (b) -Increased bus load shedding (b,p)	ATWS	-Revise training on bound control rods (b) -Install automatic ADS inhibit (b) -Install alternate boron injection system (b) -Capability to remove power to bus on trip breaker failure (p) -Install Westinghouse ATWS mitigation (p)
Coolant injection systems	-Replace ECCS pump motor with air-cooled motors (b) -Align LPCI or CS to CST on loss of suppression pool (b) -Align fire water for reactor vessel injection (b) -Revise HPCI and RCIC actuation trip setpoints (b) -Revise procedures to inhibit ADS for non-ATWS (b) -Improve procedures and training for switch to recirculation (p) -Increased training for feed and bleed (p)	RCP Seal LOCAS	-Evaluate or replace RCP seal material (b,p) -Independent seal injection or charging pump for station blackout (p) -Operator training on tripping pumps on loss of cooling (p) -Review HPSI dependency on CCW (p)
		SG tube rupture	-Revise procedure for higher inventory of water in the BWST or refill BWST -Procedures and training to isolate affected SG
Decay heat removal	-Add hard pipe vent (b) -Portable fire pump to improve isolation condenser makeup (b) -Install new AFW pump or improving existing pump reliability (p) -Refill CST when using AFW (p) -Align firewater pump to feed steam generator (p)	Internal flooding	-Increased flood protection of components (b) -Periodic inspection of cooling water piping and components (b,p) -Revise procedures for inspecting flood drain and flood barriers (p) -Install water-tight doors.
Containment performance	- Provide alternate power source to hydrogen igniters (b) -Enhance communication between sump and cavity (p) -Inspect piping for cavity flooding systems (p) -Revise procedures to use PORVs to depressurize vessel after damage (p)	Interfacing systems LOCA	-Review surveillance procedure re isolation valves (b,p) -Modify procedures to depressurize PCS (p) -Revise training for ISLOCA (b,p)
		Miscellaneous	Incorporate IPE insight into the operator training program (b,p)

Generally these changes are to improve AC-independent core cooling systems during prolonged operation. Such as isolation condenser improvements in the early BWR designs, e.g., better valve reliability, additional guidance for isolation condenser use during an extended blackout, and providing prolonged fire water capability for the isolation condensers during a long station blackout condition. Similarly, some improvements ensure better reliability of AC-independent coolant injection systems in later BWR plants (i.e., HPCI and RCIC).

Some licensees have a switch to bypass RCIC high steam tunnel temperature trips. Some licensees are evaluating improvements to prevent seal LOCAs from loss of seal cooling which are most important for W plants, but B&W licensees identified improvements related to alternate seal flow capability under loss of power conditions. The use of high temperature seals is noted for some W plants. Many PWR IPEs identify AFWS improvements. These include additional backup water supplies such as the firewater system and redundant pump cooling capability. Other reliability

Applications of PSA

improvements are identified in a few of the plants, including the ability to operate AFWS manually even under loss of DC power using one AC-independent core cooling system thereby increasing the chances of preventing core damage due to loss of secondary cooling. Other examples of PWR improvements are procedural and design improvements to deal with internal flooding. PWR licensees identified changes to deal with steam generator tube ruptures, interfacing system LOCAs, and other miscellaneous system weaknesses including better procedural guidance for dealing with steam generator tube ruptures and improved valve status checking for lessening the potential for interfacing system LOCAs.

The IPE Program, while identifying few plant-specific severe accident vulnerabilities susceptible to low-cost fixes, served as a catalyst for improving the overall safety of nuclear power plants. Furthermore, improvements at one plant may be applicable to another plant.

11.1.9 Risk-Based Regulation

The purpose of regulation is to protect the public from the risk of nuclear power. PSA make it possible to express the risk numerically. However, NRC regulations have been proscriptive to achieve an unknown risk level. Clearly too much regulation that destroys the industry is not desirable and too little may fail to protect the public. A possible solution is the use of PSA in regulations. Such has been resisted because of the uncertainties; on the other hand there are uncertainties in proscriptive regulation but no attempt is made to express them quantitatively. The following condenses material from Murphy (1996) to reflect NRC thinking on this subject.

The Atomic Energy Act, 40 years ago, required that an operating license should pose "no undue risk to the health and safety of the public." Prior to PSA, reliance was placed on defense-in-depth using redundancy and diversity. With the development of PSA, the NRC began to use insights from risk assessments - not directly in the regulations, but to prioritize technical and generic issues and to help identify Unresolved Safety Issues (SECY-78-616, 1978).

Generic Issue Resolution

This pioneering effort was followed by quantitative guidelines to prioritize Generic Safety Issues (SECY-83-221, 1983) designed to shift work from risk-free items to concentrate work on higher risk items. Experience with these guidelines, led to revision (SECY-93-108, 1993) that removed some conservatism. By 1996, over 400 issues affecting safety were prioritized (NUREG-0933).

For example, discussions on the conditions for license renewal beyond the original 40-year operating license will exceed the number of allowed metal fatigue cycles. Twennty more years of fatigue cycles would result from license renewal. Extensive research from Japan and the U.S (NUREG/CR-6336) showed that the fatigue life in a simulated light water reactor (LWR) environment is shortened compared with air at room temperature. Thus, the lower-bound ASME Design Curve for S-N (strain amplitude vs cycles to failure) in the ASME Code Section III was not correct in a reactor environment. A risk-based study resolved the effect on plant safety resulting from exceeding the cumulative useage factors for plant components with new fatigue design curves (NUREG/CR-5999). Then the data were fitted with analytical models (NUREG/CR-6336) to calculate the mean and standard deviation for the fatigue life. Assuming a log-normal distribution, the prohahility of failure was calculated. Finally, a method was developed by the staff (Hrabal, 1995)[7] for application to reactor coolant system components from which the effect on CDF could be calculated. Conclusions are:

- The fatigue failure risk in the primary piping system and reactor vessel in the upper 1/3 of the core is negligible.
- The risk from fatigue failure for the reactor vessel in the lower 2/3 of the core is negligible.

Regulatory Analysis and Safety Goal Considerations

These are addressed ind:The Backfit Rule (10 CFR 50.109, and the NRC Safety Goal Policy Statement (SECY-89-102). The Backfit Rule applies not to the regulated industry, but to the NRC staff. It says that backfitting is required if it will result in substantial increase in safety and the direct and indirect costs of backfitting are justified. (This limitation does not apply if the modification is necessary for compliance with regulations).

The Safety Goal Policy Statement was published to define acceptable radiological risk from nuclear power plant operation, and by implication provide a de minimus risk to be assured without cost considerations. Safety beyond the minimum requires cost-benefit analysis. Since being promulgated, bulletins and generic letters have heen imposed to enhance safety, under the provisions of 10 CFR 50.109, the Backfit Rule.

Individual plants, with extra safety features, may exceed the Safety Goals. The five plants analyzed in NUREG-1150 meet or exceed the Safety Goals. This approach is contained in the Regulatory Analysis Guidelines (NUREG/CR-0058).

Severe Accidents

Risk-based information provides a foundation for regulation of severe accidents. Early PRAs, with large uncertainties, indicated risk that was above or below the Safety Goals depending on containment performance. Consequently the NRC developed an Integration Plan for Closure of Severe Accident Issues (SECY-88-47) with six main elements to this plan: 1) individual plant examinations (IPE), 2) containment performance improvements, 3) improved plant operations, 4) severe accident research, 5) external event considerations, and 6) accident management.

Based on PRAs and information on containment loading, licensees with BWR Mark I containment were requested to install hardened vents (NRC, 1989). All other licensee were requested to consider the need for other potential enhancements on a plant-specific basis as part of their IPE.

The severe accident research program improved public risk assessment, reduced uncertainties, and the reliance on subjective expert opinion. To close two severe accident issues in NRC's Severe Accident Research Plan (NUREG-1365): Mark I Liner Attack and Direct Containment Heating (DCH) were addressed with a new approach using the Risk Oriented Accident Analysis Method (ROAAM) (Theofanous, 1994, 1989). The resolution of the Mark-I Liner Attack issue constitutes the first full demonstration of ROAAM. It emphasizes the determinism and provides a basis for synergistic collaboration among experts through a common communication frame.

Applications of PSA

The Mark-I analysis (NUREG/CR-6025) was controversial and manifested prior polarization. Follow-up work was carried out at RPI (on melt release scenarios), ANL (on spreading), SNL (on coriumconcrete interactions), and ANATECH (on liner structural failure by creep) resulted in a consensus (NUREG/CR-6025).

For DCH, the original document (NUREG/CR-6075) was with criticism, but consensus was achieved in the supplimentary report that proposed a basis for extrapolation to all PWRs with Zion-like features. Resolution of DCH for ice-condenser plants, and CE and B&W plants was also achieved.

11.1.10 Practical Application of PSA: Utility Experience and NRC Perspective

EPRI NP-5664 is a study based on interviews of personnel at 10 utilities and 15 NRC personnel regarding the usefulness of PSA (they use the term PRA - probabilistic risk assessment). The general utility motivation for using PSA is to demonstrate an acceptably low level of risk to the NRC. Some utilities applied PSA to individual systems, functions, or issues. These smaller programs served to train a PSA cadre and introduce PSA to other utility personnel and management.

Specific direct benefits cited by utility personnel are:

1. Submittal to the NRC demonstrating an acceptably low level of risk.
2. Submittal to NRC to obtain relief from specific backfit requirements.
3. Supporting information for interactions with the NRC.
4. Internal use for design or operational improvements.
5. Internal use for managing the plant modification process.
6. Internal use for training and staff development.

All utilities interviewed said the benefits were worth the cost. The benefits varied; some utilities used their PSA to satisfy the NRC and have received few additional benefits since. Others have continuously used and continue to benefit from their PSA. Some utilities cited detrimental impacts of their PSA program which are presented below.

Benefits
1. Specific Direct Beneficial Impacts on Design and Operation

- Identification of hitherto unrecognized deficiencies in design or operation.
- Identification or a more cost-beneficial alternative to internally planned modification or activity.
- Exemption from an NRC requirement that would not have improved safety in a cost-beneficial manner.
- Replacement of an NRC proposed modification with a significantly more cost-beneficial modification.
- Support of procedure or technical specification improvement.
- Support of analyses designed to improve operator performance.

2. An Improved Design Control Process

- Use in support of safety evaluations.
- Use in prioritization of plant modifications.

3. Improved Staff Capabilities

- Improved plant knowledge.
- Support of operator training.

4. Improved Ability to Interact with the NRC

- Support for licensing.
- Protection from NRC sponsored studies.
- Enhanced credibility with the NRC.

Detriments
1. Erroneous identification of a safety problem.

- Some utilities experienced a PSA, submitted to the NRC, being misused to identify apparent safety problems. However the PSA also provided the defense with no adverse impacts.

2. Providing the NRC with new issues that have little safety importance.

- This concern results from emphasis by the NRC on legitimate but minor issues in the PSA, especially in seeking an operating license. Utilities with plants having near-term operating cited examples of PSA-caused new issues prior to licensing. The PSA also provided their defense.

General

- PSA is becoming increasingly important in NRC decisions regarding safety.
- Utilities need in house PSA expertise for plant modifications, training, and communications within the utility and with the NRC.
- A level 1 PSA with a simplified containment analysis is usually sufficient.
- There is some concern that PSA will be used to erode safety margins.

Applications of PSA

11.2 PSA of the CANDU (Heavy Water Power Reactor)

The Canadian Deuterium Uranium reactor fissions with natural uranium, hence, no dependence on national or international fuel enrichment facilities that are needed to enrich uranium to about 3% U-235 to achieve criticality with light water moderation.

Canada has supplied itself and many nations of the world with CANDUs: Argentina - 3, Canada - 21, China - 2, India - 14, Korea - 4, Pakistan - 1, and Romania 3.

11.2.1 Reactor Description

Figure 11.2-1 shows the layout of a CANDU. The uranium fuel is supported horizontally by uniformly spaced tubes in the calandria (5) which

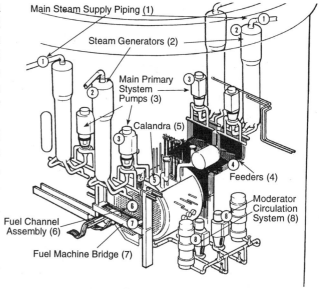

Fig. 11.2-1 Sketch of the CANDU

is filled with a heavy water moderator. The longer slowing down distance of the D_2O moderator results in an relatively large lattice spacing compared with an LWR. Fuel bundles are in sealed horizontal pressure tubes (6) with an gap to allow the passage of the heavy water pressurized coolant which is used to generate steam (2) for electric power production. The nuclear heat is removed by D_2O coolant which is pumped through the pressure tubes. CANDU completely separates the moderator (8) from the cooling system (3). The moderator is pumped through the calandria and through a heat exchanger in a non-boiling loop nearly at atmospheric pressure. The primary coolant is pumped through 10.4 cm diameter fuel-containing pressure tubes, and through the steam generators at a reactor outlet temperature of about 310° C and a pressure of 1.02 MPa (145 psi) to generate steam to power a turbo-electric generator. The relatively cool heavy water moderator in the calandria is a large heat sink that minimizes accidents. Reactivity control consists of variable neutron absorbers and absorber "adjuster" rods that are located in the reactor. The reactor is shut down by scram rods that fall from above the calandria into the reactor. A backup independent scram system rapidly injects neutron poison into the moderator. The calandria is located inside a concrete reactor vault which is filled with light water. This water shields the concrete walls from fast neutron irradiation and heating. It maintains the calandria shell at a constant temperature. The steel-water end shields of the reactor are located in openings in the vault wall, forming part of the vault enclosure.

On-Line Fuel Handling and Control

A CANDU feature is the short fuel bundles which are fed into and removed from the reactor while at power. This is the reason the CANDUs are world leaders in plant availability. The reactor is refueled by two remotely controlled fueling machines, one at each end of the horizontally tubed reactor to insert new fuel and remove spent fuel. New fuel is brought into the Reactor Building from a storage area in the Service Building via the main air lock and loaded into the fueling machines. The spent fuel is transferred under water, through a canal and transfer lock, to a 10-year capacity spent fuel bay in the Service Building. The fuel is supplied in bundles 50 cm long by 10.4 cm dia. Each bundle consists of 37 zirconium alloy tubes, or elements, which contain pellets of natural uranium dioxide. Each of the fuel channels in the reactor contains 12 bundles. The plant control is automatic using dual computers.

Safety Summary

Pressure-tubes allow the separate, low-pressure, heavy-water moderator to act as a backup heat sink even if there is no water in the fuel channels. Should this fail, the calandria shell itself can contain the debris, with the decay heat being transferred to the water-filled shield tank around the core. Should the severe core damage sequence progress further, the shield tank and the concrete reactor vault significantly delay the challenge to containment. Furthermore, should core melt lead to containment overpressure, the concrete containment wall will leak and reduce the possibility of catastrophic structural failure (Snell, 1990).

The Canadian licensing philosophy requires that each accident, together with failure of each safety system in turn, be assessed (and specified dose limits met) as part of the design and licensing process. In response, designers have provided CANDUs with two independent dedicated shutdown systems, and the likelihood of anticipated transients without scram is negligible.

11.2.2 CANDU-2 PSA

Canada's first risk assessment was for a CANDU-2, the Darlington Probabilistic Safety Evaluation (DPSE) (Ontario Hydro, 1987); King, 1987; and Raina, 1987) to verify the design of this four unit station. DPSE is a level 3 PSA that models internal events using licensing-type conservative assumptions. Principal objectives were: a) to provide design verification using probabilistic methods; b) to identify initiating events and accident sequences that dominate public risk and economic risk to the utility; c) to provide information for operating procedures and training for accident situations, and d) to provide system reliability

Table 11.2-1 CANDU-2 PSA Results	
Damage	Freq./y
Loss of core structural integrity	3.8E-6
Large LOCA and loss of ECI	2.2E-6
Small LOCA and loss of ECI	8.1E-5
Fuel cooling last at least 1 hr. after shutdown	4.7E-4
Large LOCA w. early stagnation	3.0E-5
Large LOCA w. potential for delayed release after steam blowdown	1.0E-4
Single channel event; e.g., end-fitting failure w. fuel ejection to vault	2.3E-3
Single channel event; e.g., in-core break with fuel ejection into the moderator	3.2E-3
Loss of cooling to irradiated fuel in fueling machine	2.0E-3
Small LOCA w.o. significant fuel failures	2.3E-2

and event sequence assessments required as part of the licensing process.

DPSE, like all subsequent Ontario Hydro PSAs, used the SETS code consequently it is constructed entirely of fault trees, although simple event trees were drawn to describe the accident sequences which were then redone as high-level fault trees for ten Fuel Damage Categories (FDCs) and five Plant Release Categories (PRCs) into which the system-level fault trees were merged to get the results in Table 11.2-1.

The DPSE found an acceptably low risk that was dominated by releases during normal operation. The mean risk of the station was estimated to be 9E6 Sv/y to the individual at the site boundary and 7E4 person-Sv/yr to the surrounding population to a distance of 100 km. The consequences of events beyond the design basis were not calculated but their frequency was predicted to be 4E-6 /reactor-y which is acceptably low. The mean economic risk was estimated to be about 10 M$/reactor-yr.

11.2.3 CANDU-6 PSA

The Pickering A Risk Assessment (PARA) (Ontario Hydro, 1995) is also a level 3 PSA for 1 of the 4 units at Pickering. A difference between PARA and DPSE is that sequences beyond the design basis were modeled using the MAAP-CANDU codes with best estimate assumptions. Other parts of the analysis used licensing-type conservative assumptions.

The scope of PARA was to assess public health and economic risk from internal initiating events and LOOP. Extreme natural phenomena were excluded because they were judged unlikely to pose significant additional risk to the public over and above that arising directly from the event itself and because of the uncertainty in assessing risks from these causes. The study did include the effects of internal environmental conditions such as high or low ambient temperature, steam and flooding, but fire was not included because of large uncertainty in probabilistic treatment.

The PARA process involved:

- Systematic review of plant design and operating experience to identify potential initiating events;
- Use of event trees to identify mitigating systems and their failure modes of interest;
- Fault tree analysis of mitigating systems to develop logic models of how system failure may occur;

Table 11.2-2 CANDU-6 PSA Results

Damage	Freq./y
Rapid loss of core structural integrity	5.0E-7
Slow loss of core structural integrity w. LOCA1 w. other LOCAs	1.3E-4 5.6E-5 7.4E-5
Moderator required as a heat sink <0.5 h after reactor trip	7.4E-6
Moderator required as a heat sink 0.5 to 24 h after reactor trip	1.2E-5
Moderator required as a heat sink >24 h after reactor trip	4.0E-5
Large LOCA with stagnation	2.3E-5
Single channel fuel release with sufficient containment pressure rise to initiate automatic containment button-up	7.5E-4
Single channel fuel release with insufficient containment pressure rise to initiate automatic containment button-up	7.0E-3
Intermediate LOCA w. successful ECI	2.6E-2

- Linking event tree and fault tree models to generate sequences of failure events using Boolean logic; and
- Calculation of the frequencies of occurrence of the sequences.

The process was iterative because of the need to modify conservative assumptions as the assembly process proceeded. The results are summarized in Table 11.2-2.

The Pickering risk is estimated to be 2E-4 Sv/y to an individual at the site boundary and about 4E-3 person-Sv/y to the surrounding population. Both of which are less than DPSE in spite of age of station, the higher demographics, and the inclusion of beyond design basis events. The mean site economic risk was estimated to be 1.5 M$/reactor-y.

11.2.4 CANDU-9 PSA

In addition to the PSAs that the utility has prepared, the vendor, AECL, has prepared a CANDU-6 PSA (level 2- Allen, 1990), CANDU-3 PSA (level 1 - AECL, 1989), Wolsong 2/3/4 PSA (level 2 - AECL, 1995), and CANDU-9 PSA (level 1 - AECL, 1996, Jaitly, 1995).

CANDU-9 is the next generation of CANDUs. It is a larger version of the CANDU producing 870 MWe, the CANDU-9 to complement the mid-size CANDU-6 with updated proven technology, a modified CANDU-6 station layout, improved construction methods and operational features. Standardization, a feature of CANDU reactors, is emphasized in CANDU-9 in the key components (steam generators, coolant pumps, pressure tubes, etc.) being the same design as those proven in service at CANDU power stations.

A key part of the CANDU 9 design program was the review by the AECB that explored and resolved issues during design. PSA in the design process ensures that safety related requirements are addressed early to reduce retrofits as well as its role in safety assessment and licensing.

Lessons learned from earlier PSAs incorporated in the CANDU 9 design are:

- Improved feedwater reliability via two independent sources of high pressure auxiliary feedwater, one seismically qualified, the other uses an auxiliary diesel-driven pump to cope with LOOP
- Improved emergency core cooling (ECC) reliability by reduction in number of valves, auto change-over to recirculation phase, sustained low pressure conditioning, four heat exchangers and four pumps
- Relocation of service water pumps and the electrical power distribution system for immunity from steam and boiler feedwater line break
- Redundant service water supplies to the shutdown cooler for improved shutdown cooling reliability
- Automatic trip of heat transport pump on high bearing temperature to prevent bearing damage service water is lost and the operator fails to take corrective action
- Improved service water reliability by elimination of rubber expansion joints in the recirculated cooling water system and minimization of their use in the raw service water systems

Applications of PSA

- Automatic reactor power reduction in the event of low flow or high temperature in the end shield cooling system
- Four onsite power diesel generators that start automatically on loss of offsite power to reduce the frequency of the loss of station AC by an order of magnitude compared with CANDU-6. This with the diesel driven auxiliary feedwater pump provides ample recovery time
- Moderator make-up capability from the reactor building floor to allow fuel cooling in the event of feeder stagnation break and failure of the ECC system
- Rapid bolt-up and fill-up provisions for the heat transport system to allow the steam generators to function as a heat sink in the event of problems with the shutdown cooling heat sink when the heat transport system is drained to the header level
- Improved independence between process systems
- A very large reserve water tank, located near the top of the reactor building to store inventory for the low pressure recirculating phase of the emergency core cooling system, and provide a source of passive inventory make-up to the heat transport, boilers, moderator and shield tank. It is an integral part of the end shield cooling system

Completion of the PSA to higher levels depends on the sale of a CANDU-9. The plant is expected to meet the following PSA acceptance criteria.

1. Individual accident sequences leading to severe core melt shall have a frequency < 1E-6/y.
2. Individual accident sequences requiring the moderator as a heat sink shall have a frequency < 1E-5/y.

11.3 Research and Production Reactor PSAs

In the U.S., the Department of Energy operates or has operated reactors that are used for nuclear physics, materials studies, and isotope production. Their safety has been assessed by PSA.

11.3.1 Advanced Test Reactor PSA

The Reactor

The Advanced Test Reactor (ATR) Eide (1990); Atkinson (1993); Atkinson (1995); and Atkinson (1996) is located at the Test Reactor Area within the boundaries of the INEL. It began operation in 1968 at a power of 250-MWt to study the effects of irradiation on reactor material samples. Its fully enriched core (93% U-235) is light water cooled and reflected. The ATR is significantly different from a PWR, having smaller core (89 lb), high power density (1 MW/ltr), low primary coolant system pressure (355 psi) and temperature (125° F), higher ratio of coolant weight to power (2,400), and confinement (leak rate 10%/day) rather than containment. The confinement has a lower overpressurization limit than an LWR, but does not have to be vented during a severe accident because of the low pressure and low power of ATR. Its decay heat is 13 MW at 10 seconds after shutdown and 1.3 MW one day later (~7% of an LWR).

PSA

The PSA began with ATR familiarization as it was in May 1989. This included: plant design, operating practices and procedures, test and maintenance procedures and training programs. Initiating event identification and quantification was an extensive process because of the unique nature of the plant. Both comprehensive engineering evaluation and a master logic diagram were used to identify 152 initiating events which were grouped into 22 initiating event categories for event tree development (Table 11.3-1). Several of the initiating event categories are unusual compared with LWR PRAs. Power-related events are: loss of diesel power (both the normally-running diesel and the automatic starting backup diesel) and loss of diesel-commercial power (loss of the bus powered by the normally-running diesel). The loss of all AC power initiator includes both commercial (offsite) power and all three diesels, including the normally-running diesel. Local fault events include localized flow blockages in the core and localized fuel faults. A lobe power imbalance initiator occurs if the individually-controlled core section powers (lobe powers) stray outside a predetermined safety envelope but the total core power is acceptable.

Table 11.3-1 ATR Initiating Events

Event	Freg/y
Normal/spurious shutdown	21.9
Low primary coolant system (PCS) flow	0.075
Low PCS pressure	0.05
High PCS pressure	0.05
Loss of heat sink	0.025
Loss of commercial power	0.9
Loss of all AC power	1.3E-4
Loss of diesel commercial power	0.075
Loss of diesel power	0.63
Loss of instrument air	0.025
Small LOCA	0.025
Large LOCA	5E-4
Interfacing system LOCA (demineralized water)	3.3E-8
Interfacing system LOCA (firewater)	1.5E-8
Reactivity insertion (fast ramp)	0.1
Reactivity insertion (slow ramp)	0.1
Reactivity insertion (LOOP ramp)	1.1
Lobe power imbalance	0.1
Large reactivity insertion (experimental loop rupture)	5E-4
Very large reactivity insertion	3E-7
Local faults	0.025
Fuel storage channel draining	4.2E-7

Initiating event category frequencies were estimated using a variety of methods. ATR historical experience was used to estimate 13 frequently occurring initiating events. Three categories were quantified with system fault trees. Large LOCAs and experiment loop ruptures used commercial power plant estimates modified for ATR piping characteristics. The remaining initiating events were quantified from ATR information. A small event tree with large fault trees was developed for each of the 22 initiating event categories from which 227 accident sequences led to core damage. Fault trees were developed for the 13 systems modeled in the ATR PRA. The fault trees are called "large" because support systems are included in the trees for systems requiring them. The fault trees use about 1,200 basic events and 800 logic gates.

Dependency analysis consists of: 1) explicit modeling of known dependencies within system fault trees and 2) unknown dependence modeling using the Greek Letter Methodology (EPRI NP-3967) consisting of a general beta factor of 0.1, and a plant-specific 0.2 for the diesel generators. A gamma factor of 0.5 was used to account for third component dependence.

The ATR PRA human reliability analysis (HRA) was divided into three subtasks: 1) screening analysis of human errors before an accident (miscalibration and failure to restore components following test or maintenance), 2) screening analysis of operator error during an accident, and 3) refined analysis of dominant human errors. The HRA probabilities were obtained using the Accident Sequence Evaluation Program Nominal HRA technique (NUREG/CR-4772) with discussions with operators without detailed walk-throughs and talk-throughs. Using an interim accident sequence quantification, five dominant human errors were chosen for refinement. The refined HRA included a combination of THERP (NUREG/CR-1278 and Hannaman, 1984) including detailed walk-throughs and talk-throughs with operators for events in the context of the accident sequences.

Generic and plant-specific component failure information was collected to develop a generic database from past PRAs (Eide, 1990b). This data was supplemented with ATR-specific data on test and maintenance outages, selected pumps and valves diesel generators, and scram system components. Component failure rates were obtained using a Bayesian update with the generic value as the prior and ATR data. The system fault trees were created, checked, and plotted using the IRRAS. Accident sequences were quantified using a microcomputer version of SETS. Accident sequence cutset manipulations were performed with SARA code with truncation at 1.0E-10/yr. This was followed by uncertainty and sensitivity analyses.

Two separate study teams completed the ATR Level 1 PRA. This was not planned but resulted from personnel changes. The combined teams provided redundant coverage of ATR knowledge and PRA knowledge. Quality assurance used detailed guidelines for the tasks including close cooperation between the ATR and PRA personnel. Three formal reviews were used during the PRA: 1) as tasks were performed, 2) as tasks were completed, and 3) external peer review of completed tasks and documentation. When the second PRA team took over the project, the work done by the first team was reviewed.

Unavailabilities ranged from 1.8E-5 for the scram system to >0.1 for several systems requiring significant operator action. The firewater injection system (used for low pressure feed and bleed and for coolant injection for LOCAs) is 5.0E-5 because it is nearly passive operation. Quantification of the 227 accident sequences leading to core damage resulted in 70 sequences with cutsets greater than 1.0E-10/yr.

Results

The mean core damage frequency from all internal events is 1.8E-4/yr, with an error factor (95% percentile divided by the median) of 5.0. The percentage contributions to the core damage frequencies are: Large Reactivity Insertion; 28%; Large LOCA, 28%; Reactivity Insertion LOOP ramp, 12%; Spurious/normal shutdown, 11%; Loss of commercial power, 10%; and others, 5%.

11.3.2 High Flux Beam Reactor PSA[e]

The Reactor

The HFBR at Brookhaven National Laboratory is a heavy water moderated and cooled reactor designed to provide an intense beam of neutrons to the experimental area. In addition using thimbles contained within the vessel, it provides isotopic production, neutron activation analysis, and material irradiations. It began operation in 1965 at a power of 40 MW to be upgraded to 60 MW in 1982.

The HFBR core uses fully-enriched (93%) uranium oxide-aluminum cermet curved plates clad in aluminum. The core height is 0.58 m and the diameter is 0.48 m or a volume of 103.7 ltr. The U-235 weighs 9.83 kg supported by a grid plate on the vessel bottom. The coolant flow is downward, hence, flow reversal is necessary for natural circulation. It operating temperature and pressure are 60° C and 195 psi. There are 8 main and 8 auxiliary control rod blades made of europium oxide (Eu_2O_3) and dysprosium oxide (Dy_2O_3), clad in stainless steel that operate in the reflector region. The scram system is the winch-clutch release type to drop the blades into the reflector region. Actuation of scram causes a setback for the auxiliary control rods which are driven upward by drive motors.

The two primary pumps with AC and DC pony motors flow 18,000 gpm of heavy water through the core for cooling. Following 60 MW operation, for 3 minutes, forced flow is needed to remove the decay heat. Upon loss of pony motors and primary pumps, forced flow is established through the shutdown cooling water system consisting of two pumps in parallel - one pump running and the other on standby.

The PSA

In May 1988, a Level-1 PSA was undertaken as part of the general risk assessment at DOE facilities. Revision 0 was completed, and reviewed by BNL, DOE and contractors. The revised report was available July 1990 (Azarm, 1990). The broad objective of the HFBR PRA program is to enhance the safety and operational activities throughout the remaining lifetime of the reactor.

The tasks performed were:

1. plant familiarization and identification of initiating events
2. accident sequence delineation,
3. system analysis,
4. accident sequence quantification,
5. human reliability analysis
6. plant specific (HFBR) data base development,
7. Importance, Sensitivity, and Uncertainty (ISU), and
8. analysis of assessment

---

[e] Summarized from Azarm (1990) and Chu (1990).

Applications of PSA

Initiating Event Identification

Initiating events, in this study, initiate plant scram or setback. Other initiators, such as refueling discharge accidents, do not necessarily cause a reactor shutdown but may lead to minor fuel damage and radioactive releases. The list of initiators for nuclear power plants has little relevance for HFBR because of size and design differences. A list of HFBR-specific initiators was developed from: a list prepared with the HFBR staff, the FSAR, the plant design manual, the procedures manual, technical specifications, monthly operating reports, and the HFIR PRA (Johnson, 1988).

The goal was identification of a complete list of initiators based on the evidence accumulated by the operational experience of the HFBR, other research reactors, and the engineering evaluation of the plant design. A Master Logic Diagram was constructed to approach completeness that resulted in 73 initiators. The were categorized as: 14 transients, 3 classes of loss of offsite power, 3 Loss of Coolant Accidents, 1 Beam Tube Rupture, and 5 unprotected initiators of minor core damage. Table 11.3-2 summarizes the method of PSA quantification.

Accident Sequence Delineation

Event Sequence Diagrams (ESDs) and Event Trees (ETs) were used to logically display the plant response to various initiators in a two step process:

1. An ESD was developed for all initiator categories not involving LOCA. This generic ESD was then specialized for each transient category and the categories of loss of-offsite power. Four additional ESDs were developed; three for different sizes of LOCAs and one for beam tube rupture. No ESD was developed for special initiators.

2. The ESDs were then translated into associated event trees. A generic event tree was developed for all initiators not involving LOCA. The generic transient event tree for each category of the transient initiators and loss of offsite power were specialized by: the impact of the initiators on the safety and support systems, from the success criteria of the mitigating systems, and the initiator-specific human actions which were modeled in the fault trees.

Table 11.3-2 Quantification of HFBR PSA

Initiating event ID	Search monthly reports, FSAR, PRAs, etc.
Initiating event quantification	Bayesian analysis, Thomas model, Expert judgment
Fault tree and event tree	Fault tree linking, beta factor, IRRAS 2.0 Truncation limit: systems 1E-8, sequences 1E-10
Component failure data	Bayesian
Maintenance unavailability	Expert judgment, comparison with other PSAs
Core damage frequency	IRRAS 2.0, in-house code
Uncertainty analysis	IRRAS 2.0, in-house code, LHS, TEMAC
Importance measures	Fussell-Vesely, Uncertainty

Fault Tree and Event Tree Quantification

Ten front line systems and 5 support systems were modeled using about 700 basic events. A generic transient event tree with 215 sequences was developed and quantified for 14 different transients and three loss-of-offsite-power events. Four LOCA event trees were developed and

412

quantified depending on break size and location. All event trees and event sequence diagrams were reviewed by the reactor operating staff for accuracy.

Fault trees were developed using the IRRAS 2.0 code (Russell, 1988) which allows definition of individual sequences in an event tree, and generation of their cutsets, but does not generate cutsets for total core damage frequency. An in-house code was developed to combine the cutsets of various sequences. Because IRRAS 2.0 was preliminary, use was also made of the SETS code (Worrel, 1985). The uncertainty calculation of IRRAS 2.0 was compared with the results of LHS (Iman, 1984), TEMAC (Iman, 1986), and an in-house code.

Expert Judgment

Expert judgment was used for: beam tube rupture frequency, human reliability analysis, and maintenance unavailability. The beam tube rupture frequency was estimated metallurgical, structural analysis, and, radiation damage to materials experts. Initially, probabilistic models were attempted and used because the model parameters could be quantified. Finally, a probability distribution with very large uncertainty was estimated by one of the experts. Human actions were identified in the fault tree and event trees. The tasks were analyzed by system and human reliability analysts reviewing the procedures, step-by-step with reactor operators, and plant walkdowns for familiarity. Quantitative estimates of the human error probabilities were judged a human reliability analyst and compared with tasks modeled in PRAs of commercial reactors. Maintenance unavailability was judged by a panel consisting of three Plant Operation and Maintenance Department staff.

Data Analysis

Bayesian analysis was the main tool for quantification of initiating event frequencies, component failure data, and the beta factor for failure of control rods. Bayesian analysis requires a prior distributions and data. Prior distributions were derived from generic data or from fault trees. For example, the prior for frequency of LOCA used Thomas (1981) with field data to generate a posterior distribution. Component failure data was aggregated from three generic data sources as the prior knowledge. Plant-specific data was collected for key components of the plant for the Bayesian analysis. Transient initiating events were modeled with simple fault trees as needed and quantified with data from other PRAs to obtain means of the prior distributions and updated with plant specific data to obtain the posteriors.

Results and Discussion

At 60 MW, the core damage frequency is $5.1E-04/y$. The proportion of accident classes is: LOCA, 55%; beam tube rupture, 20%; ATWS, 13%; LOOP, 10%; and other transients, 2%. The core damage frequency is dominated by large LOCA and beam tube rupture. Core damage from large LOCA occurs from the inability to mitigate the initiating event in a radioactive environment that results from the minor core damage resulting from lack of 3 minutes of forced flow through the core. Beam tube rupture results in core uncovery because of pressurizing the vessel by bulk boiling to force coolant out of the break. Other contributors are loss of offsite power and anticipated transient without scram.

At 40 MW operation, the core damage frequency is 3.7E-04/y. The proportion of accident classes is: LOCA, 50%; beam tube rupture, 27%; ATWS, 17%; LOOP, 4%; and other transients, 2%. Three minutes of forced flow are not required and large LOCAs with break size smaller than 2.8 inches can be mitigated.

11.3.3 HFIR PSA

Reactor

The High Flux Isotope Reactor (HFIR) is a high neutron flux density isotope production and research reactor in operation at ORNL since 1965 for transuranic and cobalt isotope production, materials irradiation and neutron scattering research. It is a 85-MWt flux trap reactor using water cooling (468 psig and 158° F, outlet) with a beryllium moderator. The peak thermal flux in the flux trap is 5E15 n/cm^2-sec is the highest in the world. The reactor core is 17.5 in. dia and 24 in. high with a 5 in. dia. target hole in its center for the flux trap. The core has 9.6 kg of 93% enriched U-235 arranged in two concentric, cylindrical elements. The inner element contains 171 and the outer 369 involute aluminum-clad fuel plates. Both elements of the core are replaced every 24 days. The moderator is about 1 ft thick. Control is achieved by four safety plates arranged in a cylinder around a solid control cylinder. These control cylinders are located between the core and the Be reflector. Insertion of any one of the five control elements renders the core subcritical. The reactor core is contained in an 8-foot-diameter pressure vessel that is 19 ft high located near the bottom of pool of 85,000 gal. of water for shielding and to prevent core uncovery. A flow rate of 16,000 gpm is provided by 3/4 (three-out-of-four) main AC cooling pumps with DC motor backup. Pressure is provided by (½) pressurizer pumps whose speed is controlled by controlling the slip in a hystersis clutch.

HFIR operated for 20 years until shutdown in November 1986 because of vessel embrittlement concerns. The were addressed in Cheverton (1987) and by a PSA assessment (Johnson, 1988); it returned to service 2.5 years later with a power reduction to 85 MW (from 100 MW), an operating pressure reduction to 468 psig (from 750 psig), and an annual hydrostatic test at 885 psig. In addition, a system was added to automatically depressurize the vessel in the event of low inlet temperature. To limit overcooling, logic was added to automatically prevent post-scram overcooling by tripping the main secondary pumps in response to a scram signal.

Internal Events PSA

In 1988, an internal events PRA was published (Johnson, 1988) this was followed by an external events analysis. The results were reported by Johnson (1991), and Flanagan (1990). The basic approach to risk is that of Kaplan (1981) that asks the questions: "What can happen, and How likely is it? and what are the consequences?" These are organized as triplets to characterize the risk

Scenarios were identified by: 1) a master logic diagram, 2) a dependency matrix, 3) event sequence diagrams, and 3) event trees (no fault trees). Systems analyses and operator action assessments provided information for the quantification of the scenarios. Eight initiating event categories and 21 specific initiating events were defined. The categories and the 21 events are shown in Table 11.3-1. Also shown in Table 11.3-1 are the mean initiator frequency, the CDF of that initiator, and the percentage of each initiator to the total internal CDF.

Several initiating events are unique: 1) large LOCA (6), 2) flow blockage events (4a, 4b, and 4d), and 3) fuel defects (4e). The analysis of these required special tools not found in commercial power plant PRAs.

Large LOCA is a break >2-in. equivalent diameter, i.e., beyond the design basis, thus, the initiating event frequency is the core damage frequency. The initiating event frequency is low because of operating a low pressure and system with annual hydro testing at twice the operating pressure. It is designed for 3x the operating pressure; it is free from stress corrosion because of water chemistry, and the piping >2 in is regularly inspected. Thomas (1981) was used to estimate the large break LOCA frequency.

There are no information for modeling and quantifying the flow blockage scenarios, therefore the method of Kaplan (1989) was used with the HFIR designers, operators, and engineers as the knowledge experts. The first step defined where debris[f] could be located that could cause a flow blockage. The second step divided the primary loop into debris locations: Ll encompasses the core outlet to the stainer, L2 from stainer to core outlet, and L3 includes the locations of external debris. The material types were defined as: 1) elastomers (gaskets and seals), 2) small metal parts, 3) large metal parts, 4) corrosion and erosion products, and 5) plastic, glass, paper/fabrics. Detailed event trees were formulated for the HFIR experts to evaluate the branch point split fractions. Several iterations were performed to reach a consensus. The initial results indicated a CDF of 2.6E-2/y. Several modifications were identified, and the evaluation process was repeated to find 9E-5/y.

The high power density and narrow flow channels could cause fuel hot spots and flow starvation. Fuel experts at ORNL and ANL identified five types of defects: three fuel inhomogeneities and two assembly errors to arrive at 4D, for fuel defects is 2.1×10^{-5}/yr.

External Events PSA

The external events PSA was based on standard methods used for commercial reactor PSAs. Fire risk was estimated from commercial nuclear power plant data combined with industrial fire information. The seismic hazard was evaluated using a combination of the EPRI and LLNL (NUREG/CR-5250) databases. Wind hazards were analyzed by EQE, Inc., using NRC-based methodology.

Results

The results are summarized in Table 11.3-3 for both internal and external results.; the CDF from all causes is 7.4E-4/y of which internal events contribute 3.1E-4 and external events 2.9E-4. This number has been reduced by design, administrative, and operational procedure changes.

---

[f] Internal debris is loose parts of the system (gaskets, bolts, small pieces of metal, etc.); external debris enters the system unintentionally (tools, rags, cigarette packages, watch crystals, etc.).

Applications of PSA

The dominant internal event accident scenarios were used in conjunction with conservative source terms (100% of the core melts, releasing 100% of the noble gases, 100% of iodine and cesium, and 1% solids), and site-specific meteorological data to estimate off-site consequences to be well below the 10CFR100 siting guidelines.

11.3.4 K-Reactor PSA

11.3.4.1 The Reactor

The K, L, P, and C reactors are heavy water cooled and reflected for the production of, primarily, weapons-grade plutonium and tritium.[g] All are shutdown; the K-reactor was the last one used. They are located at the 300 square miles Savannah River Site (SRS) in South Carolina. The PRA was applied to the Mark 22 core design in the K Reactor as it existed in 1987. The Mark 22 charge is composed of relatively light-weight, highly enriched uranium-aluminum alloy tubular fuel in which plutonium is produced. Tritium is produced by including tubular lithium targets. The reactor is cooled by six primary loops with redundancies in pump motors (AC and DC pony motors) and in electric power supplies. The six primary loops are cooled by twelve heat exchangers that are supplied by cooling water from a large basin that is supplied by water from the Savannah River.

The reactor contains positions for 600 principal fuel and target assemblies. Assemblies are arranged in a triangular array on 7-in centers and are suspended from the water plenum. Other principal lattice positions are used for control rod housings, spargers, gas ports, and pressure-relief tubes. Spargers promote moderator circulation and provide injection ports for soluble neutron poison from the Supplementary Safety System (SSS).

Table 11.3-3 HFIR Risk

1a Category	init./ y	CDF/y	%
1b Manual scram	21.4	2.2E-5	7.1
1c Inadvertent control rod drop	3.16	3.7E-6	1.2
1c Inadvertent scram	0.1	3.6E-7	0.1
2a LOOP	0.44	5.6E-5	18
2b Loss of preferred feeder	0.41	7.4E-6	2.4
2c Loss of switchgear DC	3.8E-3	3.9E-7	0.1
3a Runaway pressurizer pump	0.13	4.8E-6	1.5
3b Loss of running pressurizer	1	1.6E-6	0.5
4a Loose parts L1	1.5E-5	1.5E-5	4.8
4b Loose parts L2E	2.0E-5	2.1E-5	6.3
4c Loose parts L2MCP	3.4E-5	3.4E-5	11
4d Loose parts L3	2.0E-5	2.0E-5	6.4
4e Loose parts L4	2.1E-5	2.1E-5	6.8
5 Small LOCA	4.6E-5	1.6E-5	5.1
6 Large LOCA	3.3E-5	3.3E-5	11
7 Beam tube failure	5.9E-4	2.1E-6	6.8
8a Reactivity insertion	0.16	1.3E-5	4.2
8b Degraded secondary	0.23	2.0E-6	0.6
8c Loss of instrument air	0.081	1.7E-5	5.5
8d Degraded primary flow	0.27	2.2E-5	7.1
Total internal CDF		3.1E-4	
Fire		1.4E-5	
Wind and Tornado		2.9E-4	
Seismic		1.2E-4	
Other external		2.0E-6	
Total external CDF		2.9E-4	
Grand total CDF		7.4E-4	

---

[g] Note that the only other plutonium production site in the U.S. was at Richland, Washington where graphite-reflected light water cooled reactors were used.

Interspersed among the principal lattice positions are 162 secondary positions for 66 safety rods, instrument rods, tie bolts connecting the water plenum to the top shield, and special components. The safety rods use a winch-cable system that releases when the clutch is deenergized to drop the safety rods into the core.

The vessel is a 15 ft high cylinder with a diameter of 16.25 ft constructed of half-in thick, type 304 stainless steel plate. Six outlet nozzles are uniformly spaced around the circumference at the bottom, and six tapered inlet nozzles are uniformly spaced around the circumference of the water plenum at the top of the vessel. The primary water system PWS contains 40,000 gal of D_2O: 24,000 gal in the reactor tank and 16,000 gal in the piping system. Each of the six primary loops is pumped at 25,000 gpm from outlet nozzles at the bottom of the reactor tank through a motor operated valve to the circulating pump using a 3-ton flywheel to extend coast-down. The pump in each loop discharges to two parallel heat exchangers. D_2O flows through the tube side of the HXs and cooling light water (H_2O) flows counter currently through the shell side. Each HX is normally supplied with 14,000 gpm of cooling water. The process water pressure is higher than cooling water pressure to prevent in-leakage of H_2O while allowing slightly radioactive process water to leak into the cooling water.

It is important to note safety differences between the SRS reactors and LWRs. Since the SRS reactors are not for power production they operate at a maximum temperature of $90°C$ and about 200 psi pressure. Thus, there are no concerns with steam blowdown, turbine trip, or other scenarios related to the high temperature and pressure aspects of an LWR. On the other hand, uranium-aluminum alloy fuel clad with aluminum for the SRS reactors melts at a much lower temperature than LWR fuel, and has lower heat capacity. This requires the SRS reactors to act faster than an LWR to maintain adequate core cooling in an accident

11.3.4.2 Internal Events PSA (Tinnes, 1990)

Because of these major differences, an LWRs PRA cannot be adapted to make an SRS PRA which must be made from basics. All phases of the analysis are as unique to SRS.

The SRS Level 1 PRA is divided into eight principal task areas (Table 11.3-4). The tasks analyze progression of an accident for the assumed 24 hours duration to represent the plant response based on operational experience.

The initiating event task was a detailed and systematic search for accident initiators that fail barriers to radioactive material release using dendograms (hierarchical trees). Dendograms define barriers and their failure modes in terms of basic physical phenomena (e.g., melt, chemical damage, mechanical damage, etc.). The output lists primal initiators in five different classes leading to core melt (Table 11.3-5).

Table 11.3-4 SRS Methodology
Initiating Event Identification and Analysis
Event Tree Development
System Fault Tree Development
Electric Power Model Development and Analysis
Component Fault Tree Development
Component and HRA Database Development
Sequence Analysis and Compilation of Results
Uncertainty Analysis

Applications of PSA

Accident progression scenarios are developed and modeled as event trees for each of these accident classes. System fault trees are developed to the component level for each branch point, and the plant response to the failure is identified. Generic subtrees are linked to the system fault trees. An example is "loss of electric power" which is analyzed in a Markov model that considers the frequencies of losing normal power, the probabilities of failure of emergency power, and the mean times to repair parts of the electric power supply.

Plant design features and response strategies, including operator actions as specified by SRS procedures, are incorporated directly into the models. Actual plant experience from incident reports and operator logs, is used to compute SRS component reliability data. The SRS experience base of 110 reactor-years of operation is used and supplemented by generic industry data only when necessary.

Table 11.3-5 SRS Accident Classes

Transient: An upset of the reactor power-to-flow ratio.
Primary Loss of Coolant Accident: A leak in the primary (D2O) cooling system.
Secondary Loss of Coolant Accident: A leak in the light water secondary cooling system which leads to a Loss of Pumping Accident.
Loss of Heat Sink: Failure of heat transfer between the primary and secondary cooling systems requiring emergency cooling.
Loss of River Water: Failure of the river water supply to the once-through secondary cooling system requiring conservation of the water on hand.

The accident sequence frequencies are quantified by linking the system fault tree models together as indicated by the event trees for the accident sequence and quantified with plant-specific data to estimate initiator frequencies and component/human failure rates. The SETS code solves the fault trees for their minimal cutsets; the TEMAC code quantitatively evaluates the cut sets and provides best estimates of component/event probabilities and frequencies.

Uncertainty estimates are made for the total CDF by assigning probability distributions to basic events and propagating the distributions through a simplified model. Uncertainties are assumed to be either log-normal or "maximum entropy" distributions. Chi-squared confidence interval tests are used at 50% and 95% of these distributions. The simplified CDF model includes the dominant cutsets from all five contributing classes of accidents, and is within 97% of the CDF calculated with the full Level 1 model.

11.3.4.3 External Events PSA (Brandyberry, 1990)

Seismic Analysis

There are four parts to seismic risk assessment: 1) seismic hazard assessment, 2) seismic fragility analysis, 3) system modeling, and 4) risk quantification. Seismic hazard estimates the frequency of ground motion at the site. Uncertainty in the seismic hazard is characterized by a family of hazardcurves with a probability normalized-weight assigned to each curve which is a measure of belief that a curve represents the true site hazard. The seismic hazard was estimated based on the methodology of EPRI's Seismicity Owners Group seismic hazard evaluation project.

The fragility analysis evaluates the conditional fraction of failure of plant structures and equipment as a function of ground motion. The seismically initiated failure of plant components is expressed in terms spectral acceleration at 5.0 Hz which is between the fundamental frequency of the reactor building (2.0 Hz), and the fundamental frequencies of the equipment. A logic model was developed to describe the performance of the reactor system, in a seismic event. A seismic event tree indicates the response of major structures and safety systems to earthquake ground motion; a fault tree defined the seismically-initiated and random component failures.

The quantification of the CDF was performed in two steps: 1) the plant logic model was quantified by evaluating the seismic event tree using component fragility information to produce a fragility curve for each core melt sequence, which defined the conditional fraction of times the sequence occurred as a function of ground motion level. A total plant level fragility was then determined that quantified the conditional fraction of times core melt occurred as a function of ground motion level for all seismic sequences. 2) The frequency of core melt was estimated by combining the sequence fragility curves generated in the first step with the frequency that ground motion levels occur to determine the frequency of occurrence of each sequence as well as the total CDF. A point estimate of the seismic risk was determined and an uncertainty analysis was performed. To quantify the uncertainty in the seismic risk, the uncertainty in each part of the analysis was propagated through the analysis. In the quantification process, the uncertainty in component capacities was propagated through the plant logic model to quantify the uncertainty in the sequence fragility curves. The product of this evaluation was a probability distribution on the conditional failure fraction as a function of ground motion level. Combining the uncertain sequence fragility curves with the family of hazard curves gives the probability distribution of CDF.

The failure of circulation of D_2O in sequence 4 was dominated by relay chatter that caused closure of rotovalves in the D_2O system. These fast acting rotovalves could close off D_2O flow to the core.

Failure of cooling water piping in sequence 17 was dominated by soil consolidation failure of underground piping. The pipe is subject to non-ductile cracking during extremely cold weather.

Failure of the 25 million gallon cooling water basin in sequence 16 was dominated by failure of the Building 186 cooling water reservoir from soil consolidation.

Loss of river water supply to the cooling water reservoir and in sequence 2 was dominated by failure of operating personnel to respond to the alarm for loss of cooling water to the cooling water reservoir. Values estimated from one minute to several hours, and for various stress levels were estimated.

Failure of the integrity of the D_2O system in sequence 5 resulted from a small leak in the D_2O system and failure of the operating personnel to isolate the leak which dominated the sequence. The probability of the small leak that challenges of the tank used to pressurize seals for the main circulating pumps was estimated to be 0.5 @ 0.21 g.

Operator error probabilities were estimated using NUREG/CR-4910 normalized to errors determined in the internal events analysis. This allowed for: varying number of personnel, amount of time available, and stress level. When more pessimistic values were substituted for best estimate values, the calculated core melt frequency increased by a factor of at least three.

Operator error probability under stressful conditions depends upon the time to complete a sequence of events. The total time available for limiting sequences was 7.9 hours which is the time to completely drain the cooling water basin. Sequence (6) required the most operator actions (11) was for shutting down the reactor, starting emergency electrical generators, and recirculating cooling water effluent.

LLNL was contracted to use the results of the Seismic Hazard Characterization Project (SHCP), (NUREG/CR-5250) to calculate the seismic hazard at the SRS using methods similar to the Seismic Owners Group-Electric Power Research Institute (SOG/EPRI) and LLNL the two seismic hazard estimates for the SRS are different. The SHCP (Savy, 1988) seismic hazard results are typically within a factor of 5. Results of the seismic analysis are given in Table 11.3-6.

Fire Analysis

The fire analysis was assisted by SNL using methods described in NUREG/CR-4840 and the extensive operating history and fire experience. For 94 reactor-years from 1958 and 1987, 20 significant fire events were recorded. Hence, frequencies for the reactor building and the diesel generator buildings of 0.12 and 0.03 /y, respectively. A control room fire has never occurred at a SRS reactor.

The extensive operating data allowed the quantification of the core melt frequency with both plant specific data and commercial fire data (NUREG/CR-4586) using Bayesian updating. (NUREG/CR-4586).

There are important features that caused the fire-induced core melt frequency at K-Reactor to be lower than other reactors: 1) The cooling water system can sustain a complete loss of pumping power and still supply sufficient cooling water for decay heat removal in a gravity feed mode. Critical cooling water system isolation valves cannot shut with the cooling water pumps running, and their cabling is separated from cabling for coolant system pumps. 2) The process water pump DC motor power is independent of the control room. There are no plant areas where a fire can fail any more than three of the six main cooling pumps, and three or more DC driven pumps can remove the decay heat. 3) There are several plant areas where fire-induced damage can totally fail the ECCS system, but are widely separated from DC power cable routing for the main pumps. 4) No fire cause was identified for defeating all shutdown systems (safety rods and SSS, poison injection).

Weather Related Events

The risk of reactor building flooding from external sources is negligible due to the siting of the reactor. The only significant flooding scenario involves dam failures upstream of SRS on the Savannah River destroying the River Water Pump houses and depriving the reactor areas of the normal source of makeup water to the 25 million gallon basins.

Ice and hail, snow, etc. can cause a loss of off-site power. The reactor building was designed to withstand blast pressure of 1,000 lbs/ft^2 can withstand tornado missile impact (Sharp, 1986). A tornado could damage the reactor by hitting the river water pump houses similar to the flooding scenario.

Other Risks

Transportation accidents were analyzed. An aircraft could directly impact the reactor. The K-Reactor is >20 miles from an airport, not on an airway and built with blast resistant construction. The nearest public highway is 2 miles away, the nearest pipeline 17 miles away, the nearest public railway is 2.8 miles away and there is no ship traffic on the Savannah River. A study of the risks from onsite transportation of hazardous substances postulated that any release would cause control room evacuation. These also were evaluated to be negligible.

Another possibility is dropping into the heat exchanger bay the 100 ton heat exchanger to cause a primary or secondary LOCA. If irradiated fuel were still in the reactor, melting could occur. Heat exchangers have been removed with hot fuel in the reactor (at least twice since 1971). This scenario was estimated to be ~1E-4/y and procedures were changed to eliminate it.

11.3.4.4 Results

The point estimate core damage frequencies for the K-Reactor for both internal and external initiators are given in Table 11.3-6.

Internal accidents, alone, have 267 sequences with a mean value (2.3E-4) slightly higher than the point estimate. The 5% and 95% confidence values are 1.7E-5 and 1.0E-3/ reactor-year, respectively.

The Primary LOCA sequences result from expansion joint breaks (51/59) and large break LOCAs (8/59). Response to LOCAs in the primary system involves shutdown, emergency injection makeup of lost coolant inventory, leak control and/or isolation, water removal from the building to avoid flooding of pumps, and direct core cooling in the event of pump flooding.

Table 11.3-6 Point Estimate CDFs for the K-Reactor

Class of Accident	Init. /y	CDF/y	%
Primary LOCA	5.6E-3	1.2E-4	59
Secondary LOCA	3.3E-3	4.9E-5	24
Transients	2.5	1.6E-5	8
Loss of River Water	1.2E-3	1.0E-5	5
Loss of Heat Sink	1.2E-4	9.8E-6	4
Total Internal Events		2.1E-4	100
Seismic		1.2E-4	
Crane Failure		<1E-4	
Fire		1.4E-7	
Transportation		<1E-7	
Total External Events		2.2E-4	
Grand Total		4.3E-4	

The Secondary LOCA sequences are related to large breaks (23/24) with small breaks and expansion joint (< 1%). Response to Secondary LOCA involves shutdown, leak control and/or isolation, water removal from the building to avoid flooding of pumps, and direct core cooling in the event of pump flooding.

Transients are 8% of the CDF and have no subclasses of initiators that dominate. It is assumed for all classes of transients (i.e., reactivity-addition, flow reduction, and coolant heat up transients) that an automatic shutdown (Safety Rod System or SSS ink injection) is required to prevent severe core melt. Once shutdown, process water (D_2O) and cooling water flows (H_2O) continue to cool the reactor core. Process water flow to the reactor core is maintained by the six on line diesel generators that drive the six primary cooling pumps. Cooling water flow is maintained to the process water heat exchangers by gravity flow from the 25-million-gallon cooling water reservoir.

The Loss of River Water is initiated by loss of plant power grid and river water pump house equally. Response to loss of river water involves event recognition, shutdown, and water conservation. Loss of grid includes recovery modeling based on industry-wide (utility) experience.

The Loss of Heat Sink has three subclasses: total loss of primary system circulation (5%), loss of secondary cooling from effluent header failures (1%), and total loss of secondary cooling from inlet

Applications of PSA

header failures (<1%). Response to loss of heat sink initiators involves shutdown and direct core cooling.

From these results, the CDF form internal events is dominated by accidents involving pipe breaks. Primary and secondary LOCAs are 83% of the CDF. The five most dominant sequences are: 1) primary LOCA bellows break, and failure to inject emergency coolant (32%); 2) large secondary LOCA pipe break and failure to inject emergency coolant following flooding (21%); 3) primary LOCA bellows break and over throttling of emergency coolant (9%); 4) transient and failure to shutdown (8%); and 5) a large primary LOCA plenum inlet pipe break in a line containing an ECCS path with throttling of emergency coolant (7%).

11.3.5 N-Reactor PSA

The Reactor

The DOE N-Reactor is one of the plutonium production reactors located on the Hanford Reservation near Richland, Washington. It is graphite moderated, pressurized water reactors that in addition to production of special nuclear materials also provided steam to turbine generators owned by the Washington Public Power Supply System for electric power production. It began operation in 1963, was put into standby status in 1988 and closed because of similarities to Chernobyl.

The reactor consists of 1,003 horizontal zircaloy pressure tubes through which water, channeled through 16 inlet risers, flows to cool the zircaloy-clad metallic uranium fuel elements[h] through 16 outlet risers to the hot leg manifold. Each of these contains a check valve before the manifold. Piping runs from the hot leg manifold to 12 horizontal steam generators configured as six pairs, but only ten are normally operating. One reactor coolant pump is associated with each pair of steam generators. The high pressure injection system (HPI) injects into the high pressure side of each pump. The return lines from the steam generators run to the cold leg manifold, from which 16 inlet lines, each with a check valve, become the 16 inlet risers completing the loop. The pressurizer surge line is connected to the hot leg manifold. In comparison with a commercial LWR, the coolant inventory of N Reactor is large.

The ECCS injects into the core through the 16 inlet risers, each containing separate ECCS valving that includes multiple check valves. It is possible for emergency cooling to fail to individual sections of the core by failing to supply coolant to one or more of the inlet risers while maintaining flow to the rest. The ECCS pumps are individually diesel driven and can deliver coolant to the core if the primary system pressure is less than 330 psi. The ECCS coolant flows through 8 valves connected to two separate ECCS manifolds that open when the pumps start to effectively divide the core into halves. The normal operating pressure at the inlet of the core is 1,716 psi, hence depressurization of the primary system is required for ECCS to function. Depressurization is achieved by opening the 8 valves. The primary system inventory flows into two separate dump manifolds then flows through 4 valves (4 valves per manifold) to the liquid effluent retention facility

---

[h] Earlier reactors at this site used aluminum clad because it dissolves in sodium hydroxide.

(LERF). A high pressure flush line connects the bottom of all of the outlet risers allowing the possibility of ECCS flow from one side of the core to the other side by reverse flow through part of the core. The flush line consists of two 10 inch manifolds connected with a 6 inch crossover line. The ECCS is a once-through cooling system; waste ECCS water is retained in the LERF for later processing and ultimate disposal.

The N-Reactor primary system pressure tubes are supported by a Lincoln log assemblage of graphite moderator blocks through which a second, independent cooling system flows - the graphite and shield cooling system (GSCS). The pressure tubes contain the fuel elements consisting of two concentric fuel tubes, i.e., three cooling channels per pressure tube: 1) an annulus between the pressure tube and the outer fuel, 2) an annulus between the outer and inner fuels, and 3) a center channel inside the inner fuel. The water inventory inside the core is significantly smaller per unit energy produced than typical of an LWR since the reactor is moderated by the graphite stack. However, the large thermal capacity of the graphite acts to stabilize core temperatures under accident conditions. Previous detailed analyses showed that the GSCS can limit fuel damage under certain conditions even if normal and emergency core cooling are completely lost. In addition, decay heat in one region of the core can be removed (to a certain extent) by coolant in another region of the core in the event that ECCS coolant fails to reach one region of the core. The graphite stack is also normally enveloped in a helium "blanket" to prevent carbon-oxygen interaction. Two diverse reactor protection systems provide for rapid reactor shutdown. The first is a system of 84 horizontal control rods and the second is a system of boron-carbide balls that can be released into the core from hoppers located above the core.

The reactor building is a 5 psi filtered confinement designed to allow steam released in a rapid primary system blowdown to escape through pressure-actuated passive vents that are reclosed by redundant blocking valves. Small pressurizations are relieved through a filtered stack. The steam pressure relief valves actuation is passive, other confinement components are actuated by complex logic signals from timers and pressure sensors. This logic is to ensure that any primary system blowdown is complete before closing of steam vents and opening of the filtered release path to protect filters and building from the damaging effects of confinement overpressure. Vacuum breakers in the confinement open to prevent building underpressure (>-2 psi) and collapse from steam condensation.

The PSA

The PSA (Miller, 1990, Wyss, 1990a, 1990b) consisted of three steps: 1) issues important to safety were identified by "brainstorms" constructed as an accident progression event tree, 2) deterministic calculations were performed on the issues when information was not available from previous calculations or similar systems, and 3) information from step 2 was used to elicit expert judgement of the issues identified in step 1.

Its unique design suggests several accident scenarios that could not occur at other reactors. For example, failure to supply ECC to 1/16 of the core due to the failure of an ECC inlet valve. On the other hand, some phenomena of concern to other types of reactors seem impossible (e.g., core-concrete interactions). The list of phenomena for consideration came from previous studies, comments of an external review group and from literature review. From this, came the issues selected for the accident progression event tree (APET) according to uncertainty and point estimates.

A issue addressed was the hydrogen generation rates and the possibility of a hydrogen explosion and the small building failure pressure. Another issue was the amount of fuel damage that could occur during degraded emergency cooling which could occur: from complete loss of the ECCS, by ECCS valving that blocked a single riser, or by common cause failure of the inlet riser valves or the emergency dump tank valves. These complexities in the primary system thermal-hydraulics make uncertainties in estimating fuel damage. N-Reactor features raised accident progression questions such as the "banana" that divides the zone I confinement from a zone III confinement area (zone I is the reactor building, zone III is a lower level of confinement). A lower portion of this wall extends into a pool of water where the underwater conveyor belt moves spent fuel from the reactor to the examination and storage facility. Is possible that during confinement pressurization, the water level on the zone I side of the wall may be depressed enough to allow gas venting under the wall. Such issues were analyzed according to availability of information and importance of the phenomena to the overall results.

Accident Calculations

Calculations of fuel damage used integrated severe accident code developed at SNL, MELCOR with a new core package to specifically model the N-Reactor. MELCOR allowed modeling of a complete accident scenario from initiating event to the radionuclide release (if any) from the confinement. In particular was the fairly high likelihood T4 transient sequence involving the failure-to-open of 1 of 2 check valves in the emergency core cooling system. The MELCOR calculation assumed that the primary coolant pumps and reactor were tripped at time zero and the ECCS valves opened within 67 seconds (based on RELAP5). The ECCS functioned normally except for flow to the highest power density fuel.

Another complex problem is how much fuel can fail if only a portion of the ECC is supplied. This was addressed by an upper bound calculations using RELAP5 and engineering judgment. The detailed calculations showed that peak temperatures did not occur during the blowdown because water flashed to steam but in the reflood stage of the accident. Reflood of the upper core levels took longer than reflood of the lower levels, the pressure drop in the reflooded portions continually decreased as the fraction reflooded increased, and the lower pressure drop yielded lower forces to drive the reflood process for the remaining tubes. From this it appears that fuel heat up during the blowdown phase would not fail in any fuel. During the reflood phase, the ECC will initially be reflooding a relatively small portion of the core with a high water flow rate with a high pressure drop across the core that will drive a large steam flow through the upper, unflooded regions. However, as more of the core is flooded the pressure differential will become lower, thus driving a smaller steam flow through the unflooded regions which may not prevent fuel damage. The maximum level of reflood that would provide an adequate pressure differential for steam cooling of the remaining unflooded region was calculated with MELCOR to find the maximum fuel temperature and pressure drop for steam-filled tubes for varying steam flow rates, and to characterize the pressure drop that develops in reflooded tubes for varying levels of ECC flow per tube. Combining the information from these relatively simple calculations, determines an upper bound estimate for a given fuel failure temperature and ECC flow rate.

Hydrogen production, transport and explosion was addressed with MELCOR Version 1.8.0 and HECTR 1.5N. HECTR 1.5 is a lumped-parameter containment analysis code for calculating containment atmosphere pressure temperature response to combustion. HECTR 1.5N is a modification of this code for the N-Reactor where a major concerns is the possibility of a hydrogen explosion during an accident in a building only capable of 5 psi. The hydrogen production was calculated according to the ECCS and GSCS cooling available in the first 2 hours and in the first 24 hours. In 5 calculations with the same basic set of assumptions, hydrogen production was maximized by conservatively assuming an unlimited steam supply for oxidation that did not cool the fuel.

The possibility of pressure relief through the banana wall was addressed with several MELCOR calculations assuming instantaneous confinement pressurizations of: 4, 4.5, 5, and 6 psi, a small LOCA, and nitrogen injection from the hydrogen mitigation system. These calculations showed the response of the banana wall to various confinement pressurizations.

Expert judgement was used to estimate uncertainty distributions for the parameters using the issue definitions with the accident calculations. Expert judgement elicitation used the method of NUREG/CR-4550. Independent panels were not convened; the panels consisted of project members and experts familiar with N Reactor, but the fundamental structure of NUREG/CR-4550 was retained (i.e., elicitation training and debiasing of the experts, issue decomposition, aggregation of independent elicitations). Typically, participants were asked to consider an issue or parameter. Each analyst established his realistic bounds for the parameter's value. Then, a cumulative probability function (CPF) for the parameter was developed. This might be a well defined probability distribution with mean, standard deviation, and shaping factors, or it might be an empirical CPF characterized by parameters for probability quartiles. The elicitations from analysts were weighted equally and 2 or 3 individual distributions were arithmetically averaged to yield an aggregated CDF for the parameter.

Event Trees and Fault Trees

An accident progression event tree (APET) (Wyss, 1990) modeled the primary and confinement system responses to the accidents found in the Level 1 analysis. It consisted of 87 events to model the core damage state, confinement venting, and isolation status. The accident progression phenomenology includes hydrogen and graphite issues, filter effectiveness, pressure suppression, and the formation of air and liquid pathways for the release of radionuclides from the plant. The APET results were generated by the EVNTRE code and analyzed by a parametric source term model (NSOR) to determine the radioactive source term for each accident progression pathway found in the APET analysis. These source terms were then grouped and analyzed for consequences using MACCS. The risk-results were finally calculated by PRAMIS. These codes (or, in the case of NSOR, types of codes) were used in NUREG-1150 as well as Latin Hypercube Sampling (LHS) and Monte Carlo uncertainty analysis techniques to produce an integrated uncertainty analysis for the entire PSA.

Results

The mean frequencies of events damaging more than 5% of the reactor core per year were found to be: Internal Events: 6.7E-5, Fire: 1.7E-5, Seismic: 1.7E-4, and total: 2.5E-4. Thus, within the range of U. S. commercial light water reactors The core damage frequency itself, is only part of the story because many N-Reactor accident sequences damage only a small fraction of the core. The

Applications of PSA

pressure tube design makes it possible to lose emergency cooling to parts of the core while successfully cooling the remainder. In addition, the GSCS is shown to provide sufficient cooling to prevent fuel damage in large regions of the core. Thus, the PRA showed that more than 95% of all internal core damage accidents involve failure of less than half of the fuel in the reactor. This significantly reduces the radionuclide inventory for release for further reduction by the confinement systems. The uniqueness of the N-Reactor results in sparse data questionable applicability of generic failure data. A key insight is active systems can inadvertently interfere with passive systems unless precautions are taken.

The accident progression analysis showed:

- The confinement maintains its integrity in more than 85% of internally initiated core damage accidents. About half of these failures are from mispositioning of a single switch to fail both the ECCS and confinement isolation.
- About half of the confinement failures are caused by hydrogen combustion in an ECCS effluent dump tank. Inerting the atmosphere in this tank would further reduce the confinement failure rate.
- The gaseous effluent filtration system was found to operate successfully in 95% of the internally initiated core damage accidents in which the confinement maintains its integrity.
- A fire in either of two critical areas would disable all front line and emergency core cooling systems, but not affect confinement isolation and effluent filtration.
- A seismic event three times or more the 0.25 g safe shutdown earthquake was the only significant mechanism found to cause oxidation of the graphite moderator stack.

The inventory of long-lived fission products is far less than an LWR because of the short exposure of the fuel to minimize Pu-240 production. But the health effects from an accident are comparable because it primarily results from short lived radionuclides.

N Reactor accidents are expected at lower fuel temperatures than LWR accidents. The large thermal capacity of the graphite moderator stack, the low melting point of the fuel (1,407°K) and the GSCS contribute to lower accident temperatures which retains heavy metals in the fuel.

The public consequences were significantly lower public than expected for a typical commercial LWR primarily because of the large distance to the public (>6 miles) and dispersal afforded by the height of the stack. The mean frequencies per year for offsite early fatalities are Internal Events: 9.7E-13, Fire: 3.8E-9, Seismic: 2.2E-8, and Total: 3.1E-9. For offsite latent cancer fatalities the results are: Internal Events - 4.1E-3, Fire - 4.2E-3, Seismic - 6.0E-2, Total - 6.8E-2. For onsite early fatalities the results are Internal Events: 1.0E-6, Fire: 8.3E-5, Seismic: 8.6E-4, and Total: 9.4E-4. For onsite latent cancer fatalities the results are Internal Events: 2.7E-4, Fire: 2.3E-4, Seismic: 3.0E-3, and Total: 3.5E-3.

11.3.6 Omega West Reactor PSA

The Omega West Reactor (OWR) is a water-cooled pool reactor operated by for the DOE by Los Alamos National Laboratory (LANL) for neutron irradiation experiments at a power of 8 MWt about 40 hrs per week. It is cooled by water at 40°C, atmospheric pressure flowing at 3500 gpm. It is powered by 33 fuel elements of the Materials Testing Reactor (MTR) type. These elements tested at the MTR such that 31 fuel elements operated at 30 MWt. Thus Omega West operates at 21% of the power at which the fuel has been tested.

A simple Level 1 PRA was prepared by Idaho State University College of Engineering (Neill, 1990a, Neill, 1990b) commensurate with the simplicity of OWR. Gradations of fuel damage were not defined. Accident scenarios were evaluated according to: the probability of fuel element damage or the probability of shutdown without fuel element damage. Operators were assumed not to scram in response to instrument readings. Nevertheless, the results of binary consequences is useful for establishing the relative probability of accidents covered by protective systems to those not covered.

Initiating Events

The OWR facility procedures, schematics, systems inspections, and staff interviews were studied including after which a Master Logic Diagram (MLD) constructed that identified 24 independent, generic for fuel damage events. Of these, 12 were sequential in that installed protective systems could intervene and prevent fuel element damage by shutdown or emergency cooling. The remaining 12 events were treated as leading directly to fuel element damage. The initiating event probabilities were quantified, using chi-squared estimation with one degree of freedom (no failures) to estimate the mean time between occurrences at 50 and 95 percentile confidence levels for initiating events over 30 years (15.6 and 132 years respectively). The mean time between occurrences were inverted to yield aggregate failure frequencies which were prorated equally over all 24 generic initiating events for frequencies at the 50% and 95% confidence levels. A log-normal probability density was assumed for the distribution of the individual initiating event frequency and a mean value and error factor were calculated. This substitution of log-normal distribution for chi-squares was a computational simplification. A Poisson probability distribution was assumed for the number of initiating events occurring over a one-year mission time. Each initiating event probability is generic in that it describes a group of failures that lead to a particular event without specifying any details of the failure. The MLD showed that all protective systems were challenged by either 1, 2, 3, or 5 initiating events.

Event and Fault Trees

The OWR protective systems were modeled with event tree diagrams for the time sequence following an initiating event to fuel damage or safe shutdown. Fault trees were used to find the probability of failure of each protective system in a particular event tree.

Results

Three steps were involved: 1) protective system fault tree top event probabilities were calculated by the IRRAS2 computer code; 2) probabilities applicable to a given event tree were combined with the initiating event probability, to give the probability of a fuel damage state or safe shutdown; 3) fuel damage probabilities from all event trees were combined for the total probability of fuel damage from the protected sequences. The probabilities of fuel damage according to initiating event type is shown as probability density functions in Figure 11.3-1.

11.4 Chemical Process PSAs

Unfortunately chemical process plant PSA have no requirements for public disclosure, consequently examples of their PSAs are hard to find. However the following are some examples from the open literature. They are similar to nuclear PSAs except less elaborate with more empahsis on consequences than on probabilities.

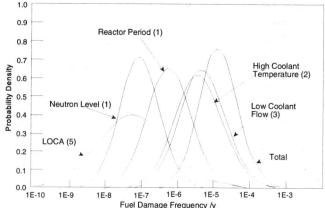

Fig. 11.3-1 Omega West Results (Numbers in parenthesis are number of initiating events)

In addition to these are studies prepared before President Carter stopped the GESMO (Generic Environmental Statement for Mixed Oxide) that addressed the chemical processing of fissionable material for the nuclear fuel cycle. Some references are Cohen (1975), Schneider (1982), Erdmann (1979), Fullwood (1980), and Fullwood (1983).

11.4.1 Canvey Island

11.4.1.1 Background

The Site

Canvey (1978) is a study of the risks of a petrochemical complex that already involved plants operated by Shell UK Oil, Mobil Oil Co. Ltd., Calor Gas Ltd., United Refineries, Occidental Refineries Ltd., Texico Ltd., and British Gas Corporation. In March 1976 the Health and Safety Commission agreed to a request to investigate the risks to health and safety associated with various installations, both existing and proposed, on Canvey Island and the neighboring part of Thurrock (Figure 11.4-1). The investigation followed an exploratory public inquiry recommending revoking the permission given in 1973 to United Refineries Limited to build an oil refinery on Canvey Island. This report recommended a detailed risk study because of the exceptional concentration of hazards and the demography. The investigating team consisted of personnel of the Safety and Reliability Directorate of the U.K. Atomic Energy Authority with expertise in hazard evaluation, the District Inspector of Factories for South Essex, and a Senior Chemical Inspector of Factories.

The investigating team was charged to realistically assess the risk at individual installations before considering interactions that may arise from fires, explosions and the release of airborne toxic substances and other interactions between installations.

The area is about 9 miles from west to east, along the north bank of the Thames, and 2.5 miles north to south at its widest, across Canvey Island. The area is divided by Holehaven Creek which separates Canvey Island from the mainland of Thurrock to the west. It is flat, low-lying salt marshland protected by sea walls. Canvey Island was reclaimed from the sea in the early 17th century. The population is about 33,000. New housing is being built and older property is steadily being replaced. Figure 11.4-1 shows the residential areas on the eastern half of the island, light industrial development is towards the southwest. Texaco Limited and London and Coastal Oil Wharves Limited tank storage, and the British Gas Corporation methane terminal front on the Thames. Occidental Refineries Limited and United Refineries Limited have requested permits to construct oil refineries on the west side of the island (at the time of the report's publication).

The part of the mainland within the area covered by the investigation is not residential, but primarily industrial. The two large oil refineries belonging to Shell UK Oil and Mobil Oil Company Limited are located on the riverside. The liquefied petroleum gases cylinder (LPG) filling plant of Calor Gas Limited is north of the refineries. Fisons Limited's ammonium nitrate plant is in the western part of the area.

Manor Way, A1014 provides road access to the mainland; a rail line for freight runs from the main line through the Shell UK Oil refinery to Thames Haven, and to the Mobil Oil distribution terminal at Coryton. A deep water channel provides access to the ten jetties of the two refineries. The control room and pumping station for the U.K. oil pipeline is located at Coryton. Oil products are transported by pipeline from Texaco Ltd on Canvey Island and from the two oil refineries on the mainland. Helicopters from the site landing pad survey the pipelines.

The Plants of Concern

British Gas Corporation Methane Terminal imports and stores about fifty shipments of 12,000 tonnes of liquefied natural gas (LNG) gas from Algeria each year, using two specially designed ships. The fully refrigerated liquid gas is pumped ashore from the ships to above-ground and in-ground storage tanks with 100,000 tonnes total capacity. Some LNG is road transported, but generally it leaves the terminal in vapor form by pipeline. Occasionally LNG-carrying ships are commissioned which involves cooling the cargo tanks with LNG from the terminal. The terminal is also used for the storage of liquefied butane brought by ships. About 200 people work at the terminal.

Texaco Limited, Canvey Island tank farm on the west side of the methane terminal is used to import petroleum products by sea via the two jetties, store and distribute them by tanker-trucks and sea, but mostly by the U.K. oil pipeline. The storage capacity is 80,000 tonnes of petroleum products employing about 130 people.

London and Coastal Oil Wharves Limited bulk stores, in tanks, a variety of petroleum products and other substances that are flammable or toxic. Some tankage is leased to Texaco Limited which is connected into the UK oil pipeline for transfer from the site. Products also are received by road and sea. The storage capacity is more than 300,000 tonnes with about fifty people employed.

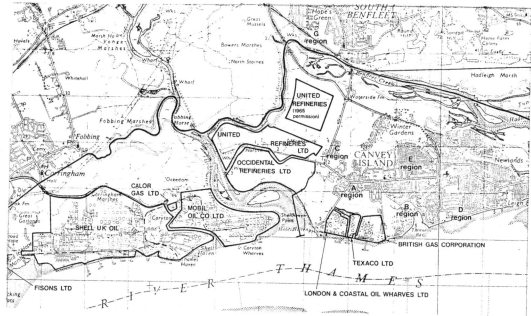
Fig. 11.4-1 Map of the Canvey Island/Thurrock Area

Mobil Oil Company Limited, Coryton occupies three adjoining sites on the mainland of Thurrock bordering the Thames and Holehaven Creek: the oil refinery, a bulk distribution depot associated with the refinery, and a large research and technical services laboratory. About 800 people are employed. Crude oil is imported by sea, stored in tanks for distillation into components which are either transferred directly to storage tanks, or chemically reacted into other products. From the storage tanks holding more than 1,500,000 tonnes, the products are distributed by pipelines, sea, road, and rail.

Calor Gas Limited, Coryton occupies a small site employing about 100 people to fill cylinders with LPG which is supplied by pipelines from both of the nearby refineries and stored in its seven tanks 350 tonnes total capacity. Large numbers of portable cylinders of all sizes, full and empty, are on the site totalling about 500 tonnes of LPG. These are transported by road; gases are transported in tank trucks.

Shell UK Oil, Shellhaven has the largest sites, about two miles along the Thames by about one mile across at its widest part and employs about 900 people. Crude oil, imported by sea, is transferred into storage tanks for distillation into components that are transferred into storage tanks, or chemically reacted into other products. The capacity of the storage tanks for hydrocarbons is more than 3,500,000 tonnes. The products are shipped by pipelines, sea, road, and rail. A large storage tank is used for storage of up to 14,000 tonnes of liquefied ammonia which is imported exclusively by sea and shipped from the area by sea, road, and occasionally rail.

Fisons Limited, Stanford-le-Nope is at the western extremity of the area and employs about 80 people in the production of strong solutions in water of ammonium nitrate from anhydrous ammonia. The liquefied ammonia is received from outside of the area in rail tankcars, stored, and partly refrigerated in a large spherical tank. The strong ammonium nitrate solution, produced on site, by reacting ammonia and nitric acid is stored in two large heated tanks. It is transported off-site in special tank trucks.

Explosives are trans-shipped at Chapmans Anchorage in the River Thames off the eastern end of Canvey Island. There is also a specified anchorage for ships carrying explosives at the opposite end of the area. This more sheltered anchorage for emergencies, such as bad weather, is well outside the area of the investigation being at least one mile south of Mucking Creek.

Ships in passage travel fairly close to the jetties and sea walls of the river bank within the area. All vessels within the estuary are navigated under requirements of the Thames Navigation Service of the Port of London Authority.

11.4.1.2 Methodology

All members of the investigating team were appointed to be inspectors by the Health and Safety Executive. This enabled them to make the necessary inquiries. All facilities and their hazardous substances in the area were familiar to the HM Factory Inspectorate. The facility operators, local authorities at county and district level, the Port of London Authority, and many local organizations and members of the public provided information, some as a result of advertisements in the local newspapers informing readers of the scope of the investigation that invited submissions. The investigating team was reassured to find that the managements were responsive and knowledgeable of operational safety, codes of practices and design and construction of plant and facilities for hazardous materials processing, handling, or storage. However, none of the companies had systematically examined and documented those few potentially serious events which might cause accidents among people in the surrounding community. In the absence of this documentation the first major task of the investigating team was to identify and analyze from first principles the routes by which single or multiple accidents could affect the community.

To carry out the assessment, the investigating team proceeded to:

1. Identify any potentially hazardous materials, their location and the quantities stored and in process
2. Obtain and review the relevant material properties such as flammability and toxicity
3. Identify the possible ways in which failure of plants might present a hazard to the community
4. Identify possible routes to selected failures typically involving: operator errors, equipment failure, plant aging, corrosion, over-filling, impurities, fire, explosion, missiles, and flooding
5. Quantify the probability of selected failures and their consequences

For probabilities and consequences, the investigating team used data they had and information from many sources, including:

1. U.K. industries, including oil, chemical and other process industries, and transport
2. Government agencies concerned with fire, and road, rail, sea and air transport
3. Professional and industry organizations
4. International safety conferences e.g. loss prevention in the process industries, ammonia plant safety, and hazardous materials spills
5. Insurance organizations e.g., Lloyds, Det Norsk Veritas, and Fire Protection Association
6. Foreign and international agencies, e.g., U.S. Coast Guard, U.S. Department of Transportation, OECD, and EEC
7. Specialized research laboratories
8. Experts

Several special projects were initiated for this study:

1. History of identified storage tanks and their possibility of failure
2. Probability of particular storage tanks or process vessels being hit by missiles caused by fires or explosions by fragmentation of rotating machines or pressure vessels, or transport accidents
3. Effects of vapor cloud explosions on people, houses, engineering structures, etc.
4. Evaporation of LNG from within a containment area on land or from a spill on water
5. Special problems of frozen earth storage tanks for LNG and the effect of flooding
6. Study of possible failures in handling operations
7. Benefits and practicality of evacuation
8. Sea-wall and its chance of being breached by subsidence, explosion or ship impact, timing consideration in flood
9. Statistics of ship collisions and their severity, groundings, etc.
10. Reliability analysis of fluid handling practices, ship to shore, and store to road vehicles and pipelines
11. Toxicology of hazardous substances
12. Lethal range for releases of toxic or explosive materials such as an ammonia spilt on water or land, or explosion or conflagration from a cloud of methane or liquefied petroleum gas

Other projects, mostly of more general application, were commissioned as the investigation developed. For example, the explosive behavior of clouds of liquefied petroleum or liquefied natural gas. Whether ammonia would rise as a light gas or slump from evaporative cooling. Potential hazards were identified early in the study stemming from the handling, processing, or storage of flammable liquids, ammonia, LPG, LNG, and hydrofluoric acid (HF). The degree of hazard, the extent of travel, and the number of people at risk, are all related to the large quantities of these materials on the various sites.

Risk

Risk depends on toxicity, design, and the competence of management at each site. Thus, management is responsible for hardware, work practices, and the selection, training, duties, and interrelationships of personnel. It must not only audit the plant but also the organization and its competence.

In the absence of company documentation of the hazard potential of their plants and procedures and methods it was necessary for the investigating team to identify and analyze from first principles the routes by which a single or multiple accident could affect the community. The assessment of risk in a complex situation has uncertainties that are best expressed numerically. The team should quantify the probability and consequences of various types of accidents. This enables qualitative (large or small) expression of risk as well as numerical representation. To say that the probability is 1 in 10,000 of killing 100 people as compared with a probability of 1 in 100 of killing 1,000 people is much more meaningful than saying the chances of these types of accidents occurring are "very remote" or "quite high." The team made best estimates of particular events expressed as 1 in 10,000 per year.

Quantitative assessment requires historical data which may be suspect for two reasons. There is the possibility that there are latent accidents not in the database. It is possible that past accidents have been rectified and will not recurr. In the absence of data, judgment based on experience and speculation must be used. Notwithstanding this weakness, the quantitative approach was adopted. The investigating team identified situations that could cause a number of public casualties. Events limited to the employees or which might cause single off-site casualties were not included in the assessment.

Typical events that are considered are fire, explosion, ship collision, and the failure of pressurized storage vessels for which historical data established the failure frequencies. Assessment of consequences was based partly on conservative treatment of past experience. For example the assessment of the number of casualties from the release of a toxic material was based on past history conditioned by knowledge of the toxicology and the prevailing weather conditions. An alternative used fault trees to estimate probabilities and identify the consequences. Credit is taken in this process for preventative measures in design, operation, and maintenance procedures. Historical data provide reliability expected from plant components and humans.

Both these methods have weaknesses. The historical data, implicit in both, may not be appropriate for the component in the assumed environment for the particular circumstances considered and may be irrelevant because of design rectification based on knowledge of previous failure. The broad approach may then be unduly pessimistic. On the other hand, the fault tree may fail to identify a primary cause which may have been missed by the plant designer to underestimate the probability of failure. The fault tree approach also takes credit for the preventative measures which may not be present in practice. The broad approach is likely to overestimate the risks because of insufficient account of preventative measures. The PSA team used the broad approach, recognizing that more accuracy may be attained by detailed industry studies.

Confidence Limits

Confidence limits for the conclusions cannot be expressed simply because of the complexity of hazard assessments. The estimation of uncertainties is itself a process subject to professional judgment. However, the team's estimates of probability are believed to be realistic, but may be pessimistic by a factor of perhaps two or three, but less than a factor of ten. Uncertainties also exist in the estimation of the consequences of any particular accident sequence e.g. in the health and debilitating effects of exposure to varying concentrations of toxic substances.

Risk Comparison

Accident statistics show that individuals in an industrialized society are exposed to widely differing levels of risk according to age, sex, occupation, location, and recreation. There is not and cannot be equality in the levels of risks, from natural or man-made sources. For example, there are 18,000 accidental deaths per year in Great Britain (1976) of which 7,000 are traffic accidents or a probability of 1.3E-4 /yr if equally shared. The risk is not equally shared; it is 3 to 4 times lower for those who do not ride in cars; it is very much higher for young males between 18 and 25. Similarly, the risk of dying from an accident at home is 1.2E-4 /yr again the risk is not shared equally - 70% is carried by the very young and the elderly. The risk of a fatal accident at work is 0.3E-4 /yr. The chance of fatal injury is 11 times greater for workers in some occupations than it is for those in others.

These figures relate to the risks to an individual. Societal risk is more concerned with a single accident causing many deaths than with many accidents causing the same number of deaths. Part of this concern stems from public belief in the rectifiability of high consequence accidents. In practice, the overall risk to the individual is affected slightly by the risk from high consequence accidents. Most of the 18,000 accidental deaths in Great Britain arise from accidents in which only a few people die at a time. In some countries natural occurences kill thousands even hundred thousands, but little can be done to rectify nature and continue to live in these locations and the probability is not perceived to be high.

11.4.1.3 Major Hazards

Attention focused on events that cold release a quantity of flammable, explosive or toxic vapors, hence, operations involving toxic liquefied gases such as ammonia and hydrogen fluoride, and flammables such as LNG and LPG.

Ammonia

Ammonia, when released is a toxic gas with little flammability. It is imported by sea into the 14,000 tonnes capacity tank at Shell UK Oil where the refrigeration maintains the temperature below the boiling point of the gas (33° C). Three ways were identified whereby several hundred tonnes of liquid ammonia could be released into the river to vaporize and disperse. The worst accident would have an accompanying explosion or fire on an ammonia carrier berthed at the unloading jetty. Next in order of severity is a ship collision and spillage into the river near the unloading jetty. The consequences of a collision between ships occurring within the area but not near the jetty were also calculated.

The team suggested mitigations. For example, the consequences of an ammonia release could be markedly reduced by a reliable and quick-acting water spray system. The probability of damage due to ship collisions could be reduced by a factor of five by a speed limit of eight knots. Evacuation of the potentially affected is practicable and effective.

Liquefied ammonia is delivered in rail tankcars to Fisons Limited for storage in a 1,900 tonnes spherical tank at -6° C. Several hundred tonnes of liquefied ammonia could be released on land if either of the two storage tanks, at Shell UK Oil and at Fisons Limited failed. The consequences of failure of the Shell tank would be minimal, because a high concrete wall to contain the contents and limit the heat transfer and consequently the rate of evaporation of the liquid. Such protection has not been provided. Because of the storage under pressure there are numerous ways the tank could fail from material defect to missile. The spillage of 50 to 100 tonnes, could kill people if not promptly evacuation.

Hydrogen Fluoride

Hydrogen fluoride is at present stored in bulk and used at only the Shellhaven refinery, but future use is envisioned. Hydrogen fluoride boils point 19° C but stored and handled as a liquefied gas. Its vapor is highly irritating and toxic. A cold cloud will be denser than air in the early stages when released but will become neutral or buoyant as it warms with dispersion. This assessment assumes negative buoyancy

Hydrogen fluoride is a component of the reaction to make the detergent alkylate plant at the Shellhaven refinery. The plant has 20 tonnes of hydrogen fluoride in several vessels operating above the boiling point of hydrogen fluoride. In addition, the plant has two 40 tonnes capacity store tanks to contain hydrogen fluoride to make up process losses that is delivered to the plant in tank trucks. It is assessed that the design and construction would withstand the overpressure resulting from an unconfined vapor cloud explosion. A water spray for the storage tanks and operating plant would mitigate a release.

Liquefied Natural Gas

LNG is predominantly methane, not toxic but flammable in air. Liquid below -161° C is delivered to the methane terminal in shiploads of 12,000 tonnes and is transferred into either four in-ground tanks, each of about 20,000 tonnes capacity or eight above-ground tanks, six with 4,000 tonnes capacity and two with 1,000 tonnes capacity. The LNG is refrigerated below the boiling point of the gas mixture.

If LNG were released and cloud ignited the accident would be serious. If the cloud did not ignite on site but drifted to a populated area and then ignited, the number of casualties would be be much greater. Ignition sources are strictly controlled at the terminal so on-site ignition was not considered

An explosion in the engine room of an LNG carrying ship berthed at the jetty could release vapor cloud, whose explosion would have serious consequences. No amelioration was suggested by the team. Release of cargo, from ship collision was discounted because of the eight knots speed limit.

Spillage of LNG within the containment walls of the aboveground tanks was considered as was roofs collapse of the in-ground storage tanks if the river were to overtop the sea wall causing a substantial release of vapor as the tanks flooded. This effect could be eliminated by higher containment walls. Risks from the commissioning of new LNG-carrying ships and from frozen ground LNG storage tanks were examined without identifying major problems.

Liquefied Petroleum Gas

LPG is a mixture of flammable hydrocarbons which are gas at normal temperature but liquid under pressure or when cooled below the boiling point at atmospheric pressure. Two mixtures are in common use, commercial propane and commercial butane. Large quantities are stored and handled at: British Gas Corporation methane terminal, Shell UK Oil, Mobil Oil Co. Ltd, and Calor Gas Ltd. The last also fills and handles large numbers of portable LPG cylinders.

There are three aboveground tanks at the British Gas Corporation methane terminal for storage of liquid butane by refrigeration below the boiling point of $0°$ C. Two tanks have a capacity of 5,000 tonnes each, third's capacity is 10,000 tonnes. A pipeline from the tanks crosses the area. A substantial quantity of LPG could be released from any one of the three refrigerated storage tanks to form a large flammable cloud that could ignite and explode, causing casualties in the vicinity of the terminal. The pipeline from these tanks could fail and cause casualties.

At the Shell UK Oil refinery the storage capacity for LPG is 5,000 tonnes. The gases are stored at ambient temperature. Most of the storage capacity of four spherical tanks for butane is 3,200 tonnes. The total capacity of the four spherical tanks for propane is 1,600 tonnes with each tank holding about 400 tonnes. In addition, three horizontal cylindrical tanks store about 400 tonnes of mixed liquefied petroleum gases, each tank holding about 135 tonnes.

At the Mobil Oil Co. Ltd. refinery, the storage capacity for liquefied petroleum gases is about 4,000 tonnes, stored under pressure at ambient temperature, and it is proposed to double this in the near future. Later it is intended to add to these tanks, a single tank in which about 5,000 tonnes of the liquefied gas would be stored, fully refrigerated, at a temperature lower than the boiling point. Fifteen tanks are in use at present, of which ten are relatively small horizontal cylindrical tanks and the others are spherical tanks ranging in capacity from about 250 tonnes to one of about 1,000 tonnes.

Calor Gas Ltd. stores a small quantity of liquefied gas because it can be delivered readily by pipelines from the neighboring oil refineries. The five tanks each storage about 60 tonnes capacity at ambient temperature. Three are used for propane and two for butane. The total quantity of gas contained in filled cylinders averages 500 tonnes.

The possibility that the storage tanks could fracture has been calculated. At the Shell UK Oil refinery, the storage vessels are located in two groups, one close to the vehicle filling position. Consideration is given to an explosion in the filling area sending a shock wave to the other storage area, failing the tanks to generate a large vapour cloud to drift off-site to a populated area before ignition with serious effects.

The possibility missile impact failing a storage tank at the Shell site was assessed. Similarly missile impact failure was assessed at the Calor Gas Limited site. An assessment was made of a major fire at this LPG cylinder filling plant violently projecting some of the filled cylinders from the site to impact on the Mobil plant, however it seems that the missiles lack sufficient energy to penetrate the storage units but pipes may be fractured. Large clouds of vapor could be generated if LPG were released from an explosion on a ship transferring LPG at one of the Shell or Mobil jetties or the Occidental jetty, or from a collision between an LPG-carrying ship near the unloading jetty and another ship. The consequences of these events would be serious, and little action could be taken to lessen the probability of the explosion, although observance of the eight knots speed limit should reduce the probability of ship collisions.

Table 11.4-1 Frequency /10,000 Years of Exceeding Maximum Casualties for Existing Facilities													
Max. Casualties	10		1500		3000		4500		6000		12000		18000	
Status*	a	b	a	b	a	b	a	b	a	b	a	b	a	b
Shell UK Oil	7.8	3.3	5	1.9	3.6	1.3	2.7	0.9	1.8	0.5	1.2	0.3	0.8	0.2
Mobil Oil Co. Ltd.	1.4	0.3	1.1	0.2	0.7	0.1	0.3	0.1						
British Gas Corp.	14.7	4	5.4	1.5	3	1	1.6	0.7	0.4	0.4	0.2	0.2	0.1	0.1
Fisons Ltd.	3.6	1	2.4	0.6	2	0.5	1.1	0.3	0.8	0.1	0.3	0.1	0.1	
Texaco Ltd. and London & Coastal Oil Wharves	3.9		3.1		1.5		0.4							
Total	31.4	8.6	17	4.2	10.8	2.9	6.1	2	3	1	1.7	0.6	1	0.3
Frequency /10,000 of Exceeding Maximum Casualties for Proposed Facilities														
Mobil Oil Extension	2.8	0.4	1.7	0.2	1.1	0.1	0.7	0.1	0.3		0.2		0.1	
Occidental Refinery	2.9	0.9	2.2	0.7	1.7	0.5	1.4	0.4	1.1	0.3	0.8	0.2	0.7	0.2
United Refinery	1.4	0.9	0.9	0.6	0.5	0.3	0.2	0.1						
Texaco Ltd. and London & Coastal Oil Wharves	9		7.3		3.6		0.9							
Total	16.1	2.2	12.1	1.5	6.9	0.9	3.2	0.6	1.4	0.3	1	0.2	0.8	0.2

* Columns labeled "a" are the assessed exceedance probabilities before safety improvements; Columns labeled "b" are the assessed exceedance probabilities after safety improvements

Flammable Liquids

Large quantities of flammable liquids are storaged on Canvey Island at Texaco Ltd and at London and Coastal Oil Wharves Ltd., and at the two operating refineries on the mainland of Shell UK Oil and Mobil Oil Co. Ltd. Proposed oil refineries for Canvey Island would also store large quantities of flammable liquids. It is conceivable that an explosion could damage the storage tanks, releasing the flammable substances which might ignite. If many tanks failed, their contents could overflow the dikes and burning liquid could flow in drainage channels to ignite houses. Such accidents include the explosion of vapor released in the atmosphere from an explosion of an LNG-carrying ship berthed at the methane terminal or of an LPG carrier berthed at the Occidental jetty. Other possible causes include a process plant explosion at the proposed Occidental refinery, or spontaneous failure of LPG storage tanks at proposed refineries. Other possibilities are an explosion of a barge or ship carrying explosives stranded on the opposite shore. These probabilities were assessed. The severest consequence of such a fire could be prevented on Canvey Island by constructing a simple secondary containment wall around the Texaco Limited and London and Coastal Oil Wharves Limited installations and around proposed oil refineries.

Ammonium Nitrate

Ammonium nitrate is made at Fisons Ltd and stored in solution in two heated tanks to prevent crystalization. One tank holds 3,100 tonnes and the other 6,200 tonnes of 92% aqueous ammonium nitrate solution. Trains laden with oil refinery products from the Mobil and Shell refineries pass the factory on a near embankment. A derailment could spill and ignite hydrocarbons from a rail tank car to explode an ammonium nitrate storage tank. Suggestions were made to mitigate or prevent such a domino effect.

11.4.1.4 Societal Risk

Many of the accidents considered in the investigation could occur without causing any significant public casualties. However, if the conditions at the time of the accident were sufficiently unfavorable, the number of deaths among the public could range from tens up to thousands (Table 11.4-1). Table 11.4-1 is the summarized population risk assessed by the study team. It is in frequency per 10,000 years of an accident at the indicated facility that causes casualties exceeding the indicated limit. Reference should be made to Canvey (1978) for details.

11.4.2 PSA of a Butane Storage Facility (Oliveira, 1994)

Introduction

A deactivated butane pressurized storage facility with two 750 m^3 and two 1,600 m^3 spheres was considered for partial activation of the two smaller spheres. A safety review showed that aspects of the facility did not comply with with current standards of PETROBRAS regarding layout and separation. A PSA was performed with the following objectives:

- Assess the public and workers risks imposed by operation of the facility as it is;
- Compare the results to existing international acceptability criteria;
- Propose and evaluate risk reduction measures; and
- Evaluate whether it is acceptability with risk reduction measures, or move the facility to a recently refurbished LPG storage facility.

PSA

Risks were expressed as triplets: <what can go wrong | frequency | consequences>. The first element of the triplet was found using accident records and a PHA. The databases used were MHIDAS (1992) (>5000 accidents) and ACCIDATA (>1,500 mostly Brazil). The PHA was performed by personnel from REDUC (facility operator) and PRINCIPIA (the PSA vendor). About 170 basic initiating events (ruptures of pipes, flanges, valves, spheres, pumps and human actions) were grouped into 12 initiators by equivalent diameter, pressure, flow type and rupture location.

The second triad element was determined from international frequencies for the basic initiating events. One fault tree for sphere rupture was constructed and evaluated using the fault tree program FTW (1991).

The third triad member was evaluated from the accident scenarios involving failure or success of existing protection systems, daytime and nighttime conditions, wind speed and direction, and ignition sources. There were 1,474 different accident scenarios were evaluated, involving BLEVE's, flash and pool fires and unconfined vapor cloud explosions (TNO, 1980 and Colambrander, 1983). The consequences of each scenario were calculated for the public and workers. A risk assessment program VULNER+ was developed by PRINCIPIA to perform scenario frequency and consequence calculations. The accident scenarios were generated by the program from information provided by the analyst. A square grid overlaid the site. Each square was associated with the population contained therein and the ignition sources. Cells were subdivided for greater resolution as needed.

The fatality conditional probability at each cell center is associated with the scenario frequency to produce the value of the individual risk at that cell. The program interpolates to draw the contours of individual risk around the facility. F-N curves and average societal risks are found by coupling results to the population in the cell. The ranked scenarios help in proposing risk-reduction measures.

Risk Results As the Facility Is

Contours of individual risks at frequencies of: 1E-4/y, 1E-5/y, and 1E-6/y were obtained, at approximated radii of 200, 500 and 700 m around the facility. The F-N curve for the public showed >10 in 2E-4/y, and >100 in 7E-5/y, with a maximum of 275 fatalities. The existing facility public risk is 2.1E-2 fatalities/y.

Risk Reduction

Table 11.4-2 shows risk reduction measures and their impact on public risk, their cost, and the cost per death averted. From this table, with the exception of the last one (because of high cost), the measures are practical for implementation.

Conclusions

It was concluded that:

1. The operation of existing facility imposes an unacceptable social risk to the public based on criteria used in the Netherlands and Denmark.

2. After the implementation of all risk-reduction measures, the public risk is not acceptable (Denmark criteria), mainly due to low frequencies/high consequences (highly populated area). After these

Table 11.4-2 Risk Reductions and Costs

Measure	Risk reduction fatal./y	Cost $US	$/death averted
Test liquid blockage valves every 3 months	5.1E-5	0	
Water injection directly into the manifold eliminating a valve and reducing the valve size for drainage	3.3E-3	2.8E4	1.1E6
Elevate and slope ground under spheres	1.3E-2	4.6E5	5.0E6
Valves removal at the pump discharge	1.3E-4	1.0E4	1.1E7
Explosion-proof control room and operator always present in control room	1.1E-3	1.0E5	1.3E7
Only one product receiving line; removal of other	1.1E-4	1.1E4	1.4E7
Nafta pipeway relocation	6.3E-3	9.5E5	2.0E7
Removal of neighboring liquid fuel tanks	4.1E5	3.1E6	1E10
All risk reduction measures	2.4E-2	4.6E-6	2.6E7

Applications of PSA

measures, the public area subjected to levels higher than 1.0E-5 (the maximum value for acceptability in England) is relatively small being mainly employees of adjacent companies.

3. Because the risks were not reduced by the risk-reduction measures to an acceptable level, relocation of the butane storage to the refurbished LPG facility is attractive.

11.4.3 Comparative Applications of HAZOP, Facility Risk Review, and Fault Trees

Montague (1990) addresses the problem of selecting the appropriate analysis method for chemical process safety. He describes three methods and provides an example of each. Chapter 3 describes the methods (except facility risk review), hence, we go to examples of each.

11.4.3.1 HAZOP of an HF Alkylation Unit

System Description

A large oil refinery had a failure in their alkylation unit resulting in a significant release of hydrofluoric acid (HF). This incident and others in refineries during recent years prompted members of the hydrocarbon processors to turn to HAZOP to better understand their risks.

Figure 11.4-2 shows process flows for an HF alkylation unit. The three sections are: 1) reaction, 2) settling and 3) fractionation. In the reaction section isobutane feed is mixed with the olefin feed (usually propylene and butylene) in approximately a 10 or 15 to 1 ratio. In the presence of the HF acid catalyst the olefins react to form alkylate for gasoline blending. The exothermic reaction requires water cooling. The hydrocarbon/HF mixture goes to the settling section where the denser HF is separated by gravity and recycled back to the reaction section with fresh acid to compensate for losses. The hydrocarbon is fractionated in several towers where it is

Table 11.4-3 HF Alkylation HAZOP Results

Hardware
1. Designate/install emergency isolation valves in the alkylation unit and install remote switches in the alkylation unit control room so that operators can close these valves during emergencies
2. Install remote switches in the alkylation unit control room to shut down the major pumps in the alkylation unit during an emergency
3. Designate critical process alarms in the alkylation unit and ensure that these alarms are installed independent of equipment control loops
4. Review the design bases for unit relief valves and the flare system to ensure that they are still valid
5. Evaluate ways to prevent backflow in critical lines between process equipment
6. Evaluate making the alkylation unit control building a "safe haven" during hazardous material releases administrative action items
Administrative Action
1. Review the acid unloading system and its procedures to ensure that adequate safety precautions are included for minimizing potentially hazardous incidents during acid unloading operations
2. Establish/enhance consistent programs to periodically inspect and test the engineered defenses that are designed to protect against hazardous upsets in the alkylation unit
3. Investigate ways to improve the control of replacement materials installed in the alkylation unit during turnarounds and routine maintenance
4. Investigate ways to reduce the potential for external impacts to damage process piping while the alkylation unit is operating
5. Evaluate the acid relief neutralization system in HF alkylation units to ensure its adequacy for neutralizing design basis relief valve discharges and unit ventings
6. Enhance refinery emergency procedures by specifically addressing accidental releases of acid

separated into a recycle isobutane stream and three products: alkylate, propane and n-butane. This section also purifies the products by removing the small amounts of HF in the product stream and returning the acid to the reaction section.

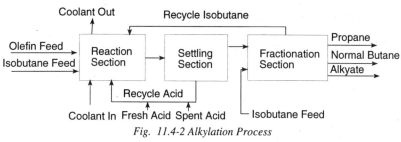

Fig. 11.4-2 Alkylation Process

HAZOP

The HAZOP team consisted of a HAZOP leader, scribe, process engineer, senior operator, and a safety engineer. Previously, the HAZOP leader collected information on the process design (e.g., P&IDs, operating procedures, emergency procedures and incident reports) for distribution to each team member. The leader sectioned the process and identified process deviations prior to the HAZOP.

The actual HAZOP required a 1-week meeting and a 3-day meeting. During the first meeting, the team "HAZOPed" each section of the process, identified process upsets with significant consequences, and made recommendations for safety improvements. In the second meeting, a month later, the HAZOP team resolved open issues from the first meeting and examined sections of the process not reviewed in the first meeting.

The HAZOP took 3 calendar months to complete. However, preliminary recommendations for system safety improvements were available after the second HAZOP meeting. The HAZOP required about 2 staff-months of effort. In addition, the HAZOP leader and scribe spent another 3 staff-months in preparation and documentation.

Results

Sixty recommendations were made for improving the safety of the HF alkylation unit. The suggestions ranged from minor piping changes to installing remote control isolation valves to changing emergency procedures. The recommendations can be divided into: hardware and administrative. Hardware requires new or modifies existing hardware; administrative changes require modifying procedures, initiating or enhancing programs - usually at less cost than hardware. Table 11.4-3 lists 12 of the 60 suggestions for HF alkylation.

11.4.3.2 Facility Risk Review (FRR) of a Mining Operation

FRR Procedures

A facility risk review (FRR) is intermediate between a qualitative HAZOP and a quantitative risk assessment (QRA) achieved by broad probability and consequence classifications. Although not a risk assessment, an FRR uses PSA to get optimum risk cost-benefit.

QRA is generally resource intensive. FRR, is much less although it uses the framework of QRA for screening and ranking systems by relative risk using order-of-magnitude estimates of the frequencies and consequences of events. It incorporates plant experience and industry data to estimate the potential for future losses.

An FRR consists of the following:

- Failure modes and effects analysis
- Consequence assessment
- Frequency assessment
- Risk categorization
- Recommendations

Background and System Description

Three years ago, a fire in a large mining operation resulted in tens of millions of dollars in equipment damage and a half-billion dollar loss from months of lost production. Previously, the facility had several multimillion dollar accidents, and near misses. After each accident, investigations were made and corrective actions taken. However the large fire precipitated a risk evaluation of the entire mining complex to find latent accidents.

Table 11.4-4 Recommendations from FRR of Mining Operation

Mining
- Develop and emphasize maintenance work plans on large mining equipment
- Emphasize and improve fire protection for mine equipment areas,
- Ensure that the fire protection systems in the major equipment electrical rooms are adequate and are periodically tested
- Stop bringing the explosives truck through process and administrative areas
Ore processing
- Evaluate the current operability of critical items for the conditioning drums
- Consider providing remote shutdown capability for the charge pumps
- Evaluate capability to respond to major process leaks in the first or second stage pump areas
- Consider a foam fire protection system for the first stage pump area
- Consider moving the chemical storage tank to an outdoor location
- Develop a contingency plan for responding to collapse or blockage of the Area 2 sewer line from the building to the decant pond
Refinery area
- Consider installing remotely operated isolation valves in the bottoms piping connected to vessels containing large inventories of hot solvents
- Consider a test program for periodically checking critical instruments and trip devices
- Consider a test procedure requiring periodic cycling of MOVs
- Consider a policy of independent inspection and verification of work performed by maintenance
- Review the instrumentation/trip designs for all furnaces in the refinery area
- Evaluate whether the interlocks on the swing valves are adequate to prevent operators from inadvertently mispositioning them
- Review the compressor building ventilation system to ensure that it has adequate capacity to maintain acceptable temperatures in the compressor bays
- Consider a test procedure to periodically check the operation of an important steam flow control valve and the position of its associated bypass valve
Utilities area
- Evaluate alternatives for protecting the facilities from events that can result in plant freezeup
- Evaluate alternatives to upgrade the raw water system valve box and piping
- Evaluate the integrity of the old cooling water loop piping, and identify ways to protect against its failure
- Improve the current capability to isolate steam lines leaving the powerhouse to ensure that the steam system can be protected from piping ruptures outside the powerhouse
- Provide an emergency response plan for flammable gas leaks in the powerhouse, including methods for rapidly isolating potential leaks or ruptures

The mining facility consists of four major operations: 1) mining, ore extraction and transportation, 2) ore processing - treatment with chemical solvents to remove minerals, 3) removing impurities and 4) utilities e.g., electricity, steam, water, air and natural gas provided by the facility's power plant and by off-site suppliers. Accidents in any of these operations can stop the whole process.

The FRR Analysis

Each of four teams for each mining operation consisting of a risk analyst and engineers and operators performed the FRR. The risk analyst provided the FRR expertise and directed the analysis of the operation; facility engineers and operators provided operational knowledge and estimated the likelihood of equipment failures and their impact on facility operations.

Following the FRR procedure, each team performed an FMEA, to identify costly accidents. Each accident was assigned a frequency category and the cost of equipment damage, outage time and percentage reduction in facility output was estimated. The frequency and cost were combined to estimate the accident risk. From the risk and the insights gained while performing the analysis, the FRR teams identified recommendations for reducing the likelihood of multimillion dollar accidents.

The FRR required 8 calendar months to complete. It was performed by a team of three risk analysts who expended 2 staff-years of effort conducting the analysis. During the FRR, the mining facility provided 15 people who collectively expended about 4 staff-months of effort providing the risk analysts with facility design information and estimating accident impacts.

Results

The FRR teams identified more than 1,600 potential accidents of which 1,000 were considered significant enough to be assigned a frequency and consequence category. These accidents were then arrayed by frequency and dollar cost to show 165 accidents of greatest concern. The FRR teams recommended actions to reduce the likelihood of these 165 accidents. The risks expressed as M$/y for the four operations are mining: 3.2, ore processing: 2.9, refinery: 46.1, and utilities: 8.5, to give a total risk of 60.8 M$/y. Some recommendations are given in Table 11.4-4.

11.4.3.3 Quantitative Risk Assessment of a Chemical Reactor

Background and System Description

A chemical company was considering a larger reactor to expand production. Because of industry reported accidents with this process, a HAZOP was performed that identified the potential for a violent reaction with deadly consequences. A QRA was conducted to better understand the risk of expansion and to determine additional safety needs.

The reactor system (Figure 11.4-3) consists of a 5,000-gallon vessel, an agitator and pumps for circulating cooling water through the reactor jacket. Chemicals X and Y are fed at a controlled rate to the reactor, where they react exothermically in the presence of a catalyst to form the product. If the temperature gets too high, the reaction may runaway.

Causes of high temperature (identified in the HAZOP) included loss of cooling, loss of agitation. and a high ratio of catalyst to chemical feed. Proposed safeguards include 1) an emergency shutdown system that automatically shuts off the feeds to the reactor, increases cooling, and dumps a reaction quenching poison into the vessel if required and 2) a manually controlled dump line to a quench tank.

Applications of PSA

QRA

The QRA was conducted by risk analysts and design personnel to determine the probability of explosive releases of the chemical. Fault tree analysis identified several combinations of equipment failures and operator errors that could cause the top event (reactor explosion). Failure data were obtained from plant experience and industry databases to quantify the fault trees to estimate the frequency of reactor explosions. The fault trees suggested several safety improvements. These suggestions were tested by incorporating them into the fault tree models to assess their risk reduction. Sensitivity studies were performed to assess the robustness of the analysis.

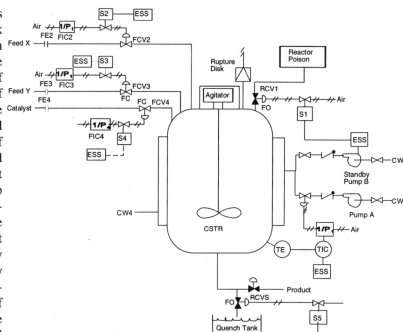

11.4-3 Proposed New Chemical Reactor

Parallel to the frequency assessment, the consequences of the scenarios were calculated for the effects on the public and plant personnel of chemical release through the rupture disc and fire and explosion. Gas dispersion models provided the toxic effects of chemical releases, fire, or unconfined vapor cloud explosion. The affected areas were plotted on demographic maps to determine the number of people affected.

The QRA required 5 calendar months by two risk analysts. A total of 6 staff-months were required for completion. During the QRA, the

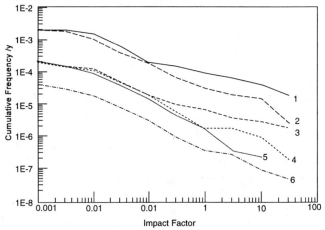

Fig. 11.4-4 Case Studies of QRA Results

chemical company provided an additional 1 staff-month in gathering information and reviewing results.

Results

Suggestions for mitigation were tested using the consequence models to determine their efficacy. Sensitivity studies were performed to test assumptions on the frequency/ consequence curves. Case 1 in Figure 11.4-4, the base proposed reactor design, shows that an explosion affecting off-site personnel (consequences greater than impact factor 10) has a frequency of about once every 10,000 years. Alternate reactor designs, Cases 2-6, could lower this by about 1000.

These results showed management that the proposed larger reactor design had an unacceptably high risk. Design modifications can substantially reduce risk, none eliminate the possibility of adversely affecting the public. Only by reducing the reactor volume or by reacting the chemicals with a solvent (economic penalty) could consequences be reduced to on-site only. Table 11.4-5 shows some results of the sensitivity studies.

Table 11.4-5 Frequency (1/y) vs Assumptions

Assumption	High	Low
Reactor volume (1E3 - 1E4 gal.	4.3E-5	8.5E-3
Reactor pressure (5-50 psig)	1.1E-4	1.5E-4
Extra cooling capacity (10 - 100%)	8.1E-3	1.3E-2
Emergency System Effectiveness (100 -70%)	2.9E-4	8.5E-3
Reactor mixture dilution (50%/50%)	2.7E-7	8.5E-3

Summary of QRA of the Chemical Reactor

1. The variation in risk between the designs analyzed is influenced more by differences in the consequences of the accidents than by differences in the accident frequencies.

2. The risk of a violent reaction causing a chemical release increases as the design volume of the reactor increases.

3. Improving the capability to detect a runaway reaction and terminate it (by emergency protection) significantly reduces risk.

4. The risk of a violent reaction decreases if the chemicals are mixed with an inert solvent that provides a heat sink and reduces the chemical inventory.

5. High-pressure can increase risk because of greater overpressure from a violent reaction.

11.4.3.4 Comparison of HAZOP, FRR and QRA

Montegue (1990) compares the three methods as presented in Table 11.4-6.

11.4.4 Probabilistic Safety Analysis of an Ammonia Storage Plant

Introduction and Facility Description

The following PSA of a refrigerated ammonia storage facility is from Papazoglou (1990a, 1990b). An objective of the work was to test PSA as it had been applied to nuclear power

Applications of PSA

Table 11.4-6 Comparison of Some PSA Methods

Hazards and Operability (HAZOP)	Facility Risk Review	Quantitative Risk Analysis
Focuses on one area or operation, and many types of hazards	Broad review of many areas or operations, many types of hazards	Focuses on a specific type of hazard
Involves many company personnel for short periods of time (ties up several people at one time)	Involves a few company personnel for short periods of time (ties up one or two people at a time)	Involves a few company personnel for short periods of time (ties up one or two people at a time)
Provides a qualitative measure of risk	Provides a coarse quantitative measure of risk	Provides an absolute quantitative measure of risk
Good screening of hazards/risks	Inefficient screening of hazards/risks	Inefficient screening of hazards/risk
Subjective evaluation of recommendation effectiveness	Subjective evaluation of recommendation effectiveness	Cost/benefit analysis of recommendations

plants for applicability to chemical process facilities. Related papers are Papazoglou (1992, 1996).

The facility stores ammonia to supply nearby fertilizer manufacturers. It is transported to the plant by ship, transferred and stored in the tank for transfer as needed. The facility mainly consists of a storage tank, a refrigeration system, a control system, and a pipe connecting the tank with the plants. Ammonia is transported and stored as a refrigerated liquid (-33° C) at atmospheric pressure.

The storage facility operates in three phases:

1. Transfer of liquid ammonia from the ship using a special pipeline to the thermally insulated storage tank. The transfer, requiring 20 hours, uses the ship's pumps; there are five transfers each year.
2. The ammonia is in refrigerated storage up to 1,646 hours
3. Ammonia is unloaded from the storage tank to either of the two fertilizer plants using three discharge pumps for 86 hours.

During each of the three phases a refrigeration system, consisting of three (3) compressors, two (2) condensers and one (1) separation drum, ensures that the ammonia is at its storage temperature.

PSA

The methodology and the procedures consist of the following seven major steps:

1. Hazard Identification, the main sources of ammonia that could release are identified and the initiating events (IE) that can cause accidents leading to the release of ammonia are determined. Three methods used for IE identification were: Master Logic Diagram, checklists, and HAZOP
2. Accident Sequence Modeling using a logic model for the facility was developed. The model included all initiators of potential accidents and the response of the installation to these initiators. Specific accident sequences are defined in event trees consisting of an initiating

event group, specific system failures or successes and their timing, and human responses. Accident sequences result in plant damage states that may or may not release ammonia. System failures are modeled in fault trees of component failures and human errors. Nineteen event trees were developed and thirteen were quantified.

3. Data Acquisition and Parameter Estimation determines frequencies of the initiating events, component unavailability and probabilities of human actions were estimated from plant history. If insufficient, generic values were used including generic data from the nuclear industry (IAEA, 1988). In addition meteorological data and data on the population distribution around the plant were gathered and processed.

4. Accident Sequence Quantification estimates the IE frequency. Specifically, the plant model built in the Step 2 is quantified by data from Step 3 according to Boolean algebra. Quantification may be a point-value calculation in which all parameters are deterministic, or as uncertain values known by their distribution function.

5. Ammonia Release Categories Assessment consists of grouping accident sequences of the same failure mode into a plant-damage state. A plant-damage state is characterized by the quantity and type of release corresponding to the plant physical conditions at the time of release.

6. Consequence Assessment calculates the atmospheric dispersion of the released ammonia, the assessment of the individual dose at each point around the site, and the construction of a dose/response model. Each ammonia release category might lead to different concentrations, hence, different doses and probabilities of consequences. An uncertainty analysis is performed for the distributed parameters for various model assumptions.

7. Integration of Results - consists of integration of the model with results from Steps 4, 5 and 6, to determine consequences, probabilities and range of uncertainties. Two types of results were calculated in this study: individual risk as a function of distance from the installation, and isorisk curves. Both are risk indices and are related to the probability of death from ammonia inhalation.

Individual Risk

The risk to an individual from operation of the facility was quantified as: isorisk contours (Figure 11.4-2 - loci of equal individual fatality frequency points) and maximum individual risk vs distance curves for the most exposed individual vs distance from the site (Figure 11.4-3). Uncertainty analyses were performed (not shown).

The five types of accidents significantly contributing to the risk are failure and their mean frequency are:

1. Piping between the ship and the tank during loading phase (3.8E-3/y),
2. Ammonia storage tank from ammonia overpressure (1.1E-3/y),
3. Ammonia storage tank from an earthquake (1.3E-3/y),
4. Piping between the storage tank and the fertilizer plants (5.9E-4/y), and
5. Storage tank from underpressure (1.0E-5/y).

The largest contributor to individual risk is the failure of the piping between the ship and the storage tank during the loading operation because of the amount, the mechanism and the dispersion of the ammonia released. Next is storage tank failure from overpressure, and from an earthquake. Smaller contributions are from failure of piping between the storage tank and the fertilizer plants, and the tank failure from underpressure. The total mean frequency of a substantial ammonia release from the storage installation is 6.8E-3/y. Thus, on the average there will be one substantial release of ammonia every 150 years of operation. There is a 70% probability (one sigma) that the frequency of ammonia release will be less than its mean.

Ammonia Release Categories

A release category is a grouping of specific types of releases with similar characteristics. Here, two general types of ammonia release were defined:

Category 1. Release of gaseous ammonia primarily from failure of Tank DK-101 by three plant-damage states: overpressure, underpressure and seismic load. For all three it was assumed that:

- The tank fails at the welding seam between the curved roof and the cylindrical plate wall with break area, A.
- All gaseous ammonia in the tank at rupture escapes. This amount depends on the liquid depth, H.

The duration of the release depends on the break area among other things. For a given quantity of ammonia, there is a critical area A_o, below which gaseous ammonia is dispersed as a buoyant plume, above which it is dispersed as a heavier-than-air plume. The liquid ammonia depth in the tank, the size of the break, and the duration of release are treated as random variables. Tank failure from overpressure and seismic load are characterized by triplets of heights, H, equivalent areas, A, and duration of release corresponding to a particular release category. Tank failure from underpressure was associated with minimum ammonia in the tank, the break size and duration of release. In all cases, evaporation of the remaining liquid ammonia in the tank was considered.

Category 2. Release of liquid ammonia from pipe failure for two plant-damage states involved the assumptions:

- Ammonia is released at a constant rate (loading or unloading) for time T.
- The released ammonia forms a pool of refrigerated liquid which evaporates by heat transfer from the soil. A constant mass value was assumed for the evaporation rate and a heavier-than-air gas dispersion model was used.

The duration of the release was treated as a random variable for the uncertainty calculations.

Consequence Calculations

The atmospheric dispersion model for dense ammonia vapor evolves a slice of the plume, from the source to receptor (Kaizer, 1989;

$$P_o = 0.5*[1+erf(P-5)/1.4142] \quad (11.4\text{-}1)$$
$$P = -35.9+1.85*ln(c^2*t) \quad (11.4\text{-}2)$$

Technica, 1988; Wheatley, 1988; Ziomas, 1989). Plume dispersion consists of two phases: the gravity-slumping and passive dispersion period. Continuous releases with any time-profile and puff releases as of short duration continuous releases are treated using the release category and the weather conditions. Weather is also treated as a random variable. The probability of death of an individual (P_o) exposed to an ammonia cloud is calculated according to equation 11.4-1, where erf is the error function and P (equation 11.4-2) is the probit value of ammonia, c is the concentration of ammonia in ppm, t is the exposure time assumed to be one-half hour. These calculations are integrated in the computer code DECARA (Papazoglou, 1990b).

11.5 Problems

1. If a nuclear power plant loses its connection to the offsite load, it must shutdown because it cannot be cutback sufficiently that its electrical output matches its "hotel" load. Outline a PSA study to determine the risk reduction that might be achieved by switching in a dummy load to avoid shutdown and keep the plant online.
2. If the containment holds, nuclear power plants present no risk to the public. Overpressurization of the containment is the failure mode that could allow direct release of radioactivity to the public. Design a risk reduction investigation of the benefits of releasing the gas pressure through an offgas processing system that removes the particulates.
3. Many nuclear plants are connected to rivers such as Indian Point. Discuss accident sequences that could result in a PWR releasing significant radioactivity into a river.
    a) Outline the PSA steps that would be necessary to determine the probabilities and consequences.
    b) How could such an accident be mitigated?

Chapter 12
Appendix: Software on the Distribution Disk

The distribution disk located at the Butterworth-Heinemann website http://www.bh.com/ contains the following software for use on an IBM or IBM-clone personal computer operating in DOS, Windows 3.1 or Windows 95. The programs are small, operate in the DOS mode with minimal system requirements. It is recommended that a directory (folder) be constructed on your hard drive and copy them to this directory. No guarantees are made; use them if you find them useful. They are:

1. ANSPIPE Calculates pipe break probability using the Thomas Model
2. BETA Calculates and draws event trees using word processor and other input
3. BNLDATA Failure rate data
4. FTAPSUIT Set of codes for finding fault tree cutsets, evaluating fault trees, calculating importances and performing uncertainty analysis
5. LAMBDA Code for estimating failure rate for confidence intervals using a Chi-Squaredand an F-Number estimator. Also includes Gamma and Normal calculators.
6. UNITSCNV Program for units conversion and relativity calculations

12.1 ANSPIPE

Is an IBM-PC code to estimate pipe break and leak probabilities by implementing Thomas's procedures (Thomas, 1981). ANSPIPE aids the engineer guiding the inputting of information needed for a parametric study, predicting the leak and break probabilities, indicating the probabilities according to the specified pipe sizes of large, medium, small and small-small and according to whether they are submerged in pool water or not. It presents this information on the CRT screen, prints it or graphs it as a histogram. Figure 12.1-1 shows the main menu. The first selection

```
             MAIN ANSPIPE MENU
========================================
1  Set Default File
2  Read in Previous Pipe Segment Definitions
3  Define the Pipe Segments
4  Compute and Display the Pipe Failure Probabilities
5  Plot the Results
6  Print Results including Setup Data
7  Record Pipe Segment Definitions on a Disk
8  Quit
     Select the Number of the Task to be Performed
```

Fig. 12.1-1 Main Screen for ANSPIPE

Appendix

creates and a default file for the large, medium and small pipe diameters, the type of steel, pipe wall thickness, number of circumferential welds per segment, replications, whether or not there are longitudinal welds, and the internal pipe pressure.

If the default file previously was defined, selection 1 is skipped and the file is read in using selection 2. Selection 3 defines a segment with a title, the number of segment replications, type of material, internal pressure, diameter, the ASME critical wall thickness, and the segment length. Selection 4 computes the pipe failure probabilities, selection 5 plots the results, selection 6 prints the results, and 7 saves the definitions to a disk. The program has edit capability if changes are needed in an approximately correct setup, or a data file may be prepared and edited with an ASCII text editor. Both output and input files may be labeled for storage and retrieval or optionally the default name may be accepted.

12.2 BETA

Event trees exhaustively explore the combinations of system failures that may follow from accident initiator through dendritic diagrams connecting initiator with plant damage state and by calculating the frequency of each accident sequence. They are the central element of PRA, but require supplementary text to relate the sequences to plant systems, and explain the logic and state assumptions. These tasks are accomplished by the integration of a popular word processor and the event tree analysis code BETA (Fullwood and Shier, 1990). Using the word processor, you write text that describes each sequence to the necessary depth and prepare an Event Table which indicates, by symbols, the operability of each system in the accident sequence and the judged fuel damage state. The matrix of symbols so formed is copied to another file, edited to remove spaces, additional information is added and the file is saved in ASCII format to provide input to BETA. BETA, which may be used for further editing, displays or plots on HP laser or Epson printer, the event tree specified by the Event Table logic and computes, displays and/or prints the end or intermediate state probabilities according to damage criteria. It also provides for changing nodal probabilities within the event tree according to the preceding sequences ("binary conditionals"). The size of event trees is essentially unlimited by its ability to branch and link with other event trees. Work sessions may be saved to floppy or hard disk for recall.

Selections 1 to 5 allow data modification and supplementation. Selection 6 displays the event tree. The tree will be

```
BETA Main Menu
==================================
 1: Edit Accident Sequence Logic
 2: Edit Base Case System Failure
    Probabilities
 3: Edit System Identification
 4: Edit Problem Title
 5: Edit Binary Conditionals
 6: Display the Event Tree
 7: Display Accident Sequence Probabilities
 8: Print Accident Sequence Probabilities
 9: Save Data File
10: Read in New Data File
11: Link this Event Tree to Secondary Tree(s)
12: Leave the Analysis Session

   Select the Number Identifying Your Job
```

Fig. 12.2-1 The Main Beta Menu

Appendix

correctly displayed, although it may be in parts depending upon how the events are input. Selection 7 displays the accident sequence probabilities; selection 8 prints them, and selection 9 saves all of the information to a floppy or hard disk. Selection 10 reads in a new data file; selection 11 links to a secondary tree and 12 exits the program..

12.3 BNLDATA

This folder contains three files: bnlgener.xls, bnlgener.wq1, and referen.txt. The database, bnlgener is an eclectic collection of 1,311 failure rates from 31 references. The database is provided as a spread sheet in the Microsoft Excel 4 format (xls) and in Corel Quatro-Pro format (wq1). This information is not provided in ASCII because of format scrambling and to use the search capabilities of either of the two spreadsheet programs. The references that are cited in the spreadsheet are provided in ASCII as file: referen.text.

12.4 FTAPSUIT

Fault tree or equivalent analysis is key to PSA. Small logical structures may be evaluated by hand using the principles of Chapter 2 but at some point computer support is needed. Even for simple structures, uncertainty analysis by Monte Carlo methods requires a computer. However, most of the codes are proprietary or a fee is charged for their use. Fortunately a set of codes was made available[a] for use on a personal computer. The suite consists of the following FORTRAN programs:

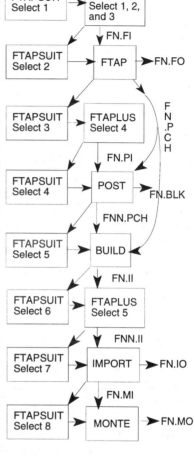

Fig. 12.4-1 Procedure for Using FTAPSUIT

1. FTAP give it fault tree or equivalent input as linked Boolean equations, it will give the cutsets (Section 2.2), but does not numerically evaluate the probabilities.
2. PREPROCESSOR modifies the FTAP punch file output for common cause and dependent analyses (conditional probabilities), remove complemented events, correct for mutually exclusive events.

---

[a] These codes were made available by Dr. Howard Lambert, FTA Associates, 3728 Brunell Dr., Oakland, CA 94602, phone (510) 482-1492.

3. BUILD accepts a file from FTAP or PREPROCESSOR that specify events whose probabilities must be added for quantitative analysis.
4. IMPORTANCE accepts the output from BUILD and calculates importance measures (Section 2.8) and provides input to the uncertainty analysis.
5. MONTE performs a Monte Carlo analysis using uncertainties in the data to estimate uncertainties in the calculation of the system and subsystem failure probability.

The programs were written for a main frame computer using card input. To make the set of codes easier to use, I wrote auxiliary codes to organize the procedures and permit the user to provide input in comma-delimited format. This organizes the "deck" input and assures proper column placement of the data.

Figure 12.4-1 shows the sequence of actions. Selections from FTAPSUIT are in the left column. The right column uses three selections from FTAPLUS and the rest are from the previous FORTRAN programs. Although the suite of programs can be run from the floppy disk, for reasons of speed and problem size, it is recommended to run from a hard disk. In DOS, type MD "directory name," enter, and cd\"directory name," enter. Type copy a:*.* and the disk will be copied to that directory. Type FTAPSUIT and the screen shown in Figure 12.4-2 is presented.

First, system information must be input. From FTAPSUIT, type 1 and enter. This starts FTAPLUS with its menu as shown in Figure 12.4-3.

Select 1 and type a title for the problem; the default title is "Station Blackout from Loss of Offsite Power," unless you want to use this enter your title.

Select 2 to enter a less than 7 character file name. This name is used throughout the suite with the extender being changed as shown in Figure 12.4-1. (In this figure F. stands for file name.) It is advisable to keep the name short because it is requested for the FORTRAN programs. The default designator is "DG" (diesel generator).

Select 3 to input the linked logic sequences using a comma delimiter, no comma at the end, and enter to indicate the end of the line. Type # and hit enter to mark the end of input. For example, the test logic is: BLACKOUT,*,LOSP*NO-AC; NO-AC,*,A,B,C; A,+,DG-A,DG-A-M; B,+,DG-B,DG-B-M; C,+,DG-C,DG-C-M;#; where the semicolon designates enter. Notice these equations are linked with BLACKOUT being the top event. This completes the preparation of FTAP input; selection 6 exits FTAPlus to return to the FTAPSUIT menu (Figure 12.4-2). Figure 12.4-1 shows file FN.FI was created. This is an ASCII file which can be edited, if needed, with a text editor.

The next step is to run FTAP by selection 2 from the FTAPSUIT menu shown in Figure 12.4-2. It asks you to enter the file name that was assigned in selection 2 from the FTAPlus menu. Do not give an extender. A "successful completion" sign indicates proper FTAP execution. At this point, the cutsets have been found and can be seen by reading file FN.FO with a text editor. While this qualitative analysis is useful, usually a quantitative analysis is needed. Figure A-1 shows that the qualitative analysis can be used as-constructed by going to BUILD via file FN.PCH or POSTPROCESS can be used to modify the operation and indicate conditional probabilities.

The postprocessor reads the punch file from FTAP and generates an output file in a format which can be read by BUILD. It can execute commands not found in FTAP. Features of the postprocessor include:

Appendix

- Conducting common cause and dependent event analysis
- Dropping complemented events and performing the subsequent minimization
- Generating block files (i.e., a set of Boolean equations) for subsystems
- Eliminating mincutsets with mutually exclusive events

Selection 3 from the FTAPSUIT menu (Figure 12.4-1) runs FTAPlus for preparing input for

```
Master Program for FTA Associates
Codes

1. Prepare Input for FTAP
2. Run FTAP
3. Prepare Input for POSTPROCESSOR
4. Run POSTPROCESSOR
5. Run BUILD
6. Prepare Input for IMPORTANCE
7. Run IMPORTANCE
8. Run MONTE
H. Help File
Q. Quit

Input designation of Task, Press ENTER
```

Fig. 12.4-2 Menu for FTAPSUIT

```
FTA Associates FTAPlus Main Menu
==================================

1: Provide Title for FTAP
2: Provide File Name
3: Input FTAP Tree Logic and Save
4: Input Data for POSTPROCESSOR and Save
5: Input FT Data for IMPORTANCE and Save
6: Leave the Analysis Session

Select the Number Identifying Your Job
```

Fig. 12.4-3 FTAPlus Menu

```
FTA Associates PostProcessor Input
==================================

1: Eliminate Complemented Events
   (NOCMPL)
2: Print Mincuts in FTAP Format (PRTDNF)
3: Same as 2 and Boole Equa. for Top Event
   (FRMBLK)
4: SORT
5: Mutually Exclusive Input (MEX)
6: Substitute Cutsets for Dependencies (SUB)
7: Exit Menu

Select the Number Identifying Your Job
```

Fig. 12.4-4 PostProcessor Menu

the postprocessor (Figure 12.4-4). To print the contents of the blockfile, the postprocessor input must contain either the command PRTDNF or FRMBLK (selections 2 and 3 respectively). The command PRTDNF (print equation in disjunctive normal form) allows you display the contents of a punch file in readable form. FRMBLK (form block) is required to generate an FTAP input file. The command NOCMPL drops complemented events (i.e., assumes that they are true) and performs the subsequent minimization. The command SORT sorts cutsets according to order. It is necessary to sort the punch file to run IMPORTANCE if cutsets are not listed according to order in the punch file (which occurs when probabilistically culling in FTAP using the IMPORT instruction).

MEX and SUB are optional commands. MEX command must precede the SUB command. The MEX command sets to false, basic events that are pairwise mutually exclusive. For the SUB command, the user identifies sets of basic events within cutsets (original sets) that are to be replaced by another set of basic events (substituted sets). The SUB command accounts for basic events within cutsets that have either a common cause or statistical dependency.

After selecting the MEX command (selection 5) the comma delimited input consists of groups of basic events that are pairwise mutually exclusive. To extend the group to the next and subsequent lines, the first field must be blank. Up to 100 names per group are allowed with a new group starting when a non-blank character is encountered in the first field. The number of pairs that are mutually exclusive is: $n!/[(n-2)! *2!]$, where n is the number of basic events in the group. An example for the plant blackout fault tree is: DG-A-M,DG-B-M,DG-C-M; (where ";" means enter).

Selection 6 (Figure A-4) executes the SUB command. The comma-delimited input alternates between the cutset to be replaced and the cutset replacing it. For example for the plant blackout problem, the input is: DG-A,DG-B,DG-C;DG-A,DG-B/A,DG-C/A&B;DG-A,DG-B;DG-A,DG-B/A;DG-A,DG-C;DG-A,DG-C/A;DG-B,DG-C;DG-B,DG-C/B. The actual input to BUILD alternates lines with * and ** to identify lines to be replaced with replacing lines, but this bookkeeping is done automatically by FTAPlus. The original set of basic events and the substituted set cannot exceed 99 basic events. The SUB command completes the postprocessor input. Select 7 to exit the postprocessor menu and 6 to exit FTAPlus for the FTAPSUIT menu.

Select 4 from this menu and POST (postprocessor) begins by requesting the data input file name (e.g., dg.pi - prepared by FTAPlus), the punch file name (e.g., dg.pch), the block file name (e.g., dg.blk) and the name to be assigned to the new punch file (e.g., dgn.pch). Two outputs are produced: FN.BLK and FNN.PCH.

Select 5 from the FTAPSUIT menu and BUILD starts by requesting the punchfile name (e.g., dgnpch or dgpch depending on whether you want to run the original fault tree the postprocessed fault tree. It requests the name to be given the importance input file (e.g., dg.ii). The last request regards the type of input. Option 5 is pure probability, option 1 is failure rate. BUILD generates file FN.II for editing.

Selection 6 goes to FTAPlus, from which menu 5 is selected, to provide the information needed for IMPORT. This program presents event names as given in FTAP and POST and requests failure rates or probabilities (e-format), uncertainty (f-format), and an event description. Each time this set is given, it presents the next event name until all events are specified. Then it returns to the main FTAPlus menu to be exited by selecting 6 to go to the FTAPSUIT menu.

Selection 7 causes IMPORT (importance) to start by requesting the name of the input file FN.II (e.g., dgn.ii). The importance output is contained in the FN.IO file. The output needed by MONTE is FN.MI.

Selection 8 from the FTAPSUIT menu runs MONTE (Monte Carlo). It requests the name of the input file, FN.MI (e.g., dgn.mi). The Monte Carlo analysis is contained in file FN.MO.

12.5 Lambda

To run Lambda go to the directory containing the distribution files and type "Lambda." A banner announces the program. The next screen explains the program that does several calculations (Figure 12.5-1).

Selection 1 performs a one-sided, fractional confidence bound of the failure rate based on Chi-Squared estimation (Section 2.5.3.1). When this selection is made, it asks you to specify your confidence bound, the number of component years of experience, and the number of failures. For example, to estimate the failure rate for a component that has had no failures in 7,015 component-years of experience with 50% confidence: input "1" (for chi-squared), input 0.5 (50% confidence), input 7015 (component years of experience), and input "0" (number of failures). The program answers: $\lambda(50\%)$ = 9.883E-5/years. Rerun the program for 90% confidence and get: $\lambda(90\%)$ = 3.282E-4/years.

Selection 2 is a similar calculation using the F-Number method (Section 2.5.3.2). Selection 3 calculates the integral over the Chi-Squared distribution. When selected, you are requested to input the upper limit of integration and the number of degrees of freedom. Selection 4 is the inverse calculation and is part of the Selection 1 calculation. You are requested to input the confidence, and the degrees of freedom. The answer is the inverse cumulative Chi-Squared. Selection 5 is part of the Selection 2 calculation. You input the confidence limit (equation 2.5-34) and the nu values to get the F-Number. Selection 6 gives the integral over the normal distribution from -∞ to the limit you specify (Abramowitz and Stegun, 1970 - 26.2-17). For example if zero is the limit then the integral is over half of the area

```
      Lambda - Statistical Functions
   ==================================
   1) Upper Bound Failure Rate (Inverse Chi2)
   2) Upper Bound Probability (Inverse
        Binomial)
   3) Cumulative Chi-Squared
   4) Inverse Cumulative Chi-Squared
   5) Inverse Cumulative F-number
   6) Cumulative Gaussian (normal)
   7) Inverse Cumulative Gaussian (normal)
   8) Gamma and Factorial Integer and Half
        Integer Functions
   9) Exit
            Select your job number
```

Fig. 12.5-1 Menu for Lambda

```
Brookhaven National Laboratory Microcomputer Club
-------------------------------------------------
 1   Length              13   Electrical Charge
 2   Area                14   Magnetic Field Int.
 3   Volume              15   Magnetic Field Density
 4   Mass                16   Radioactivity
 5   Force               17   Dose
 6   Time                18   10's Power Prefixes
 7   Power               19   Temperature
 8   Energy              20   Constants
 9   Pressure            21   Relativity
10   Viscosity           22   Quit
11   Velocity
12   Acceleration
+------------------------------------------------
Select the Number of the Type of Units or Constants?
```

Fig. 12.6-1 Menu for the Units Conversion Program

Appendix

and the answer is 0.5. Selection 7 gives the value of 1, the cumulative integral from the specified limit to ∞ (Abramowitz and Stegun, 1970, equation 26.2.23). Finally, selection 8 gives values of integer and half integer factorials and gamma functions.

12.6 UNITSCNV

To run UNITSCNV go to the directory containing the distribution files and type "unitscnv." A banner announces the program. The next screen explains the program. This is followed by the main menu (Figure 12.6-1). Units are grouped by type. For example, to change from one kind of pressure units to another, select 9. Another screen comes up to identify the units you are in. It then asks the quantity. When this is provided, it presents another screen displaying the quantity of the units provided and the equivalent in other units. Selections 18 and 20 do no calculations; they only present information. Selection 19 calculates temperature; selection 21 performs the calculations of special relativity.

Chapter 13

Glossary of Acronyms And Unusual Terms

accident - An unplanned event with undesirable consequences.

accident Analysis - The process of systematically indentifying hazards, analyzing the conditions under which these lead to undesirable consequences, evaluation of the consequences and recommendations for preventing these accidents.

accident Event Sequence - The sequence of events leading to undesirable consequences.

ACGIH - American Conference of Governmental Industrial Hygienists, Inc., 1330 Kemper Meadow Dr., Ste 600, Cincinnati, OH 45240, Phone: (513) 742-2020 • Fax: (513)742-3355

ACRS - Advisory Committee on Reactor Safety.

Adiabatic - No heat transfer between systems.

ADS - Automatic Depressurization System.

advection - Spatial variation with temperature of a medium through which vapor is moving.

AE - Architect Engineer.

AEC - Atomic Energy Commission.

AECB - Atomic Energy Control Board, the nuclear licensing authority in Canada.

AECL - Atomic Energy of Canada Ltd.

AFS - Auxiliary Feedwater System.

AFW - same as AFS.

Aggregate Threshold Quantity - The total amount of a hazardous chemical contained in interconnected or nearby vessels that may be adversely affect an accident.

AIChE - American Institute of Chemical Engineers, 345 East 47th Street, New York, NY 10017, Phone (212) 705-7319.

airborne release fraction - the fraction of hazardous material released by an accident that is suspended in air for air transport to public, worker or undesired receptors.

ANL - Argonne National Laboratory.

ANS - American Nuclear Society, La Grange Park, IL.

AOT - Allowed Outage Time.

API - American Petroleum Institute, 1220 L Street. N.W., Washington, D.C. 20005.

ASEP - Accident Sequence Evaluation Program.

ATWS - Anticipated Transient without Scram.

B&W - Babcock and Wilcox (a PWR manufacturer).

BCL - Battelle Columbus Laboratories.

BFR - Binomial Failure Rate (model of common cause system interactions).

BNL - Brookhaven National Laboratory, PSA analysts.

Glossary of Acronyms and Unusual Terms

BR - The Business Roundtable, 200 Park Avenue, New York, NY 10166.
BRP - Big Rock Point, an early, small BWR.
burnup - Depletion of nuclear fuel by fission.
BWR - Boiling Water Reactor (GE reactor having no steam generator separate from the reactor).
BWST - Borated Water Storage Tank.
CAD- Compter aided drawing.
C&EN - Chemical and Engineering News.
Catastrophic Release - A major uncontrolled emission, of hazardous material that presents serious danger to workers or the public.
CCDF - Cumulative Complementary Distribution Function
CCPS - Center for Chemical Process Safety, part of the AIChE, 345 East 47th Street, New York, NY 10017, Phone (212) 705-7319.
CDC - Control Data Corporation (a former computer manufacturer).
CDF - Core Damage Frequency.
CE - Combustion Engineering (a PWR manufacturer).
CFR - Code of Federal Regulations.
choked flow - Flow at sonic velocity - the maximum, pressure-independent flow.
CMA - Chemical Manufacturers Association; 2501 M Street NW, Washington, DC 20037.
CMR - Christian Michelsen Research - a Norwegian PSA consulting company.
completeness - The degree to which all accidents have been considered.
component - An elemental unit of a system that is considered in no greater detail, i.e., a"black box."
corium - A mixture of core and reactor material.
CPI - Chemical process industry.
Cray Research - a manufacturer of computers.
CS - Containment Spray.
CST - Core spray tank.
DBA - Design Basis Accident.
DBE - Design Basis Earthquakes.
DBF - Design Basis Flood.
DBT - Design Basis Tornado.
DCH - Direct Containment Heating.
delphi - A procedure for quantitative estimation based on a consensus of experts.
dense gas - Gas with density greater than the ambient surrounding air.
dependency - An interaction between systems degrading their separately analyzed reliability.
DHR - Decay Heat Removal.
DG - Diesel Generator.
dike - A barrier to prevent a liquid release from spreading beyond this boundary.
DNV - det Norsk Veritas a Norwegian PSA consulting company.
DOC - U.S. Department of Commerce.
DOE - U.S. Department of Energy.
DOL - U.S. Department of Labor, Occupational Safety and Health Administration.
DOT - U.S. Department of Transportation.

Glossary of Acronyms and Unusual Terms

doubleton - Double failure.
Dow - Dow Chemical Company, Midland, Michigan 48674.
DPD - Discrete Probability Distribution.
DPV - Depressurization valves.
ECC - Emergency Control Center.
ECCI - Emergency Core Cooling Injection.
ECCS - Emergency Core Cooling System.
EFWS - Emergency Feedwater System.
EI - Energy Incorporated (a PSA services supplier).
Employee - According to 29CFR1910.119, "Process Safety Management of Highly Hazardous Chemicals," an hourly, salaried, or contract person who works at a facility and is subject to the process hazards.
entrainment - The mixing of gas/vapor cloud, plume, jet, aerosol with the surrounding air due to advection, momentum exchange induced by shear flow, molecular transport across a flow boundary, or turbulent mixing.
EPA - U.S. Emergency Protection Agency, Washington, D.C. 20460.
EPRI - Electric Power Research Institute.
EPZ - Emergency Protection Zone.
EQE International - A PSA services supplier (acquired PL&G) http:\\www.eqe.com.
ERIN - Engineering & Research Incorporated, a PSA services provider.
Event - An occurrence involving process, equipment, or human performance either internal or external to a system that causes system upset.
Facility - The buildings, containers, or equipment that contain a process.
FEMA - Federal Emergency Management Agency, Washington DC.
Flammable Gas - A gas that, at ambient temperature and pressure, forms a flammable mixture with air at ambient temperature and pressure.
Flammable Liquid - A liquid with a flash point below $100°$ F except mixtures where such liquids account for 1 percent or less of the total volume.
flashing - The sudden evaporation of a liquid whose temperature is above boiling at the new pressure after release.
FMEA - Failure Modes and Effects Analysis.
FMECA - Failure Modes, Effects and Criticality Analysis.
F-N curve - A cumulative frequency number affected curve.
FSAR - Final Safety Analysis Report (Chapter 15 is the accident analysis section in the FSAR standard format).
GAO - U.S. General Accounting Office.
GDCS - Gravity driven cooling system (SBWR feature).
GE - General Electric (a BWR manufacturer).
GSA - U. S. General Services Administration
guillotine break - Pipe break with the ends displaced allowing unimpeded flow from each.
hazard - Something that has the potential to cause illness, injury, or death to people, property or to the environment, without regard for the likelihood.

HAZWOPER - Hazardous Waste Operations and Emergency Response (29CFR1910.120).
HEP - Human Error Probability.
HFBR - High Flux Beam Reactor, a research reactor at BNL.
Highly Hazardous Chemical - Toxic, reactive, flammable, or explosive substances, as defined in Appendix A of 29 CFR 1910.119, "Process Safety Management of Highly Hazardous Chemicals."
Hot Work - Work involving welding, cutting, brazing or other spark-producing operations. It also means work in a radiation field.
HPCI - High Pressure Cooling Injection.
HPSI - High Pressure Safety Injection.
HRA - Human Reliability Assessment.
HTML - hypertext markup language. The language of the Internet.
HVAC - Heating, Ventilation and Cooling
hypertext - Cursor-sensitive areas (hot buttons) which when clicked by the left mouse button causes a jump to another program, a location in text or display of additional information.
I&C - Instrumentation and Control.
IC - Isolation condenser - used with some BWRs and the SBWR.
IDCOR - Industry Degraded Core Rulemaking Program.
incident - An unplanned event that may or may not result in injuries and/or loss.
INEL - Idaho National Engineering Laboratory, PSA analysts and research.
insolation - Heating (of a plume) by the sun.
IPE - Integrated Plant Examination.
IPPSS - Indian Point Probabilistic Safety Study (PSA).
IPRDS - In-Plant Reliability Data System.
IREP - Interim Reliability Evaluation Program.
IRRAS - Integrated Reliability and Risk Analysis System.
IRWST - In-containment refueling water storage tank.
ISLOCA - Interfacing system LOCA.
JCO - Justification for Continued Operation.
LANL - Los Alamos National Laboratory, Los Alamos, NM 87545.
LER - Licensee Event Report (see Reg. Guide 1.16).
LLNL - Lawrence Livermore National Laboratories, P.O. Box 808, Livermore, CA 94551.
LNG - Liquified natural gas (methane).
LOCA - Loss of Cooling Accident.
LOOP - Loss Of Off-site Power.
LPCI - Low Pressure Cooling Injection.
LPG - Liquified petroleum gas (propane or butane).
LPIS - Low Pressure Injection System.
LWR - Light Water Reactor.
MAGNOX - English CO_2 gas cooled reactor.
math notation - "*" means numerical or Boolean multiplication. nEx means $n*10^x$, i.e., n with x zeros after it; nE-x means $n*10^{-x}$ or n with a decimal point and x-1 zeros before it.

Glossary of Acronyms and Unusual Terms

MFT - Modular Fault Tree.

mincut - The least combinations of component failures that will cause system failure.

min-max - Partially optimized solution; the best of a bad situation; a saddle-point.

mission time - The time during which the component or system must work. Often this is the time since it was last known to work (since test). This is multiplied by the failure rate to give the probability of working when needed.

MORT - Management Oversight Review Technique.

MSDS - Material safety data sheets. Information provided by chemical manufacturers that list the hazards.

MSIV - Main Steam Isolation Valve.

MTR - Materials Test Reactor (at INEL).

MWe - Million watts of electric power.

MWt - Million watts of thermal power.

Near Miss - (Precursors) An event that did not result in an accidental release of a highly hazardous material but would have if another failure occurred.

negative buoyancy - a plume with more density than the surrounding air.

NERC - National Electrical Reliability Council.

neutral buoyancy - A plume with the same density as the surrounding air.

NGSGC - National Geophysical and Solar-Terrestrial Data Center.

NIOSH - National Institute for Occupational Safety and Health. Web site: http://www.cdc.gov/niosh/homepage.html.

NN - Nearest neighbor - usually referring to a control rod configuration.

NIST - National Institute of Standards and Technology, Gaithersburg, MD 20899.

Normally Unoccupied Remote Facility - A facility that is operated, maintained, or serviced by workers who visit the facility only periodically to check its operation and to perform necessary operating or maintenance tasks. No workers are regularly or permanently stationed at the facility. Such facilities are not contiguous with, and must be geographically remote from, all other buildings, processes, or persons. If workers spend more than 1 hour at a facility each day, that facility is not considered to be normally unoccupied.

NPRDS - Nuclear Plant Reliability Data System.

NRC - U.S. Nuclear Regulatory Commission.

NSC - National Safety Council, 44 Micigan Ave., Chicago, IL 60611-3991.

NUMARC - Nuclear Utility Management and Research Consortium.

NUS - Formerly, Nuclear Utility Services, a PSA services supplier, now part of Scientech.

OECD - The Organization for Economic Cooperation and Development of the United Nations.

ORNL - Oak Ridge National Laboratory.

OSHA - Occupational Safety and Health Administration (part of the Department of Labor).

P - (capital P) probability.

p - (lower case p) probability density.

PC - Personal Computer.

PCS - Power Conversion System.

PG&E - Pacific Gas and Electric, an electric and gas utility.

PHA - Preliminary Hazards Analysis.

PL&G Inc. - Formerly Pickart Lowe and Garrick, formerly PLG, acquired by EQE in 1997, a PSA services supplier.

PMF - Probable maximum flood.

PNL - Pacific Northwest Laboratory, technical analytical DOE laboratory including PSA.

Population dose - The sum of the doses received by a population. If each person received a dose, and the population is N, it is d*N.

PORV - Pilot (or power) Operated Relief Valve.

positive buoyancy - a plume with less density as the surrounding air.

PRA - probabilistic risk assessment/probabilistic safety assessment, synonymous with PSA. The PRA Procedures Guide is NUREG/CR-2300. PRA was the original term but became probabilistic safety assessment to avoid the negative implication of "risk."

PrHA - Process Hazards Analysis.

Price Anderson - A law providing insurance coverage for nuclear power in the United States.

P RHR HX - Passive residual heat removal heat exchanger.

PRA - Procedures Guide the name of

Process - An activity that involves highly hazardous materials reacting chemically, their use, storage, manufacturing, handling, or movement

Process Hazard - An inherent chemical or physical characteristic with the energy potential for damaging people, property, and the environment.

Process Hazards Analysis (PrHA) - The application of analytical methods to identify and evaluate process hazards to determine the adequacy or control.

Process Safety Management - The application of management principles, methods, and practices to prevent and control accidental releases of process chemicals or energy.

PSM Rule - The U.S. Occupational Safety and Health Administration's rule "Process Safety Management of Highly Hazardous Chemicals," 29 CFR 1910.119.

PWR - Pressurized Water Reactor (a reactor with steam generators separate from the reactor).

QA - Quality Assurance.

QRA - Quantitative Risk Assessment

Qualitative risk - Risk expressed without numbers.

Quantified risk - Risk expressed numerically.

RBD - Reliability Block Diagram.

RCIC - Reactor Cooling Isolation Condenser.

RCP - Reactor Cooling Pump.

RCS - Reactor Cooling System.

Replacement in kind - A replacement that satisfies design specifications.

resuspension - Re-entrainment of particulate into a wind field for dispersion.

RHR - Residual Heat Removal System.

risk - The quantitative or qualitative expression of possible loss that considers both the probability that a hazard will result in an adverse event and the consequences of that event.

RMIEP - Risk Methodology Integration Evaluation Program.

ROAAM - Risk Oriented Accident Analysis Method.
RPS - Reactor Protection System (scram system).
RPV - Reactor pressure vessel.
RSS - Reactor Safety Study (reported in WASH-1400).
RSSMAP - Reactor Safety Study Methodology Applications Program.
RWST - Refueling Water Storage Tank.
SAIC - Science Application International Corporation, a PSA services provider.
SARA - Systems Analysis and Risk Assessment code - now part of IRRAS.
SBWR - Simplified Boiling Water Reactor - second generation reactor.
Scientech - A PSA services supplier.
scram - Immediate reactivity shutdown using control rods.
SDS - Safety Depressurization System.
SG - Steam Generator
SGTS - Standby gas treatment system.
singletons - Single components whose failure could fail a system.
SIS - Safety Injection System.
SNL - Sandia National Laboratories, Albuquerque, NM.
sonic flow - Flow at sonic speed. Below sonic speed, the velocity of flow is a function of the differential pressure; at sonic speed, flow does not depend on pressure differential.
source term - Amount and characteristics of hazardous material released. Source for the dispersion calculation.
squib - An explosively operated valve.
SRC - Westinghouse Savannah River Corporation, Aiken SC
SRV - Safety relief valves.
SSC - Structures, systems and components.
ST - Surveillance Testing.
sublimation - Passing from a solid to a gas without going through the vapor phase.
Supercomponent - Set of components all of which must function for the set to function.
Sv - Sievert radiation health effects (1J/kg tissue) equivalent to 100 REM.
system - A collection of components to perform a purpose.
TCC - Texas Chemical Council, 1402 Nueces Street, Austin, TX 78701-1534.
Tech. Specs. - Technical Specifications (lOCFR50.36).
THERP - Techniques for Human Error Rate Prediction.
Threshold Quantity - As defined in 29 CFR 1910.119, the minimum amount of a toxic, reactive, or flammable chemical judged by OSHA as being capable of causing a catastrophic event. The threshold quantity triggers application of the rule's requirements.
tonnes - A metric ton, i.e., 2,200 lb.
train - a series of linked components. The failure of one fails the train.
Transient - A passing disturbance that in itself is not damaging but the process upset may be damaging.
VIA - Value Impact Analysis, a form of cost benefit analysis that includes attribute evaluations in addition to dollars.

Glossary of Acronyms and Unusual Terms

$\underline{W}$ - Westinghouse (a PWR manufacturer - pronounced "W bar").

WASH-1400 - The report of the Rasmussen Study that effectively started the use of probabilistic safety assessment.

ZIP - Zion 1 and 2 and Indian Point 1 and 2 PSAs.

zircalloy - An alloy of zirconium used for fuel cladding because of the low neutron capture cross section.

Chapter 14

References

14.1 Nuclear Regulatory Commission Reports Identified by NUREG Numbers

0046   "Anticipated Transients without Scram for Light Water Reactors, Vol.1-3, December 1978.
0492   Haasl, D. F. et al., "Fault Tree Handbook," January 1981.
0626   "Generic Evaluation of Feedwater Transients and Small Break Loss-of-Coolant Accidents in GE-Designed Operating Plants and Near-Term Operating License Applications," January 1980.
0666   Baranowsky, P.W., A.M. Kolaczkowski, and M.A. Fedele, "A Probabilistic Safety Analysis of DC Power Supply Requirements for Nuclear Power Plants," April 1981.
0700   "Guidelines for Control Room Design Reviews," September 1981.
0737   "Clarification of TMI Action Plan Requirements," November 1980.
0739   "An Approach to Quantitative Safety Goals for Nuclear Power Plants," October 1980.
0771   Rasedag, W. F. et al., "Regulatory Impact on Nuclear Reactor Accident Source Term Assumptions," June 1981.
0772   "Technical Bases for Estimating Fission Product Behavior during LWR Accidents," June 1981.
0773   Blond, R., "The Development Severe Reactor Source Terms: 1957-1981," November 1982.
0880   "Safety Goals for Nuclear Power Plant Operation" May 1983.
0930   Emrit, R. et al., "A Prioritization of Generic Safety Issues," NUREG-0933 suppliment.
0956   "Reassessment of the Technical Bases for Estimating Source Terms," Draft, May 1985.
1032   Baranowsky, P.W., "Evaluation of Station Blackout Accidents at Nuclear Power Plants," May 1985.
1037   Bagchi, G., "Containment Performance Working Group Report," Draft, 1985.
1050   Murphy, J. A., "Probabilistic Risk Assessment (PRA) Reference Document," September 1984.
1079   Speis, T. P., "Containment Loads Working Group Report," Draft, 1985.
1150   "Reactor Risk Reference Document," February 1987.
1365   "Revised Severe Accident Reseach Plan," 1989.
1560   "Individual Plant Examination Program: Perspectives in Reactor Safety and Plant Performance," 1996.
75/014 WASH-1400, "Reactor Safety Study: An Assessment of Accident Risks in Commercial Nuclear Power Plants," October 1975.

References

14.2 Nuclear Regulatory Commission Contractor Reports Identified by NUREG/CR Numbers

0165 Carlson, D. D. and J. W., Hickman, "Value-Impact Assessment of Alternate Containment Concepts," SNL, July 1978.

0400 Lewis, H. W. et al., "Risk Assessment Review Group Report to the U.S. Nuclear Regulatory Commission," September 1978.

0809 Varnado, G. B. and N. R. Ortiz,."Fault Tree Analysis for Vital Area Identification," SNL, June 1979.

1201 Jackson, J. F. and M. G. Stevenson, "Nuclear Reactor Safety Quarterly Progress Report," LANL, September 1979.

1205 Sullivan, W. H. and Poloski, J. P., "Data Summaries of Licensee Event Reports of Pumps at U.S. Commercial Nuclear Power Plants, January 1, 1972-April 30, 1978," EG&G, January 1980.

1278 Swain, A. D. and H. E. Guttmann, "Handbook of Human-Reliability Analysis with Emphasis on Nuclear Power Plant Applications," SNL, August 1983.

1321 Boyd, G.J., et al., "Final Report Phase 1 Systems Interaction Methodology Applications Program," April 1980

1331 Hubble, W. H. and C. F. Miller, "Data Summaries of Licensee Event Reports of Control Rods and Drive Mechanisms at U.S. Commercial Nuclear Power Plants, January 1, 1972-April 30, 1978," EG&G, January 1980.

1362 Poloski, J. P. and W. H. Sullivan, "Data Summaries of Licensee Event Reports of Diesel Generators at U.S. Commercial Nuclear Power Plants," EG&G. March 1980.

1363 Hubble, W. H. and C. F. Miller, "Data Summaries of Licensee Event Reports of Valves at U.S. Commercial Nuclear Power Plants. Volume 1. Main Report. January 1, 1976-December 31, 1978," EG&G, May 1980.

1711 Wooton, R. O. and H. I. Avci, "MARCH (Meltdown Accident Response Characteristics) Code Description and User's Manual," BCL, October 1980.

1730 Sams, D. W. and M. Trojovsky, "Data Summaries of Licensee Event Reports of Primary Containment Penetrations at U.S. Commercial Nuclear Plants," EG&G, September 1980.

1740 Trojovsky, M. and S. R. Brown, "Data Summaries of Licensee Event Reports of Selected Instrumentation and Control Components at U.S. Commercial Nuclear Power Plants, January 1, 1976-December 31, 1981, Revision 1," EG&G, July 1984.

1859 Lin, J.J. et al., "Systems Interaction: State-of-the-Art Review and Methods Evaluation," January 1981.

1879 Hall, R.E. et al.,"Sensitivity of Risk Parameters to Human Errors in Reactor Safety Study for a PWR," BNL, January 1981

1901 Buslik, A.J. et al., "Review and Evaluation of System Interaction Methods," January 1981.

1930 Miller, B. and R. E. Hall, "Index of Risk Exposure and Risk Acceptance Criteria," February 1981.

References

2139	Sutter et al., "Aerosols Generated by Free Fall Spills of Powders and Solutions in Static Air," 1981
2211	Samanta, P. K. and S. P. Mitra, "Modeling of Multiple Sequential Failures During Testing, Maintenance, and Calibration," BNL, December 1981.
2254	Bell, B. J. and A. D. Swain, "Procedure for Conducting a Human- Reliability Analysis for Nuclear Power Plants," SNL, May 1983.
2285	Rivard, J. B. et al., "Interim Technical Assessment of the March Code," SNL, November 1981.
2300	"PRA Procedures Guide: A Guide to the Performance of Probabilistic Risk Assessments for Nuclear Power Plants Review," ANS/IEEE, January 1983.
2497	Minarick, J. W. and C. A. Kukielka, "Precursors to Potential Severe Core-Damage Accidents: 1969-1979 A Status Report," Vols. 1 and 2, ORNL, June 1982.
2534	Beare, H. N. et al., "Criteria for Safety-Related NPP Operator Actions: Initial BWR Simulator Exercises," November 1982.
2554	Bell, B. J. and H. D. Swain, "A Procedure for Conducting a Human Reliability Analysis for Nuclear Power Plants," May 1983.
2635	Carlson D. D. et al., "Interim Reliability Evaluation Program Procedures Guide," SNL, January 1983.
2728	Carlson, D. D. (Principal Investigator), "Interim Reliability Evaluation Program Procedures Guide," SNL.
2815	Papazoglou, I. A. et al., "Probabilistic Safety Analysis Procedures Guide," BNL, January 1984.
2886	Drago, J. P. et al., "In-Plant Reliability Database for Nuclear Plant Components: Interim Data Report, The Pump Component," ORNL, December 1972.
2934	Kolb, G. J. et al., "Review and Evaluation of the Indian Point Probabilistic Safety Study," SNL, December 1982.
2989	Battle, R.E., and D. J. Campbell, "Reliability of Emergency AC Power Systems at Nuclear Power Plants," July 1983
3010	Hall, R. E. et al., "Post-Event Human Decision Errors: Operator Action Tree/Time Reliability Correlation," BNL, November 1982.
3028	Papazoglou, I. A. et al., "Review of the Limerick Generating Station Probabilistic Risk Assessment," BNL, February 1983.
3085	Curry, J. J. et al., "IREP: Analysis of Millstone Point Unit 1 Nuclear Power Plant," SAI, May 1983.
3092	Beare, A. N. et al., "Criteria for Safety-Related NPP Operator Actions: Initial Simulator Field Data Calibration," February 1983.
3098	Borkowski, R. J. et al., "In-Plant Reliability Database for Nuclear Plant Components: Interim Report - The Valve Component," ORNL, December 1983.
3154	Borkowski, R. J. et al., "The In-Plant Reliability Database for Nuclear Plant Components: Interim Report - The Valve Component."
3226	Kolaczkowski, A. M. and A. C. Payne, Jr., "Station Blackout Accident Analyses (Part of NRC Task Action Plan A-44)," SNL, May, 1983.

3248	Winegardner, W. K., "Studies of Fission Product Scrubbing within Ice Compartments," PNL, May 1983.
3518	Embry, D. E. et al.,"SLIM-MAUD: an Approach to Assessing Human Error Probabilities Using Structured Expert Judgment," BNL, March 1984.
3519	Voska, K.J. and J.N. O'Brien," Human Error Probability Estimation Using Licensee Event Reports," BNL, July 1984.
3626	Siegel, A.I. et al., "Maintenance Personnel Performance Simulation (MAPPS) Model: Description of Model Content, Structure and Sensitivity Testing," 1984.
3634	Kopstein, F. F. and J. J. Wolf, "Maintenance Personnel Performance Simulator (MAPPS) Model User's Manual," ORNL, July 1985.
3688	Comer, M. K. et al., "enerating Human Reliability Estimates using Expert Judgment," SNL, November 1984.
3837	Samanta, P. K. et al., "Multiple-Sequential Failure Model: Evaluation of and Procedures for Human Error Dependency," BNL May 1985.
4022	Selby, D. L. et al., "Pressurized Thermal Shock Evaluation of Calvert Cliffs Unit 1 Nuclear Plant," September 1985.
4183	Selby, D. L. et al., "Pressurized Thermal Shock Evaluation of H. B. Robinson Unit 2 Nuclear Power Plant," September 1985
4314	Waller, R. A., "A Brief Survey and Comparison of Common Cause Failure Analysis," June 1985
4350	"Probabilistic Risk Assessment Course Documentation," Volumes 1 through 7, August 1985
4550	M. T. Drouin, F. T. Harper and A. L. Camp, "Analysis of Core Damage Frequency from Internal Events: Methodology Guidelines," Vol. 1, September 1987.
4550	Wheeler, T. A. et al., "Analysis of Core Damage Frequency from Internal Events: Expert Judgment Elicitation," Vol. 2. 1989
4586	"User's Guide for a Personal Computer-Based Nuclear Power Plant Fire Database," August 1986.
4598	Iman, R. L. and M. J. Shortencarier, "A User's Guide for the Top Event Matrix Analysis Code (TEMAC)," August 1986.
4639	Gertman, D. I. et al., "Nuclear Computerized Library for Assessing Reactor Reliability (NUCLARR)," 1990.
4639	Reece, W. J. et al., "Nuclear Computerized Library for Assessing Reactor Reliability (NUCLARR), Part. 2 Human Error Probability Data (HEP)," 1994.
4658	Ballinger and Hodgson, "Aerosols Generated by Spills of Viscous Solutions and Slurries," 1986.
4772	Swain, A. D., "Accident Sequence Evaluation Procedure (ASEP) Huan Reliability Analysis Procedure," 1987.
4834	Weston, L. M., D.W. Whitehead, and N. L.Graves, "Recovery Actions in PRA for the Risk Methods Integration and Evaluation Program (RMIEP): Volume I Development of the Databased Method," June 1987.

4840	Bohn, M. P., and J. A. Lambright, "Recommended Procedures for Simplified External Event Risk Analyses," February 1988.
4910	Budnitz, R. J., H. E. Lambert, and E. E. Hill, "Relay Chatter and Operator Response after a Large Earthquake," August 1987.
4997	Ballinger, M.Y. et al., "Methods for Describing Airborne Fractions of Free Fall Spills of Powders and Liquids," Pacific Northwest Laboratory, January 1988.
5032	Iman, R. L. and S. C. Nora, "Modelilng Time to Recovery and Initiating Event Frequency for Loss of Off-Site Power Incidents at Nuclear Power Plants," January 1988.
5213	Woods D. D., H. E. Pople, and E. M. Roth, "The Cognitive Environment Simulation (CES)," 1990.
5250	Bernreuter, D. L. et al., "Seismic Hazard Characterization of 69 Nuclear Plant Sites East of the Rocky Mountains, Methodology, Input Data, and Commparisons with Previous Results," Vol 1-8, January 1989.
5423	Theofanous, T G. et al., "The Probability of Liner Failure in a Mark I Containment," 1991
5534	Wells, J. E. et al., "Talent Analysis Linked Evaluation Technique (TALENT), 1991.
5964	Russell, K. D. et al., "SAPHIRE Technical Reference Manual: IRRAS/SARA, Volume 4.0, June 1993.
5999	Majumdar, S. et al., "Interium Fatigue Design Curves for Carbon, Low Alloy, and Austensic Stainless Steels in LWR Environments," 1993.
6025	Theofanous, T. G. et al., "The Probability of Mark I Containment Failure by Melt-Attack of the Liner," 1993.
6075	Pilch, M. M, et al., "The Probability of Containment Failure by Direct Containment Heating in Zion," 1994.
6111	Fullwood, R. et al., "Integrated Systems Analysis of the Pius Reactor," November 1993.
6116	Russell, K. D. et al. "System Analysis Program for Hands-on Integration Reliability Evaluation (SAPHIRE) Version 5," Volumes 1-10, July 1994.
6336	Keisler, J. et al., "Fatigue Strain-Life Behavior of Carbon and Low-Alloy Steels," 1995.

14.3 Electric Power Research Institute Reports Identified by NP Number

768	Twisdale, L. A., et al., "Tornado Missile Analysis," 1978.
2170-3	Hannaman, G. W. et al., "Human Cognitative Reliability (HCR) Model for PRA Analysis," 1984.
2301	McClymont, A. S. and B. W. Poehlman, "Loss of On-Site Power at Nuclear Power Plants: Data and Analysis," SAIC, March 1982.
2433	McClymont, A. S. and G. McLogan, "Diesel Generator Reliability at Nuclear Power Plants: Data and Preliminary Analysis," SAIC.
2682	Von Hermann, J. L. and P. J. Wood, "The Practical Application of Probabilistic Risk Assessment," IT-Delian.
3123	Energy Inc., "GO Methodology Vol. 1 Overview Manual," June 1983.

References

> A. P. Kelley, Jr., and D. W. Stillwell, "GO Methodology Vol. 2 Application and Comparison of the GO Methodology and Fault Tree Analysis," PL&G, June 1983.
> Azizi, S. M. et al., "GO Methodology Vol. 3 GO Modeling Manual," EI, November 1983.
> Williams, R. L. et al., "GO Methodology Vol. 5 Program and Users Manual (IBM Version)," Kaman Sciences, November 1983.
> Williams, R. L. et al., "GO Methodology Vol. 6 Program and User's Manual (CDC Version)," Kaman Sciences, November 1983.

3583 Hannaman, G. W. and A. J. Spurgin, "Systematic Human Action Reliability Procedure (SHARP)," NUS Corp., June 1984.

3967 Fleming, K.N., and A. Moslem, "Classification and Analysis of Reactor Operating Experience Involving Dependent Events," 1985.

4213 Worrell, W.B., "SETS Reference Manual," 1985.

14.4 References by Author and Date

AAR, 1973, "Summary of Ruptured Tank Cars Involved in Past Accidents," Association of American Railroads, AAR Report 130.

AAR, 1972, "Analysis of Tank Car Tub Rocketing in Accidents," Association of American Railroads, AAR Report 146.

Abbitt, J. F., 1969 "A Quantitative Approach to the Evaluation of the Safety Function of Operators in Nuclear Reactors," Atomic Health and Safety Board, UK. AHSB(s) R 160.

Abramowitz, M, and I. A. Stegun, 1970 *Handbook of Mathematical Functions*, U.S. Dept. of Commerce Applied Mathematics Series 55.

ACCIDATA, 1992, "Accident Data," PRINCIPIA, Engenharia de Confiabilidade e Informatica Ltda, Brazil.

Adams, J. A., 1984,"A Slow Comeback," IEEE System p 27 April.

AECL, 1989, "CANDU-3 Conceptual Probabilistic Safety Assessment."

AECL, 1995, "Probabilsitic Safety Report, Wolsong 2/3/4."

AECL, 1996, "CANDU-9 Probabilsitic Safety Assessment."

Airline Pilots Association International, 1978, "Human Factors Report on the Tenerife Accident," March 27, 1977," Washington, D.C.

Alesso, H. P., 1982 "Review of PASNY Systems Interaction Study," LLNL, UCID 19130, April.

Alesso, H. P., 1984, "On the Relationship of Digraph Matrix Analysis to Petri Net Theory and Fault Trees," LLNL UCRL-90271 Preprint, January.

Algermissen, S. T. and D. M. Perkins, 1976, "A Probabilistic Estimate of Maximum Acceleration in Rock in the Contiguous United States," Geological Survey Report 76-416.

Allen, P. J. et al., 1990, "Summary of CANDU-6 Probabilistic Safety Assessment Study Results," Nuclear Safety 31, 2, pp 202-214.

Allison, T. L, 1992, "Fire Load Verification for RTF Room 9, 15, and 17," EPD-SR-92-1159, SRC, November.

Alvares, N. J., K. L. Foote, and P. J. Pagni, 1984, "Force Ventilation Enclosure Fires," Combustion Science and Technology, Vol.39, p 55-81.

API(1), "Management of Process Hazards," (API Recommended Practice 750).
API(2). "Improving Owner and Contractor Safety Performance," (API Recommended Practice 2220).
Apostolakis, G. E. and M. Kazarians, 1980, "The Frequency of Fires in Light Water Reactor Compartments," ANS,'ENS Topical Meeting on Thermal Reactor Safety, Knoxville TN CONF-800-403 Vol I.
APS, 1975,"Review of the Reactor Safety Study" Rev. Mod. Phys. 47 Supp 1.
Army, 1969, "Structures to Resist the Effects of Accidental Explosions," Technical Manual TM5-1300, U.S. Army, Navy, and Air Force, Washington, D.C.
Atkinson, S. A., 1996, "PSA-Operations Synergisms for Advanced Test Reactor Shutdown Operations PSA," Proceeding of the International Topical Meeting on Probabilistic Safety Assessment, Park City, Utah, pp 600-604, Sept. 29 - Oct. 3.
Atkinson, S. A., 1995, "A Full ATR PRA for Risk Assessment Reduction, and Management," DOE Risk Management Quarterly, 3, 1, p 4, January.
Atkinson, S. A. et al., 1993, "Advanced Test Reactor Probabilistic Risk Assessment," Proceeding of the International Topical Meeting on Probabilistic Safety Assessment, Clearwater Beach, FL, pp 747-753, January 26-29.
Aviation Week and Space Technology, 1982, "Air Florida 731 Voice Recorder Transcribed," February 22.
Aviation Week and Space Technology, 1984, "Maintenance Cited in Triple-Engine Out," August 13.
Azarm, M. A. et al., 1990a, "Level-1 for High Flux Beam Reactor (HFBR) Accident Sequence Delineation," ANS Topical Meeting, The Safety, Status and Future of Non-Commercial Reactors and Irradiation Facilities, Boise, ID, Sept. 31 - Oct. 4, 1990.
Azarm, M. A. et al., 1990b, "Level-1 Internal Events PRA for the High Flux Beam Reactor," BNL Report, Vol. 1 and 2, July.
Bagnol, C.W., R. A. Matzie, and R. S. Turk, 1992, "System 80+ PWR Safety Design," Nuc. Safety 33, pp 47-57, January-March.
Baker et. al., 1983, *Explosive Hazards and Evaluation*, Elsevier, NY ISBN 044-420494-0.
Baumeister, T. and L.S. Marks (editors), 1967, *Standard Handbook for Mechanical Engineers*, Seventh Edition, McGraw-Hill, New York NY.
Belore, R. and J. Buist, 1986, "A Computer Model for Predictin Leak Rates of Chemicals from Damaged Storage and Transportation Tanks," Report EE-75, Environmental Canada.
Berkowitz, L., 1969, "Seismic Analysis of Primary Piping Systems for Nuclear Generating stations," Reactor Fuel Proc. Technol. 12 p 267.
Bernero, R.M. 1984, "Probabilistic Risk Analyses: NRC Programs and Perspectives," *Risk Analysis* 4, p 287, December.
Bignell, V et al., 1977, *Catastrophic Failures*, Open University Press, U.K.
Bird,R. B, W. E. Stewart, and E. N. Lightfoot, *Transport Phenomena*, Wiley NY.
Blevins, R. D., 1985, *Applied Fluid Dynamics Handbook*, Van Nostrand Reinhold, NY.
Bloom, E.D., 1981, "A Technique to Reliability Estimate Earthquake Recurrence Intervals," EPRI NP-1857.

References

Bloom, E. D. and R. C. Erdmann, 1979, "Frequency-Magnitude-Time Relationships in the NGSDC Data File," Bull. Seismol. Soc. Am. 69 p 2085.

Bloom, E. D. and R. C. Erdmann, 1980, "The Observation of a Universal Shape Regularity in Earthquake Frequency - Magnitude Distributions," Bull. Seismol. Soc. Am. 70 p 349.

Boccio, J. L. et al., 1984, "Program Plan for Evaluating Technical Specifications (PETS)," BNL A-3230, November.

Bolt, Beranek, and Newman, Inc., 1981, "Evaluation of Proposed Control Room Improvements Through Analysis of Critical Operator Decisions," EPRI NP-1982.

Bolten, J. G., 1983, "Risk-Cost Assessment Methodology for Toxic Pollutants from Fossil Fuel Power Plants," Rand report R-2993-EPRI, June.

Borkowski, R. J. et al., 1984, "In-Plant Reliability Data Bank for Nuclear Plant Components: a Feasibility Study on Human Error Information," ORNL/TM 9066, March.

Bourne, A.E. and A.V. Green, 1978, *Reliability Technology*, Wiley, New York.

Brandyberry, M. D. and H. E. Wingo, 1990, "External Events Analysis for the Savannah River Site K Reactor," ANS Topical Meeting, The Safety, Status and Future of Non-Commercial Reactors and Irradiation Facilities, Boise ID, Sept. 31 - October 4, 1990

Brereton, S. et al., 1997, "Final Report of the Accident Phenomenology and Consequence (APAC) Methodology Evaluation: Spills Working Group," UCRL-ID-125479, August.

Briggs, G. A., 1969, *Plume Rise*, AEC Critical Review Series TID-25075.

Briscoe F. and P. Shaw, 1980, "Spread and Evaporation of Liquid," *Progress in Energy and Combined Sciences*, 6, 2, 127-140.

Britannica, 1990, *The New Encyclopaedia Britannica*, 15th Edition, Chicago.

Brown, D. F., W. E. Dunn, and M. A. Lazaro, 1997a, "CASRAM: The Chemical Accident Stochastic Risk Assessment Model: Technical Documentation," BetaVersion 0.8, ANL.

Brown, D. F., W. E. Dunn, and M. A. Lazaro,1997b, "Users' Guide for CASRAM-SC" ANL.

Brown,D. F., Dunn, W. E. and A. J. Policastro, 1994a: "Statistical Determination of Downwind Concentration Decay for the 1993 Emergency Response Guidebook," University of Illinois, Urbana, IL.

Brown, D. F., W. E. Dunn, and A. J. Policastro, 1994b: "Application of a MonteCarlo Model for Transportation Risk Assessment from DOE Facilities," University of Illinois, Urbana, IL.

Brown, S. C. and J. N. T. Martin, 1977, *Human Aspects of Man-Made System*, Open University Press, U.K.

Brunot, W.K., 1970, "Reliability of a Complex Safety Injection System from Digital Simulation," ANS Trans 12 p 169, June.

Bunz, H. et al., 1983, "NAUA Mod 4: A Code for Calculating Aerosol Behavior in LWR Core Melt Accidents" Karlsruhe, Germany KfK-3554, August.

Burdick, G. R., and N. H. Wilson, and J. R. Wilson, "COMCAN - A Computer Program for Common Cause Analysis," INEL ANCR-1314, May.

Butikofer, R. E., 1986, "Safety Digest of Lessons Learned," Amer. Pet. Inst. API publication 758.

Buzzelli, G., 1986, "Nuclear Emergency Planning Course of Age," Power, pp 35-38, March.

C&EN - Chemical and Engineering News, 1155, 16th St. NW Washington DC 20036, published semi-weekly.

Canvey, 1978, "An Investigation of Potential Hazards from Operations in the Canvey Island/Thurrock Area," Her Majesty's Stationery Office, 49 High Holburn, London, WC1V 6HB.

Casarett, L. J. et al., 1991, *Casarett and Doull's Toxicology : the Basic Science of Poisons*, 4th Edition, Pergamon Press New York NY.

Cate, C. L., and J. B. Fussell, 1977, "BACKFIRE - A Computer Code for Common Cause Failure Analysis," University of. Tennessee, May.

CCPS, 1994, *Guidelines for Evaluating the Characteristics of Vapor Cloud Explosions, Flash Fires, and BLEVES*.

CCPS,1994, *Guidelines for Preventing Human Error in Process Safety*.

CCPS, 1989a, *Guidelines for Chemical Process Quantitative Risk Analysis*

CCPS, 1989b, *Guidelines for Process Equipment Reliability Data*

CPSS, 1989c, *Guidelines for Technical Management of Chemical Process Safety*.

CCPS,1992, *Guidelines for Hazard Evaluation Procedures*.

Chan, M. K. et. al., 1989, "User's Manual for FIRIN - A Computer Code to Estimate Accidental Fire and Radioactive Airborne Releases in Nuclear Fuel Cycle Facilities," NUREG/CR-3037 (PNL-4532), PNL, February.

Cheverton, R. D. et al., 1987, "Evaluation of HFIR Pressure Vessel Integrity Considering Radiation Embrittlement," ORNL/TM-10444, September.

Chong, J., 1980, "Hazardous Gas Release Model," Air Resources Branch, EMGRESP Program Documentation, Ontario Ministry of the Environment.

Chu, T. L., Z. Musicki, and P. Kohut, 1995, "Results and Insights of Internal Fire and Internal Flood Analysis of the Surry Unit 1 Nuclear Power Plant during Mid-loop Operations," PSA 95 Seoul Korea, pp 967-973, November.

Chu, T. L. et al., 1990, "Quantification of the Probabilistic Risk Assessment of the High Flux Beam Reactor (HFBR) at Brookhaven National Laboratory," ANS Topical Meeting, The Safety, Status and Future of Non-Commercial Reactors and Irradiation Facilities, Boise, ID, September 31 - October 4, 1990.

Chung, G., N. Siu, and G. Apostolakis, 1985, "Improvements in Compartment Fire Modeling and Simulation of Experiments," Nuclear Technology, 69, p. 14.

Claybrook, S., 1992, "Comparison of FIRIN Predictions to 1986 LLNL Enclosure Fire Tests 9 and 10," Numerical Applications Inc., Personal Communications.

Clough, P.N. et al., 1987, "Thermodynamics of Mixing and Final State of a Mixture formed by the Dilution of Anhydrous Hydrogen Fluoride with Most Air," Safety and Reliability Directorate, UKAEA, Wigshaw Lane, Culcheth, Warrington, Cheshire, England, WA3 4NE, SRD R 396.

CMA(1), "Evaluating Process Safety in the Chemical Industry."

CMA(2) "Safe Warehousing of Chemicals,"

CMA(3) "CMA's Manager Gulde," First Edition, September 1991.

Cohen, B.L., 1982, "Physics of the Reactor Meltdown Accident," Nucl. Sci. Eng. 80 p 47.

Cohen, B. L. and I. S. Lee, 1979, "A Catalog of Risks" Health Physics 36, pp 707-722, June.

References

Cohen, S. C. and K. D. Dance, 1975, "Scoping Assessment of the Environmental Health Risk associated with Accidents in the LWR Supporting Fuel Cycle," Teknekron Inc. 4701 Sangamore Rd., Washington, D.C. 20016.

Colenbrander, G. W. and J. S. Puttock, 1980, "Maplin Sands experiments 1980: interpretation and modelling of liquified gas spills on the sea," in *Atmospheric Dispersion of Heavy Gases and Small Particles*, G. Ooms and H. Tennekes, editors, Springer-Verlag, Berlin.

Colenbrander, G. W. and J. S. Puttock, 1983, "Dense Gas Dispersion Behavior: Experimental Observations and Model Developments," International Symposium on Loss Prevention and Safety Promotion in the Process Industries, Harrogate, England, September.

Commonwealth Edison Co., 1981, "Zion Probabilistic Safety Study," Chicago, IL.

Covello, V. T., 1981, "Actual and Perceived Risk: A Review of the Literature," in *Technological Risk Assessment* edited by P. F. Ricci et al. NATO ASI Series E-81, Martinos Nijhoff Publisher, The Hague, Netherlands.

Cox, D. C. and P. Baybutt, 1982, "Limit Lines for Risk," Nuclear Technology 57 pp 320-330, June.

Cox, R.A., editor, 1989, *Mathematics in Major Accident Risk Assessment*, Clarendon Press Oxford.

CRBR, 1977, "CRBRP Safety Study, An Assessment of Accident Risk," CRBRP-1.

CRC Handbook, 1979 *CRC Handbook of Chemistry and Physics*, (also known as the rubber handbook), CRC Press Inc., Boca Raton, FL 33431.

Crellin G. L., 1972, "The Philosophy of Mathematics and Bayes Equation," Trans. Rel. R-21, p 131, August

Croff, A. G., 1983, "A Versatile Computer Code for Calculating the Nuclide Compositions and Characteristics of Nuclear Material," Nucl. Technol. 62, p 335, September.

Crosby, P. B., 1984, *Quality without Tears: The Art of Hassle-free Management*, McGraw-Hill, New York, NY.

Crosetti, P. A., 1971, "Fault Tree Analysis for Reactor Systems," Inst. Power Ind. 14 p 54.

Crowl, D. A., and J. F. Louvar, 1990, *Chemical Process Safety Fundamentals*, Prentice Hall, Englewood Cliffs NJ.

Deal, S., 1995, "Technical Reference Guide for FPEtool Version 3.2," NISTIR 5486, NIST, April.

Delichatsios, M. et al., 1982, "Computer Modeling of Aircraft Cabin Fire Phenomena" Factory Mutual, FMRC J. I. OGON1.BU, Norwood, MA, December.

Deming, W. E., 1986, *Out of the Crisis*, MIT Center for Advanced Engineering Study.

Diamond, G. L. et al., 1988, *A Portable Computing System for Use in Toxic Gas Emergencies*, Ontario Ministry of the Environment, ARB-150-87, ISBN 0-7729-3448-7, revised.

DOC, 1969, "Earthquake Investigation in the United States," C&GS Special Publication No. 282.

DOC, 1973, "Earthquake History of the United States," Publication 41-1.

DOE, 1996, "DOE Handbook: Chemical Process Hazards Analysis," DOE-HDBK-1100-96, February.

DOE, 1996, "Process Safety Manaagement for Highly Hazardous Chemicals," DOE-HDBK-1101-96, February.

DOE, 1994, "Preparation Guide for U.S. Department of Energy Nonreactor Nuclear Facility Safety Analysis Reports," DOE-STD-3009-94, July

DOE, 1993, "Example: Process Hazard Analysis of a Department of Energy Water Chlorination Process," DOE/EH-0340 , September.
DOL, 1986, "Safety and Health Guide for the Chemical Industry," (OSHA 9081).
DOL, 1989, "Safety and Health Program Management Guidelines."
Donovan, N. C. and A. F. Borstein, 1977, "The Problems of Uncertainties in the Use of Seismic Risk Procedures," Dames and Moore Report EE77-4, Denver, CO.
Dougherty, E. M. and J. R. Fragola, 1988, *Human Reliability Analysis: A Systems Engineering Approach with Nuclear Power Plant Applications*, Wiley New York, NY.
Dougherty, E. M., 1990, "Human Reliability Analysis: Where Shouldst Thou Turn," Rel. Eng & Sys. Saf. 29, pp 283-299.
Dow, 1987, "Fire &; Exploslon Index Hazard Classification Guide," 6th Edition, May.
Dow, 1988, "Chemical Exposure Index," May.
Drouin, M. T. et al., 1996, "Individual Plant Examinations Perspectives on Reactor Safety," PSA-96 Park City, UT, September 29 - October 3, pp 1411-1417.
Dumas, R., 1987, "Safety and Quality: the Human Dimension," Professional Safety, pp 11-14, December.
DuPont, 1988, "Hydrofluoric Acid, Anhydrous - Technical, Properties, Uses, Storage and Handling," undated bulletin, E.I. duPont de Nemours & Company, Wilmington, DL,.
Eagling, D. G., editor, 1996, "Seismic Safety Manual: A Practical Guide for Facility Managers and Earthquake Engineers," UCRL-MA-125085, September.
Egan, B. A., 1975, "Turbulent Diffusion in Complex Terrain," Lectures on Air Pollution and Environmental Impact Analyses, American Meteorological Society, Boston, MA, p 123.
Eide, S. A. et al., 1990a, "Advanced Test Reactor Level 1 Probabilistic Risk Assessment," ANS Topical Meeting, The Safety, Status, and Future of Non-Commercial Reactors and Irradiation Facilities, Boise ID, Sept. 31 - October 4, 1990.
Eide, S. A. et al., 1990b,"Generic Component Failure Database for Light Water and Liquid Sodium Reactor PRAs," INEL, EGG-SSRE-8875.
Eidsvik, K. J., 1980, "A Model for Heavy Gas Dispersion in the Atmosphere," Atmospheric Environment, 14, 769-777.
EPA, 1996, "Exposure Models Library and Integrated Model Evaulation System," EPA/600/C-92/002, Third Revision, Office of Research and Development, CD-ROM.
EPA, 1992, "Work Book of Screening Techniques for Assessing Impacts of Toxic Air Pollutants," EPA-454/R-92-024, December.
EPA, 1988, "Review of Emergency Systems," 1988, Office of Solid Waste and Emergency Response, June.
EPA, FEMA and DOT, 1987, "Technical Guidance for Hazards Analysis, Emergency Planning for Extremely Hazardous Substances," December.
Erdmann, R. C. et al., 1976, "ATWS: A Reappraisal Part II, Evaluation of Societal Risks Due to Reactor Protection System Failure" Vol II BWR Risk Analysis EPRI NP-265, August.
Erdmann, R. C., F. L. Leverenz, H. Kirch, "WAMCUT, A Computer Code for Fault Tree Evaluation," EPRI-NP-803, June.

References

Erdmann, R.C. et al., 1079, "Status Report of the EPRI Fuel Cycle Risk Assessment," EPRI NP-1128.

Etherington, H., editor, 1958, *Nuclear Engineering Handbook*, McGraw-Hill, NY.

Evans, R. A. 1975, "Statistical Independence and Common-Mode Failures," IEEE Trans. Rel., R-24, p 289.

Farmer, F. R., 1967, "Reactor Safety and Siting: A Proposed Risk Criterion," Nuclear Safety 8, 539.

Farmer, F. R., 1967, "A New Approach in Containment and Siting of Nuclear Power Reactors," IAEA report ST3/PUB/154, Vienna.

Fauske, H. K. and M. Epstein, 1988, "Source Term Considerations in Connection with Chemical Accidents and Vapor Cloud Modeling," *Proceedings of the International Conference on Vapor Cloud Modeling*, CCSP, 251-273.

Fiksel, J. et al., 1982, "Development, Application and Evaluation of a Value-Impact Methodology for Prioritization of Reactor Safety R & D Projects," EPRI NP-2530, August.

Fisher, H. G. et al., "The Design Institute for Emergency Relief Systems (DIERS) Project Manual IV," AIChE, New York..

Fleming, K. N. et al., 1975, "A Reliability Model for Common Mode Failures In Redundant Safety Systems," Proceedings of the Sixth Annual Pittsburgh Conference on Modeling and Simulation, April.

Fleming, K. N. et al., 1979, "A Methodology for Risk Assessment of Major Fires and Its Application to an HTGR Plant," General Atomic GA-A15401.

Fragola, J. R. et al., 1983, "A Systematic Approach to Human Error Categorization in Nuclear Power Plant Tasks" ANS Trans. 44 p 176, June.

Fragola, J. R., 1983, "A Resource-Limited Model for Operator Response to Multiple Contemporaneous Events," ANS Trans. 45 November.

Fragola, J.R. and E.P. Collins, 1985, "Human Reliability Data Framework Development and Application," ANS Trans. 49 p 137 (quoted information is in a supplement).

Fried, L. E., a, "CHEETAH 1.39 User's Manual," UCRL-MA-117541 Rev 3, LLNL.

Fried, L. E., b, "READ ME CHEETAH 1.40," LLNL.

Fryer, L. S. and G. D. Kaiser, 1979, "DENZ - A Computer Program for the Calculation of the Dispersion of Dense Toxic or Explosive Gases in the Atmosphere," UKAEA, Report SRD R152.

Fullwood, R. and K. Majumdar, 1983, "On the Use of Fault Trees for Control Room Review" ANS Trans. 44, p 166, June.

Fullwood, R. and R. C. Erdman, 1974, "On the Use of Leak Path Analysis in Fault Tree Construction for Fast Reactor Safety," CONF-740401-P3.

Fullwood, R. and R. E. Hall, 1988, *Probabilisitic Risk Assessment in the Nuclear Power Industry*, Pergamon Press, Oxford.

Fullwood, R. and W. Shier, 1990, "PRA Using Event Tables and the Brookhaven Event Tree Analyzer (BETA)," *The Role and Use of Personal Computers in Probabilistic Safety Assessment and Decision Making*, Elsevier, NY, ISBN 1-85166-501-3, pp 79-92/.

Fullwood, R. et al., 1977, "Application of the Bayes Equation to Predicting Reactor System Reliability," Nucl. Technol. 34 p 341, August.

Fullwood, R, et al., 1984, "Value-Impact Analysis of Selected Safety Modifications to Nuclear Power Plants," EPRI NP-3434, March.

Fullwood, R. R. 1978, "Human Factors Considerations at Vermont Yankee Related to a Stuck-Open Relief Valve" SAI-083-79-PA, April.

Fullwood, R. R. and K .J. Gilbert, "Data and Assessment of Human Factors Impact on Nuclear Reactors," ANS Trans 26 p 385, June 1977.

Fullwood, R. R. and R.C. Erdmann, 1983, *Risks Associated with Nuclear Material Recovery and Waste Processing*, Progress in Nuclear Energy, Pergammon Press Ltd.

Fullwood, R. R. and R. R. Jackson, 1980, "Partitioning-Transmutation Program Final Report: VI Short Term Risk Analyis Reprocessing, Refabrication, and Transportation," ORNL/TM6986, and ORNL/Sub-80/31048/1.

Fullwood, R. R. et al., 1976, "ATWS: A Reappraisal Part I: An Examination and Analysis of WASH-1270, Technical Report on ATWS for Water-Cooled Power Reactors," EPRI NP-251, August.

Fussell, J. B. 1975, "Computer Aided Fault Tree Construction for Electrical Systems," *Reliability and Fault Tree Analysis*, SIAM, Philadelphia, PA, p 37.

Fussell, J. B. and W. E. Vesely, 1972, "A New Method for Obtaining Cutsets for Fault Trees," Trans. ANS, 15, p. 262.

Fussell, J. B. et al., 1974, "MOCUS - A Computer Program to Obtain Minimal Cutsets" Aerojet Nuclear ANCR-1156.

Gallucci, R. H. V., 1980, "A Methodology for Evaluating the Probability for Fire Loss of Nuclear Power Plant Safety Functions," Ph.D. Thesis at Rensselaer Poly. Inst., Troy, NY.

GAO, 1983, "Report to the Honorable Richard L. Ottinger, Chairman, Subcommittee on Energy Conservation and Power Committee on Energy and Commerce," U.S. House of Representatives.

GAO, 1985, "Probabilistic Risk Assessment: An Emerging Aid to Nuclear Power Plant Safety Regulation," GAO/RCED-85-11, June.

Garrick, B. J. et al., 1967, "Reliability Analysis of Nuclear Power Plant Protection Systems," Holmes and Narver HN-190, May.

Garrison, 1989, *Large Property Damage Losses in the Hydrocarbon-Chemical Industries: A Thirty Year Review*, 12th ed., Marsh and McLennan Protection Consultants, Chicago.

Generic Letter 88-20 "IPE for Severe Accident Vulnerabilities at Nuclear Power Plants," NRC, December 1,

German Risk Study, 1981, EPRI NP-1804-SR, April.

Gertman, D. L., and H. S. Blackman, 1994, *Human Reliability and Safety Analysis Data Handbook*, Wiley, New York, NY.

Gifford, F. A., 1984, "Statistical Properties of a Fluctuating Plume Dispersion Model" <u>Advances in Geophysics</u> 6 Academic Press, NY, p 117.

Gifford, F. A., 1972, "Atmospheric Transport and Dispersion Over Cities" Nucl. Safety 13, p 391, September-October.

Gilluly, J. et al., 1954 *Principles of Geology*, Freeman, San Francisco, CA, p. 456.

References

Glasstone, S. and P. J. Dolan, 1977, *The Effects of Nuclear Weapons*, 3rd Edition, U.S. Superintendent of Documents, WADC.

Green, A.E., 1982, *High Risk Technology*, Wiley, New York.

Gregory, W. S. and B. D. Nichols, 1991, "EXPAC User's Manual: A Computer Code for Analyzing Explosion-Induced Flow and Material Transport in Nuclear Facilities," LA-11823-M, LANL, July.

Gregory, W. S., et. al., 1991, "Fires in Large Scale Ventilation Systems," Nuclear Engineering Design, 125, pg 403-410.

Gregory, W. S, et al., "FIRAC-PC User's Manual - DRAFT," LANL Engineering and Safety Analysis Group (N6)

Gregory, W.S., et al., 1989, "FIRAC Code Predictions of Kerosene Pool Fire Tests," (Unpublished Report), LANL, May.

Gritzo, L. A. et al., 1995a, "Heat Transfer to the Fuel Surface in Large Pool Fires," Transport Phenomenon in Combustion, S. H. Choa (ed.), Taylor and Francis Publishing, Washington, DC.

Gritzo, L. A. et al., 1995, "Wind-Induced Interaction of a Large Cylindrical Calorimeter and an Engulfing JRE Pool Fire," Symposium on Thermal Science & Engineering in Honor of C. L. Tien, Berkeley, CA, November 14.

Guarro, S. and D. Okrent, 1984, "P 456 The Logic Flowgraph: A New Approach to Process Failure Modeling and Diagnosis for Disturbance Analysis Applications," Nucl. Technol. 67, p 348, December.

Gull, 1990, "An Analysis of Nuclear Incidents resulting from Cognitive Error," *11th Advances in Reliability Technology Symposium*, University of Liverpool, Elsevier, April.

Gumbel, E. J., 1958 *Statistics of Extremes* Columbia Press, New York.

Gupta, I. N., 1976, "Attenuation of Intensities Based on Isoseismals of Earthquakes in Central United States," Earthquake-Notes 47, p 13.

Gutenberg, B. and C.F. Richter, 1942, "Earthquake Magnitude, Intensity, Energy and Acceleration,"" Bull. Seismol. Soc. Am. 32, p 163.

Hall, R. et al., 1983, "Control Rod Trip Failures: Salem 1, The Cause, The Response and Potential Fixes," ANS Trans. Topical Meeting, Jackson Hole, WY, April.

Hanna, S. R. and D. Strimaitis, 1989, *Workbook of Test Cases for Vapour Cloud Dispersion Modes*, CCPS, New York.

Hannaman, G. W. and A .J. Spurgin, and Y. D. Lukic, 1984, "Human Cognitive Reliability Model for PRA Analysis," NUS Corp. NUS-4531 (Rev. 3).

Hannerz, H., 1983, "Towards Intrinsically Safe Light Water Reactors," ORAU/IEA-83-2(M)-Rev. Institute for Energy Analysis.

Haugen, E.B. 1980, *Probabilistic Mechanical Design*, Wiley, New York.

Havens, J. A., and T. O. Spicer, 1985, "Development of an Atmospheric Dispersion Model for Hearier-than-Air Gas Mixtures," Vol. 1-3 CG-D-2385, U.S. Coast Guard, May.

Hazzon, M. J. and E. A. Warman, 1986, "A Rational Approach to Emergency Planning" AMS France, November.

Henley, E. J. and H. Kunamoto, 1981, *Reliability Engineering and Risk Assessment*, Prentice Hall, Englewood Cliffs, NJ.

Henry, R. E. and H. K. Fauske, 1971, "The Two-Phase Critical Flow of One-Component Mistures in Nozzles, Orifices, and Short Tubes," *J. Heat Transfer*, pp 179-187, May.

Heskestad, G. and J. P. Hill, 1986, "Experimental fires in Multi-Room/Corridor Enclosures," NIST, NBS-GCR-86-502, Gaithersburg, MD 20899.

Hinga, K R., 1982, "Disposal of High-Level Radioactive Waste by Burial in the Sea Floor," Env. Sci. Technol. 16.

Hinkley, P.L., 1975, "Work by the Fire Research Station on the Control of Smoke in Covered Shopping Centers," Proceedings of the Conseil International du Batiment (CIB) W14/168/76 (UK) 24p., September.

Hockenbury, R. W. and M. L. Yeater, 1980, "Development and Testing of a Model for Fire Potential in Nuclear Power Plants," NUREG/CR-1819.

Holden, P. L. and A. B. Reeves, 1985, "Fragment Hazards from Failures of Pressurized Liquified Gas Vessels," *Assessment and Control of Major Hazards, I Chem. E Symposium Series* No. 93 ISBN 0-85295-189-2.

Holen, J., M. Brostrom, and B. F. Magnussen, 1990, "Finite Difference Calculation of Pool Fires," Proc. of 23rd Int. Symp. Combustion, pp.1677-1683.

Hollnagel, E., 1993a, *Human Reliability Analysis: Context and Control*, Academic Press, London.

Hollnagel, E., 1993b, *Reliability of Cognition: Foundations of Human Reliability Analysis*, Plenum Press, New York.

Homann, S. G., 1994, "HOTSPOT: Health Physics Code for the PC," UCRL-MA-106315, LLNL, March.

Huff, J. E., 1985, "Multiphase Flashing Flow in Pressure Relief Systems," Plant Operations, $\underline{4}$, 191-199.

Hymes, I., 1983, "The Physiological and Pathological Effects of Thermal Radiation," UKAEA Safety and Reliability Directorate, Report SRD R275, Culcheth U.K.

IAEA, 1988, "Component Reliability Data for Use in Probabilistic Safety Assessment," TecDoc-478, Vienna.

IAEA, 1980, "Atmospheric Dispersion in Nuclear Power Plant Siting - A Safety Guide," Safety Series No. 50-SG-S3 UNIPUB, NY.

IEEE Std 500-1984, *IEEE Guide to the Collection and Presentation of Electrical, Electronic, Sensing Component and Mechanical Equipment Reliability Data for Nuclear Power Generating Stations*, IEEE, NY.

Ille, G. and C. Springer, 1981, "The Evaporation and Dispersion of Hydrazine Propellants from Ground Spills," CEEDO-TR-83-30, ADA059407.

Ille, G. and C. Springer, 1978, "The Evaporation and Dispersion of Hydrazine Propellants from Ground Spills," CEEDO-TR-78-30.

Iman, R. L. and M.J. Shortencarrier, 1986, "A User's Guide for the Top Event Matrix Analysis Code (TEMAC)," NUREG/CR-4958, August.

References

Iman, R. L. and M. J. Shortencarrier, 1984, "A FORTRAN 77 Program and User's Guide for the Generation of Latin Hypercube and Random Samples for Use with Computer Model," NUREG/CR-3624, March.

Iman, R. L. and W. J. Conover, 1983, *A Modern Approach to Statistics*, Wiley, N Y

Indian Point 2 and 3, 1982, "Indian Point Probabilistic Safety Study," Consolidated Edison and NY Power Authority.

Inhaber, H., 1979, "Risk of Energy Production," Science, 203 pp 718-723.

Ito, V., N. Siu, G. Apostolakis, 1985, "COMPBRN III-A Computer Code for Modelling Compartment Fires," UCLA Report - ENG-8524, November.

Jaitly, R K., 1995, "The CANDU-9 Probabilistic Safety Assessment Program," PSA'95, Seoul, Korea, pp 33-38, November 26-30.

Johnson, D. H. et al., 1991, "The High Flux Isotope Reactor Probabilistic Risk Assessment and Its Application in Risk Assessment," Probabilistic Safety and Management PSAM1, Beverly Hills, CA pp 357-362, February 4-7.

Johnson, D. H. et al., 1988, "The High Flux Isotope Reactor Probabilistic Risk Assessment," Final Report PLG-0604, Pickard Low and Garrick, Inc. January.

Johnson, W. B. et al., 1975, "Gas Tracer Study of Roof-Vent Effluent Diffusion at Millstone Nuclear Power Station," Atomic Industrial Forum AIF/NESP 0076.

Johnson, W. G., 1980 *MORT Safety Assurance System*, Marcel Dekker Inc. NY, N.Y.

Joksimovich, V, 1984, "A Review of Plant Specific PRA's," Risk Analysis $\underline{4}$ p 255, December.

Joksimovich, Y. et al., 1983, "A Review of Some Early Large-Scale Probabilistic Risk Assessments," EPRI NP-3265, October.

Jones, W. W. and R. D. Peacock, 1994 "Refinement and Experimental Verification of a Model for Fire Growth and Smoke Transport," 2nd IAFSS Meeting.

Jones, W. W., 1994, "Modeling Smoke Movement Through Compartmented Structures," *Journal of Fire Sciences*.

Joschek, H. I., 1981, "Risk Assessment in the Chemical Industry," ANS/ENS Topical Meeting on Probabilistic Risk Assessment , ANS, Sept.

Juran, J. M., 1979, *Juran's Quality Control Handbook*, 3rd Edition, McGraw-Hill, NY.

Kaiser, G. D., 1986, "Implications of Reduced Source Terms for Ex-Plant Consequence Modeling and Emergency Planning," Nuclear Safety 27,3, pp 369-384, July-September.

Kaizer, G. D., 1989, "A Review of Models for Predicting the Dispersion of Ammonia in the Atmosphere," Plant Operation Progress 8, p1.

Kameleon and Kameleon, Fire User's Manuals available from SINTEF.

Kaplan, I. and B.J. Garrick, 1981, "On the Quantitative Definition of Risk," Risk Analysis $\underline{1}$, p1.

Kaplan, S., 1989, "Expert Information vs Expert Opinion: Another Approach to the Problem of Eliciting/Combining/Using Expert Opinion in PRA," PSA-89, p. 593.

Kazarians, M. and G. E. Apostolakis, 1978, "Some Aspects of the Fire Hazard in Nuclear Power Plants," Nucl. Eng. Design 47, p 157.

KBERT, 1995, "User's Guide for the KBERT 1.0 Code," Sandia National Laboratories SAND95-1324, June

Kelly, A. P. Jr. and D. W. Stillwell, 1983, refer to EPRI NP-3123 Vol 2.

Kelly, J. E. et al., 1976, "ATWS. A Reappraisal Part II, Evaluation of Societal Risks due to Reactor Protection System Failure Vol.3 PWR Risk Analysis," EPRI NP 265, August.

Kendall, H. W. Study Director, 1977, *The Risks of Nuclear Power Reactors* Union of Concerned Scientists, Cambridge, MA, August.

Kenney, W. F., 1993, *Process Risk Engineering*, VCH Publishers, Florham Park, NJ.

King, F. K. et al., 1987 "The Darlington Probabilisitic Safety Evaluation - A CANDU Risk Assessment," Canadian Nuclear Society 8th Annual Conference, Proceedings, pp 151-178.

Kirwan, B., 1992., "Human Error Identification in Human Reliability Assessment," Applied Ergonomics, 23, pp 299-318 and pp 371-381.

Kletz, T., 1991, *An Engineer's View of Human Error*, 2nd Edition, Institution of Chemical Engineers, Rugby U.K.

Kletz, T., 1994, *What Went Wrong: Case Histories of Process Plant Disasters*, 3rd Edition, Gulf Publishing Co., Houston.

Kouts, H., 1986, "The Chernobyl Accident," BNL-52033, September.

Kunkel, B. A., 1983, "A Comparison of Evaporative Source Strength Models for Toxic Chemical Spills", AFGL-TR83-0307, ADA139431.

Lambert, H. E., 1975, "Fault Trees for Decision Making in Systems Analysis," LLNL UCRL-51829.

Lambright, J. et al., 1989, "Fire Risk Scoping Study," Sandia National Laboratories, SAND 88-0177, NUREG/CR-5088, December.

Lapides, M. W, 1976, "Nuclear Unit Productivity Analysis," EPRI SR-46, August.

Lazaro, M. A. et al., 1997, "Model Review and Evaluation for Application in DE Safety Basis Documentation of Chemical Accidents: Modeling Guidance for Atmospheric Dispersion and Consequence Assessment," ANL/EAD/TM-75, September.

Lees, F. P., editor, 1980, *Loss Prevention in the Process Industries*, Butterworth, London.

Lees, F.P.,1986, *Loss Preventlon in the Process Industries*, Volumes I and II, Butterworth, London

Leung, J. C., 1986, "A Generalized Correlation for One-Component Homogenous Equilibrium Flashing Choked Flow," AIChE Journal U32U, 10, pp1743-1746.

Leverenz, F. L. and H. Kirch, "Users Guide for the WAM-BAM Computer Code," EPRI-217-2-5, January.

Leverenz, F. L. and R. C. Erdmann, 1979, "Comparison of the EPRI and Lewis Committee Review of the Reactor Safety Study," EPRI NP-1130.

Leverenz, F. L. et al., 1978, "ATWS: A Reappraisal Part III Frequency of Anticipated Transients," EPRI NP-801.

Levine, S. and N.C. Rasmussen, 1984, "Nuclear Plant PRA; How Far Has It Come?" *Risk Analysis* 4, pp 247-254, December.

Lewis, E. G., 1977, *Nuclear Power Reactor Safety*, Wiley, NY.

Lisk, K. C., 1972 *Nuclear Power Plant Systems and Equipment*, Industrial Press, NY.

Long, W. T., 1975, "Go Evaluation of a PWR Spray System," EPRI 350-1, August.

Lorenzo, D. K., 1990, *A Manager's Guide to Reducing Human Errors: Improving Human Performance in the Chemical Industry*, CMA, Washington, DC.

Luckas, W. J. et al.,"A Human Reliability Analysis for the ATWS Accident Sequence at the Peach Bottom Atomic Power Station," BNL Technical Report A3272, May 1986.

References

M&M, 1997, "Large Property Damage Losses in the Hydrocarbon-Chemical Industries: A Thirty-year Review," J&H Marsh and McLennon, New York, NY.

MacArthur, C., 1981, "Dayton Aircraft Cabin Fire Model Version 3," Vols 1 and 2 Univ. of Dayton. Research Inst.

Mackay, D. and R. S. Matsugo, 1973, "Evaporation Rates of Liquid Hydrocarbon Spills on Land and Water," *Canadian Journal of Chemical Engineering*, 51, 434-439

Mann, N., R. Shafer and N. Singpurwalla, 1984, *Methods for Statistical Analysis and Life Data*, Wiley, New York.

Marshall, A. W. and I. Olkin, 1967, "A Multivariate Exponential Distribution," *J. Am. Stat. Assoc.* p 30.

Matthews, S.D., 1977, "MOCARS: A Monte Carlo Simulation Code for Determining Distribution and Simulation Limits," TREE-1138, July.

Mattson et al., 1980, "Concepts, Problems and Issues in Developing Safety Goals and Objectives for Commercial Nuclear Power," Nuclear Safety 21, pp 703-716, November-December.

McCandless, R. J. and J. R. Redding, 1989, "SBWR: The Key to Improved Safety Performance and Economics," Nuc. Eng. Int., pp 20-24, Nov.

McCormick, N. J., 1981, *Reliability and Risk Analysis*, Academic Press, NY.

McIntyre, B. A., and R. K. Beck, 1992 "Westinghouse Advanced Passive 600 Plant," Nuc. Safety 33, pp 36-37, January-March.

Melhem, G., et al. 1997, "Explosions and Energetic Events (EEE) Modeling Guidance for Accident Consequence and Safety Analysis," Draft Report, A.D. Little Inc., January.

MHIDAS, 1992, "Major Hazard Incident Data Service, Safety and Reliability Business," UKAEA, distributed by Silverplate Information Inc.

Mill, R. C. (Ed.), 1992, *Human Factors in Process Operations*, Institute of Chemical Engineers, Rugby, U.K.

Miller, L. A., et al., 1990, "Application of Phenomenological Calculations to the N-Reactor Probabilistic Risk Assessment," ANS Topical Meeting, The Safety, Status and Future of Non-Commercial Reactors and Irradiation Facilities, Boise, ID, Sept. 31 - October 4, 1990.

Mishima, J., 1993, "Recommended Values and Tedmical Bases for Airborne Release Fractions (ARFs), Airborne Release Rates (ARRs), and Respirable Fractions (RFs) for Materials from Accidents in DOE Fuel Cycle, Ex-Reactor Facilities," Revision 2, Draft DOE report, April.

Mitler, H. and H. Emmons, 1981, "Documentation for CFCV, the Fifth Harvard Computer Fire Code," Harvard University, Cambridge, MA, October.

Montague, D. F., 1990, "Process Risk Evaluation - What Method to Use?" Rel. Eng. & Sys.Safety, 29, pp 27-53.

Moore, C. V., 1967, "The Design of Barricades for Hazardous Pressure systems," Nuc. Eng. Des. $\underline{5}$, pp 1550-1566.

Morehouse, J. H. et al., 1983, "Value-Impact Analysis of Recommendations Concerning Steam Generator Tube Degradations and Rupture Events," SAIC for NRC/NRR Contract NRC-03-82-131, February.

Mulvihill, R. J. 1966, "A Probabilistic Methodology for the Safety Analysis of Nuclear Power Reactors," USAEC report PRC-R-657, February.

Murphy, J. A., 1996, "Risk-Based Regulation: Practical Experience in Using Risk-Related Insights to Solve Regulatory Issues," PSA 96, pp 945-948.

Murphy, J. R. and L .J. O'Brien, 1977, "The Correlation of Peak Ground Acceleration Amplitude with Seismic Intensity and Other Physical Parameters," Bull. Seismol. Soc. Am. 67 p 877.

Myrtle Beach, 1979, "Conference Record of the 1979 IEEE Standards Workshop on Human Factors and Nuclear Safety," IEEE Cat. No. TH0075-2, December.

Myrtle Beach, 1981, "Conference Record for the 1981 IEEE Standards Workshop on Human Factors and Nuclear Safety," IEEE Cat. No. TH0098-4.

Nadeau, M. L., 1995a, "Transmittal of Software Management and QA Package for the HOTSPOT Codes," (HOTSPOT Version 7.01, Fidler version 7.0), ECS-EST-95-0077 SRC, August.

Nadeau, M. L., 1995b, "Evaluation of the HOTSPOT Computer Codes, (HOTSPOT Version 7.01, Fidler version 7.0)," S-CLC-G-00080 SRC August.

Nadeau, M. L., 1995c, "Operation and Management of the HOTSPOT Computer Codes," (HOTSPOT Version 7.01, Fidler version 7.0)," TP-95-003, SRC, August.

National Research Council, 1972, "The Biological Effects of Ionizing Radiations," National Academy of Sciences.

Neill, D. T. et al., 1990a, "Summary of Omega West Reactor, Level 1, Probabilistic Risk Assessment," ANS Topical Meeting, The Safety, Status and Future of Non-Commercial Reactors and Irradiation Facilities, Boise, ID, September 31 - October 4.

Neill, D.T. et al., 1990b, "Omega West Reactor, Level 1, Probabilistic Risk Assessment," Idaho State University, College of Engineering, LANL, LA-UR-90-962.

Nelson, H. E. and S. Deal, 1992, "Comparison of Four Fires with Four Fire Models," Proceedings of 3rd International Symposium, IAFSS, Edinburg, Scotland, pp. 719-728.

Nichols, B. D. and W. S. Gregory, 1986, "FIRAC User's Manual: A Computer Code to Simulate Fire Accidents in Nuclear Facilities," NUREG/CR-4561 (LA-10678-M), LANL, April.

Nichols, B. D. et al., "Fire-Accident Analysis Code (FIRAC) Verification," LA-UR-86-2860, LANL, 19th DOE/NRC Nuclear Air Cleaning Conference.

Nicolette,V. et al., 1989, "Observations Concerning the COMPBRN III Fire Growth," ANS/ENS Conference on PRA, Pittsburgh, PA, April 12,and SAND 88-2160.

Notarianni, K. A. and W. D. Davis, 1993, "Use of Computer Models to Predict Temperature and Smoke Movement in High Bay Spaces," NIST NISTIR 530.

NRC, 1984, "Regulatory Analysis Guidelines of the U.S. Nuclear Regulatory Commission," NUREG/BR-0058, May.

NRC, 1989, "Installation of a Hardened Wetwell Vent," Generic Letter 89-16.

NSAC-60, 1984, "Oconee PRA: A Probabilistic Risk Assessment of Oconee Unit 3," EPRI/Duke Power Co., June.

NSC, 1983 "Accident Investigation: A New Approach,"

O'Donnell, E. P., 1982, "Safety Goals and Risk Analysis: Status of Regulatory Development," Joint ASME/ANS Nuclear Engineering Conference, July.

Oil Insurance Association, 1971, "Report on Boiler Safety."

References

Okrent, D., 1980, "Comment on Societal Risk" *Science* 208, pp 372-375, April.

Okrent, D., 1981, *Nuclear Reactor Safety* University of Wisconsin Press, Madison.

Oil Insurance Association, 1971, *Report on Boiler Safety*.

Oliveira, L. F. S., et al., 1994, "Quantitative Risk Analysis of a Butane Storage Facility," PSAM-II, San Diego, CA, March 20-25.

Oliveira, L. F. S. et al., 1991, "Risk Assessment of a Marine Terminal for Hazardous Chemicals," PSAM-I, Beverly Hills, CA, February.

Ontario Hydro, 1995, "Pickering NGS A Risk Assessment."

Ontario Hydro, 1987, "Darlington Probabilistic Safety Evaluation."

Opschoor, Ir. G.,1978 "Evaporation", Chapter 5 of TNO *Yellow Book*, Methods for Calculation of the Physical Effects of Escape of Dangerous Materials, 1st Edition, Volume III, February.

Orvis, D. D. et al., 1985, "Review of Selected Topics from PRA Studies: System Dependencies, Human Interactions, and Containment Event Trees," EPRI NP-3838, May.

OSHA, 1996, "Process Safety Management of Highly Hazardous Materials," 29CFR1910.119.

Otway, H. J. and R. C. Erdmann, 1970, "Reactor Siting and Design from a Risk Viewpoint," Nucl. Eng. Design 13, p 365.

Otway, H. J. et al., 1970, "A Risk Analysis of the Omega West Reactor," LASL LA-4449, July.

Pande, P. K. et al., 1975, "Computerized Fault Tree Analysis: TREEL and MICSUP," Operations Research Center, U.C. Berkeley., ORC 75.3.

Papazoglou, I. A. et al., 1996, "SOCRATES: a Computerized Toolkit for Quantification of the Risk from Accidental Releases of Toxic and/or Flammable Substances," *Int. J. Envir. & "Pollution* 6, 4-6, pp 500 -533.

Papazoglou, I. A. et al., 1992, "On the Management of Severe Chemical Accidents DECARA: A Computer code for Consequence Analysis in Chemical Installations Case Study: September 14, 1998 Ammonia Plant," *J. Haz. Mat.* 31, pp 135-153.

Papazoglou, I.A., et al., 1990a, "Probabilistic Safety Analysis of an Ammonia Storage Plant,"" PSAM-I, Beverly Hills California, February 4-7.

Papazoglou, I.A. et al., 1990b, "Probabilistic Safety Analysis of an Ammonia Storage Plant,"" DEMO report, NCSR "DEMOKRITOS", July.

Peacock, R D., et al., 1993a, "CFAST, The Consolidated Model of Fire Growth and Smoke Transport", NIST Technical Note 1299, NIST.

Peacock s R. D. et. al., 1993b, "Verification of a Model of Fire and Smoke Transport," *Fire Safety Joumal*: 21(2), pp. 89-129.

Perkins, D. M. et al., 1980, "Probabilistic Estimates of Maximum Seismic Horizontal Ground Motion on Rock in The Pacific Northwest and the Adjacent Outer Continental Shelf," U.S. Geological Survey No.80-471.

Peters, T. J., and R. H. Waterman, 1982, *In Search of Excellence*, Harper & Row, NY.

Peterson, R. J. et al., 1981, "Performance-Based Evaluation of Safety Parameter Display: Detection," EG&G, Sd-B-81-004, November.

Philadelphia Electric, 1983, "Severe Accident Risk Assessment," Limerick Generating Station," Report No. 4161.

Phillips, L. D., P. Humphreys, and D. E. Embrey, 1983, "A Socio-Technical Approach to Assessing Human Reliability (STAHR), TR 83-4, July.

Portier, R. W. et al., 1992, "A User's Guide for CFAST Version 1.6," NISTIR 4985, NIST.

Post, L. (editor), 1994, "HGSystem 3.0 Technical Reference Manual," Report No. TNER.94.059, Shell Research Limited, Thornton Research Centre, P.O. Box 1, Chester, England, Shell Internationale Research Maatschappij B.V.

Potash, L. M. et al., "Experience in Integrating the Operator Contributions of the PRA of Actual Operating Plants, "Proceeding of the ANS/ENS September 1981 Topical Meeting on Prbabilisitic Risk Assessment, Port Chester, NY pp 1054-1063, ANS.

Powers, D. A. et al., 1985,"VANESA, A Mechanistic Model of Radionuclide Release and Aerosol Generation during Core Debris Interaction with Concrete," NUREG/CR-4308.

Powers, G. J. and F. C. Thompkins, 1974, "Fault Tree Synthesis for Chemical Processes," AICE Journal 20, March.

Powers, G. J. and S. A. Lapp, 1976, "Computer-Aided Fault Tree Synthesis," Chem. Eng. Progr. 72 p 89, April.

Press, F. and R. Siever, 1974, *Earth*, p 411, Freeman, San Francisco.

Quintiere, J,. 1977, "Growth of Fire in Building Compartments," ASTM Special Tech. Pub. 614.

Raina, V. M. et al., "System Modelling Techniques and Insights from the Darlington Probabilistic Evaluation Study," PSA '87 CONF-870820 Vol. 2 TUEV Rhineland.

Raj, P. K., 1981, "Models for Cryogenic Liquid Spill Behavior on Land and Water," J. Haz. Mat. 5, 111-130.

Raj, P. K. and J. A. Morris, 1987, "Source Characterization and Heavy Gas Dispersion Models for Reactive Chemicals," AFGL-TR-88-0003 (I), ADA200121.

Rasmussen, J., 1979, "On the Structure of Knowledge: A Morphology of Mental Models in a Man-Machine Context," Riso-M-2192, Riso Nat. Lab., Denmark.

Rasmussen, J., 1989, "Chemical Process Hazard Identification," Rel. Eng. & Sys. Saf. 24, p 11-20.

Reg. Guide 1.145, 1983, "Atmospheric Dispersion Models for Potential Accident Consequence Assessments at Nuclear Power Plants," USNRC, February.

Restrepo, L. F., et al., 1996, "Accident Phenomenology and Consequence Methodology Evaluation: Fire Working Group," Draft Report, Sandia National Laboratory, April.

Ringhals, 1983, "Ringhals 2 Probabilistic Safety Study," Swedish State Power Board.

Roberts, A. F., 1981, "Thermal Radiation Hazards for Releases of LPG from Pressurized Storage," Fire Safety Journal 4, p 197-212.

Rothbart, G. et al., 1981, "Verification of Fault Tree Analysis," Vols. 1 and 2 EPRI NP-1570, May.

Rowsome, F. H. III, 1976, "How Finely Should Faults be Resolved in Fault Tree Analysis?" ANS/CNA, Toronto, Canada, June 18.

Rubber Handbook, 1979, *Handbook of Chemistry and Physics*, Robert Weast (editor), CRC Press Inc., Boca Raton, FL 33431.

Ruger, C., J. L. Boccio, and M. A. Azarm, 1985, "Evaluation of Current Methodology Employed in Probabilistic Risk Assessment of Fire Events at Nuclear Power Plants," BNL report A-3710, February.

References

Russell, K. D. et al., 1988, "Integrated Reliability and Risk Analysis System (IRRAS) User's Guide Version 2.0," NUREG/CR-5111, November

Russell, K. D. et al., 1987, "Integrated Reliability and Risk Analysis System (IRRAS) User's Guide Version 1.0," NUREG/CR-4844, February.

Sacks, I., 1978, "Techniques for the Determination of Potential Adversary Success with Tampering," LLNL report MC 78-9280, October.

Sacks, I. J. et al., 1977, "Target Identification Procedure for Plutonium Reprocessing Facilities," LLNL UCRL 79215-R1, June.

Sacks, I. J. et al., 1983, "Systems Interaction Results from the Digraph Matrix Analysis of the Watts Bar Nuclear Power Plant High Pressure Safety Injection Systems," LLNL UCRL-53467, December.

Sagendorf, J. F., 1974, "A Program for Evaluating Atmospheric Dispersion Calculations Considering Spatial and Temporal Meteorological Variations," NOAA Tech. Memo ERL-ARL-44.

Salem, S. L. et al., 1980, "Issues and Problems in Inferring a Level of Acceptable Risk," R-2561-DOE, August.

Salvatori, R., 1970, "Systematic Approach to Safety Design and Evaluation," IEEE Trans. NS-18, $\underline{1}$ p 495, February.

Samanta, P. K. and S. M. Wong, 1985 "AOT Risk Analysis and Issues: Limerick Emergency Cooling Systems," BNL A3230-2-25-85, February.

Sattison, M. B., et al., 1987, "System Analysis and Risk Assessment System (SARA) User's Manual Version 2.0," EGG-SPAG-7532, January.

Savy, J. B., 1980, "Seismic Hazard Analysis of the Savannah River Site," UCID-21596, November.

Sax, N. et al., 1987, *Hazardous Chemicals Desk Reference*, Van Nostrand Reinhold, New York, NY.

Sax, N. et al., 1989, *Dangerous Properties of Industrial Materials*, 7th Edition, Van Nostrand Reinhold, New York.

Schmidt, E. and P. Silveston, 1959, "Natural Convection in Horizontal Layers," *Heat Transfer*, $\underline{55}$ 20, American Institute of Chemical Engineers, Chicago.

Schneider, K. J., Coordinator, 1982, "Nuclear Fuel Cycle Risk Assessment," Vol 1 and 2, PNL-4306.

SECT-93-106, 1993, "Revised Guidelines for Prioritization of Generic Safety Issues," USNRC.

SECY-89-102, 1990, "Implementation of the Safety Goals," USNRC, memorandum from S. J. Chalk to J. M. Taylor.

SECY-88-147, 1987, "Integration Plan for Closure of Severe Accident Issues," USNRC.

SECY-83-221, 1983, "Prioritization of Generic Safety Issues," USNRC.

SECY-80-283, 1980, "Report of the Task Force on Interim Operation of Indian Point," USNRC.

SECY-78-616, 1978, "Unresolved Safety Issues," USNRC.

Schotte, W., 1987, "Fog Formation of Hydrogen Fluoride in Air", Ind. Eng. Chem. Res., 26, pp 300-306,

Schotte, W., 1988, "Thermodynamic Model for HF Fog Formation", letter to C. A. Soczek, dated August 31, 1988, E.I. DuPont de Nemours & Company, duPont Experimental Station, Engineering Department, Wilmington, DE.

Shapiro, A. H., 1953, *Dynamics and Thermodynamics of Compressible Flow*, 1, Wiley, New York.
Sharp, D., 1986, "Tornado Hazards to Production Reactors," SRL DPST-86-579, January.
Shrivastava, P., 1987, *Bhopal: Anatomy of a Crisis*, Ballinger Pub. Co., Cambridge, MA.
Singh, A. J., G. W. Perry and A. N. Beare, "An Approach to the Analysis of Operating Crew Responses for Use in PRAs," Proceedings of PSA '93, Clearwater Beach FL, Jan. 27-29 pp 294-300, ANS.
Siu, N. 1983, "COMPBRN - A Computer Code for Modeling Compartment Fires," NUREG/CR3239, UCLA - ENG-8257
Siu, N., 1982, "COMPBRN: A Computer Code for Modeling Compartment Fires," UCLA-ENG-8257, August.
Siu, N. and G. Apostolakis, 1982, "Probabilistic Models for Cable Tray Fires," Reliability Engineering 3, p 213.
Siu, N., 1980, "Probabilistic Models for the Behavior of Compartment Fires," UCLA-ENG-8090.
Sizewell-B, 1982, "Sizewell B Probabilistic Safety Study," Westinghouse Electric WCAP-9991.
Smith, A. M. and I. A. Watson, 1980, "Common Cause Failures: A Dilemma in Perspective," Reliability and Maintainability Conf. pp 127-142.
Smith, C. L. and R. D. Fowler, 1996, "SAPHIRE Recover Cut Sets editor: a technique for automating the manipulation of PRA cut sets," Rel. Eng. & System Safety, 54, pp 47-52,
Smith, P. D. et al., 1980, "An Overview of Seismic Risk Analysis for Nuclear Power Plants." LLNL UCID-18680.
Snell, V. G., et al., 1990, "CANDU Safety under Severe Accidents," Nuclear Safety, 31, 1, pp 20-36, January-March.
Solomon, K. D. and S. C. Abraham, 1979, "The Index of Harm: A Measure for Comparing Occupational Risk across Industries," Rand report R-2409-RC, June.
Solomon, K. D. and W. G. Kastenburg, 1985, "Estimating the Planning Zones for the Shoreham Nuclear Reactor, A Review of Four Safety Analysis," Rand note N-2353-DOE September.
Spicer, J. W. and J. Havens, 1989, "User's Guide for the DEGADIS 2.1 Dense Cloud Dispersion Model," EPA-4504-89-019, November.
Stamatelatos, M. et al., "Value-Impact Methodology for Decision Makers," EPRI NP-2529, August.
Stamm, J. et al., 1996, "Safety Monitor Implementation Project at Wolf Creek, Callaway, and Comanche Peak Stations," PSA '96, Park City, UT, p 12-19, Sept. 29 - Oct. 3.
Starr, C., 1969, "Social Benefit versus Technological Risk," Science 165 pp 1232-1238.
Starr, C., 1971, "Benefit-Cost Studies in Sociotechnical Systems," in *Perspectives on Benefit-Risk Decision Making*, National Academy of Engineering p 17.
Starr, C., 1980, "Risk Criteria for Nuclear Power Plants: A Pragmatic Proposal" ANS/ENS International Conference, Washington, DC, November 16-21.
Start, G. E. and L. L. Wendall, 1974, "Regional Effluent Dispersion Calculations Considering Spatial and Temporal Meteorological Variations," NOAA Tech Memo. ERL-ARL-44.
Swain, A. D., 1989, *Comparative Evaluation of Methods for Human Reliability Analysis*, GRS-71, Gesellschaft fur Reaktorsicherheit (GRS) mbH, Garchin Koln, Germany.
Swamy, M. N. S. and K. Thularisaman, 1981, *Graphs, Networks and Algorithms*, Wiley, New York.

References

Swann, C. D. and M. L. Preston, 1995, "Twenty-five Years of HAZOPs," J. Loss Prev., 8, 6, p 349-354, November.

Tatem, P. et al., 1982, "Liquid Pool Fires in a Complete Enclosure," 1982, Technical Meeting, Eastern Section of the Combustion Inst., Atlantic City, NJ, December 14-16.

Taylor, J. H. et al., 1986, "Probabilistic Safety Study Applications Program for Inspection of the Indian Point Unit 3 Nuclear Power Plant," NUREG/CR-4565, March.

TCC, "Recommended Outlines for Contractor Safety and Health,"

Technica, 1988, "WHAZAN Computer Code: User's Guide and Theory Manual."

Theofanous, T. G., 1994, "Dealing with Phenomenological Uncertainty in Risk Analysis," NUREG/CP-0138.

Thomas, H. M., 1981, "Pipe and Vessel Failure Probability," Reliability Engineering, 2, pp 83-124.

Thomas, H. M., 1979, "In-Service Inspection is Over-Rated," *Inst. Nucl. Eng. J.* pp 43-46; April/May.

Thomas, K., 1981, "Comparative Risk Perception: How the Public Perceives the Risks and Benefits of Energy Systems" in The Assessment and Perception of Risk, Royal Society, Gordon pp 35-50.

Tines, S. P., et al., 1990, "Results of the Level 1 Probabilistic Risk Assessment of Internal Events for Heavy Water Production Reactors," ANS Topical Meeting, The Safety, Status and Future of Non-Commercial Reactors and Irradiation Facilities, Boise ID, September. 31 - October 4

TNG, 1980, "Methods for the Calculation of the Physical Effects of the Escape of Dangerous Materials," Directorate General of Labour, Holland.

Travis, J. R., K L. Lam, and T. L. Wilson, 1994, "GASFLOW: A Three-Dimensional Finite-Volume Fluid Dynamics Code for Calculating the Transport, Mixing, and Combustion of Flammable Gases in Geometrically Complex Domains," LA-UR-94-2270 Vol. 1, 2, and 3 LANL, July.

Tribus, M., 1969, *Rational Description Decisions and Designs* Pergamon Press Elmsford NY.

Turner, 1970, "Workbook of Atmospheric Dispersion Estimates," EPA, AP-26.

Uehara, Y. and H. Hasegawa, 1986, "Analysis of Causes of Accidents at Factories dealing with Hazardous Materials," *5th International Symposium on Loss Prevention and Safety Promotion in the Process Industries*, 1 Ch. 23 Societie de Chimie Industrielle.

Vanderzee, C. E. and W. W. Rodenburg, 1970, "Gas Imperfections and Thermodynamic Excess Properties of Gaseous Hydrogen Fluoride," *Journal of Chemical Thermodynamics*, Vol. 2, pp. 461-478,.

Van Slyke, W. J. and D. E. Griffing, 1975, "ALLCUTS, A Fast Comprehensive Fault Tree Analysis Code," Atlantic Richfield ARH-ST-112, July.

Van Ulden, A. P., "On the Spreading of a Heavy Gas Released Near the Ground," *First International Loss Prevention Symposium*, The Hague/Delft Elsevier, Amsterdam, pp 221-226.

Verna, B. J., 1981, "Nuclear Power Experience," Vol. BWR-2, Event No. VI. F.2, XV.13 Vol PWR-2, Event No. VI.F. 32, July.

Vesely, W.E., 1969, "Analysis of Fault Trees by Kinetic Tree Theory," INEL IN-1330, October.

Vesely, W. E., 1971, "Reliability and Fault Tree Applications at NRTS," IEEE Trans NS 18 p 472, February.

Vesely, W. E., 1977, "Estimating Common Cause Failure Probability in Reliability and Risk Analyses: Marshall-Olkin Specializations," Proc. Int. Conf. Nucl. Systems Rel. Eng. & Risk Assessment, Gatlinburg, TN, June.

Vesely, W. E. and R. E. Goldberg, 1977, "FRANTIC - A Computer Code for Time Dependent Unavailability Analysis," NUREG-0193, October.

Vesely, W. E. and R. E. Narum, 1970, "PREP and KITT Computer codes for the Automatic Evaluation of a Fault Tree," INEL IN-1349.

Vesely, W. E. et al., 1983, "Measures of Risk Importance and Their Applications," NUREG/CR-3385, July.

Vijuk, R. and H. Bruschi, 1988 "AP600 Offers a Simpler Way to Greater Safety, Operability and Maintainability," Nuc. Eng. Int., pp 22-26, November.

Virolainen, R., 1984, "On Common Cause Failures Statistical Dependence and Calculation of Uncertainty: Disagreement in Interpretations of Data," Nuc. Eng. and Design 77 pp 103-108.

Wakefield, 1992, "A Revised Systematic Human Action Reliability Procedure (SHARP1), EPRI TR-101711, December.

Wall, I. B., 1974, "Probabilistic Assessment of Flooding Hazard for Nuclear Power Plants," Nucl. Saf. 15, p 399.

Waller, R. A. and V. T. Covello, 1984, *Low-Probability High-Consequence Risk Analysis*, Plenum Press, New York.

Walton, W. D. and K. A. Notarianni, 1993, "Comparison of Ceiling Jet Temperatures Measured in an Aircraft Hanger Test Fire with Temperatures Predicted by the DETACT-QS and LAVENT Computer Models," NIST, NISTIR 4947.

WASH-740, 1957, "Theoretical Possibilities and Consequences of Major Nuclear Accidents at Large Nuclear Plants," March.

Wheatley, C.J., et al., 1988, "Comparison and Test of Model for Atmospheric Dispersion of Continuous Releases of Chlorine," SRD Report R438, UKAEA, July.

Whiston, J. and B. Eddershaw, 1989, "Quality and Safety - Distant Cousins or Close Realtives," The Chemical Engineer, June.

Whitacre, C. G. et al., 1987, "Personal Computer Program for Chemical Hazard Prediction (D2PC)," U.S. Army Munitions Command, Aberdeen Proving Ground, CRDEC-TR-87021.

Wilkins, D. R. and J. Chang, 1992, "GE Advanced Boiling Water Reactors and Plant System Designs," 8th Pacific Basin Nuclear Conference, Taiwan, April.

Williams, J. C., 1989, "A Data-Based Method for Assessing and Reducing Human Error to Improve Operational Performance, *Proceedings of the 1988 IEEE Fourth Conference on Human Factors and Power Plants*, Monterey, CA, June 5-9, pp 436-450, IEEE.

Willie, R. R., 1978, "Computer Aided Fault Tree Analysis," Operations Research Center, University of California Berkeley.

References

Witlox, H. W. M., 1993, "Thermodynamics Model for Mixing of Moist Air with Pollutant Consisting of HF, Ideal Gas, and Water," Shell Research Limited, Thornton Research Center, TNER.93.021,.

Worrell, R. B. and D. W. Stack, 1974, "A SETS Users' Manual for the Fault Tree Analyst," p 36, SNL, NUREG/CT-0465, SAND 77-2051.

Worrell, R. B. 1997, Logic Analysts, 1717 Lousiana Ave., Suite 102A, Albuquerque, NM, 87110, communication with Ralph Fullwood

Wreathall, J., 1982, "Operator Action Trees, An Approach to Quantifying Operator Error Probability during Accident Sequences," NUS Report 4655, July.

Wyss, G. D. et al., 1990, "Accident Progression Event Tree Analysis for Postulated Severe Accidents at N-Reactor," SNL, SAND89-2100.

Ziomas, I. C. et al., 1989, "Design of a System for Real Time Modeling of the Dispersion of Hazardous Gas Releases from Industrial Plants," *J. Loss Prevention* 2 October.

Zion, 1982, "Zion Probabilistic Safety Study," Commonwealth Edison, Chicago, IL.

ZIP (refer to Zion 1982 and Indian Point 1982).

Zukowski, E. and T. Kubota, 1980, "Two Layer Modeling of Smoke Movement in Building Fires," Fire Material 4.

Chapter 15

Answers to Problems

15.1 Chapter 1

1. A certain insurance company requires a 30% overhead on the premiums. If the payment to your beneficiary is $100,000 and you pay $1,500/yr in premiums, what is your probability of dying in the year?

Answer: Equation 1.4-1 says R = p*C. In this case, C = $100,000, R = $1,500, with no overhead, p = 0.015. However, 30% of your premium goes to overhead only leaving $1,050 for insurance, hence, the corrected p = 0.0105 or your odds for dying in the year are about 1 in 100.

1. What is the mean frequency of deaths from nuclear power indicated in Figure 1.4.3-1?

Answer: The mean is: <C> = Σ p_i*C_i. Figure 15.1.2-1 is a redrawing of Figure 1.4.3-1 to more finely divide the scales. This is a cumulative probability plot. The probability of more than 1 fatality is 6.5E-4, >4 is 2.4E-4, >10 is, >40 is 5.9E-6, >100 is 6.0E-7. Hence the probability of 3 is 2.4E-4, 6 is 8.2E-5, 30 is 5.9E-6, 60 is 6.0E-7. The mean is 1.4E-5/(year-plant).

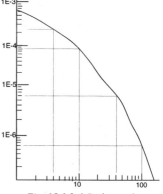

3a. Devise an alternative to logarithmic plotting that is more familiar that compresses the range of data.

Answer: Figure 1.4.3-5 shows volume scaling. Another alternative is area scaling.

Fig. 15.1.2-1 Redraw of Figure 1.4.3-1

3b. Apply your method to Zion in Figure 1.4.3-6.

Answer: The results are shown in Figure 15.1.3-1.

4a. 10CFR50 Appendix I gives the value of a person-rem as $1,000. The report, Biological Effects of Ionizing Radiation (BEIR, 1972) assesses about 10,000 person-rem per death (statistically-not exposed to an individual). Given this information, what is the cost of a death from radiation?

Answer: Given $1,000/rem and 10,000 rem/death, the cost of a human death from radiation is: $10 M.

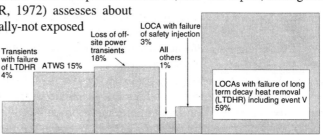

Fig. 15.1.3-1 Comparing by Area

Answers to Problems

4b. At Pickett's charge at Gettysburg, approximately 30,000 soldiers died in 30 minutes. What is the expenditure rate per minute if the same life equivalence were used?

Answer: Using the cost of a life for radiation, the cost was $ 3E11. The expenditure rate was $ 1E10/minute or 10 billion dollars/minute. Clearly if soldier lives were valued the same as radiation death, no government could afford to go to war.

5a. Discuss worker vs public risk.

Answer: Humans accept a greater voluntary risk than involuntary risk. Part of this argument is that one should be free to commit suicide whether with a gun or in "sports." A counter argument is that if the suicide is not successful, the cost of health care may fall on society. The same is true of smoking or drinking alcohol, although in the case of cigarette smoking, the manufacturers may have to pay the health costs. It seems that there is a tendency to take risk but not want to pay the cost of the risk taking. Figure 1.4.4-1 shows that the ratio of voluntary to involuntary risk is 1,000 to 1. The reason why worker risks are higher than involuntary risks is that worker risks are presumed to be voluntary. It is true that a worker understands that there are risks associated with the job, but the worker has aptitude and training that presumably makes the job higher paying and perhaps more gratifying than other alternatives.

5b. Should non-repository workers be treated as public or workers from a risk standpoint?
Answer: The repository workers accept certain risks (primarily that of a miner) as a condition of employment and the other workers accept other risks as a condition of employment. It seems reasonable that the non-repository workers should be treated as public for all risks that are not part of their conditions of employment. Similarly, repository workers should be treated as public for all risks that are not a condition of their employment.

6. Discuss why the EPZ should be independent of plant size, safety features, demography and meteorology.

Answer: The only reason I can think of for making the EPZ independent of plant size, safety features, demography and meteorology is the convenience of emergency planners. Such would take no consideration for the hazard a plant poses; it would favor hazardous activity and penalize safe activity.

7. Compare the risk of evacuation with that of radiation exposure at TMI-2.

Answer: The automobile death rate is about 1E-7/passenger mile. If 25,000 people evacuate 20 miles, this is 5E5 passenger miles, hence, the risk is $5E5*1E-7 = 0.05$ deaths. The radiation exposure is $2.5E4*5E-4*48 = 600$ person-rem. Using information from problem 4, the estimated deaths from radiation is $600*1E-4 = 0.06$. About the same. The risk from radiation may be over estimated because the radiation level was measured close to the plant; on the other hand, the traffic fatality estimate may be high because of police presence and slow driving.

8a. The government is charged by the Constitution with a responsibility for the public welfare. Discuss the government's responsibility in controlling the risks in society.

Answer: The ambiguity comes in the meaning of public welfare. It may mean improving economic and environmental conditions of the public; it may mean protecting the public from their own folly generally by prohibiting certain actions and activities. Some people wish to engage in hazardous activities without accepting the responsibility for their actions. The current tendency in the courts is to absolve them of this responsibility and place the burden on anything with "deep pockets" however nebulously connected with the hazard. Primarily this denies that the risk taker understood the risk.

8b. Tobacco is not subject to the Delaney Amendment that bans material that can be demonstrated to cause cancer in animals regardless of the quantity of material used in the tests. Why aren't all materials subject to the same rules?

Answer: Tobacco is an example (this problem was written 10 years ago before the recent court decisions). In the '1930s, it was generally acknowledged that tobacco was unhealthful. Cigarettes were called "coffin nails." There was a popular song about driving nails in you coffin. Nevertheless courts have taken the attitude that the risk was not understood, and that the tobacco companies acted to gain addicts especially among the youth. Tobacco is not subject to the Delaney Amendment because Congress chose to protect the tobacco growers. In a more ideal world, laws should apply equally to all substances regardless of economic interests. Moreover, laws should have a scientific basis in reality.

15.2 Chapter 2

1. The control rods in a hypothetical reactor have the pattern shown in Figure 15.2.1-1. Based on plant experience, it is determined that the probability of a rod failing to insert is 0.01/plant-year. Assuming that if 3 nearest-neighbor rods fail to insert, core damage will result. What is the probability of core damage from failure to scram from control rod sticking?

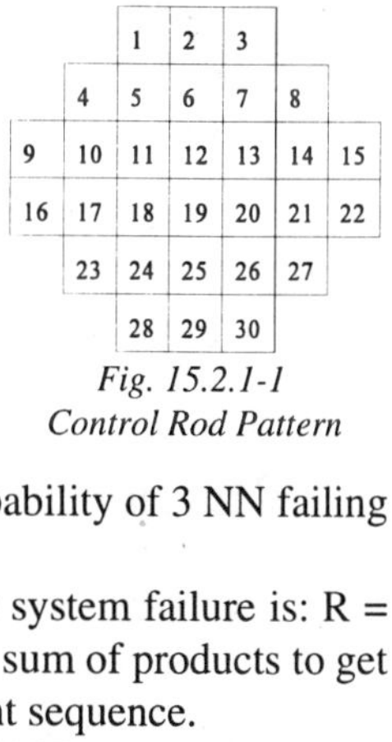

Fig. 15.2.1-1 Control Rod Pattern

Answer The counting technique uses pivot points. The 3 NN unique combinations for 1 are: 1,4,5; 1,5,6; 1,2,6. 2 has 6 combinations, 3 has 3. 4, 5, 6, 7 and 8 have 4. 9, 10, 11, 12, 13, 14, 17, 18, 19, 20, 24 and 25 have 2. 15, 16, 21, 23, and 26 have 1. The total number of unique 3 NN combinations is 61; the probability of 3 rods failing is 1E-6, hence the probability of 3 NN failing on a scram request is 6.1E-5.

2. The Boolean equation for the probability of a chemical process system failure is: R = A*(B+C*(D+E*(B+F*G+C). Using Table 2.2-1, factor the equation into a sum of products to get the mincut representation with each of the products representing an accident sequence.

Answer: The cutsets are: A*B + A*C*D + A*C*E. The terms A*C*E*B and A*C*E*F*G are contained by the A*C*E term.

3. The probability that a system will fail is the sum of the following probabilities: 0.1, 0.2, 0.3, 0.4, and 0.5. What is the probability of failure?

Answer: The easiest way is to calculate the probability of not failing and subtract it from 1. P = 1-(.9*.8*.7*.6*.5) = 0.8488.

4. The probabilities for the equation in Problem 2 are: A = 1E-3/y, B = 4E-3, C = 7E-4, D = 0.1, E = 0.5, F = 0.25 and G = 0.034. What is the value of R? What is the probability of each of the sequences?

Answer: The probabilities of the sequences are A*B = 1E-3/y* 4E-3 = 4E-6/y; A*C*D = 1E-3/y*7E-4*0.1/y = 7E-8/y; A*C*E = 1E-3/y*7E-4*0.5 = 3.5E-7/y. R = 4.42E-6/y.

5. Using the information and result of Problem 4, calculate the Birnbaum, Inspection,

Answers to Problems

Fussell-Vesely, Risk Reduction Worth Ratio, Risk Reduction Worth Increment, Risk Achievement Worth Ratio and Risk Achievement Worth Increment for each of the components A through G. Do your results agree with the equivalences in Table 2.8-1?

Answer: R = A*B + A*C*D + A*C*E. The Birnbaum importance of A = 4E-3 + 7E-4*0.1+7E-4*0.5 = 4.42E-3. Similarly the other Birnbaum importances are: B = 1E-3, C = 6E-4, D = 7E-7, E = 7E-7, F = 0, G = 0. Inspection importance is: A = 4.42E-6, B = 4E-6, C = 4.2E-7, D = 7E-8, E = 3.5E-7, F = 0, G = 0. Fussel-Vesely is Inspection Importance divided by risk, hence: A = 1, B = 0.9, C = 9.5E-2, D = 1.6E-3, E = 0.7.9E-3, F = 0, G = 0. The Risk Reduction Worth Ratio is the ratio of risk with the component failed (set to 1) to the risk unfailed: A = 1000, B = 226, C = 136, D = 1.14, E = 1.08, F = 0, G = 0. The Risk Reduction Worth Increment is the risk with the component failed minus the risk. A = 4.4E-3, B = 1E-3, C = 6E-4, D = 6.3E-7, E = 3.5E-7, F = 0, G = 0. The Risk Achievement Worth Ratio is the ratio of the risk with the component unfailed (set to 0) to the risk A = 0, B = 0.095, C = 0.9, D = 0.98, E = 0.92, F = 1, G = 1. The Risk Achievement Worth Increment is the risk minus the failed component: A = 4.4E-6, B = 4E-6, C = 4.2E-7, D = 7E-8, E = 3.5E-7, F = 0, G = 0. Yes the results agree with the table. Spot checking: RRWI(C) = Birnbaum(C) - Inspection(C) = 6E-4.

6. In Section 2.5.4, we found the availability of a repairable emergency generator (EG) by Markov methods. If a plant requires that two identical, independent EGs must both work for time T for success. What is the probability of this? Assume the failure rates are λ_1, λ_2, and the repair rates are μ_1, μ_2.

Answer: Equation 2.5-42 gives the probability of an EG operating: $P_1 = \{\mu_1 + \lambda_1 * \exp[-(\lambda_1+\mu_1)*T]\} /(\lambda_1+\mu_1)$. The independent probability of the other working in the same time period is $P_1 = \{\mu_2 + \lambda_2 * \exp[-(\lambda_2+\mu_2)*T]\} /(\lambda_2+\mu_2)$. The probability of both working at the same time is $P_{combined} = P_1 * P_2$.

15.3 Chapter 3

1a. Transform the event tree of Figure 3.4.5-2 into several fault trees for each plant damage state noting that end points 1, 3 and 6; 2, 4 and 7; 5, 8, 9, and 10 lead to the same damage state.

Answer: End points 1, 3, and 6 are success paths which are not modeled by a fault tree. The fault tree for slow melt is Figure 15.3.1-1.

1b. Find the cutsets of each fault tree. Answer: The equations are: Slow Melt = PB*ECR, and Melt = PB*EC1 + PB*RS.

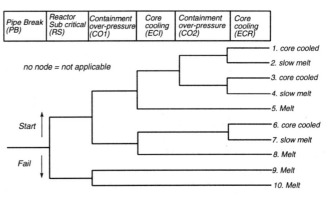

Fig. 3.4.5-2 Large LOCA as a Function Event Tree (adapted from NUREG/CR-2300)

1c. Assuming the following failure rates: PB = 3.1E-2, RS = 1.0E-5, CO1 = 2.3E-2, ECI

496

= 3.2E-3, CO2 = 0.1, ECR = 2.5E-2, calculate the Birnbaum, Inspection and Risk Achievement Worth Ratio for each system including the initiator.
Answer: Slow Melt = 3.1E-2*2.5E-2 = 7.75E-4/y. Melt = 3.1E-2*(3.2E-3+1E-5) = 1E-4/y. Let Slow Melt Birnbaum of PB be represented by B(SM, PB) = 2.5E-2, B(SM, ECR) = 3.1E-2. Similarly for Melt, B(M, PB) = 3.2E-3, B(M, ECI) = 3.1E-2, B(M, RS) = 3.1E-2. For Inspection Importance: I(SM, PB) = 7.75E-4, I(SM, ECR) = 7.75E-4, I(M, PB) = 9.9E-5, I(M, ECI) = 99E-5, I(M, RS) = 3.1E-7. For Risk Achievement Worth Ratio: RAWR(SM, PB) = 32, RAWR(SM, ECR) = 40, RAWR(M, PB) = 32, RAWR(S, ECI) = 320, and RAWR(S, RS) = 320

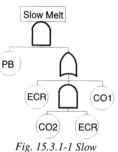

Fig. 15.3.1-1 Slow Melt Fault Tree

2a. Construct a Reliability Block Diagram (RBD) of the simple injection system shown in Figure 3.4.4-6. Answer: The RBD is shown in Figure 15.3.2-1.

2b. Discuss the relative merits of these models and the fault tree model of the same system.
Answer: The RBD is a clear logical figure in which water flows from left to right. Redundancies such as Valves B and C are shown as parallel paths. The RBD uses positive, i.e., success logic. It is anachronistic to introduce negative logic into an RBD such a Valve Closes. A fault tree looks nothing like the flow diagram. The logic is reversed to look for failure. There is no success path. Its large set of symbols allows easy expression of the most complex systems.

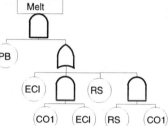

Fig. 15.3.1-2 Melt Fault Tree

Fig. 15.3.2-1 RBD of a Simple Injection System

2c. Assuming the following failure probabilities: pump = 0.01, SIS = 1.5E-3, valves A through E = 4.3E-3, evaluate each of your models to determine the probability of this system functioning on demand. Calculate the Birnbaum, Inspection and Risk Achievement Worth Ratio for each of the components using which ever model you prefer. (Once you get the cutsets, it does not matter which model was used as long as it is a correct system model.)
Answer: The success probability of the RBD is: 9.985E-1* 9.9E-1 *9.957E-1 *9.957E-1 *9.999815E-1 *9.957E-1 = 0.9758. The importance measures are for negative logic. The success of the train is SI = SIS + P + VE + VA + (VB*VC) + VD. Insert numbers and test and get 2.42E-2 which is the same answer as by success logic. Birnbaum importance is B(SIS) = 1, B(P) =1, B(VE) =1, B(VA) =1, B(VB) = 4.3E-3, B(P) = 4.3E-3, B(VD) =1. Inspection Importance is: I(SIS) = 1.5E-3, I(P) = .01, I(VA) = 4.3E-3, I(VB) = 1.8E-5, I(VC) = 1.8E-5, I(VD) = 4.3E-3. RAWR(SIS) = 41, RAWR(P) = 41, RAWR(VE) = 41, RAWR(VA) = 41, RAWR(VB) = 1.177, RAWR(VC) = 1.177, and RAWR(VD) = 41.

2d. Rank order the systems for each importance measure. Answer: the Birnbaum rank order where [ ] encloses items of the same rank is: [SIS, P, VE, VA, VD], [VB, VC]; Inspection is: P, [VA, VD], SIS, [VB, VC]; RAWR is: [SIS, P, VE, VA, VD],[VB, VC].

2e. Why doesn't each measure result in the same ordering? Discuss which is the "best" measure.

Answer: They don't give the same result because they are not measuring the same thing. Birnbaum and RAWR are similar because the both show what happens when a component's failure rate is set to 1. This statement is only approximately true for more complex systems than the simple RBD. A basic difference is RAWR is divided by the risk but this just shifts the scale. In this case Inspection seems superior because it distinguishes between more of the components except for those that are identical with identical roles. It introduces a weighting by the failure rate of the component which gives higher priority to those with a high failure rate.

3a. In the ZIP, members of a redundant train are treated as identical and for this reason correlated. In such case, the means and variances do not propagate as independent distributions and recourse is made to Equations 2.7-24 and 2.7-26. For the "V" sequence (release bypassing containment), the mean failure rate of an MOV if 1.4E-8/hr with a variance of 5.3E-15/hr*hr. For two doubly redundant valves, what is the mean failure rate and variance? Answer: The mean of the combination is $<Q> = <q>^2$. The variance of the combination $\mu(Q) = <q>^2 + \mu(q)$. Hence $<Q> = $ 2E-16/hr². $\mu(Q) = $ 2E-16/hr² + 5.3E-15/hr² = 5.5E-15/hr².

3b. What is the beta factor under this assumption of the valves being identical?

Answer: It is $\beta = \lambda/\lambda_c$. But if the common-cause failure rate is the same as the component failure rate, $\lambda = 1$ and $\lambda_c = \infty = \lambda$. This is impossible.

3c. Discuss your opinion of the validity of the identical assumption and suggest experimental tests of the validity of such strong coupling.

Answer: It is unrealistic. No components are identical, but even if they were the causes of random failure in component are not correlated to the other component because they are defined to be random. However, components can fail at the same time from deterministic coupling such as fire, missile, common utilities etc.

4a. Prepare a FMECA for Valve A (a motor-operated valve - MOV) in Figure 3.4.4-6.

Answer: The FMEA is presented in Table 15.3.4-1.

4b. How do you handle the redundancy between Valve B and Valve C in a FMECA? Note this problem requires some understanding of MOVs and how they can fail.

Answer: You can treat them as one component in the component column and prepare the FMECA for ways they can fail together. For random failure of both valves in a mission time, estimate the failure probability as one-half the failure rate of one times the probability of ailing in the mission time of the other.

5a. Prepare a fault tree of a dual emergency power system, considering that the compressed air used for starting the diesels is a separate system which may be a common-cause failure to start the

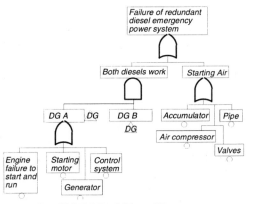

Fig. 15.3.5-1 Fault Tree of Emergency Electric Power

diesels. Answer: Figure 15.3.5-1.

5b. Using λ_1 and λ_2 for the failure rates and μ_1 and μ_2 for the repair rates, what is the frequency of both systems working? One or the other? None?

Table 15.3.4-1 Failure Modes Effects Analysis for Valve A in Figure 3.4.4-6

Component	Failure mode	Effect	Mitigation
Valve Failure	Valve body crack	Divert fluid from injection	Periodic test and inspection
	Valve flapper	Disconnect from shaft, situation ambiguous	Test or inspect by disassembly.
		Foreign material blocking the ability to open	Strainer upstream, test and inspection
	Valve seat	Fails to block flow when closed. Possible leakage of SIS fluid into system and reduced operability.	Need leakage taps on system to determine if there is leakage between series valves. Sense temperature of pipe. Inspection.
Motor Failure	Bearing	Motor cannot open valve	Shorts can be sensed by ground leakage.
	Shorted winding		Bearing failure can be detected by test
	Open winding		Open winding can be sensed by winding resistance.
Shaft coupling	Coupling slipping, sheared pin or key	Cannot open valve	Motor current when in operation, inspection
Control switches	Limit switch failure	Valve binding, in ability to move valve	Test and inspection
	Improper limit switch setting	In ability to move valve	Test and inspection

Answer: Ignore the Starting air system. The fault tree does not model repair. The frequency of both systems failing is: $\frac{1}{2}*\lambda_1*\lambda_2*t$, where t is the mission time, hence, the frequency of both systems working is $1-\lambda_1*\lambda_2*t$. The frequency of one or the other (not both) working is $1- \lambda_1+\lambda_2 + \frac{1}{2}*\lambda_1*\lambda_2*t$. (Assuming that $\lambda_1, \lambda_2 << 1$.)

5c. Compare these results with the Markov model prepared in Section 2.5.4. Under what conditions do the two methods agree?

Answer: The Markov model was prepared for one diesel having two states: working and failed. For two diesels, the probability of both failing at the same time is $\lambda = \frac{1}{2}*\lambda_1*\lambda_2*t$, where t is the mission time. If the repair rate is short $\mu<<\lambda$, the failure probability (Equation 2.5-42) is: $Q = \exp(-\lambda*t)$. When expanded it is $Q = \lambda*t$. The failure frequency is $\lambda = \frac{1}{2}*\lambda_1*\lambda_2*t$.

6. The "gedanken" experiment in Section 3.4.4.2 results in equation 3.4.4-2 suggesting the operations of summation and multiplication in the algebraic sense - not as probabilities. Since this is a simulation why are not the results correct until given the probability interpretation? Hint: Refer to the Venn Diagram discussion (Section 2.2).

Answer: It is necessary in taking the summation that overlapping areas be removed to get the probability interpretation.

7a. Refer to the discussion of Problem 7 at the end of Chapter 3. Develop a relationship that shows the improvement that may be achieved by staggered testing.

Answer: Referring to Figure 3.7-2, we assume the failure probability increases linearly from 0 to some amount with a slope a. $P_a = a*t$ and $P_b = a*(T-t_1+t)$ for the time interval $0 < t < t_1$, and $P_a = a*t$ and $P_b = a*(t-t_1)$ for the time interval $t_1 < t < T$. The mean unreliability is

$$<U> = a^2 * \int_0^{t_1} t*(T-t_1+t)*dt / \int_0^{t_1} t*dt + a^2 * \int_{t_1}^{T} t*(t-t_1)*dt / \int_{t_1}^{T} t*dt.$$ The results of integration after

considerable and factoring a cubic is $<U> = a^2*T*(T+t_1)/3$

7b. Assume that the failure rate of system A or B is 0.1/yr, test interval, T, is monthly and t_1 is one day for sequential testing, what is the numerical ratio of the risks of the two schemes?

Answer: For staggered testing, $2*t_1 = T$ and $<U_{stag}> = 2*a^2*t_1^2$. For sequential testing $30*t_1 = T$, and $<U_{seq}> = 310*a^2*t_1^2$. The ratio is $<U_{seq}>/<U_{stag}> = 155$.

15.4 Chapter 4

1. Table 4.1-5 shows in these records, there has been 3 control rod drive failures. Assume 100 plants in the U.S. with an average of 30 control rods/plant and 10.7 years of experience in this database. Estimate, the mode, 90% and 10% confidence limits for the failure rate.

Answer: Using equation 2.5-32 and Table 2.5-1, the mode, which is the 50% confidence level is: $\lambda_{50} = 7.2/(2*100*30*10.7) = 1.1\text{E-}4/\text{y}$. Similarly, $\lambda_{90} = 2.0\text{E-}4/\text{y}$ and $\lambda_{10} = 5.5\text{E-}5/\text{y}$.

2. Prepare a reliability data acquisition plan for your or a hypothetical plant. Detail the sources of information, method of recording, analysis methods, and interfaces with management, maintenance, and operations.

Answer: This can only be answered in general - it must be specific to a plant. Step 1; Determine what plant records you have for as long as the plant has been operating. Such records should include the design drawings and revisions, P&IDs, wiring diagrams, process descriptions, flow sheets, requisition sheets, test, inspection and maintenance reports and operator logs. Step 2; Decide on data handling. Undoubtedly this will involve computers. Determine what computers are available. Possibly all of the data can be handled with a personal computer that may be devoted to this specific purpose. If so find what database management software is available for this particular machine and if the software will handle the amount of data envisioned with the required speed. Especially important is the backup for the data files. It should be automatic or at least daily and store the backup on diverse media such as floppy disks, read only memory, tapes, removable hard drives. Step 3; Inputting data. Undoubtedly much data such as handwriting must be manually input. This may be done by clerical personnel on separate computers which may be networked to yours or transferred on floppy disks. Some data may be input by scanning which saves time over manual input but, especially with poor copies, the scanning will make errors that must be checked manually. The data will likely come in as ASCII which must be imported into the database manager. One way of doing this is by comma-quote formatting. Using a search and replace routine commas may be inserted between what is to become fields and quotation added to designate records. Step 4; Using the data. Between emergencies you will become familiar with the data and the database manager program to organize categories and classifications. It is good to work with the maintenance personnel to establish statistics on component failure rates. The ability to rapidly collect incidence data using various search schemes is important so when an accident occurs at the plant you can quickly provide management with related incidents, who was involved, and the corrective actions that were taken.

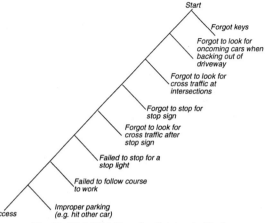

Fig. 15.4.3-1 HRA Tree for Driving to Work

3. Prepare an HRA event tree for driving from your home to work or school. Use this tree to estimate your annual error rate. Compare this with your actual rate. Explain disagreements.

Answer: The HRA tree is shown in Figure 15.4.3-1. For error of commission or omission use 1E-1. There are 8 error events, hence, the failure estimate is 1E-8/trip. Assuming 600 trips per year, the annual failure probability is 6E-6/y. For 10 years it is 6E-5. This has never occurred to me but the low probability means it is unlikely.

4. Bar-code readers (like those in grocery stores) are to identify equipment in a plant. Suggest a way that this may be used to check the temperature, acid type and quantity, and aluminum quantity for acid dissolution of aluminum.

Answer: An electronic temperature monitor could be equipped with a bar-code liquid crystal display which could be read by a portable bar-code reader. These devices have a memory so several readings may be taken before they are readout over a telephone modem to a data logging computer. The simplest way to read the acid type would be to post a label that is bar-coded to indicate the acid type. The acid quantity could be indicated by an acid level gage using a bar-code display of the level. The aluminum quality could be indicated by displaying a label in bar-code. The amount of aluminum could be determined by weight using a bar-code readout on the scales.

15.5 Chapter 6

1. A critical assembly is a split bed on which fissionable material used to mockup up a

Answers to Problems

Table 15.5.1-1 What-If Analysis of Critical Assembly

LINE/VESSEL:Critical Assembly 1   Date: 8/28/1998 Page 1 of 1

What If	Consequences	Danger Level	Comments
Radiation sensor fails?	Fail to detect critical condition and fails to initiate bed separation	High	Redundant detectors are used. In addition technicians have personal audible (cherper) monitors to alert to an incipient critical condition.
Electronic trip fails	Separation not actuated	High	Circuit is designed to fail safe
Technician leaves an object on the rail?	Moveable assembly cannot move	Could be high	Object must be located close to a wheel on the side toward the movement. Even slight movement may shutdown assembly.
Spring to separate units fails?	Failure to separate halves	High	Use multiple springs.
Drive jams?	Failure to separate halves	High	Clutch disengages with drive power.
Release mechanism fails?	Failure to separate halves	High	Release is actuated by dropping power.
Power fails?	Should get automatic separation	Low	Release is actuated by dropping power.
Technician fails to follow procedures	Should have no consequences	Low	Critical Assembly is designed to fail safe.
Earthquake	Jam separation mechanism	Unlikely	Rare occurrence.
Fire	Disable separation mechanism	Possible	No combustibles.
Flood	Cause criticality	Unlikely	No water in critical facility.

separated reactor core that is stacked half on each half. One half is on roller guides so that the two halves may be quickly pulled apart if the neutron multiplication gets too high. Use the What-If Analysis method described in Section 3.3.2 to identify the possible accidents that may occur and the qualitative probabilities and consequences. List the initiators in a matrix to systematically investigate the whole process. Don't forget human error.

Answer: See Table 15.5.1-1 for the What-If analysis and Table 15.5.1-2 for the initiator list.

2a. Removing decay heat has been the "Achilles Heel" of nuclear power. The designs shown in this section use active methods to remove the heat. Sketch and discuss a design that removes the heat passively.

Answer: Figure 15.5.2-1 shows the "percolator" scheme. At the top of the a reactor vessel, presumably the hottest location, is a diaphragm that blows out if the reactor exceeds the maximum allowable pressure. It is attached by an alloy that melts at that the maximum allowable temperature. A pipe connects the top of the pressure vessel with a large water

Table 15.5.1-2 Critical Assembly Accident Initiators

| Radiation sensor failsure |
| Electronic trip failure |
| Object blocking rail? |
| Spring failure |
| Drive jams |
| Release mechanism failsure |
| Power failure |
| Procedure failure |
| Earthquake |
| Fire |
| Flood |

reservoir. Normally the pipe is water filled to reduce heat loss and to prevent steam buildup in the pipe. If either pressure or temperature is excessive, the fuseable plug blows which ruptures the release diaphragm and pressure is relieved in the reservoir. With pressure relief, water flows into the reactor, forms steam and blows out and condenses in the reservoir. Thus keeping the reactor near atmospheric pressure and $212°F$.

2b. It would seem that the energy in the decay heat could be used for its own removal. Sketch and discuss a design that uses this property to remove the decay heat. (This problem was written before the advanced reactors were designed.)

Answer: Figure 15.5.2-2 shows an automatic depressurization valve that opens on excessive conditions to release the steam which passes through a turbine to drive an electric generator to provide makeup water and operate other cooling equipment. After the steam has cooled in the low pressure turbine it discharges under water in the cooling reservoir and condensed. The reservoir cools by evaporation.

3. When Hamlet said to Horatio, "there are many things twixt heaven and earth not conceived of in your philosophy", was he complete in the scope of his coverage? If so can this approach be used for completeness in identifying all the possible accidents?

Answer: Yes his statement is like a Venn diagram that includes set A and not set A. Therefore it is all inclusive. I have tried to use this for completeness but the problem is that it is no help in specifying systems and causes of failure.

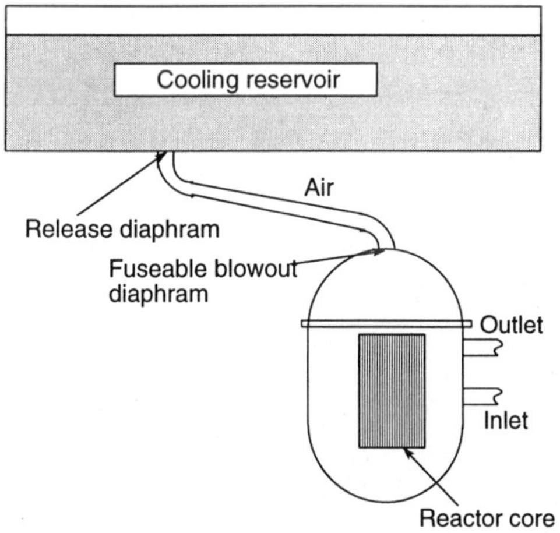

Fig. 15.5.2-1 The Percolator Scheme

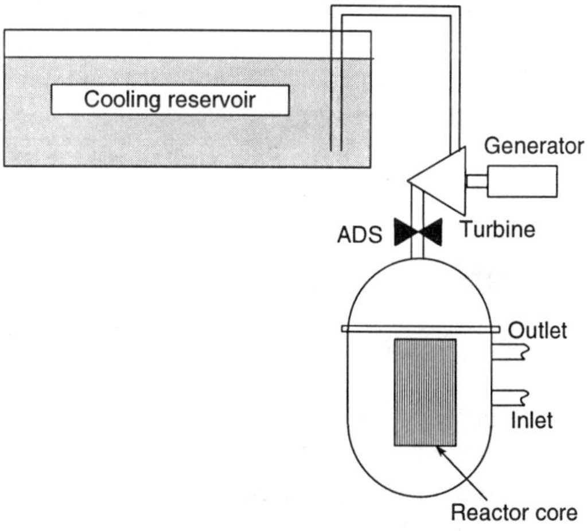

Fig. 15.5.2-2 Using Residual Heat

4. In the WASH-1400 analyses of nuclear power accidents, it was calculated that it is possible to overpressure and rupture the containment. Discuss whether this is better or worse than a pressure

Answers to Problems

relief that releases radioactivity but prevents the pressure from exceeding the rupture point. Spent fuel reprocessing plants are designed to release filtered and processed gases. Typically the final filters are large containers of sand or HEPA (high efficiency particulate air) filters. Discuss filtered containment with consideration for possible failure modes of the filter. Answer: It seems to me the argument is whether or not to design for failure of the containment. It is reasonable to design the containment with a blowout panel to blowout before the containment fails and to discharge the gases through scrubbers and HEPA filters before exhausting the stack. This is essentially the practice of chemical processing in which the effluent is hazardous and cannot be contained, therefore, it is processed to make it benign.

5. Early BWRs used an isolation condenser, although such is not specific to the direct cycle. This device removes decay heat by steam flow through a heat exchanger the other side of which is water vented to the atmosphere. Discuss the relative merits of such a boiler.

Answer: There are obvious merits in using an isolation condenser because it is being included in some of the advanced reactor designs.

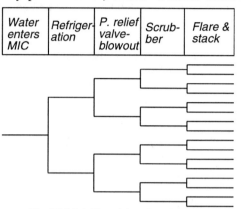

Fig. 15.6.1-1 Bhopal Accident Event Tree

15.6 Chapter 7

1. Prepare an event tree diagram of the Bhopal accident.
Answer: See Figure 15.6.1-1.

2. Use FTAPSUIT to calculate the fault tree shown in Figure 3.4.4-7. For data use $A = C = D = E = H = 1.0E-2/d$ for valve failure to open; for failure of the SIS signal: $B = 1.0E-4/d$, for a pump failing to start: $F = 2.0E-2/d$, and for a pump failing to run after it starts: $G = 5E-3/d$. Answer: The input for the fault tree is shown in Figure 15.6.2-1. Running IMPORTANCE gave 4.93E-2. The AND gate would require a mission time if these were failure rates but since they are failure per demand, the units are correct.

```
** FTAP INPUT
** Failure to Inject Fault tree
A1    +    A    B    B1
B1    *         C1   C2
C1    +         C    B    D1
C2    +         D    B    D1
D1    +         E    B    E1
E1    +         F1   F2
F1    +         F    G    B
F2    +         H    B
ENDTREE
PUNCH
*XEQ
ENDJOB
```

Fig. 15.6.2-1 Failure to Inject Fault Tree Input

15.7 Chapter 8

1. Suppose an interstate highway passes 1 km perpen-

504

dicular distance from a nuclear power plant control room air intake on which 10 trucks/day pass carrying 10 tons of chlorine each. Assume the probability of truck accident is constant at 1.0E-8/mi, but if an accident occurs, the full cargo is released and the chlorine flashes to a gas. Assume that the winds are isotropically distributed with mean values of 5 mph and Pasquill "F" stability class. What is the probability of exceeding Regulatory Guide 1.78 criteria for chlorine of 45 mg/m^3 (15 ppm).

Answer: The geometry is a line passing to within a kilometer of the plant. The probability of a release in any mile is 1.0E-8. For a plane geometry the probability that the wind is blowing toward a 0.1 km wide plant is $0.1/(2*\pi*r)$, where $r = \sqrt{(x^2+d^2)}$. Equation 8.3-1 shows an exponential dependence on r, and σ_h. I fit the equation $\sigma_h = 37*r^{0.877}$. For the denominator, I fit $\sigma_y = 140*r^{0.5}$. These are assembled into the BASIC code CHLOR to calculate the probability weighted average χ/Q. The answer from running the program is $<\chi>/Q = 1.25E-16$. However 100 tons per day are transported past the plant to give a probability of 4.15E-6 gm/y reaching the plant. The NRC criterion is 0.045 gm so the probability of 9.2E-5/y of exceeding the criterion. The computer program is included on the disk.

2a. How long can you remain in a very large flat field that is contaminated with radioactivity at a density of 1 curie/m^2 emitting gamma rays of 0.6 MeV mean energy before you exceed 10CFR100 limits for whole body?

Answer: Assume the receptor is at 1 m, Etherington, 1958 gives the total attenuation coefficient in air: $\mu_t = 0.0804$ cm^2/gm, and the energy absorption coefficient $\mu_e = 0.0311$ cm^2/gm for tissue. These coefficients must be multiplied by the density of air and tissue, respectively. Figure 15.7.2-1 depicts a radiation fallout field. Let C be the curie activity/m^2. The radiation into a unit area receptor at $z = 1$ m above the ground. The area is emitting $C*r*dr*d\theta$ gammas/s. These are attenuated in the air as $\exp(-\mu_t*R)$ and geometrically as $1/(4*\pi*R)$. The radiation received by the receptor is given by equation 15.7.2-1 which becomes 15.7.2-2 by a change of variable.

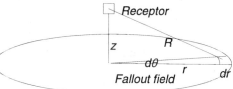

Fig. 15.7.2-1 A Radiation Fallout Field

$$G = C/2 * \int_z^\infty \exp(-\mu_t*R)*dR/R \quad (15.7.2\text{-}1)$$

$$= C/2 * \int_{\mu_t*z}^\infty \exp(-t)*dt/t \quad (15.7.2\text{-}2)$$

This integral is the well known exponential integral. Abramowitz, 1964 approximates it as: $E_1(x) = -\gamma - \ln(x) - \Sigma(-1)^n*x^n/(n*n!)$, where γ is Euler's constant $= 0.57721$. Using the density of air as 1.25 kg/m^3, I get $\mu_t*z = 9.8E-3$ which gives $E_1(9.8E-3) = 5.28$. The energy deposition rate is $3.7E6*5.28/2*0.311*0.6*1.6E-13*1E3*3600 = 0.401$ J/hr $= 0.401$ Gy/hr or 40.1 rem/hr. The 10CFR100 limit is 25 rem (whole body), therefore a person could only remain in this field 37 minutes before exceeding this limit.

2b. What whole body dose will you get from a 100 megacurie puff release passing at a perpendicular distance of 1 km traveling in a straight line at 3 mph? Use the gamma ray parameters of problem 2a and for your area, use the estimate on page 327 which is 0.072 m^2.

Answer: Imagine a 3.7E17 d/s point source moving through the sky. The dose is given by Equation 15.7.2.3. If the differential is changed from dt to dR, the equation is not a standard form.

Answers to Problems

The easiest way to evaluate it is numerically. The program CLOUD on the disk does a rectangular integration to get D = 1.6E-5 rad. I was surprised by this result. It is low because of attenuation in the air. A more accurate result would include a buildup factor but still the dose would not be significant. The finite width of the cloud would increase the dose somewhat.

$$D = 2*S*<E>*\mu_E*A*\int_0^\infty \exp(-\mu_t*R)*dt/(4*\pi*R^2) \quad (15.7.2\text{-}3)$$

$$\text{where } R = \sqrt{[(v*t)^2 + z^2]}$$

3. Suppose that the RHR system in the example in Section 8.1.2 lost half of its heat removal capability after 90 s. When would it reach 2,300°F (a temperature criterion once suggested by Steve Hanauer regarding ATWS)?

Answer: The heat to take the core from 400°F to 2,300°F is 0.07*1.75E5*1900 = 2.33E7 BTU. The heat removal rate after 90 s is Q_{cool} = 7.2E6 + 4E4*t. The early cumulative heat evolution curve in Figure 8.1-2 can be fitted to Q_{decay} = 6.67E4*t. The time to reach 2,300° F is obtained from 2.33E7 = 6.67E4*t - 7.2E6 - 4E4*t. Giving t = 18.9 minutes.

15.8 Chapter 11

1. If a nuclear power plant loses its connection to the offsite load, it must shutdown because it cannot be cutback sufficiently that its electrical output matches its "hotel" load. Outline a PSA study to determine the risk reduction that might be achieved by switching in a dummy load to avoid shutdown and keep the plant online.

Answer: This problem is related to the risk from the Loss of Off-Site Power (LOOP) initiator. It is necessary to determine the risk from LOOP, and the frequency of the LOOP initiator. Correct the LOOP initiator for the frequency that the plant could not have continued operation if it could have been connected to a load. Then a systems analysis of the dummy load systems must be performed to determine its reliability which is used to correct the LOOP initiator. Any risk associated with the dummy load must be added to the LOOP risk. These corrections give the risk change due to not having to shut the plant down when it loses its connection to the power line. Another way to avoid shutting down when the load is lost is with a steam bypass around the turbine so the turbine-generator only supplies enough power for the hotel load.

2. If the containment holds, nuclear power plants present no risk to the public. Over pressurization of the containment is the failure mode that could allow direct release of radioactivity to the public. Design a risk reduction investigation of the benefits of releasing the gas pressure through an offgas processing system that removes the particulates.

Answer: Use the plant's PSA to determine the risk of accidents that include containment failure from overpressurization. Then make a preliminary design of a vented containment that has sufficiently low impedance to the gas at the pressure predicted for the most severe accident sequences such that the containment is not damaged. This containment bypass will include iodine and HEPA filters as well as scrubbers and a discharge through a stack. Estimate the dose that the population would get using this bypass for comparison with the PSA result for ruptured containment sequences.

Put a branch in the containment event tree for the success or failure of the off-gas system with end states for the consequences of off-gas system failure.

3a. Many nuclear plants are connected to rivers such as Indian Point. Discuss accident sequences that could result in a PWR releasing significant radioactivity into a river. Outline the PSA steps that would be necessary to determine the probabilities and consequences.

Answer: Review the plant's design to determine how radioactive water could get from the plant to the river. Some ways are: i) through the heat exchanger and through the condenser, ii) from the closed circuit water into the service water, iii) from the spent fuel storage pool, and iv) from the sump. Prepare fault trees or adapt existing fault trees to determine the probability of each of these release paths. Obtain reliability data for the components that are involved and evaluate the fault trees to determine the probability of each type of failure. For those pathways with a probability of <1E-7/y, calculate the amount of release that could occur before preventive measures are taken. Compare the results with Federal regulations for river water considering the nearest intake for municipal water.

3b. How could such an accident be mitigated?

Answer: By placing redundant radiation in each release path. Such may not have sufficient sensitivity and it may be necessary to collect water samples and concentrate them by evaporation to get the needed sensitivity. Another way is to monitor the radioactivity in organisms that concentrate it; such as clams.

Index

χ/Q	323
ABB PIUS	218
ablation	316
absorption	35
accident analysis	382
acetic	259
acetylene	272
ACRS	3, 20
acrylic	283
Adam 2.1	349
ADS	215
AFTOX	349
aliphatic hydrocarbons	272
alkyds	278
ALOHA 5.2	350
ammonium dichromate	258
ammonium nitrate	262, 276
ASLB	19
ASME pressure vessel codes	2
associative	35
Atomic Energy Act of 1948	3
ATR	413
attenuation	192
ATWS	237, 386
bathtub curve	46
Bayes conjugates	51
Bayes's equation	50
Bayesian	50, 415
benzene	271
benzoic acid	262
beta factor	126
beta function	52
Bhopal	254
Birnbaum importance	62
BIT	212
black powder	275
BLEVE	345
block flow diagram	68
BNLDATA	151
boiler	2
Boolean algebra	35
Boolean logic	38
Boolean variables	37
bottom-up	105
bromine	270
Browns Ferry fire	198
buildup factor	328
BWR	208, 385
CALPUFF	349
CANDU	407
Canvey Island	431
carbon black	273
CASRAM	350
catalytic conversion	291
CCDF	8, 10
CCSP	347
central limit theorem	44
CERCLA	23
CFAST	367
Chebyshev distribution-free inequality	44
check list	77
CHEETAH	366
Chernobyl	224
chi-square	47
chlorine	264
Clean Air Act (CAA)	22, 67

Clean Water Act (CWA) 23
Combustion Engineering System 80+ . . 219
common cause 123
common cause multiparameter models . 127
commutative 35
comparison of HAZOP, FRR and QRA   448
COMPBRN 367
completeness 35, 380
Compliance 33, 75
Comprehensive Environmental Response,
    Compensation, and Liability Act . 23
computer codes for ground acceleration   192
computer codes for system reliability
    analysis 128
confidence 43, 54
confusion matrix 176
constant failure rate 52
containment 208
containment analysis 382
containment event trees 118
control room 210
convolution 56
core melt 311, 316
CRAC 332
CRACIT 325, 330
cracking 292
critique of the Reactor Safety Study 4
cut sets 8, 39, 100, 103
data adequacy 381
DC power 388
de Morgan's theorem 35, 101, 135
deflagration 341
DEGADIS 351
detonation 341
Digraph 119
direct numerical estimation 177
discrete probability distributions 56
disjoint 38
distributions 42
distributive 35
DOE 77
dose from a line source 327

dose from a point source 327
Dresden 1 208
dynamic loads 193
dynamite 276
elastomers 273
emergency control center 75
emergency planning 16, 33
Emergency Planning & Community Right-
    to-know Act (EPCRA) 23
emergency preparedness 74
EMGRESP 352
employee involvement in process safety . 67
Endangered Species Act 24
Environmental Impact Statements 25
EPA 23, 67
EPCRA 23
epoxy 281
EPZ 16
ethanol 274
ethylene 272
ethylene oxide 261
event trees 111
EXPAC 364
expectation value 43
explosions 340
exponential distribution 45
extreme value distribution 205
f-M 191
failure mode and effects analysis 94
failure on demand 48
failure with time 47
fault tree analysis 101
fault tree coding 111
fault tree symbols 102
Federal Insecticide, Fungicide and
    Rodenticide Act (FIFRA) 24
FEM3C 353
fertilizer 266
FIFRA 24
film materials 273
FIRAC/FIRIN 353, 367
fire growth modeling 200

Flixborough	251
flooding	202
flooding analysis	204
flow	337
fluorine	269
FMEA	94, 99
FMEA/FMECA	99
FPEtool	368
fractional distillation	288
fragility	190, 194
fragility curve	194, 201
freedom of Information Act (FOIA)	25
frequency	39
frequency-magnitude (f-M) relations	189
frontline	211
FSAR	20
FTAP	130
FTAPlus	240, 307
FTAPSUIT	241, 309
Fussell-Vesely importance	62
gamma	53
GASFLOW	356, 365
Gauss' distribution-free inequality	44
Gaussian diffusion	325
Gaussian distribution	43
General Electric ABWR	221
General Electric SBWR	221
generating function	49
GIDEP	151
Go	119
Gray	329
ground coupling	192
halogens	268
hazard	46
hazardous chemical and process information	27
HAZOP	70, 86
HAZWOPER	74
HFBR	414
HFIR	417
HGSYSTEM	356
hoop stress	335
hot work permit	32
HOTMAC/RAPTAD	355
HOTSPOT	364
HPSI	210, 215
HRA event tree	181
human errors	108, 381
human reliability analysis	164
hydrofluoric	258
idempotency	35
IEEE 500	153
IMES	370
impact of the IPE program on reactor safety	396
importance	61
incident investigation	33, 74
initiating events	237
inspection	32
inspection importance	62
intensity-attenuation	192
intersection	36, 37, 41
iodine	270
IPE	398
IPE flood findings	205
IPRDS	155
IRWST	218
K-reactor	419
KBERT	358
KITT	130
Kyshtym	251
Laplacian probability	40
LER	158
LER-HEP	177
Lewis Committee	4
linear risk	6
LLOCA	213
LOCA	114
lognormal distribution	45
longitudinal stress	336
LWR shutdown risk	390
M-out-of-N-Combinations	42
maintenance reports	162
management of change	32, 71, 73

Index

MAPPS 177
MARCH 319
Markov 48
mean 43
median 42
Meltdown 311
mincut 103
minimal cut set 39
minimal pathsets 39
minterms 38
MISM 359
missiles 347
MLOCA 213
MMI 190
MOCUS 130
mode 43
modified central limit theorem 45
modified Mercali intensity 190
modular fault trees 119
moments method 57
MONTE 134
monte carlo analysis 56, 59, 455, 456
MSDS 68
MSF 178
MSIV 213
MTTF 45
MTTR 46
N-reactor 425
NARC 155
National Environmental Policy Act
    (NEPA) 25
natural-circulation 218
nature of process accidents 248
NEPA 19, 25
NGSDC 189
nitric acid 267
nitrocellulose 277
nitrogen 267
nitroglycerin 275
nitroparaffin 260
normal distribution 43
NPRDS 155
NUCAP+ 146
NUCLARR 155
NUCRAC 332
NUREG 1150 156
OAT 178
Occupational Safety and Health Act 25
OJT 71
Omega West 426
Organizational factors 165
ORIGEN 319
ORPS 158
OSHA 27, 67
P&IDs 69
paired comparisons 178
parts count 98
passive safety 217
Peach Bottom fire 197
performing a detailed probabilistic safety
    analysis 301
PHA 234
PHAST Professional 360
phosphorus 261, 266
picric acid 278
PIFs 167
Poisson distribution 43
Pollution Prevention Act 25
polyesters 280
polyethylene 282
polymerization 279, 292
polyurethanes 281
positive void coefficient 226
potassium 266
precursors 389
PREP 130
pressurized thermal shock 389
PrHA 30, 68, 70
Price-Anderson 3
probability 39
probability as state of belief 41
process flow diagrams 69
process hazard analysis 27, 70
process scoping 298

prophylaxis 16	RWST 210
propylene 272	Safe Drinking Water Act (SDWA) 26
PSA management 230	safety goals 1, 14
PSAPACK 141	safety monitor 147
PSAR 19	San Onofre fire 197
pseudo-failure 52	SAPHIRE 136
pseudo-time 52	SARA 26, 75
PSFs 156	scram 210
PSM Rule 27, 67, 73	sensitivity analysis 61
PSpill 362	set 37
PWR 208, 385	SETS 131
qualitative methods of accident analysis . 76	Seveso 252
qualitative results 8	Shannon's method 37
quality assurance 32, 73	Shippingport 3, 208
quantitative methods of accident analysis 97	Sievert 330
quantitative results 10	similarities between nuclear and chemical
rad 329	safety analysis 295
RADC 153	SLAB 362
RBD 100	SLCS 214
RBMK 225	SLIM-MAUD 179
RCRA 26	SLOCA 213
Reactor Safety Study 3, 45	SMACS 194
refinery 288	source terms 317
reliability 46	special nuclear material 19
reliability block diagram (RBD) 100	spring-mass model 190
REM 14, 330	SSLOCA 213
representativeness 381	SSMRP 190
resins 260	state vector 35
Resource Conservation and Recovery Act	station blackout 389
(RCRA) 26	step prior 53
revealed preferences 12	styrene 284
RHR 212, 214	success tree 110
Richter scale 190	sulfur 265
risk 6	Superfund Amendments and
Risk Achievement Worth Ratio 63	Reauthorization Act (SARA) 26 75
Risk Achievement Worth Increment 63	support 213
risk perception 13	taxonomy of the chemical process
Risk Reduction Worth Increment 63	industry 264
Risk Reduction Worth Ratio 63	test 32
RISKMAN 143	test and maintenance 72
RMDB 152	test reports 162
rocket fuel 259	Texas City disaster 249

thermal radiation 346
thermosetting resins 280
TMI-2 .... 16, 223, 385
TNT 277
toluene 271
top-down 104
Toxic Substances Control Act (TSCA) .. 26
training 32, 71
transport analysis 382
TRAPMELT 319
truncated gamma 53
TSCA 26
TSCREEN 363
union 36, 37
unity 35
utility experience 405
UVCE 341
VANESA 319
variance 43
VDI 364
Venn diagram 37-39, 167
von Misesian Probability 40, 102, 103
VULCAN 370
WAMBAM 133
WAMCUT 134
WASH-1400 .... 4, 153, 297
WASH-740 316
Westinghouse AP-600 216
what-if analysis 81
what-if/checklist analysis 84
WinNUPRA 146
work permit 73
work authorization 73
xylene 271
yield stress 335
zircalloy 313